Lecture Notes in Artificial Intelligence 13023

Subseries of Lecture Notes in Computer Science

Series Editors

Randy Goebel
University of Alberta, Edmonton, Canada
Yuzuru Tanaka
Hokkaido University, Sapporo, Japan
Wolfgang Wahlster
DFKI and Saarland University, Saarbrücken, Germany

Founding Editor

Jörg Siekmann
DFKI and Saarland University, Saarbrücken, Germany

More information about this subseries at http://www.springer.com/series/1244

Dimitris Fotakis · David Ríos Insua (Eds.)

Algorithmic Decision Theory

7th International Conference, ADT 2021
Toulouse, France, November 3–5, 2021
Proceedings

Springer

Editors
Dimitris Fotakis (iD)
National Technical University of Athens
Athens, Greece

David Ríos Insua (iD)
Consejo Superior de Investigaciones
Cientificas
Madrid, Madrid, Spain

ISSN 0302-9743   ISSN 1611-3349   (electronic)
Lecture Notes in Artificial Intelligence
ISBN 978-3-030-87755-2   ISBN 978-3-030-87756-9   (eBook)
https://doi.org/10.1007/978-3-030-87756-9

LNCS Sublibrary: SL7 – Artificial Intelligence

This Springer imprint is published by the registered company Springer Nature Switzerland AG
The registered company address is: Gewerbestrasse 11, 6330 Cham, Switzerland

Preface

The 7th International Conference on Algorithmic Decision Theory (ADT 2021), held in November 2021, at the University of Toulouse 1 Capitole, France, has continued in the tradition established by previous ADT conferences in providing a unique opportunity for scientific exchange among researchers and practitioners coming from diverse areas of computer science, economics, and operations research. Their joint aim is to improve the theory and practice of modern algorithmic decision support. Previous ADT conferences were held in Venice, Italy (2009); Piscataway, NJ, USA (2011); Brussels, Belgium (2013); Lexington, KY, USA (2015), Luxembourg (2017) and Durham, NC, USA (2019).

ADT 2021 received 58 submissions, which were all rigorously peer-reviewed by at least three Program Committee (PC) members, in a double-blind fashion. The papers were evaluated on the basis of originality, significance, and exposition. The PC eventually decided to accept 27 papers to be presented at the conference and to be included in the proceedings. The acceptance rate was 46.5%.

The program also included three invited talks by distinguished researchers in algorithmic decision theory, namely Battista Biggio (University of Cagliari, Italy), Edith Elkind (University of Oxford, UK), and Christophe Labreuche (Thales Research and Technology and SINCLAIR AI Lab, France). In addition, ADT 2021 featured a PhD student day, co-chaired by Georgios Amanatidis (University of Essex, UK), Roi Naveiro (Institute of Mathematical Sciences, Spain), and Arianna Novaro (University of Amsterdam, Netherlands).

The works accepted for publication in this volume cover most of the major aspects of algorithmic decision theory, such as preference modeling and elicitation, computational social choice, preference aggregation, voting, fair division and resource allocation, coalition formation, stable matchings, and participatory budgeting.

We thank the authors for their interest in submitting and presenting their high quality recent work to ADT 2021, as well as the PC members and the external reviewers for their great work in evaluating the submissions. We also want to thank the Artificial Intelligence Journal, the EURO Working Group on Preference Handling, the University of Toulouse 1 Capitole, the European Office of Aerospace Research and Development (EOARD), and the AXA Research Fund (through the AXA-ICMAT Chair in Adversarial Risk Analysis) for their generous financial support. We are grateful to the University of Toulouse 1 Capitole for hosting ADT 2021. Special thanks also go to the members of the local organizing committee, Umberto Grandi (chair), Sylvie Doutre, Laurent Perrussel, and Pascale Zaraté, for their excellent organization and local arrangements work, and to Rachael Colley for her help with the conference website. Finally, we want to thank Alexis Tsoukiàs, for his invaluable advice and support, Anna Kramer at Springer for helping with the proceedings, and the EasyChair conference management system.

August 2021   Dimitris Fotakis
                                                David Ríos Insua

Organization

Program Committee

Haris Aziz	The University of New South Wales and Data61, CSIRO, Australia
Katarina Cechlarova	Pavol Jozef Šafárik University, Slovakia
Lea Deleris	BNP Paribas, France
Luis Dias	University of Coimbra, Portugal
Love Ekenberg	International Institute of Applied Systems Analysis, Austria
Ulle Endriss	University of Amsterdam, The Netherlands
Piotr Faliszewski	AGH University of Science and Technology, Poland
Angelo Fanelli	CNRS, CREM, France
Aris Filos-Ratsikas	University of Liverpool, UK
Dimitris Fotakis (Co-chair)	National Technical University of Athens, Greece
Laurent Gourves	CNRS, LAMSADE, Université Paris-Dauphine, France
Tatiana V. Guy	Institute of Information Theory and Automation, Czech Academy of Sciences, Czech Republic
Carlos Henggeler Antunes	University of Coimbra, Portugal
Maria Kyropoulou	University of Essex, UK
Jérôme Lang	CNRS, LAMSADE, Université Paris-Dauphine, France
David Manlove	University of Glasgow, UK
Evangelos Markakis	Athens University of Economics and Business, Greece
Reshef Meir	Technion-Israel Institute of Technology, Israel
Fanny Pascual	LIP6, Université Pierre et Marie Curie - Paris 6, France
Patrice Perny	LIP6, Université Pierre et Marie Curie - Paris 6, France
Hans Peters	Maastricht University, The Netherlands
Marc Pirlot	Université de Mons, Belgium
Maria Polukarov	King's College London, UK
David Rios Insua (Co-chair)	Universidad Rey Juan Carlos, Spain
Fred Roberts	Rutgers University, USA
Francesca Rossi	IBM Research, USA
Antonio Salmeron	University of Almería, Spain
Ahti Salo	Aalto University School of Science and Technology, Finland
Maria Serna	Universitat Politècnica de Catalunya, Spain
Alexis Tsoukias	CNRS, LAMSADE, Université Paris-Dauphine, France
Carmine Ventre	King's College London, UK
Toby Walsh	The University of New South Wales, Australia

Organizing Committee

Rachael Colley	University of Toulouse 1 Capitole, France
Sylvie Doutre	University of Toulouse 1 Capitole, France
Umberto Grandi (Chair)	University of Toulouse 1 Capitole, France
Laurent Perrussel	University of Toulouse 1 Capitole, France
Pascale Zaraté	University of Toulouse 1 Capitole, France

Additional Reviewers

Amanatidis, Georgios
Archbold, Thomas
Auletta, Vincenzo
Bachrach, Yoram
Bielous, Gili
Boixel, Arthur
Bouveret, Sylvain
Bredereck, Robert
Cailloux, Olivier
Eirinakis, Pavlos
Fairstein, Roy
Ferraioli, Diodato
Greco, Gianluigi
Hamm, Thekla
Haret, Adrian
Kalavasis, Alkis
Kóczy, László
Lee, Barton

Mathioudaki, Angeliki
McKay, Michael
Olckers, Matthew
Panageas, Ioannis
Papasotiropoulos, Georgios
Perrussel, Laurent
Rastegari, Baharak
Rey, Simon
Roberts, Fred
Serafino, Paolo
Spanjaard, Olivier
Sun, Zhaohong
Sziklai, Balázs R.
Terzopoulou, Zoi
Voudouris, Alexandros
Wilczynski, Anaëlle
Yang, Yongjie

Abstracts of Invited Talks

Machine Learning (for) Security: Lessons Learned and Future Challenges

Battista Biggio[1,2] (iD)

[1] University of Cagliari, Italy
[2] Pluribus One

Abstract. In this talk, I will briefly review some recent advancements in the area of machine learning security [2] with a critical focus on the main factors which are hindering progress in this field. These include the lack of an underlying, systematic and scalable framework to properly evaluate machine-learning models under adversarial and out-of-distribution scenarios, along with suitable tools for easing their debugging. The latter may be helpful to unveil flaws in the evaluation process [7], as well as the presence of potential dataset biases and spurious features learned during training. I will finally report concrete examples of what our laboratory has been recently working on to enable a first step towards overcoming these limitations [1, 3], in the context of Android [6] and Windows malware detection [4, 5].

Keywords: Machine learning · Computer security · Adversarial machine learning · Malware detection

References

1. Biggio, B., et al.: Evasion attacks against machine learning at test time. In: Blockeel H., Kersting K., Nijssen S., Železný F. (eds) ECML PKDD 2013. LNCS, vol. 8190, pp. 387–402. Springer, Heidelberg (2013). https://doi.org/10.1007/978-3-642-40994-3_25
2. Biggio, B., Roli, F.: Wild patterns: ten years after the rise of adversarial machine learning. Pattern Recogn. **84**, 317–331 (2018)
3. Biggio, B., Nelson, B., Laskov, P.: Poisoning attacks against support vector machines. In: 29th ICML. pp. 1807–1814. Omnipress (2012)
4. Demetrio, L., Biggio, B., Lagorio, G., Roli, F., Armando, A.: Functionality-preserving black-box optimization of adversarial windows malware. IEEE Transactions on Information Forensics and Security 16, 3469–3478 (2021)
5. Demetrio, L., Coull, S.E., Biggio, B., Lagorio, G., Armando, A., Roli, F.: Adversarial EXEmples: a survey and experimental evaluation of practical attacks on machine learning for Windows malware detection. ACM Trans. Priv. Secur. (2021)
6. Demontis, A., et al.: Yes, machine learning can be more secure! a case study on android malware detection. IEEE Trans. Dep. Sec. Comp. **16**(4), 711–724 (2019).
7. Pintor, M., et al.: Indicators of attack failure: Debugging and improving optimization of adversarial examples. CoRR abs/2106.09947 (2021)

Mind the Gap: Fair Division With Separation Constraints

Edith Elkind

University of Oxford, UK
elkind@cs.ox.ac.uk

Abstract. This is the extended abstract for the ADT'21 invited talk. It is based on a series of papers with Erel Segal-Halevi and Warut Suksompong [1, 2, 3].

Keywords: Cake cutting · Land division · Maximin fair share

Motivated by the social distancing rules, we consider the task of fairly distributing a divisible good among several agents under the additional assumption that every two agents' shares must be *separated*. We start by analyzing the case where the good is the [0, 1] segment (usually referred to as 'cake'). In this model, the separation constraint is captured by specifying a separation parameter s such that for every pair of agents i, j and every pair of points x, y such that x is allocated to i and y is allocated to j it holds that $|x - y| \geq s$; metaphorically, the cake is cut by a blunt knife of width s.

We observe that in this setting a proportional allocation cannot be guaranteed. We therefore focus on the solution concept of *maximin fair share*. Intuitively, we first ask each agent to determine their fair share by executing a mental experiment where they need to cut the cake into n s-separated pieces (where n is the number of agents), and are allocated the piece that they value the least; their fair share is then defined as the most they can guarantee for themselves under this protocol, and an allocation is considered fair if it provides each agent with a piece that they value at least as much as their fair share.

We show that a natural moving-knife protocol guarantees that each agent receives their fair share, i.e., maximin fair share allocations exist. However, to execute that protocol, agents need to be able to compute their fair shares, and it turns out that there is no finite algorithm that can accomplish this task. We circumvent this issue by providing an algorithm that approximates the agents' shares up to an arbitrarily small error, as well as a polynomial-time algorithm for the case where all agents have piecewise constant valuations that are specified explicitly as part of the input.

We then extend our analysis of fair division with separation to richer settings: we consider fair division of a pie (i.e., a circular cake), land (i.e, a 2-dimensional good), and graphical cake (i.e., the 'cake' formed by edges of a graph). In many of these settings, maximin fair allocation is no longer guaranteed to exist. We therefore consider its *ordinal approximation*, defined as follows. Recall that, in the definition of maximin fair share, we asked each agent to perform a mental experiment where they cut the good into n pieces. We now ask each agent to re-run that experiment, but cut the good into $k > n$ pieces, for a given value of k; we refer to the outcome of that experiment as *k-fair share*. Increasing the value of k corresponds to the agents being less ambitious in terms of what they want to receive, so we are interested in the smallest value of k such that each agent can be guaranteed their k-fair share.

It turns out that for the circular cake with separation constraints it suffices to set $k = n + 1$; however, the circular cake is more challenging than the interval cake from an algorithmic perspective. For general graphical cakes, under a mild technical assumption, we can set $k = n + f$, where f is the feedback vertex set number of the underlying graph; in particular, if the graph is a tree, each agent can be guaranteed her maximin fair share (however, somewhat surprisingly, a natural extension of the moving knife protocol to trees may fail to find a maximin fair allocation).

For land division, our results depend on the geometric shape of the land itself as well as the shapes of the agents' pieces. For instance, if each agent has to be allocated a square piece of land, we can set $k = 4n - 5$. However, if agents' pieces can be arbitrary axis-aligned rectangles (and the land itself is an axis-aligned rectangle), we get a much weaker upper bound of $k = 2^{n+2}$, and converting it into a finite algorithm comes at an additional cost.

References

1. Elkind, E., Segal-Halevi, E., Suksompong, W.: Mind the gap: cake cutting with separation. In: Proceedings of AAAI 2021
2. Elkind, E., Segal-Halevi, E., Suksompong, W.: Graphical cake cutting via maximin share. In: Proceedings of IJCAI 2021
3. Elkind, E., Segal-Halevi, E., Suksompong, W.: Keep your distance: land division with separation. In: Proceedings of IJCAI 2021

Hierarchical Decision Models with Interacting Criteria: Preference Learning, Identifiability and Explainability

Christophe Labreuche[1,2] (ID)

[1] Thales Research and Technology, Palaiseau, France
christophe.labreuche@thalesgroup.com
[2] SINCLAIR AI Lab, Palaiseau, France

Multi-Criteria Decision Aiding (MCDA) aims at comparing several alternatives on the basis of multiple and possibly conflicting criteria. In industrial applications such as Air Traffic Management, elaborate decision models need to be used. Firstly, criteria are not independent as there are often statistical correlations and/or preferential dependencies among attributes. The leading model for capturing such interactions is the Choquet integral. Secondly, the number of criteria can be relatively large. In order to have an interpretable model, the set of criteria is organized in a hierarchical way instead of in a flat way. This means that criteria are aggregated by means of several nested aggregation functions. Each node in this hierarchy has a clear meaning to the decision maker. This allows enriching the representation power of the model while reducing the complexity thanks to a smaller number of parameters.

The combine use of hierarchical models and the representation of interacting criteria is expected to bring significant added values in real applications. We will expose recent results in three directions.

- **Preference Learning.** The existing approaches in Operation Research consists in eliciting each aggregation function and each marginal utility function separately from an interaction with the decision maker. However, providing only local preference information may yield global inconsistency. The promise of preference learning is to go beyond these limitations, replacing time-consuming interactions with a user, by machine learning from a large quantity of (possibly noisy) preference data. The objective is to learn all parameters of the model simultaneously, which is challenging as the underlying optimization problem is no more convex. We will present a Preference Learning approach based on the representation of the decision model as a neural network.
- **Identifiability.** To ease learning and interpreting the parameters of the model, there should not be two hierarchical models with different parameters and possibly different hierarchies yielding the same decision function pointwise. We show identifiability for the hierarchical Choquet integral model. We are in particular able to relate structuring elements on the behavior of the model to the underlying hierarchy.
- **Explainability.** It is important in practice to explain the recommendations made by the model. Most of the time, the user does not need an in-depth explanation of the internal model mechanism, but he only wishes to understand which are the nodes in

the tree at the origin of the model outcome. This is obtained by computing an index measuring the influence of each node in the tree, on the preference between two alternatives. There are many connections with Feature Attribution (FA) in Machine Learning. Computing the level of contribution of a feature in a classification black-box model or that of a criterion in a MCDA model is indeed similar. The Shapley value is one of the leading concepts for FA. Unlike our situation, feature attribution only computes the influence of leaves in a model. We will show that the Shapley value is not appropriate on trees, when we are interested in knowing the contribution level of not only the leaves but also other nodes. We will then define a consistent value for trees.

References

1. Bresson, R., Cohen, J., Hüllermeier, E., Labreuche, C., Sebag, M.: Neural representation and learning of hierarchical 2-additive Choquet integrals. In: Proceedings of the Twenty-Eight International Joint Conference on Artificial Intelligence (IJCAI 2020), pp. 1984–1991, Yokohoma, Japan (2020)
2. Bresson, R., Cohen, J., Hüllermeier, E., Labreuche, C., Sebag, M.: On the identifiability of hierarchical decision models. In: Proceedings of the 18th International Conference on Principles of Knowledge Representation and Reasoning (KR 2021). Accepted 2021
3. Choquet, G.: Theory of capacities. Annales de l'Institut Fourier, 5, 131–295 (1953)
4. Fallah Tehrani, A., Labreuche, C., Hüllermeier, E.: Choquistic Utilitaristic Regressio. In: Decision Aid to Preference Learning (DA2PL 2014) Workshop, Chatenay-Malabry, France, November 2014
5. Figueira, J., Greco, S., Ehrgott, M., (eds.) Multiple Criteria Decision Analysis: State of the Art Surveys 2nd edn. Kluwer Acad. Publ. (2016)
6. Fürnkranz, J., Hüllermeier, E.: Preference Learning. Springer-Verlag. Heidelberg. https://doi.org/10.1007/978-3-642-14125-6(2010)
7. Grabisch, M., Labreuche, C.: A decade of application of the Choquet and Sugeno integrals in multi-criteria decision aid. Ann. Oper. Res. 175, 247–286 (2010)
8. Guidotti, R., Monreale, A., Ruggieri, S., Turini, F., Giannotti, F., Pedreschi. D.: A survey of methods for explaining black box models. ACM Comput. Surv. 51(6), Article 93 (2018)
9. Labreuche, C., Fossier, S.: Explaining multi-criteria decision aiding models with an extended shapley value. In: Proceedings of the Twenty-Seventh International Joint Conference on Artificial Intelligence (IJCAI 2018), pp. 331–339, Stockholm, Sweden, July 2018
10. Labreuche, C.: A general framework for explaining the results of a multi-attribute preference model. Artif. Intell. 175, 1410–1448, (2011)
11. Labreuche, C., Hüllermeier, E., Vojtas, P., Fallah Tehrani, E.: On the Identifiability of Models in Multi-Criteria Preference Learning. In: Decision Aid to Preference Learning (DA2PL 2016) workshop, Paderborn, Germany, November 2016

12. Labreuche, C.: Explaining hierarchical multi-linear models. In: Proceedings of the 13th international conference on Scalable Uncertainty Management (SUM 2019), Compiègne, France, December 2019
13. Labreuche, C.: Explanation with the winter value: efficient computation for hierarchical Choquet integrals. In: Proceedings of the Sixteenth European Conference on Symbolic and Quantitative Approaches to Reasoning with Uncertainty (ECSQARU 2021). Accepted 2021
14. Lundberg, S., Lee, S.I.: Unified approach to interpreting model predictions. In: Guyon, I., Luxburg, U.V., Bengio, S., Wallach, H., Fergus, R., Vishwanathan, S., Garnett, R., (eds.) 31st Conference on Neural Information Processing Systems (NIPS 2017), pp. 4768–4777, Long Beach, CA, USA (2017)
15. Shapley, L.S.: A value for n-person games. In: Kuhn, H.W., Tucker, A.W. (eds.) Contributions to the Theory of Games, vol. II, number 28 in Annals of Mathematics Studies, pp. 307–317. Princeton University Press (1953)

Contents

Coalition Formation

Stable Matchings

Participatory Budgeting

Computational Social Choice
and Preference Modelling

Aggregating Preferences Represented by Conditional Preference Networks

Abu Mohammad Hammad Ali[(✉)], Howard J. Hamilton, Elizabeth Rayner, Boting Yang, and Sandra Zilles

University of Regina, Regina, SK, Canada
{howard.hamilton,rayner3e,boting.yang,sandra.zilles}@uregina.ca

Abstract. This paper focuses on the task of aggregating preference orders over combinatorial domains, where both the individual and the aggregate preference orders are represented as Conditional Preference Networks (CP-nets). We propose intuitive objective functions for finding an optimal aggregate CP-net, as well as corresponding optimal efficient aggregation algorithms for inputs with certain structural properties.

Keywords: Preference representation · Preference aggregation

1 Introduction

Preference aggregation aims at ranking a set of outcomes so as to collectively represent the rankings for a group of individuals [11,21]. Applications include recommender systems, multi-criteria object selection, and meta-search engines.

We assume each outcome is a tuple of attribute-value pairs, allowing for a compact representation of preferences. For example, if outcomes represent movies to be ranked, and *genre* is an attribute, the rule "*genre = comedy* is preferred over *genre = drama*," entails a large number of preference expressions between specific movies. One formal model of preference relations over such outcomes is the Conditional Preference Network (CP-net) [5], which expresses user preferences of the form "given attribute-value pairs $(V_1, v_1), \ldots, (V_m, v_m)$, I prefer value w over w' in attribute W". Thus, preferences over the values of an attribute W are conditioned on the values of other attributes (here $V_1, \ldots, V_m$), called the parents of W. Not all preference relations can be expressed by CP-nets [6].

This paper is the first to treat the task of aggregating preferences represented by CP-nets as an optimization problem, with the aggregate preference relation itself represented as a CP-net, called *consensus CP-net*, that is optimally close to the collection of input CP-nets using a proposed distance measure over outcome pairs. Algorithms solving this problem, since they output CP-nets, can be applied hierarchically and may hence be useful in applications where there are

This work was supported by the Natural Sciences and Engineering Research Council of Canada (NSERC).

D. Fotakis and D. Ríos Insua (Eds.): ADT 2021, LNAI 13023, pp. 3–18, 2021.
https://doi.org/10.1007/978-3-030-87756-9_1

too many users (input CP-nets) to do a global optimization. Another advantage of outputs in the form of CP-nets is that known methods for preference optimization, constrained optimization, and outcome ordering work on them directly.

We first define three measures of dissimilarity between two CP-nets, which are in turn used as the basis of objective functions. A noncommutative dissimilarity measure we propose counts the outcome pairs on which a user expresses a clear preference that is not entailed by the consensus CP-net; a commutative variant is also defined. Both are called measures of *total disagreement*, since they take all outcome pairs into account. However, due to the lack of efficient tools for computing total disagreement, let alone minimizing it over a set of CP-nets, we first approach a different measure, *swap disagreement*, which counts the so-called *swaps* on which a user's CP-net and the consensus CP-net disagree. A swap is a pair of outcomes that differ only in a single attribute; reasoning with such pairs is often of advantage compared to reasoning with general outcome pairs.

We present an algorithm that computes an optimal consensus CP-net for any set of complete[1] input CP-nets using the cumulative swap disagreement to the inputs as an objective function. If the input CP-nets are "too dissimilar", this algorithm has exponential runtime. By contrast, it has linear runtime if the input satisfies a natural structural condition. While we formally relate swap disagreement to both notions of total disagreement, our algorithm for minimizing swap disagreement does not minimize total disagreement in general. Even for aggregating just two CP-nets of the simplest ("separable") structure, rather complex consensus CP-nets are needed to minimize noncommutative total disagreement. In the same case, swap disagreement is minimized by a separable CP-net.

A fourth proposed objective function sums the squares of attribute-wise swap disagreements, to penalize a consensus CP-net that disagrees with a user substantially in an individual attribute. A linear-time algorithm is shown to minimize this objective function for separable input CP-nets.

To sum up, the main contribution of our paper is to (i) propose four CP-net aggregation problems via four objective functions, (ii) solve one of these problems optimally (on special inputs also efficiently), (iii) discuss how to solve the other three optimally for special inputs, and (iv) show that the solution to one of these problems is equivalent to the majority CP-net used in earlier work on a completely different algorithmic problem.

2   Relation to Previous Work

When aggregating explicit preference orders, the input is a set of permutations over the outcome space [21], and the output is a best collective permutation, i.e., one that minimizes some objective function wrt the set of input permutations [2,10,11]. If outcomes are represented as tuples of values over attributes, one can instead aggregate compact preference representations [1], using models like CP-nets [5], LP-trees [4], or utility-based models. We focus on CP-nets.

---

[1] Completeness is a limiting condition to be defined in Sect. 3.

One approach to aggregating CP-nets is using an mCP-net [19, 20] – a set of partial CP-nets as opposed to a single CP-net with aggregate preferences. Outcome ordering or optimization queries require applying a voting rule on the set of partial CP-nets. For every query, a new round of voting is carried out. By comparison, our approach, while limited to complete CP-nets so far, builds a consensus CP-net with a single round of voting, upon which all subsequent queries can be answered using the consensus CP-net, without the need to store the input CP-nets. Our algorithms have advantages over that by Cornelio et al. [9]. The latter represents the aggregate preferences as a Probabilistic CP-net (PCP-net) [3]. As a result, the notions of (and algorithms for) outcome optimality and dominance must be modified to the probabilistic setting.

Lang [15] presents an algorithm, that, given acyclic input CP-nets, elicits votes sequentially over attributes, choosing the winning value at each step. While our approach is also sequential over attributes, it assigns conditional preference rules for each attribute. Lang's approach thus finds a consensus outcome, while ours outputs a consensus CP-net. This enables us to answer dominance queries in addition to optimization queries. Lang discusses preference aggregation, but not with the focus on an objective function. Instead, his aim is to ensure certain fairness properties for elections. Xia et al. [23, 24] show that sequential voting may lead to paradoxical outcomes. To avoid this problem, they assume that all voters agree on a linear order over the attributes. Our approach makes no such assumption. Methods for sequential voting over attributes are described in detail in [16] and extended to constrained CP-nets in [13].

Voting over general CP-nets is possible by a hypercube-wise composition of voting rules [7, 8, 17, 22]. However, the size of the hypercube is exponential in the number of attributes. While one of our methods also builds exponentially large CP-nets in the worst case, it runs in polynomial time for CP-nets that are structurally close (refer to Definition 2). Interestingly, the hypercube from [8] and [17] is identical to that induced by the output CP-net for one of our objective functions, and it is already mentioned in [7] that this CP-net can be built directly from the input CP-nets, without the complex hypercube approach. Further, the optimal outcome using one of the approaches from [9] equals that of the hypercube, or our output CP-net. Thus, three different approaches of multiagent reasoning yield identical results.

The primary difference between our work and previous studies on preference aggregation is in our treating CP-net aggregation as an optimization problem for a family of objective functions, which are based on pairwise distance between CP-nets. We propose three new distance measures, two of which satisfy all properties of distance functions. Loreggia et al. [18] also propose two distance functions for CP-nets, both approximations to the Kendall Tau Distance extended for partial orders [12], but they do not focus on aggregating CP-nets.

3   Definitions and Terminology

Following the notation in [5], a CP-net N is a directed graph (V, E), where $V = \{V_1, \ldots, V_n\}$ is a set of n binary attributes. Each vertex $V_i \in V$ represents

Fig. 1. A very simple CP-net N.

an attribute, with $Dom(V_i) = \{v_i, v_i'\}$ denoting the set of possible values of V_i. An edge $e \in E$ from V_j to V_i indicates that the user's preferences over V_i depend on the value assigned to V_j, where a preference over V_i is a total order over the values in $Dom(V_i)$. We refer to V_j as a parent of V_i. The set of parents of V_i in a CP-net N is denoted by $Pa(N, V_i)$.

For each V_i, the preference order over $Dom(V_i)$ is given in a Conditional Preference Table (CPT), denoted $CPT(N, V_i)$. For example, the entry $v_i \succ v_i'$ is read "v_i is preferred over v_i'." If $|Pa(N, V_i)| = k$, then $CPT(N, V_i)$ has up to 2^k entries, called CPT rules. In practice, CP-nets might be incomplete, i.e., some CPTs do not have the maximum possible number of CPT rules. In this paper, we exclusively consider *complete* CP-nets, i.e., the CPT for any attribute V_i with k parents has 2^k CPT rules, one for each assignment of values to $Pa(N, V_i)$.

An assignment of values to each attribute in a set is called an instantiation of the set. The set of all instantiations of a set $X \subseteq V$ is denoted by $Inst(X)$, where $Inst(\emptyset)$ contains only the empty tuple. An outcome o is any element of $Inst(V)$. For any outcome $o \in Inst(V)$ and any $V_i \in V$, we use $o[V_i]$ $(\in Dom(V_i))$ to denote the value of V_i in o, so that we can think of o as the tuple $(o[V_1], \ldots, o[V_n])$.

Preferences over attributes are interpreted under a *ceteris paribus* assumption, which is natural in many real-life applications [5]. A CPT rule for attribute V_j, in the form $\beta : v_j \succ v_j'$, where $\beta \in Inst(Pa(V_j))$, then means that, if two outcomes are identical except for their value in V_j *and* their instantiation of the parent attributes of V_j coincides with β, then the one with value v_j is preferred over the one with value v_j' in V_j. The transitive closure of these preferences given by a CP-net N yields N's associated preference relation over the outcome space.

For example, consider the CP-net N in Fig. 1 with the set of attributes $\{A, B\}$, $Pa(N, A) = \emptyset$, $Pa(N, B) = \{A\}$, $Dom(A) = \{a, a'\}$, $Dom(B) = \{b, b'\}$, $CPT(N, A) = \{a' \succ a\}$, and $CPT(N, B) = \{a' : b' \succ b, a : b \succ b'\}$. Here $CPT(N, B)$ dictates that when A is assigned a, $b \succ b'$ and thus $ab \succ ab'$. Similarly, we get $a'b \succ ab$ and $a'b' \succ a'b$. The transitive closure of these orderings yields the preference order over the entire set of outcomes: $a'b' \succ a'b \succ ab \succ ab'$. Note that, in general, even complete CP-nets induce only a partial order over the outcome space, with some outcome pairs not being comparable.

The term *swap over* V_i refers to a pair of outcomes differing on the value of exactly one attribute, namely V_i. A *separable* CP-net is one in which no attribute has any parents. The *size* of a CP-net is the total number of its CPT rules.

4   Disagreement and Objective Functions

Suppose N^* is a proposed consensus CP-net for a set of CP-nets over the same attribute set V. How to quantify the disagreement between N^* and an individual input CP-net N, depends on one's interpretation of incomparability.

One option would be to interpret incomparability as indifference. In particular, if the CP-net N entails neither $o \succ o'$ nor $o' \succ o$, one would not count the pair (o, o') when measuring the disagreement between N and N^*. If, however, N entails $o \succ o'$ or $o' \succ o$, while N^* finds (o, o') incomparable, one would count (o, o') as a disagreement, since the user expresses a preference that would be neglected by N^*. Formally, we define the resulting *noncommutative disagreement* between two CP-nets for a given (o, o') as follows.

$$\delta_{nc}(N, N^*)(o, o') = \begin{cases} 0 & \text{if } N \text{ does not order } (o, o') \\ 0 & \text{if } N \text{ and } N^* \text{ order } (o, o') \text{ the same way} \\ 1 & \text{otherwise} \end{cases}$$

This disagreement notion makes our restriction to complete CP-nets less limiting: any incomplete input CP-net N is subsumed by a complete CP-net N' such that $\delta_{nc}(N, N')(o, o') = 0$ for all (o, o'), rendering the incomplete and the complete CP-nets equivalent for our purposes.

A second option is to interpret incomparability of two outcomes as explicitly forbidding to favor one over the other, yielding the following *commutative disagreement* measure.

$$\delta_c(N, N^*)(o, o') = \begin{cases} 0 & \text{if neither } N \text{ nor } N^* \text{ orders } (o, o') \\ 0 & \text{if } N \text{ and } N^* \text{ order } (o, o') \text{ the same way} \\ 1 & \text{otherwise} \end{cases}$$

In what follows, X denotes the set of all outcome pairs (excluding duplicates, i.e., X has only one of (o_1, o_2) and (o_2, o_1)) and $X_{swap} \subseteq X$ is its subset of swaps.

Definition 1. The *noncommutative total disagreement* of two CP-nets N and N^* is defined as

$$\Delta_{nc}^{total}(N, N^*) = \sum_{(o,o') \in X} \delta_{nc}(N, N^*)(o, o').$$

The *commutative total disagreement* Δ_c^{total} is defined analogously using δ_c instead of δ_{nc}. The *swap disagreement* of N and N^* is defined as

$$\Delta^{swap}(N, N^*) = \sum_{(o,o') \in X_{swap}} \delta_{nc}(N, N^*)(o, o').$$

While Δ_{nc}^{total} is noncommutative, swap disagreement commutes for complete CP-nets. In the latter, swaps are always comparable [6] as they differ in exactly one attribute, whose CPT determines the preference order for the pair. Thus,

$$\Delta^{swap}(N, N^*) = \sum_{(o,o') \in X_{swap}} \delta_c(N, N^*)(o, o'),$$

i.e., δ_{nc} and δ_c can be exchanged in the definition of Δ^{swap}.

When given a tuple $T = (N_1, \ldots, N_t)$ of input CP-nets, we will try to construct a CP-net N^* that minimizes the cumulative total disagreement values over all N_i, i.e., the sum

$$f_{\alpha,T}^{total}(N^*) = \Sigma_{1 \leq i \leq t} \Delta_\alpha^{total}(N_i, N^*),$$

where $\alpha \in \{nc, c\}$. Note that $f_{nc,T}^{total}(N^*) \leq f_{c,T}^{total}(N^*)$.

For reasons detailed below, we will first focus on finding a best consensus in terms of *swap disagreement*, i.e., minimizing the cumulative swap disagreement

$$f_T^{swap}(N^*) = \Sigma_{1 \leq i \leq t} \Delta^{swap}(N_i, N^*). \tag{1}$$

We drop the subscript T from $f_{\alpha,T}^{total}$ and f_T^{swap} when T is clear from the context.

Note that the objective functions studied here may in general have multiple optimal solutions for any given input. The focus of our work is to find one such optimal solution – not all of them. We leave the latter problem for future work.

5   Why Swap Disagreement?

One might prefer to minimize $f_{\alpha,T}^{total}$ rather than f_T^{swap}, since total disagreement captures differences between preference orders over the full outcome space. However, we also consider swap disagreement for several reasons.

First, the literature offers substantially more tools for addressing swap disagreement than for addressing total disagreement. This is partly because deciding whether a CP-net entails $o_1 \succ o_2$ is NP-hard for an arbitrary outcome pair (o_1, o_2) but has an efficient solution when (o_1, o_2) is a swap [5]. In particular, swap disagreement can be computed in polynomial time. Second, every complete CP-net orders every swap, which means that both Δ_{nc}^{total} and Δ_c^{total} yield the same notion of swap disagreement when restricted to only swaps. Third, a preference on a swap can immediately be attributed to the CPT of the swapped attribute, which allows us to decompose the problem of optimizing swap disagreement into subproblems separated by attributes.

Interestingly, when comparing two separable CP-nets, our measures of total disagreement become functions of the swap disagreement, as Theorem 1 shows.

Theorem 1. *Suppose an input CP-net N is separable, and N^* is an arbitrary consensus CP-net that differs from N in exactly $m \leq n$ CPTs. Then*

$$\Delta^{swap}(N, N^*) \leq m \cdot 2^{n-1} \text{ and}$$

$$\Delta_{nc}^{total}(N, N^*) \leq \sum_{l=1}^{n} \left(\binom{n}{l} - \binom{n-m}{l} \right) \cdot 2^{n-l}.$$

Moreover, if N^* is separable, these two inequalities are equalities, while

$$\Delta_c^{total}(N, N^*) = \sum_{l=1}^{m} \binom{m}{l} \cdot 2^{n-l} + 2 \sum_{l=1}^{m} \sum_{k=1}^{n-m} \binom{m}{l} \binom{n-m}{k} \cdot 2^{n-l-k}.$$

Proof. (Sketch.) Let $P \subseteq V$ be the set of attributes on which N and N^* differ, $|P| = m$. Since each CPT in the separable CP-net N orders exactly 2^{n-1} swaps, the first inequality is obvious. If N^* is separable, then all m attributes in P order all of their 2^{n-1} swaps differently, yielding $\Delta^{swap}(N, N^*) = m \cdot 2^{n-1}$.

For the second inequality, a counting argument (details omitted) shows that exactly $\sum_{l=1}^{n} \binom{n}{l} \cdot 2^{n-l}$ outcome pairs are comparable in N.

Now let (o_1, o_2) be an arbitrary pair of outcomes that differ on exactly l attributes (the "swapped" attributes) such that (o_1, o_2) is comparable in N. Let $\ell \leq l$ be the number of swapped attributes whose CPTs differ in N and N^*. For (o_1, o_2) to be ranked differently in N^* than in N, at least one of these ℓ attributes must be in P. That means, of the $\binom{n}{l} \cdot 2^{n-l}$ pairs with swapped attributes that are comparable by N, at least $\binom{n-m}{l} \cdot 2^{n-l}$ will be ranked the same way in N and N^*, because none of their l swapped attributes are among the m crucial ones. Thus at most $(\binom{n}{l} - \binom{n-m}{l}) \cdot 2^{n-l}$ outcome pairs with l swapped attributes contribute to $\Delta_{nc}^{total}(N, N^*)$. Summing up over all choices for l yields the desired upper bound. When N^* is also separable, all outcome pairs counted in this sum will actually be ordered differently by N and N^*, making the bound tight.

Finally, the formula for $\Delta_c^{total}(N, N^*)$ is obtained by adding to $\Delta_{nc}^{total}(N, N^*)$ the number of pairs for which N^* entails an order, while N does not. (Details are omitted due to space constraints.)   □

6   Minimizing Cumulative Swap Disagreement

This section describes a simple algorithm M^{swap} that, given a tuple of arbitrary complete CP-nets over an attribute set V, constructs a consensus CP-net N^* that minimizes f^{swap}. Coincidentally, this algorithm is known to solve the problem of finding non-dominated outcomes [7]. While M^{swap} does not generally run in polynomial time, we will define an interesting class of inputs on which it does.

Algorithm M^{swap} works in two steps. First, for each input CP-net N_i and each attribute V_j, it replaces $Pa(N_i, V_j)$ by $\bigcup_{i \leq t} Pa(N_i, V_j)$ without changing the semantics of the input CP-nets. (When adding k parents to $Pa(N_i, V_j)$, each original CPT rule in $CPT(N_i, V_j)$ is replaced by 2^k rules with the same preference order - one for each context over the k added parents). Now, given any j, all input CP-nets have the same parent set for V_j. Second, M^{swap} aggregates the preferences over the modified input CP-nets. For each instantiation of the (joint) parent set of V_j, it lists in $CPT(N^*, V_j)$ the majority preference over v_j and v'_j among the t modified input CP-nets. A useful third step would be to remove irrelevant parents (as defined in [14]) from the consensus CP-net, yielding a more compact representation of the final consensus CP-net. This third step takes time linear in the size of N^* and quadratic in $|V|$, but it is not essential to our formal results and thus not mentioned any further.

Theorem 2. *For any tuple of complete CP-nets, M^{swap} outputs a complete consensus CP-net N^* that minimizes f^{swap}.*

Proof. Since the modified input CP-nets are semantically equivalent to the original ones, it suffices to show that the CP-net N^* output by M^{swap} minimizes f^{swap} wrt the modified CP-nets as input.

The latter is easy to see from the definition of f^{swap}, which is the cumulative pairwise swap disagreement between each individual input CP-net and the consensus CP-net. The CPTs for any attribute V_j in all the augmented input CP-nets contain rules for the same parent instantiations, and each such rule decides on one of two possible orders of a fixed set of swaps. To minimize the cumulative pairwise swap disagreement, one must simply order each of these fixed sets of swaps as the majority of input CP-nets do. □

Example 1 shows that the first step of M^{swap} may add an exponential number of rules to a CPT. In particular, there are instances for which each optimal solution *must* set the union of all parent sets of V_j in all input CP-nets as the parent set of V_j (see the proof of Theorem 3).

Example 1. For each even $n \geq 4$, consider a tuple $T^n = (N_1^n, \ldots, N_{n-1}^n)$ of $n - 1$ input CP-nets over $V = \{V_1, \ldots, V_n\}$. For each i, $Pa(N_i^n, V_1) = \{V_{i+1}\}$, and $\text{CPT}(N_i^n, V_1)$ has rules (i) $v_{i+1}' : v_1' \succ v_1$ and (ii) $v_{i+1} : v_1 \succ v_1'$. Note that the union of all parent sets of V_1 across all input CP-nets is $\{V_2, \ldots, V_n\}$. Further, for each i and each $j \neq 1$, let $Pa(N_i^n, V_j) = \emptyset$. While each original input CP-net contains only $(n+1)$ CPT rules, each modified one contains $2^{n-1}+n-1$ preference rules, since the parent set for V_1 becomes $\{V_2, \ldots, V_n\}$, and each modified CPT for V_1 contains one rule for each of the 2^{n-1} instantiations of this parent set. Thus the size of the consensus CP-net N^* output by M^{swap} is $2^{n-1} + n - 1$. In general, the exponential size of the optimal consensus in this case is unavoidable, cf. Theorem 3.

Theorem 3. *There are tuples of CP-nets T over n attributes such that T is of size $O(n^2)$, while any optimal consensus CP-net for T wrt f^{swap} is of size $\Omega(2^n)$.*

Proof. The claim is witnessed by $T = T^n$ from Example 1, for which the quadratic input size was already shown. For each even $n \geq 4$, consider a tuple $T^n = (N_1^n, \ldots, N_{n-1}^n)$ of $n - 1$ input CP-nets over $V = \{V_1, \ldots, V_n\}$. For each i, $Pa(N_i^n, V_1) = \{V_{i+1}\}$, and $\text{CPT}(N_i^n, V_1)$ has rules (i) $v_{i+1}' : v_1' \succ v_1$ and (ii) $v_{i+1} : v_1 \succ v_1'$. Note that the union of all parent sets of V_1 across all input CP-nets is $\{V_2, \ldots, V_n\}$. Further, for each i and each $j \neq 1$, let $Pa(N_i^n, V_j) = \emptyset$. Thus, the input T^n to M^{swap} contains only $(n - 1)(n + 1)$ CPT rules in total $(n+1$ for each input CP-net). However, the output contains $(n-1)(2^{n-1}+n-1)$ preference rules, since the modified parent set for V_1 becomes $\{V_2, \ldots, V_n\}$, and each CPT for V_1 in N^* contains one rule for each of the 2^{n-1} instantiations of this parent set. Then the size of the CP-net output by M^{swap} is $2^{n-1} + n - 1$.

It remains to verify that any optimal consensus CP-net for T^n wrt f^{swap} is of size $\Omega(2^n)$. Note that a CPT for an attribute V_i in an optimal consensus CP-net for T^n wrt f^{swap} must assign, for each possible instantiation γ of the parents of

V_i, the preference order over $\{v_i, v_i'\}$ that would be preferred by the majority of input CP-nets under the same instantiation γ.

Let N^* be any acyclic CP-net such that there exists some $V_k \in \{V_2, \ldots, V_n\} \setminus Pa(N^*, V_1)$. For simplicity, we assume that $Pa(N^*, V_1) = \{V_2, \ldots, V_n\} \setminus \{V_k\}$ (in case V_1 has fewer than $n - 2$ parents in N^*, we can simply add parents that are de facto irrelevant.)

Consider the instantiation

$$\gamma^k = v_{j_1}' v_{j_2}' \ldots v_{j_{\frac{n}{2}-1}}' v_{j_{\frac{n}{2}}} \ldots v_{j_{n-2}},$$

where $\{j_1, \ldots, j_{n-2}\} = \{2, \ldots, n\} \setminus \{k\}$. This instantiation γ^k assigns values to all parents of V_1 in N^*. Given γ^k, clearly $\frac{n}{2} - 1$ input CP-nets from the tuple T^n have the preference $v_1' \succ v_1$ for attribute V_1, and the remaining $\frac{n}{2} - 1$ input CP-nets from the tuple T^n have the preference $v_1 \succ v_1'$ for V_1.

Next, consider two instantiations extending γ^k, namely $\gamma^k v_k'$, which extends γ^k by assigning the value v_k' to V_k, and $\gamma^k v_k$, which extends γ^k by assigning the value v_k to V_k. With $\gamma^k v_k'$, more than half of the input CP-nets in T^n have the preference $v_1' \succ v_1$ for attribute V_1. With $\gamma^k v_k$, more than half of the input CP-nets in T^n have the opposite preference, namely $v_1 \succ v_1'$, for attribute V_1. Thus, the following modification of N^* would yield a decrease in f^{swap}: add V_k as a parent to attribute V_1, and assign the different majority preference orders for $\gamma^k v_k'$ and $\gamma^k v_k$ over V_1 correspondingly to the CPT of V_1.

This means that N^* is not optimal. In particular, any optimal consensus CP-net for T^n wrt f^{swap} has V_k as a relevant parent of V_1, and, by analogy, all attributes in $\{V_2, \ldots, V_n\}$ as relevant parents of V_1. Hence, the CPT for V_1 in such optimal consensus CP-net is of size $\Omega(2^n)$. $\qquad\square$

Remark. One might raise the question whether the size of the optimal consensus CP-net in this proof is exponential only because our representation of complete CP-nets is unnecessarily large. When limiting ourselves to complete CP-nets, it would be sufficient in each CPT to list only the instantiations that lead to the less frequent preference rule. For example, the CPT

$$\{ab : c \succ c', \; a'b : c \succ c', \; ab' : c \succ c', \; a'b' : c' \succ c\}$$

could be stored in the more compact form $\{a'b' : c' \succ c\}$, implicitly stating that all non-listed instantiations entail the opposite preference ordering.

However, it is not hard to see that this more compact representation of complete CP-nets still requires size $\Omega(2^n)$ for any optimal consensus CP-net N for the tuple T^n wrt f^{swap}. To this end, note that the preference $v_1' \succ v_1$ occurs in exactly half of the rules in $CPT(N, V_1)$ for an optimal consensus CP-net N. Thus, even an implicit form of CPT has at least 2^{n-2} entries.

Any two input CP-nets in Example 1 are dissimilar in the sense that their preferences on V_1 depend on two disjoint parent sets. Intuitively, in such a situation, preference aggregation is difficult, e.g., when user 1 bases their preference over movie genres on the production year, while user 2 bases their preference over movie genres on the country of origin, and further users on yet further attributes. It therefore seems reasonable to limit our aggregation task to cases in which, for any input CP-net, the parent set of any attribute V_j misses a constant number of parents that are relevant for V_j in other input CP-nets.

Definition 2. Let $T = (N_1, \ldots, N_t)$ be a tuple of CP-nets. The *maximum parent difference* of T is defined as

$$\max_{V_j \in V} \{|\bigcup_{i \leq t} Pa(N_i, V_j)| - \min_{i \leq t} |Pa(N_i, V_j)|\}.$$

This immediately yields the following result.

Corollary 1. *Let k be a constant. Given a tuple T of complete CP-nets with maximum parent difference at most k, M^{swap} runs in time linear in the sum of the sizes of the CP-nets in T and quadratic in $|V|$, and yields a complete CP-net N^* that minimizes f^{swap}.*

When all input CP-nets are separable (implying a maximum parent difference of 0), M^{swap} adds no parents, so the output CP-net N^* is also separable. Theorem 1 then allows us to efficiently calculate $f_{nc}^{total}(N^*)$ and $f_c^{total}(N^*)$, using the number of CPTs in which N^* differs from each individual input CP-net. Note though that it is possible for two separable CP-nets N, N' to both minimize f^{swap}, while $f_{nc}^{total}(N) \neq f_{nc}^{total}(N')$ and $f_c^{total}(N) \neq f_c^{total}(N')$.

As an example, suppose n is even and consider the simple case of only two input CP-nets N_1 and N_2, both of which are separable and differ in all CPTs. Suppose $CPT(N_1, V_i) = \{v_i \succ v_i'\}$ when $1 \leq i \leq n/2$, and $CPT(N_2, V_i) = \{v_i \succ v_i'\}$ when $n/2+1 \leq i \leq n$. Then the CP-net N^* output by our algorithm is equal to each of N_1 and N_2 in exactly $n/2$ CPTs. By Theorem 2, it minimizes f^{swap} for (N_1, N_2). However, N_1 also minimizes f^{swap}; in fact, every separable CP-net minimizes f^{swap} for this particular input. For $f_{nc}^{total}(N^*)$ and $f_c^{total}(N^*)$, the consensus N_1 should be preferred over the consensus N^*: the value of $f_{nc/c}^{total}(N_1)$ is given by the formulas for $\Delta_{nc/c}^{total}$ using $m = n$ (as $\Delta_{nc/c}^{total}(N_1, N_1) = 0$), while $f_{nc/c}^{total}(N^*)$ is twice the value of $\Delta_{nc/c}^{total}$ with $m = n/2$ (for the disagreements with each of the two input CP-nets). It is easy to verify that $f_{nc}^{total}(N_1) = f_c^{total}(N_1) < f_{nc}^{total}(N^*) < f_c^{total}(N^*)$ when $n = 4$.

Corollary 2 will show that even for just two separable CP-nets, in most cases no separable CP-net minimizes f_{nc}^{total}.

7   Consensus of Two Separable CP-Nets

While the definitions of our commutative and noncommutative measures differ just slightly, the difficulty of minimizing cumulative disagreements based on the noncommutative measure become evident when looking at the seemingly simple case of just two input CP-nets. Such inputs have a trivial solution when optimizing f_c^{total}, as is seen from Proposition 1, but a less trivial solution when optimizing f_{nc}^{total}, even when both input CP-nets are separable.

Algorithm 1: Finding optimal consensus for f_{nc}^{total}

    Input   : A pair $T = (N_1, N_2)$ of separable CP-nets
    Output: CP-net N^* minimizing f_{nc}^{total}

1 $P_1 = \{V_i \mid \text{CPT}(N_1, V_i) = \text{CPT}(N_2, V_i)\}$
2 $P_2 = V \setminus P_1$
3 **if** $P_1 = \emptyset$ **or** $P_2 = \emptyset$ **then** $\{N^* = N_1; \text{break }\}$
4 **for** $V_i \in P_1$ $\{$ $\text{CPT}(N^*, V_i) := \text{CPT}(N_1, V_i)$ $\}$
5 **for** $V_i \in P_2$ **do**
6   Set $Pa(N^*, V_i) = P_1$
7   **for** all $\beta \in Inst(Pa(N^*, V_i))$ **do**
8   **if** β *contains an even number of parents assigned 0* **then**
9   Add $\beta : r$ to $\text{CPT}(N^*, V_i)$ for $r \in \text{CPT}(N_1, V_i)$
10   **end**
11   **else**
12   Add $\beta : r$ to $\text{CPT}(N^*, V_i)$ for $r \in \text{CPT}(N_2, V_i)$
13   **end**
14   **end**
15 **end**

Proposition 1. *Given a pair $T = (N_1, N_2)$ of CP-nets, both N_1 and N_2 minimize f_c^{total}.*

The proof is straightforward and omitted due to space constraints. It will turn out that Proposition 1 does not hold for f_{nc}^{total}. To optimize the latter, we propose Algorithm 1.

If the input CP-nets N_1 and N_2 are identical or differ in *all* CPTs, Algorithm 1 outputs one of them. Otherwise, it constructs a consensus CP-net N^* whose CPT for any attribute V_i is defined as follows. When N_1 and N_2 have equal CPTs for V_i, this CPT is copied to N^* (line 4). If the CPTs of N_1 and N_2 differ, V_i's parent set in N^* is the set of attributes for which N_1 and N_2 have identical CPTs (line 6). The corresponding CPT is then set so that all parents are relevant and one half of the rules side with N_1, while the other half side with N_2. For simplicity, Algorithm 1 assumes that $\{v_i, v_i'\} = \{0, 1\}$ for all i.

The running time of Algorithm 1 (and the size of its output) is exponential in the size of P_1 and linear in all other input parameters. In other words, the more similar the two input CP-nets are (as long as they are not identical), the more complicated the constructed consensus CP-net is, which seems counter-intuitive. To prove the correctness of Algorithm 1, we first formulate an obvious fact.

Lemma 1. *Given input $T = (N_1, N_2)$, a CP-net N^* minimizes f_{nc}^{total} if it fulfills the following conditions for each outcome pair $(o_1, o_2) \in X$: (i) if N_1 and N_2 both entail $o_1 \succ o_2$, then N^* entails $o_1 \succ o_2$; (ii) if N_1 and N_2 entail opposite orders on (o_1, o_2), then N^* entails some order on (o_1, o_2); (iii) if exactly one of N_1, N_2 entails an order over (o_1, o_2), then N^* entails the same order over (o_1, o_2).*

Theorem 4. *Given a pair (N_1, N_2) of separable CP-nets, Algorithm 1 yields a complete consensus CP-net N^* that minimizes f_{nc}^{total}.*

Proof. If $P_2 = \emptyset$, the claim is trivial. If $P_1 = \emptyset$, then any pair ordered by N_1 is ordered in the opposite way by N_2 and vice versa. Thus, any CP-net that orders all outcome pairs that are ordered by N_1 (and hence also by N_2) is optimal. No matter in which way such an outcome pair is ordered, it counts as one disagreement (either with N_1 or with N_2). In particular, N_1 is an optimal consensus CP-net wrt f_{nc}^{total}.

When neither P_1 nor P_2 is empty, let N^* be the output of Algorithm 1 and let $(o_1, o_2) \in X$. We will show that the three conditions formulated in Lemma 1 are fulfilled.

(i) is true, since $o_1 \succ o_2$ holds in both input CP-nets iff o_1 and o_2 differ only in attributes in P_1, where always o_1 has the value preferred in the two (identical) CPTs in N_1 and N_2. These CPTs stay in N^*, which thus also entails $o_1 \succ o_2$.

For (ii), note that opposite orders on (o_1, o_2) can only be entailed if o_1 and o_2 differ only in attributes in P_2. Let β be the instantiation of P_1 that occurs (identically) in o_1 and o_2. Depending on the number of zeroes in β, then N^* orders (o_1, o_2) either as N_1 does (line 9) or as N_2 does (line 12).

To obtain (iii), consider any $(o_1, o_2) \in X$ that is ordered by N_1 but not by N_2 (the opposite case is handled analogously). W.l.o.g., suppose N_1 entails $o_1 \succ o_2$ and let $V^\times$ be the set of all attributes on which o_1 and o_2 differ. It is not hard to verify that $P_i \cap V^\times \neq \emptyset$ for $i = 1, 2$. All attributes in $P_1 \cap V^\times$ are assigned the more preferred value (for both N_1 and N_2) in o_1, while all attributes in $P_2 \cap V^\times$ are assigned the more preferred value for N_1 in o_1. We now claim that N^* entails $o_1 \succ o_2$. Let β_2 be the instantiation of P_1 in o_2. If β_2 has an even number of zeroes, then the CPT rules for N^* in P_2 dictate that an outcome $\hat{o}$ with $\hat{o} \succ o_2$ is obtained by changing all of o_2's values in $P_2 \cap V^\times$ to the values in o_1, since N^* uses the same CPT rule as N_1 for the instantiation β_2 of P_1. Now N^* entails $o_1 \succ \hat{o}$, since o_1 results from $\hat{o}$ by flipping the values in $P_1 \cap V^\times$, which invokes only CPT rules on which N_1, N_2 and N^* all agree. Thus, by transitivity, N^* entails $o_1 \succ o_2$. If β_2 has an odd number of zeroes, a similar argument is obtained after first flipping one of the values of o_2 in $P_1 \cap V^\times$, which yields a more preferred outcome in N^* that also has an even number of zeroes in its instantiation of P_1. $\square$

An obvious, dual statement to Lemma 1 is the following.

Lemma 2. *Let T be any pair of CP-nets. If there exists a CP-net N^* that satisfies conditions (i)–(iii) from Lemma 1, then every minimizer of f_{nc}^{total} for T must satisfy these conditions.*

As a consequence, Proposition 1 does not hold for f_{nc}^{total}:

Corollary 2. *Let N_1, N_2 be separable CP-nets that differ in at least one and at most $n - 1$ CPTs. Then no separable CP-net N^* minimizes f_{nc}^{total}.*

Proof. (Sketch). The proof of Theorem 4 showed that, for such N_1 and N_2, there always exists a CP-net N^* meeting the conditions from Lemma 1. By Lemma 2, then every minimizer of f_{nc}^{total} for T satisfies these conditions. A standard argument (details omitted) shows that every separable CP-net violates condition (iii) and hence is sub-optimal wrt f_{nc}^{total}.   □

8   Quadratic Objective for Swaps

For each N_i in a tuple $(N_1, \ldots, N_t)$ of input CP-nets, one can decompose the swap disagreement between N_i and N^* in terms of the swap disagreement for each attribute. d_{ij} denotes the pairwise swap disagreement between CP-nets N_i and N^* restricted to only the swaps in attribute V_j. The objective function from Eq. (1) can then be written as

$$f^{swap}(N^*) = \sum_{i=1}^{t} \Delta^{swap}(N_i, N^*) = \sum_{i=1}^{t} \sum_{j=1}^{n} d_{ij}.$$

In this section, we propose a quadratic objective function based on the sum of the squares of the d_{ij}:

$$f_q^{swap}(N^*) = \sum_{i=1}^{t} \sum_{j=1}^{n} d_{ij}^2.$$

This quadratic objective function takes an attribute-oriented view. In particular, compare two situations, namely (a) the consensus CP-net disagrees with a single user's CP-net in ℓ swap pairs over the same attribute V_j, (b) the consensus CP-net has disagreements in a total of ℓ swap pairs distributed over several users and/or several attributes. While f^{swap} would regard both consensus CP-nets equally good, the squared distance would penalize the consensus CP-net in situation (a) more than in situation (b). This objective function captures the idea that a user might be more likely to accept incorrectly ranked items when they are scattered than when they occur within a certain category of items (e.g., related to swaps over a specific attribute).

Minimizing f_q^{swap} is more involved than minimizing f^{swap}. Our approach applies only to input tuples of separable CP-nets. In linear time, Algorithm 2 builds a consensus CP-net minimizing f_q^{swap} for such inputs. For each $V_j \in V$, it computes the fraction α of occurrences of the rule $v_j \succ v_j'$ among the input CP-nets and adds enough parents to assign $v_j \succ v_j'$ to (roughly) a fraction of α of the swaps over V_j.

Theorem 5. *Given a tuple T of separable CP-nets, Algorithm 2 runs in linear time and outputs a complete consensus CP-net N^* that minimizes f_q^{swap}.*

Proof. We only prove the optimality wrt f_q^{swap}. Suppose that for V_j, t' of the input CPTs are $\{v_j \succ v_j'\}$ and $t - t'$ are $\{v_j' \succ v_j\}$. The claim is trivial if

---

Algorithm 2: Finding optimal consensus for f_q^{swap}

    Input   : A tuple $T = (N_1, \ldots, N_t)$ of separable CP-nets
    Output: A consensus CP-net N^*

1 Set N^* to a separable CP-net with empty CPTs
2 for *each attribute* $V_j \in V$ **do**
3 $t' = \#$ of occurrences of rule $v_j \succ v_j'$ in T
4   **if** $t' = t$ **then** $\mathrm{CPT}(N^*, V_j) = \{v_j \succ v_j'\}$
5   **else if** $t' = 0$ **then** $\mathrm{CPT}(N^*, V_j) = \{v_j' \succ v_j\}$
6   **else**
7   Reduce $\frac{t-t'}{t}$ to $\frac{r}{s}$ where r and s are co-prime
8   Find $a, b \in \mathbb{N}$ s.t. $1 \leq b < 2^a \leq \min\{s, 2^{n-1}\}$ that minimize $|\frac{2^a-b}{2^a} - \frac{r}{s}|$
9   Add a parents to $Pa(N^*, V_j)$, and build $\mathrm{CPT}(N^*, V_j)$ with exactly
            $2^a - b$ rules $v_j' \succ v_j$ and b rules $v_j \succ v_j'$
10   **end**
11 end

$t' \in \{0, t\}$. Suppose $0 < t' < t$. Fix a consensus CP-net N^* and let z_1 (resp. z_2) be the proportion of all swaps of V_j on which the t' (resp. $t - t'$) CP-nets disagree with N^*. (Note $z_1 + z_2 = 1$.) If c is the number of swaps over V_j, then

$$\sum_{i=1}^{t} d_{ij}^2 = t'(cz_1)^2 + (t - t')(cz_2)^2.$$

This is minimized when $z_1 = \frac{t-t'}{t}$. The remaining proof details are omitted. $\square$

    This result suggests a weakness of f_q^{swap}. Which parents are introduced in line 9 of the algorithm and which parent instantiations to use in which of the two preference orders for $Dom(V_j)$ plays no role – only their numbers matter. Hence there is no meaningful relationship between the preferences over $Dom(V_j)$ and the parent attributes of V_j; the parents and their instantiations just serve as dummies in order to allow for enough rules in the CPTs. The same, in fact, applies to all non-linear norms used in place of squared distance.

9   Conclusions

The simple algorithm M^{swap} was proven optimal in aggregating any input tuple of complete CP-nets in terms of cumulative swap disagreement. Under intuitive structural conditions on the input, it was shown to run in polynomial time. Conditions like these may well translate into notions of "groupability" in the context of clustering users – if a group of user preferences are too dissimilar, we may want to refrain from aggregating them and instead split them into two or more clusters. One would then aggregate preferences for each cluster separately. The fact that M^{swap} outputs a CP-net presents the possibility of using it for hierarchical clustering. The output of M^{swap} corresponds to the preference relation constructed by earlier methods that targeted only non-dominated outcomes. This implies the equivalence of two seemingly different algorithmic problems.

While we established formal relationships between swap disagreement and both versions of total disagreement, the latter tend to be more complex to minimize than the former. We further demonstrated that aggregation wrt non-linear norms may result in unintuitive consensus CP-nets. Future research might hence focus on generating interesting suboptimal solutions. Another problem to be addressed is the aggregation of CP-nets with non-binary attributes.

References

1. Airiau, S., Endriss, U., Grandi, U., Porello, D., Uckelman, J.: Aggregating dependency graphs into voting agendas in multi-issue elections. In: IJCAI (2011)
2. Bachmaier, C., Brandenburg, F., Gleißner, A., Hofmeier, A.: On the hardness of maximum rank aggregation problems. J. Discrete Alg. **31**, 2–13 (2015)
3. Bigot, D., Zanuttini, B., Fargier, H., Mengin, J.: Probabilistic conditional preference networks. arXiv (2013)
4. Booth, R., Chevaleyre, Y., Lang, J., Mengin, J., Sombattheera, C.: Learning conditionally lexicographic preference relations. In: ECAI, pp. 269–274 (2010)
5. Boutilier, C., Brafman, R., Domshlak, C., Hoos, H., Poole, D.: CP-nets: a tool for representing and reasoning with conditional ceteris paribus preference statements. J. Artif. Intell. Res. **21**, 135–191 (2004)
6. Brafman, R., Domshlak, C., Shimony, S.: On graphical modeling of preference and importance. J. Artif. Intell. Res. **25**, 389–424 (2006)
7. Brandt, F., Conitzer, V., Endriss, U., Lang, J., Procaccia, A.: Handbook of Computational Social Choice. Cambridge University Press, Cambridge (2016)
8. Conitzer, V., Lang, J., Xia, L.: Hypercubewise preference aggregation in multi-issue domains. In: IJCAI (2011)
9. Cornelio, C., Grandi, U., Goldsmith, J., Mattei, N., Rossi, F., Venable, K.: Reasoning with PCP-nets in a multi-agent context. In: AAMAS, pp. 969–977 (2015)
10. Dinu, L., Manea, F.: An efficient approach for the rank aggregation problem. Theor. Comput. Sci. **359**(1–3), 455–461 (2006)
11. Dwork, C., Kumar, R., Naor, M., Sivakumar, D.: Rank aggregation methods for the web. In: TheWebConf, pp. 613–622. ACM (2001)
12. Fagin, R., Kumar, R., Mahdian, M., Sivakumar, D., Vee, E.: Comparing partial rankings. SIAM J. Discrete Math. **20**(3), 628–648 (2006)
13. Grandi, U., Luo, H., Maudet, N., Rossi, F.: Aggregating CP-nets with unfeasible outcomes. In: O'Sullivan, B. (ed.) CP 2014. LNCS, vol. 8656, pp. 366–381. Springer, Cham (2014). https://doi.org/10.1007/978-3-319-10428-7_28
14. Koriche, F., Zanuttini, B.: Learning conditional preference networks. Artif. Intell. **174**(11), 685–703 (2010)
15. Lang, J.: Vote and aggregation in combinatorial domains with structured preferences. In: IJCAI, vol. 7, pp. 1366–1371 (2007)
16. Lang, J., Xia, L.: Sequential composition of voting rules in multi-issue domains. Math. Soc. Sci. **57**(3), 304–324 (2009)
17. Li, M., Vo, Q., Kowalczyk, R.: Majority-rule-based preference aggregation on multi-attribute domains with CP-nets. In: AAMAS, pp. 659–666 (2011)

18. Loreggia, A., Mattei, N., Rossi, F., Venable, K.: A notion of distance between CP-nets. In: AAMAS, pp. 955–963 (2018)
19. Lukasiewicz, T., Malizia, E.: Complexity results for preference aggregation over (m) CP-nets: Pareto and majority voting. Artif. Intell. **272**, 101–142 (2019)
20. Rossi, F., Venable, K., Walsh, T.: mCP nets: representing and reasoning with preferences of multiple agents. In: AAAI, vol. 4, pp. 729–734 (2004)
21. Sculley, D.: Rank aggregation for similar items. In: Proceedings of the 2007 SIAM International Conference on Data Mining, pp. 587–592. SIAM (2007)
22. Xia, L., Conitzer, V., Lang, J.: Voting on multiattribute domains with cyclic preferential dependencies. In: AAAI, vol. 8, pp. 202–207 (2008)
23. Xia, L., Lang, J., Ying, M.: Sequential voting rules and multiple elections paradoxes. In: Proceedings of TARK, pp. 279–288 (2007)
24. Xia, L., Lang, J., Ying, M.: Strongly decomposable voting rules on multiattribute domains. In: AAAI, vol. 7, pp. 776–781 (2007)

Measuring Nearly Single-Peakedness
of an Electorate: Some New Insights

Bruno Escoffier[1,2], Olivier Spanjaard[1](✉), and Magdaléna Tydrichová[1]

[1] Sorbonne Université, CNRS, LIP6, 75005 Paris, France
{bruno.escoffier,olivier.spanjaard,magdalena.tydrichova}@lip6.fr
[2] Institut Universitaire de France, Paris, France

Abstract. After introducing a new distance measure of a preference profile to single-peakedness, directly derived from the very definition of single-peaked preferences by Black [4], we undertake a brief comparison with other popular distance measures to single-peakedness. We then tackle the computational aspects of the optimization problem raised by the proposed measure, namely we show that the problem is NP-hard and we propose an integer programming formulation. Finally, we carry out numerical tests on real and synthetic voting data. The obtained results show the interest of the proposed measure, but also shed new light on the advantages and drawbacks of some popular distance measures.

Keywords: Computational social choice · Structured preferences · Distance measure to single-peakedness

1 Introduction

The study of *structured preferences* in social choice [7,21] starts from the observation that, although the opinions of individuals on candidates in an election are heterogeneous, the voters often agree on the way the candidates are related to each other, more precisely on the *ideological proximities* between them. Various preference structures can be considered to model these proximities, among which are single-peaked preferences [4] and its extensions (see e.g. [18]), as well as Euclidean models where the ideological positions of voters and candidates are viewed as points in an Euclidean space [8].

A preference structure is also called a *domain restriction* in social choice theory, because it restricts the domain of possible preferences for the voters by assuming a consistency of the preferences with the proximities between candidates. Domain restrictions often make it possible to overcome social choice paradoxes (such as the famous Arrow's impossibility theorem [1]), and impact the computational complexity of determining the winner of an election, as well as the complexity of manipulating an election. For more details regarding these computational aspects, we refer the reader to the survey by Elkind et al. [12].

Another issue addressed in the survey is that of recognizing structures in preferences, i.e., determining whether a set of preferences has a given structural property and, if yes, returning the corresponding structure (left-right axis,

© Springer Nature Switzerland AG 2021
D. Fotakis and D. Ríos Insua (Eds.): ADT 2021, LNAI 13023, pp. 19–34, 2021.
https://doi.org/10.1007/978-3-030-87756-9_2

graph or positions in the Euclidean space). The work we present here deals with this recognition problem. We focus on the most well-known domain restriction, that of single-peaked preferences. More precisely, we introduce a new distance measure to single-peakedness for an election.

As emphasized by Feld and Grofman [15], the assumption that preferences are perfectly single-peaked is indeed very strong if the alternatives are candidates in an election (the case of numerical alternatives, such as tax levels, is obviously different). We recall that preferences are said to be single-peaked if 1) all voters agree on a left-right axis on the alternatives, and 2) the preferences of all voters decrease along the axis when moving away from their most preferred alternative to the right or left. Single-peakedness in the strictest sense thus requires that no individual preference deviates (even slightly) from the single-peakedness condition. Given an axis A, the number of rankings consistent with A (i.e., such that condition 2 holds) is 2^{m-1}, over $m!$ possible rankings in total, where m is the number of alternatives. The proportion of consistent rankings within all possible rankings thus quickly becomes tiny when m increases ($2^{m-1}/m! \approx 0.01$ for $m = 7$), as well as the likelihood that no voter deviates from this subset of preferences. This observation is corroborated by the numerical tests carried out by Sui et al. [20] on 2002 Irish General Election data in Dublin West and Dublin North, where the best axes explain only 2.9% and 0.4% of voters' preferences.

Conitzer [7] distinguishes between two interpretations of nearly single-peakedness (see e.g. [13] for a systematic study of nearly single-peaked electorates): an interpretation where preferences are said nearly single-peaked if only a few voters' preferences deviates from a given axis A and the other voters' preferences are perfectly single-peaked w.r.t. A (the numerical tests reported above corresponds to this interpretation); another interpretation where one allows all voters' preferences to deviate to some extent from a given axis A. The distance measure we propose in this paper falls under the second interpretation, which has been less studied and tested than the first one.

Given an axis A on the candidates and a set $\mathcal{P}$ of preferences, the idea is to measure how far from single-peakedness w.r.t. A each individual preference is. Put another way, each preference in the electorate partially fits with the axis (according to a non-binary measure), and one sums up the degrees of fitness of preferences in $\mathcal{P}$ to obtain the "degree of single-peakedness" of $\mathcal{P}$ w.r.t. A. More precisely, one defines a *distance* to single-peakedness, i.e., the degree is 0 if $\mathcal{P}$ is single-peaked w.r.t. A. We are thus seeking a procedure that returns both a degree of single-peakedness of a profile and an axis that witnesses the obtained value. These outputs allow the analysis of a political landscape, by answering the questions: How close to single-peakedness is an electorate? How the voters perceive the ideological proximities between candidates?

Related Work. While recognizing perfectly single-peaked preferences is a polynomial time problem [3,10], determining the distance to single-peakedness (according to various measures) is often NP-hard. Various notions of nearly single-peakedness are present in the literature. We briefly review here notions that do not relax the assumption of a one-dimensional axis on all the candidates, which excludes other approaches that return, e.g., an axis on a subset of candidates [13], an axis on clone sets [9], or multiple axes [2,20]. Most of them have been introduced and/or studied by Faliszewski et al. [14], Erdélyi et al. [13] and Elkind and Lackner [11]. Faliszewski et al. [14] studied *k-voter deletion single-peakedness*, also known as *partial single-peakedness* in economics [19]. One says that an electorate is k-voter deletion single-peaked consistent if all but k of the voters preferences ("maverick" voters) are consistent with a common axis on the candidates. The smallest number k such that there exists an axis w.r.t. which the electorate is k-voter deletion single-peaked can be viewed as a distance to single-peakedness. Erdélyi et al. [13] as well as Bredereck et al. [6] have proved that determining this distance is NP-hard. Elkind and Lackner [11] have proposed a polynomial time 2-approximation algorithm for this distance, and have established fixed-parameter tractability results (complexity $\mathcal{O}^*(1.28^k)$ if $k < n/2$, and $\mathcal{O}^*(2.08^k)$ if $k \geq n/2$, where n is the number of voters).

Erdélyi et al. [13] introduced *k-local candidate deletion single-peakedness*. They first defined single-peaked consistency of a partial preference (linear order on a subset of candidates) w.r.t. an axis A on all candidates: a partial preference is single-peaked w.r.t. A if it is single-peaked w.r.t. the axis obtained from A by removing the missing candidates. Then they say that an electorate is k-local candidate deletion single-peaked consistent if, by removing at most k candidates from each preference, one obtains a set of partial preferences that are single-peaked with respect to a common axis. As above, the smallest k for which the property holds can be viewed as a distance. Here again, the authors have proved that determining this distance is an NP-hard problem.

The class of distance measures that is the closest to our work is that of swap distances. Erdélyi et al. [13] introduced *k-global swaps single-peakedness*, where k is the number of swaps of consecutive candidates that need to be performed in the preferences to make the election single-peaked. Following Faliszewski et al. [14], they also considered a "local budget" for swaps, i.e., they allow up to k swaps *per vote*. They call *k-local swaps* this notion of nearly single-peakedness. For both notions, Erdélyi et al. [13] have proved that computing the smallest k enabling to make the election single-peaked is NP-hard. Finally, let us mention the notion of *PerceptionFlip$_k$ single-peakedness* [14]. An electorate is PerceptionFlip$_k$ single-peaked if there exists an axis A such that, for each voter, the axis A can be transformed into an axis A' by at most k swaps of consecutive candidates in A so that the voter's preference is single-peaked with respect to A'. Erdélyi et al. [13] have proved that k-local swaps single-peakedness and PerceptionFlip$_k$ single-peakedness are equivalent, in the sense that an electorate is k-local swaps single-peaked iff it is PerceptionFlip$_k$ single-peaked.

Our Contribution. The originality of the distance measure we introduce in the paper is that it directly follows from the very definition of Black's single-peakedness condition. For a given axis on the candidates, it consists in counting the number of violations of the single-peakedness condition in the preferences. We give a formal definition in Sect. 2, as well as some insights on the differences between this measure, k-voter deletion single-peakedness and k-global swap single-peakedness. We tackle computational complexity in Sect. 3: we prove that, as for most of the proposed measures in the literature, computing an axis at minimum distance to an electorate is NP-hard for our measure. We nevertheless propose an exact method to compute such an axis, that turns out to be efficient in practice. Then, in Sect. 4, we present the results of numerical tests on both real and synthetic election data, to evaluate the relevance of the returned axes on the candidates, providing also comparisons with the other notions of nearly single-peakedness. Some proofs are missing due to lack of space.

2 Definition and Comparison with Voter Deletion and Global Swap

2.1 Definition

We start by recalling some basic terminology of social choice theory. Given a set $C = \{c_1, c_2, \ldots, c_m\}$ of candidates and a set V of n voters ($|V| = n$), each voter $v \in V$ ranks all the candidates from the most to the least preferred one. This ranking is called the *preference relation* of v. The (multi)set P of preference relations of all the voters in V is called a *profile*. The couple (C, P) is called an *election*. The definition of a *single-peaked* profile states as follows:

Definition 1 (Single-Peakedness). *Let an* axis A *be a total order* $\lhd_A$ *over a set* $C = \{c_1, \ldots c_m\}$ *of candidates. Let* $>_v$ *denote the preference relation (total order) of a voter v over C. Let c^* denote the most preferred candidate of v (also called the* peak *of v), i.e., $c^* >_v c$ for all $c \neq c^*$. The preference $>_v$ is single-peaked with respect to A if for any $c_i, c_j \in C$, if $c_j \lhd_A c_i \lhd_A c^*$ or $c^* \lhd_A c_i \lhd_A c_j$ then $c^* >_v c_i >_v c_j$ holds. A profile P is said to be single-peaked with respect to A if every vote is single-peaked with respect to A.*

Definition 2 (Betweenness Relation). *The* betweenness relation *induced by an axis A is the relation R_A defined by:*

$$R_A = \{(c_i, c_j, c_k) \in C^3 : c_i \lhd_A c_j \lhd_A c_k \text{ or } c_k \lhd_A c_j \lhd_A c_i\}.$$

Put another way, $(c_i, c_j, c_k) \in R_A$ means that c_j is between c_i and c_k on the axis A (note that c_i, c_j and c_k do not need to be consecutive on A). The notion of A-*forbidden triple* that we introduce now will make it possible to measure the consistency of a profile with an axis:

Definition 3 (A-Forbidden Triple). *Let c^* be the peak of a voter v. If $c^* >_v c_i >_v c_j$ and $(c^*, c_j, c_i) \in R_A$, then the triple $T = (c^*, c_i, c_j)$ is called A-forbidden in v.*

Counting the number of A-forbidden triples in a profile P (by summing over all $>_v \in P$) amounts to counting the number of violations of the definition of single-peakedness w.r.t. A. Note that the number of A-forbidden triples per voter is upper bounded by $(m-1)(m-2)/2$ as c^* is unique.

Example 1. Let us consider the following profile with 5 candidates $\{1, 2, \ldots, 5\}$ and 20 voters with the following preferences:

- $5 > 4 > 3 > 2 > 1$ for 1 voter (type I);
- $1 > 2 > 3 > 4 > 5$ for 10 voters (type II);
- $1 > 5 > 3 > 2 > 4$ for 9 voters (type III).

Let us consider the axis A: $1 \lhd_A 2 \lhd_A 3 \lhd_A 4 \lhd_A 5$. The voter of type I is single-peaked with respect to A, so there is no A-forbidden triple for her. Voters of type II are also single-peaked w.r.t. A, so the A-forbidden triples only occur with voters of type III. There, for each of them there are 4 A-forbidden triples: $(1, 3, 2)$ (as $1 > 3 > 2$ for the voters, but $1 \lhd_A 2 \lhd_A 3$), $(1, 5, 2)$, $(1, 5, 3)$ and $(1, 5, 4)$. Then the number of A-forbidden triples in the profile is 36.

More generally, we define a new notion of nearly single-peakedness, that also entails a distance measure to single-peaked profiles. Let $FT(\mathcal{P}, A)$ denote the number of A-forbidden triples in $\mathcal{P}$, and $FT(\mathcal{P}) = \min_A FT(\mathcal{P}, A)$.

Definition 4 (k-Forbidden Triples Single-Peakedness). *We say that a profile $\mathcal{P}$ is k-forbidden triples single-peaked consistent if $FT(\mathcal{P}) \leq k$.*

We will denote by $A_{FT}(\mathcal{P})$ an optimal axis, i.e., $A_{FT}(\mathcal{P}) \in \arg\min_A FT(\mathcal{P}, A)$.

Example 2 (Example 1, Continued). Now, let us consider the following axis B: $4 \lhd_B 2 \lhd_B 1 \lhd_B 3 \lhd_B 5$. Voters of type II are single-peaked w.r.t. B. For voters of type III there is only one B-forbidden triple $((1, 5, 3))$, and there are 4 forbidden triples for the unique voter of type I. In total, there are 13 B-forbidden triples, much smaller than 36 A-forbidden triples. Indeed, 19 voters are single-peaked or very close to being so w.r.t. B. Actually, B is an optimal axis, i.e., $FT(\mathcal{P}) = 13$ and $A_{FT}(\mathcal{P}) = B$.

2.2 Forbidden Triples, Voter Deletion and Global Swap

Let us highlight some differences between k-forbidden triples single-peakedness and other notions of nearly single-peakedness. We focus on the notions of k-voter deletion single-peakedness (we denote by VD the corresponding distance measure), k-global swaps single-peakedness (GS), and of course k-forbidden triples single-peakedness (FT). They all result in a single axis on all candidates (contrary to, e.g., multi-dimensional single-peakedness [20] and k-candidate deletion single-peakedness [13]). Formally, $VD(\mathcal{P}, A) = \sum_{>\in\mathcal{P}} \delta(>, A)$ where $\delta(>, A) = 0$ if $>$ is single-peaked w.r.t. A, otherwise 1, and $GS(\mathcal{P}, A) = \sum_{>\in\mathcal{P}} d_{swap}(>, A)$, where $d_{swap}(>, A)$ is the minimum number of swaps of consecutive candidates to make $>$ single-peaked w.r.t. A.

A major difference between VD, on the one hand, and FT and GS, on the other hand, lies on the fact that the latter ones are much smoother. Namely, they quantify how far a preference is from being single-peaked with respect to an axis, by a distance which lies from 0 (single-peaked) to $\Theta(m^2)$. On the opposite side, VD only looks if a preference is single-peaked or not, so the distance is 0 or 1. This may prevent VD to find interesting axis, with respect to which almost all the preferences are almost single-peaked. We illustrate this point on Example 1.

Example 3 (Example 1, Continued). There is no axis compatible with voters of type II and III, so the axis $1 \lhd_A 2 \lhd_A 3 \lhd_A 4 \lhd_A 5$ is the (unique, up to reversal) axis compatible with voters of type I and II. Thus, it is optimal for VD. As pointed before, this axis is not satisfactory, as almost half of the voters have preferences very far from being single-peaked with respect to this axis. The axis B optimal for FT seems to better fit nearly single-peakedness, as among the 20 voters, 10 are single-peaked w.r.t. it and 9 are very close to being single-peaked.

GS and FT are intuitively closer to each other than VD, but still have some important differences. A first difference is computational, and will be dealt with in the next sections: while both of them are NP-hard, FT is much easier to compute in practice. FT and GS have also qualitative differences, and we illustrate this through a property for which they behave differently.

Let us consider the following *unpopularity property*. It states that beyond a certain level of unpopularity a candidate c can hardly be viewed as intermediate between others, and thus there should be an optimal axis where c is at an extremity. Let us say that a candidate is *unpopular* if she is never ranked in first position, and ranked last by at least $\lceil n/2 \rceil$ voters[1].

Property 1 (Unpopularity). Let $(\mathcal{C}, \mathcal{P})$ be an election. A distance d verifies the unpopularity property if for any unpopular candidate c there exists an axis A minimizing $d(\mathcal{P}, A)$ where c is at an extremity.

It is well-known that, if a profile is single-peaked, then such an unpopular candidate is indeed necessarily at an extremity of any compatible axis [10]. Interestingly, we show that dealing with nearly single-peakedness, among the considered measures, FT is the only one for which the unpopularity property holds.

Theorem 1. *FT satisfies the unpopularity property, while GS and VD do not.*

Proof (sketch). We prove that the property holds for FT.

Let c be a candidate never ranked in first position, and ranked in last position by at least half of the voters. Let A be an arbitrary axis such that c is not one of its extremities. Let us denote by m_1 the number of candidates on the left of c in A, and by m_2 the number of candidates on the right of c. We define two axes A_l and A_r obtained from A by putting c respectively on the extreme left position for A_l, and on the extreme right position for A_r. We prove that at least one of the axis A_l, A_r is at least as good as A. To do so, for each voter v, we count down the difference of the number of forbidden triples with respect to A and with respect to A_l and A_r. It consists in counting for each of axes the number of triples involving the candidate c. In fact, as c is never ranked first and the restrictions of A, A_l and A_r on $\mathcal{C} \setminus \{c\}$ lead in the same axis, we observe that the

---

[1] As in any axiomatic approach, the specific situation considered here does not need to often happen in practice: it is a thought experiment in which one considers a hypothetical situation and examines whether the measure would behave well in such a case.

triple (c_i, c_j, c_k) (with c_i, c_j, c_k different from c) is forbidden with respect to A if and only if it is forbidden with respect to A_l (resp. A_r). Let v be an arbitrary voter and c^* the peak of v. Four configurations are possible:

(i) v ranks c in last position and c^* is on the left of c in A;
(ii) v ranks c in last position and c^* is on the right of c in A;
(iii) v does not rank c in last position and c^* is on the left of c in A;
(iv) v does not rank c in last position and c^* is on the right of c.

The Table 1 expresses $FT(>_v, A_l)$ and $FT(>_v, A_r)$ in function of $FT(>_v, A)$ - if the exact value can not be given, the upper bound (representing the worst case) is given. For v of type (i), (c^*, c, c') is forbidden (w.r.t. A) if and only if c' is on the right of c in A (there are m_2 such positions). The candidate c is not involved in any forbidden triple with respect to an axis A_l or A_r, as it is placed on the extremity. The same reasoning applies to v of type (ii). For a v of type (iii), in the worst case c is not involved in any forbidden triple with respect to A, but moving it on the left (resp. right) extremity will create up to $m_1 - 1$ (resp. m_2) new forbidden triples. We reason the same way for type (iv).

Table 1. Values of $FT(>_v, A_l)$ and $FT(>_v, A_r)$ in function of $FT(>_v, A)$, according to the type of v.

Type	$FT(>_v, A_l)$	$FT(>_v, A_r)$
(i)	$FT(>_v, A) - m_2$	$FT(>_v, A) - m_2$
(ii)	$FT(>_v, A) - m_1$	$FT(>_v, A) - m_1$
(iii)	$\leq FT(>_v, A) + m_1 - 1$	$\leq FT(>_v, A) + m_2$
(iv)	$\leq FT(>_v, A) + m_1$	$\leq FT(>_v, A) + m_2 - 1$

Assume that $m_1 \leq m_2$. We prove that A_l is always at least as good as A, i.e., $FT(\mathcal{P}, A_l) \leq FT(\mathcal{P}, A)$, which is written:

$$\sum_{>_v \in \mathcal{P}} FT(>_v, A_l) \leq \sum_{>_v \in \mathcal{P}} FT(>_v, A).$$

Thanks to Table 1 it is sufficient to prove that:

$$n^{(i)} m_2 + n^{(ii)} m_1 \geq (n^{(iii)} + n^{(iv)}) m_1 - n^{(iii)}$$

with n^t the number of voters of type t. By assumption, $n^{(i)} + n^{(ii)} \geq \frac{n}{2}$. As we assume that $m_1 \leq m_2$, the inequality holds all the time.

If $m_1 \geq m_2$, we prove in the same manner that A_r is always at least as good as A. □

We will present several other qualitative differences between the measures in the experimental section.

3   Computational Aspects

Given a preference profile $\mathcal{P}$, we first note that determining $FT(\mathcal{P}, A)$ for a given axis A can be handled in polynomial time $\mathcal{O}(nm^2)$, where $n = |\mathcal{P}|$ and $m = |\mathcal{C}|$, by brute force enumeration of all triples. This complexity can be improved to $\mathcal{O}(nm\sqrt{\log m})$ (proof omitted).

Let us now focus on the complexity of determining if $FT(\mathcal{P}) \leq k$ for a given k, i.e., the following problem:

FT SINGLE-PEAKED CONSISTENCY
Input: An election $(\mathcal{C}, \mathcal{P})$ and an integer k.
Output: Yes if $FT(\mathcal{P}) \leq k$, otherwise no.

We show that, similarly to other measures such as GS or VD, this is an NP-complete problem:

Theorem 2. FT SINGLE-PEAKED CONSISTENCY *is NP-complete.*

We now provide an Integer Program (IP) formulation, that turns out to be efficient in practice in our experiments on this consistency problem. For each pair $\{c_i, c_j\}$ of candidates (with $i \neq j \in \{1, \ldots, m\}$), we introduce a binary variable x_{ij} describing their relative position on the sought axis A. More precisely, the constraints of type 1 and type 2 detailed below will ensure that: $x_{ij} = 1$ if $c_i \triangleleft_A c_j$, and 0 otherwise.

Additionally, for each voter $v \in \{1, \ldots, n\}$ and each pairwise preference $c_i >_v c_j$ with $\pi(v) \notin \{i, j\}$, where $\pi(v)$ is the index of the peak of v, we define a binary variable z_{vij} related to the triple $(c_{\pi(v)}, c_i, c_j)$. More precisely, the constraints of type 3 and type 4 detailed below will ensure that $z_{vij} = 1$ if $(c_{\pi(v)}, c_i, c_j)$ is A-forbidden in v, and 0 otherwise.

The sum of variables z_{vij} is the number of forbidden triples in the profile $\mathcal{P}$. The IP objective function is then $\min \sum_{(v,i,j) \in T} z_{vij}$, where $T = \{(v, i, j) : \pi(v) \notin \{i, j\}, i \neq j\}$.

We now detail the four types of constraints in the program:

1. For each pair $\{c_i, c_j\}$ of candidates, one and only one of the variables $\{x_{ij}, x_{ji}\}$ equals 1.
2. For each tuple (c_i, c_j, c_k), if $x_{ik} = 1$ and $x_{kj} = 1$ then $x_{ij} = 1$ (because $c_i \triangleleft_A c_k$ and $c_k \triangleleft_A c_j \Rightarrow c_i \triangleleft_A c_j$).
3. For each (v, i, j) such that $\pi(v) \notin \{i, j\}$ and $c_i >_v c_j$, if $c_{\pi(v)} \triangleleft_A c_j$ and $c_j \triangleleft_A c_i$ then $z_{vij} = 1$ $((c_{\pi(v)}, c_i, c_j)$ is A-forbidden in v, on the right side of the peak).
4. For each (v, i, j) such that $\pi(v) \notin \{i, j\}$ and $c_i >_v c_j$, if $c_i \triangleleft_A c_j$ and $c_j \triangleleft_A c_{\pi(v)}$ then $z_{vij} = 1$ $((c_{\pi(v)}, c_i, c_j)$ is A-forbidden in v, on the left side of the peak).

Altogether, we obtain the following IP, where $T = \{(v, i, j) : \pi(v) \notin \{i, j\}, i \neq j\}$:

$$\min \sum_{(v,i,j) \in T} z_{vij}$$

$$s.t. \begin{cases} x_{ij} + x_{ji} = 1 & \forall \{c_i, c_j\} \text{ with } i \neq j & (1) \\ x_{ij} \geq x_{ik} + x_{kj} - 1 & \forall (c_i, c_j, c_k) \text{ with } i \neq j \neq k & (2) \\ z_{vij} \geq x_{\pi(v)j} + x_{ji} - 1 & \forall (v, i, j) \in T \text{ with } c_i >_v c_j & (3) \\ z_{vij} \geq x_{j\pi(v)} + x_{ij} - 1 & \forall (v, i, j) \in T \text{ with } c_i >_v c_j & (4) \\ x_{ij} \in \{0, 1\} \, \forall i, j, \quad z_{vij} \in \{0, 1\} \, \forall (v, i, j) \in T \end{cases}$$

4 Experimental Study

We carried out numerical tests[2] on real and randomly generated preference profiles in order to compare experimentally the distance measures GS, VD, FT. For optimizing the VD distance, we used the C++ code developed by Sui et al. [20], made available on the web[3]. For optimizing the FT distance, we used the Gurobi software to solve the IP formulation. Finally, for optimizing the GS distance, we used a brute force algorithm - to the best of our knowledge, no efficient algorithm is known for this problem so far. In particular, no efficient IP formulation is available in the literature for swap measures.

We study the quality of optimal axes on real data, compared to reference axes whose design is detailed below. To evaluate the quality of an axis, we use the following distance ρ between two axes A and A' defined on the same set of candidates: $\rho(A, A') = |R_A \cap R_{A'}|/|R_A|$ (note that $|R_A| = |R_{A'}|$). Put another way, we measure the proportion of the betweenness relation (see Definition 2) that is common to the optimal axis and the reference axis. We call this proportion, expressed in percentage in the sequel, *recognition rate*.

To go further and better understand the impact of the characteristics of the profiles on the numerical results, we also study the quality of optimal axes on profiles randomly generated according to diverse probability distributions for structured preferences.

4.1 Numerical Tests on Real Data

The real data sets were taken from the 2007 Glasgow city council election and a 2017 voting experiment during the French presidential election. The first data set is available on the PrefLib website[4], a library of preference data and links assembled by Mattei and Walsh [17]. The second data set comes from the website of the experiment called *Voter autrement*[5] [5].

---

[2] Tests performed on an Intel Core i7 (1.3 GHz base, 3.9 GHz turbo) with 8 GB RAM.
[3] http://www.cs.toronto.edu/~lex/code/asprgen.html.
[4] https://www.preflib.org/data/index.php.
[5] https://zenodo.org/record/1199545.

The Glasgow election was separated in 21 wards (with one list of candidates per ward). The data from 20 of them were used in our tests. Each ward involved different candidates and voters, and elected 3 or 4 councillors using the Single Transferable Vote (STV) system. This implies that some political parties had several candidates for the same voting district. In order to fit the data with our setting, we restricted ourselves to the votes (ballots) consisting of complete rankings of the candidates. The number of candidates in the Glasgow data set ranges from 8 to 11, and the number of complete votes from 320 to 1003. In the *Voter autrement* data set, one file was usable for our purpose (file `stv111.csv`, here also reporting the results of an experiment about STV), with 11 candidates from as many distinct political parties and 4068 complete votes.

Regarding the computation times, an optimal axis for VD was computed in less than 3 s (sec.) on all data sets, and generally in around 30–40 s (resp. 82 s) for FT on a Glasgow ward (resp. on the French presidential election data). Of course, the brute force algorithm used for GS was much slower: determining an optimal axis with 6 candidates and 100 voters took about 20 s, which became 2 min for 7 candidates, and 10 min for 8 candidates.

For each election (at the level of a ward or a country), we built a reference left-right axis on the candidates. To do so, we used Wikipedia as external source. The free encyclopedia provides indeed a political position (of course debatable) for each political party (e.g., left wing, centre, centre right, etc.). We assumed that the political position of an affiliated candidate corresponds to that of the belonging party, and we built an axis over the affiliated candidates based on these positions. We excluded the non-affiliated candidates from the data sets as we were not able to define a political position for them. Actually, the "Wikipedia axis" is not unique since several parties can be labeled by the same political position, or some parties can have several candidates in an election. For instance, a Wikipedia axis reads $((1,3),2,(4,5))$, where the numbers are the indices of candidates, and candidates $\{1,3\}$ as well as $\{4,5\}$ have indistinguishable political positions. This leads to a set of $2 \cdot 2 = 4$ *compatible* axes : $1 \triangleleft 3 \triangleleft 2 \triangleleft 4 \triangleleft 5$, $3 \triangleleft 1 \triangleleft 2 \triangleleft 4 \triangleleft 5$, $1 \triangleleft 3 \triangleleft 2 \triangleleft 5 \triangleleft 4$, and $3 \triangleleft 1 \triangleleft 2 \triangleleft 5 \triangleleft 4$.

Note that indistinguishable political positions do *not* mean here that the candidates share the same position on the political spectrum, but that we have a partial knowledge of the exact axis. The sets of candidates with indistinguishable political positions (as $\{1,3\}$ and $\{4,5\}$ above) are called *blocks* in the following. Given a distance measure d (in $\{VD, FT, GS\}$) and a profile $\mathcal{P}$, the recognition rate is formulated in the following manner to take into account blocks:

$$\min\{\rho(A_d(\mathcal{P}), A') : A' \text{ compatible with the Wikipedia axis}\}$$

where $A_d(\mathcal{P})$ is an optimal axis according to d.

Apart from the recognition rate, we also distinguish three classes of results for the optimal axis w.r.t. a distance:

- T (True): The optimal axis is compatible with the Wikipedia axis, e.g. $3 \triangleleft 1 \triangleleft 2 \triangleleft 4 \triangleleft 5$ for $((1,3),2,(4,5))$.

- EE (Exchanged Extremities): The optimal axis can be made compatible with the Wikipedia axis by swapping the far left and far right blocks, e.g. an optimal axis $5 \lhd 2 \lhd 6 \lhd 4 \lhd 3 \lhd 1$ for the Wikipedia axis $((1,3), 2, 6, 4, 5)$.
- F (False): The optimal axis is called false otherwise.

We distinguish class EE because the experiments revealed a difficulty in recognizing the two extreme blocks. To get an intuition of what is going on, consider a profile where the two "extreme" candidates are ranked in the two last positions by a large number of voters, in an arbitrary order, and the voters who rank one of them in first position do not want to rank anyone else. The data do not provide then much information to distinguish who is left wing and who is right wing.

The results obtained are summarized in Table 2. Note that only the results for the VD and FT measures are given in the table, because the brute force algorithm used for the GS measure was not able to compute an axis in a reasonable amount of time for more than 8 candidates. Regarding the two profiles with 8 candidates in the Glasgow data set, the results obtained with the GS measure are of class EE, while with FT one result is of class T and the other of class EE (the result is of class F in both cases with VD).

Table 2. Results on real election data. Rate means recognition rate.

d	T	EE	F	Rate	T	EE	F	Rate
	Glasgow city council				French election			
VD	2	1		17	57.25%	0	0	1	58.8%
FT	5	5		10	67%	0	0	1	74.6%

Table 2 indicates how many times each class occurs for the VD and FT measures (over 20 preference profiles for the Glasgow city council election and 1 for the French presidential election), as well as the average recognition rate. The results tend to show that the recognition ability of the FT measure is better than that of VD. When the FT measure is used, an axis perfectly compatible with Wikipedia is recognized in nearly 24% of cases; it reaches 48% if one adds the cases when the extremities are swapped.

Let us detail now in a more down-to-earth manner the results obtained on the voting data from the French election. The Wikipedia axis is $W = ((1, 2, 3), 4, 5, (6, 7), (8, 9), 10)$, with one non-affiliated candidate excluded from the voting data (for readability, the candidates are here numbered in function of their position in W). The axis A_{VD} minimizing VD is $7 \lhd 8 \lhd 5 \lhd 4 \lhd 3 \lhd 2 \lhd 1 \lhd 6 \lhd 9 \lhd 10$, the axis A_{FT} minimizing the FT measure is $8 \lhd 1 \lhd 2 \lhd 3 \lhd 4 \lhd 5 \lhd 6 \lhd 9 \lhd 7 \lhd 10$; both axes are not compatible with W, but A_{FT} is much better than A_{VD} in terms of recognition rate. While it can be objected that this result may follow from the fact that FT explicitly relates to triples while VD does not (although only triples involving the peak are used in FT, not the whole betweenness relation), note that, by swapping candidates 7 and 9 and moving candidate 8 in A_{FT}, an axis compatible with W is obtained, while many more fixes are needed in A_{VD}.

As for the recognition rate, it is slightly higher for the French election. It may reflect the fact that a national election is commonly more structured by the left-right political spectrum than a local election.

4.2 Numerical Tests on Synthetic Data

We also generated synthetic election data to deepen the analysis of the recogition abilities of the VD, FT and GS measures. The aim is to model situations where the preferences are noisy but there is a strong underlying structure. Given an axis A, each preference relation is generated in two steps:

1. A candidate c^* is drawn uniformly at random in C and an auxiliary preference relation $>_0$ of peak c^* single-peaked w.r.t. A is generated uniformly at random.
2. A preference relation $>$ is drawn from the Mallows model centered around $>_0$.

We recall that the Mallows model defines a probability distribution on rankings. A central ranking $>_0$ has the highest probability, and the probability of other rankings $>$ decreases in a Gaussian manner with the Kendall tau distance from $>_0$. The central ranking $>_0$ is often interpreted as a "ground truth", and rankings $>$ as noisy views of $>_0$. Formally, given a dispersion parameter $\theta \geq 0$, the probability $P(>)$ of a ranking $>$ is proportional to $e^{-\theta d(>, >_0)}$, where $d(., .)$ is the Kendall tau distance. If $\theta = 0$, the uniform distribution is obtained. The greater the value of θ, the higher the probabilities of the rankings around $>_0$. It is known that using the Mallows model with parameters $>_0$ and θ is equivalent to generating a binary relation R where, for each pair c_i, c_j of candidates, if $c_i >_0 c_j$, then $c_i R c_j$ with probability $p = e^\theta / (1 + e^\theta)$; if the obtained binary relation R is transitive then stop and return the corresponding ranking, otherwise repeat the process until R is transitive. For the sake of interpretability, in the tables, we give the value of p instead of θ.

We used the PerMallows R package[6] for generating rankings according to the Mallows model. For a fast generation of the profiles, the number of voters is set to 100 and the number m of candidates varies from 7 to 9. The above probability p takes its values in $\{0.7, 0.75, 0.8, 0.85, 0.9\}$. For each couple (m, p) of parameter values, 100 instances were generated and one counted the number of instances for which the axis is perfectly recognized.

As the GS brute force algorithm is not usable in practice for more than 7 candidates, we give only the results for VD and FT. However, we generated instances with 5 to 7 candidates and observed very similar results with GS and FT.

The results are reported in Table 3. It appears that the VD measure is the one for which axis A is the most often recognized. This result was quite unexpected because it is well-known, as mentioned in the introduction, that an optimal axis

---

[6] https://cran.r-project.org/package=PerMallows.

for VD explains only a few percentage of voter preferences in real election data - and this is the case in these election data. Nevertheless, the good behaviour of VD can be simply explained by the manner in which the preferences are generated here: the probability that a voter preference is perfectly compatible with A is low but is the highest among all the preferences, thus the law of large numbers plays in favor of VD, and this with all the more intensity as probability p is high.

To refine the analysis, we also studied the recognition rates for FT and VD, since it is a smoother criterion than the previous one. The results are reported in Table 4. The differences are then much narrower, which means that, for the instances where A is not perfectly recognized, the optimal axis for FT is very similar to A.

Computation Times for Optimizing the FT Distance. We also carried out some numerical tests with a greater number of candidates and/or voters to evaluate how the running time scales for initializing and solving the IP of Sect. 3. In preliminary tests, the running times did not appear to vary significantly with the value of θ used for generating the profiles, thus we set $\theta = 0$, which corresponds to a uniform distribution on the rankings. The running times obtained seem to be much more sensitive to the number of candidates than to the number of voters: on the one hand, for 100 voters, solving the IP took about 3–4 (resp. 10) minutes for 15 (resp. 20) candidates; on the other hand, for 11 candidates, it took about 50 s (resp. 2 min, 5 min) for 1000 (resp. 5000, 10000) voters.

Table 3. Percentage of profiles where the axis is perfectly recognized, w.r.t. distance measure d and probability p.

p	7 candidates		8 candidates		9 candidates	
d						
	VD	FT	VD	FT	VD	FT
0.7	39%	9%	26%	5%	12%	3%
0.75	85%	46%	74%	29%	58%	16%
0.8	100%	93%	98%	81%	91%	67%
0.85	100%	98%	100%	98%	100%	95%
0.9	100%	100%	100%	100%	100%	100%

Robustness to Similar Candidates. Another case which can make single-peaked preferences noisy is the presence of similar candidates. We say that candidates c and c' are *similar* if some voters perceive c as the left neighbour of c' on the left-right spectrum while others perceive the opposite. More generally, a subset of candidates are similar if they are consecutive on the left-right spectrum and the perception of their order changes with the voters. Such a subset of candidates

Table 4. Recognition rates w.r.t. measure d and probability p.

p	7 candidates		8 candidates		9 candidates	
d						
	VD	FT	VD	FT	VD	FT
0.7	83.5%	75%	77.5%	72%	72%	69.1%
0.75	96%	89.1%	93.7%	86.7%	58%	83.2%
0.8	100%	99.1%	99.5%	97.3%	91%	96.2%
0.85	100%	99.8%	100%	99.9%	100%	99.7%
0.9	100%	100%	100%	100%	100%	100%

is called *block* below. We studied here the robustness to the presence of similar candidates of each of the considered measures.

Let us call *weak axis* an axis where several candidates are similar, and describe such an axis with the same notation used for Wikipedia axes. We considered weak axes where the blocks contained approximately the same number of candidates - that means there were no political position shared by (considerably) more candidates than the others. In practice, we worked with the axes $((1,2),(3,4),(5,6))$ for $m = 6$ candidates, $((1,2),(3,4),(5,6),(7,8))$ for $m = 8$, and $((1,2),(3,4,5),(6,7,8),(9,10))$ for $m = 10$. For each weak axis, number of voters and number of candidates, we generated 1000 profiles[7] and computed an optimal axis according to FT and VD. For each measure, we counted the number of times the returned axis was compatible with the weak axis. Regarding the FT measure, the optimal axis was compatible with the weak axis *in all tests performed*, independently of the number of voters or candidates. In contrast, the VD measure is much less robust to the presence of similar candidates: the percentages of profiles for which the optimal axis for VD was compatible with the weak axis are given in Table 5.

Table 5. Percentages of profiles for which the optimal axis for VD was compatible with the weak axis.

#cand.	#voters			
	100	200	500	1000
6	47%	25%	23%	10%
8	17%	8%	0%	0%
10	8%	1%	0%	0%

---

[7] For a weak axis A, each preference relation in the profile is generated in two steps: (1) an axis A' compatible with A is generated uniformly at random; (2) a candidate c^* is drawn uniformly at random in $\mathcal{C}$, then a preference relation is generated uniformly at random among preferences of peak c^* single-peaked w.r.t. A'.

5   Conclusion

We have proposed a new distance measure to single-peakedness, based on counting the number of violations of Black's definition. After a brief comparison with other existing measures, we have shown that determining an optimal axis for this measure is NP-hard. We have then presented an IP formulation, and carried out numerical tests on real and synthetic data. They show that the proposed measure compares favorably to other popular measures. In particular, the IP formulation is operational while no efficient procedure is known for minimizing the number of swaps in the preferences to make them single-peaked; it is more robust to noise in preferences than minimizing the number of votes to delete.

For future work, from a computational viewpoint, one may wonder whether problems that are NP-hard in general but polynomial time on single-peaked profiles remain tractable for nearly (w.r.t. FT) single-peaked electorates. Besides, a local version of the new measure (where, instead of the sum, we minimize the *maximum* over the voters of the number of forbidden triples), together with a comparison to local versions of VD and GS, might be investigated. Also, it would be interesting to undertake the same type of approach, based on the very definition of a restricted domain, for defining distance measures to other domains, such as the single-crossing domain [16], or single-peaked preferences *on a graph* [18].

References

1. Arrow, K.: Social Choice and Individual Values. Wiley, Chapman & Hall, New York (1951)
2. Barberà, S., Gul, F., Stacchetti, E.: Generalized median voter schemes and committees. J. Econ. Theory **61**(2), 262–289 (1993)
3. Bartholdi III, J., Trick, M.A.: Stable matching with preferences derived from a psychological model. Oper. Res. Lett. **5**(4), 165–169 (1986)
4. Black, D.: On the rationale of group decision-making. J. Polit. Econ. **56**(1), 23–34 (1948)
5. Bouveret, S., et al.: Voter autrement 2017 - online experiment, July 2018. https://doi.org/10.5281/zenodo.1199545
6. Bredereck, R., Chen, J., Woeginger, G.J.: Are there any nicely structured preference profiles nearby? Math. Soc. Sci. **79**, 61–73 (2016)
7. Conitzer, V.: Eliciting single-peaked preferences using comparison queries. J. Artif. Intell. Res. **35**, 161–191 (2009)
8. Coombs, C.H.: Psychological scaling without a unit of measurement. Psychol. Rev. **57**(3), 145 (1950)
9. Cornaz, D., Galand, L., Spanjaard, O.: Bounded single-peaked width and proportional representation. In: ECAI, pp. 270–275. IOS Press (2012)
10. Doignon, J., Falmagne, J.: A polynomial time algorithm for unidimensional unfolding representations. J. Algorithms **16**(2), 218–233 (1994)
11. Elkind, E., Lackner, M.: On detecting nearly structured preference profiles. In: Proceedings of the AAAI Conference on Artificial Intelligence (2014)
12. Elkind, E., Lackner, M., Peters, D.: Structured preferences (Chap. 10). In: Trends in Computational Social Choice, pp. 187–207. AI Access (2017)

13. Erdélyi, G., Lackner, M., Pfandler, A.: Computational aspects of nearly single-peaked electorates. In: Proceedings of 27th AAAI, pp. 283–289. AAAI Press (2013). (An extended version appeared in J. AI Res. in 2017)
14. Faliszewski, P., Hemaspaandra, E., Hemaspaandra, L.A.: The complexity of manipulative attacks in nearly single-peaked electorates. In: TARK, pp. 228–237. ACM (2011). (An extended version appeared in Artif. Intel. in 2014)
15. Feld, S.L., Grofman, B.: Research note partial single-peakedness: an extension and clarification. Publ. Choice **51**(1), 71–80 (1986)
16. Jaeckle, F., Peters, D., Elkind, E.: On recognising nearly single-crossing preferences. In: Proceedings of 32th AAAI (2018)
17. Mattei, N., Walsh, T.: Preflib: a library for preferences http://preflib.org. In: Perny, P., Pirlot, M., Tsoukiás, A. (eds.) ADT 2013. LNCS, vol. 8176, pp. 259–270. Springer, Heidelberg (2013). https://doi.org/10.1007/978-3-642-41575-3_20
18. Nehring, K., Puppe, C.: The structure of strategy-proof social choice-part I: general characterization and possibility results on median spaces. J. Econ. Theory **135**(1), 269–305 (2007)
19. Niemi, R.G.: Majority decision-making with partial unidimensionality. Am. Polit. Sci. Rev. **63**(2), 488–497 (1969)
20. Sui, X., Francois-Nienaber, A., Boutilier, C.: Multi-dimensional single-peaked consistency and its approximations. In: Proceedings of the Twenty-Third IJCAI, pp. 375–382. AAAI Press, Beijing (2013)
21. Walsh, T.: Uncertainty in preference elicitation and aggregation. In: AAAI, vol. 7, pp. 3–8 (2007)

Preference Aggregation in the Generalised Unavailable Candidate Model

Arnaud Grivet Sébert[1]([envelope]), Nicolas Maudet[2], Patrice Perny[2], and Paolo Viappiani[2]

[1] Université Paris-Saclay, CEA, List, 91120 Palaiseau, France
[2] Sorbonne Université, CNRS, LIP6, 75005 Paris, France
{nicolas.maudet,patrice.perny,paolo.viappiani}@lip6.fr

Abstract. While traditional social choice models assume that the set of candidates is known and fixed in advance, recently several researchers [2,5,7,15,18] have proposed to reject this hypothesis. In particular, the unavailable candidate model of Lu and Boutilier [15] considers voting situations in which some candidates may not be available and focuses on minimising the number of binary disagreements between the voters and the consensus ranking. In this paper, we extend this model and present two new voting rules based on a finer notion of disagreement, called dissatisfaction. The dissatisfaction of a voter is defined as the disutility gap between its preferred available candidate and the candidate elected by the consensus ranking. In the first approach, called *ex ante dissatisfaction* rule, the disutility is independent of the set of available candidates whereas the second approach, called *ex post dissatisfaction* rule, assumes that the disutility depends on which candidates are actually available. We provide several results for the two rules. On the one hand, we show that the ex ante rule actually coincides with standard positional scoring rules; therefore, a consensus ranking can be computed in polynomial time. On the other hand, we exhibit strong links between ex post rule and Kemeny rule and we provide a polynomial-time approximation scheme (PTAS) for the ex post problem.

Keywords: Computational social choice · Preference aggregation · Unavailable candidate model · Polynomial-time approximation scheme

1 Introduction

Traditional social choice theory assumes that the set of candidates is well known before voting takes place. This assumption is not always valid especially in computer science applications (such as recommender systems, decision aid tools, electronic commerce applications) but also in more traditional settings, such as choosing a candidate for a job (a candidate may accept a different job after the hiring committee made its decision). In recent years, several approaches have

This work was mainly conducted while at LIP6, Sorbonne Université.

© Springer Nature Switzerland AG 2021
D. Fotakis and D. Ríos Insua (Eds.): ADT 2021, LNAI 13023, pp. 35–50, 2021.
https://doi.org/10.1007/978-3-030-87756-9_3

been proposed to address the problem of candidates' unavailability, in particular within *computational* social choice [6]. Among these approaches, strategic candidatures have been studied extensively [8,9,13]. Some works have dealt with the problem of finding robust winners when considering the addition of new candidates [7] or in a context where it is possible to "query" the availability of the candidates [5]. Oren et al. [18] study how many candidates the voters have to rank to ensure the true winner with high probability despite the unavailability hypothesis. Top-k voting may also be a way to deal with unavailability in [17].

An approach of particular interest is the *unavailable candidate model* (UCM) proposed by Lu and Boutilier [15]. It assumes that candidates may become unavailable after voters expressed their preferences; therefore there is a need to make decisions in the face of uncertain candidate availability. The optimal rankings are computed by minimisation of the expected number of disagreements over all the possible subsets of available candidates. In [15], the probability distribution on these subsets is supposed to follow a Bernoulli law whereas several other authors [2,11,12] used other distributions. Lu and Boutilier provide a clear decision-theoretic justification for producing a ranking instead of a single winner. Indeed, a ranking serves as a very compact decision policy: the winner is the output ranking's best candidate among the available ones. Yet, the binary disagreement used in [2,11,12,15] relies on strong hypotheses that can be discussed, as acknowledged by the authors: a voter is satisfied only if its favourite available candidate is elected (as in "plurality" rule) and fully unsatisfied otherwise.

We argue that the voter's satisfaction should vary more smoothly and depend on the rank it gives to the candidate declared as winner by the aggregation. In this paper, we extend the UCM by assuming that positions are associated with disutility values and compute dissatisfaction as the disutility gap between the voter's preferred available candidate and the candidate declared as winner. The goal is then to produce a ranking that minimises the expected dissatisfaction under the probability distribution on the subsets of candidates. We observe that there are two opposed ways to measure the satisfaction of the voters, either by considering the ranks of the candidates in the whole preference order of the voter (*ex ante* approach), or the ranks of the candidates within the subset of available candidates (*ex post* approach). Hence, we analyse our generalisation of the UCM from these two perspectives; we also show connections to other voting schemes.

We introduce background and notations, and review the UCM in Sect. 2. In Sect. 3, we present our framework and introduce the two different models, ex ante and ex post dissatisfactions, that we thoroughly analyse in Sects. 4 and 5 respectively. Finally we provide concluding remarks (Sect. 6).

2   Background

Throughout the paper, given a set $\mathcal{E}$, $\mathscr{P}(\mathcal{E})$ denotes the set of all the subsets (powerset) of $\mathcal{E}$ and $|\mathcal{E}|$ denotes the cardinality of $\mathcal{E}$. We define $[\![1;m]\!] := \{1,\ldots,m\}$. We now present our basic assumptions (Sect. 2.1) and we summarise the UCM (Sect. 2.2).

2.1 Basic Assumptions

We assume that there are m candidates and we call C the set of all the candidates. Given a non-empty subset of candidates $S \subseteq C$, we use R_S to denote the set of rankings (permutations) of the candidates of S. We use R without subscript to mean R_C, the set of full rankings involving all candidates in C. A ranking can be represented explicitly by the tuple that lists the candidates from the most to the least preferred; for instance, the tuple (b, c, a) denotes the ranking that ranks b in first position, c in second position and a in last position. Given a ranking r, r_i denotes the candidate ranked in i-th position by r. For example, if $r = (b, c, a)$, $r_1 = b$, $r_2 = c$ and $r_3 = a$.

We suppose that, for every voter v, the preferences of v over the candidates can be modelled by a ranking. We will then identify voters with their associated rankings and all the definitions that apply to rankings implicitly apply to voters and vice-versa. Given a ranking $r \in R$, the associated preference order is denoted as $>_r$ (and derived orders $\geq_r$, $<_r$ and $\leq_r$ have the obvious meanings). For example, a and b being two candidates, $a >_r b$ means that a is better ranked (has a *lower* rank) than b in r.

Given a ranking $r \in R$ and a candidate $a \in C$, $r(a)$ denotes the position (rank) given to a by r. If S is a non-empty subset of C, the expression $r_S(a)$ denotes the rank of a in the restriction of ranking r that considers only elements of S. In other words, $r_S(a) = 1 + |\{b \in S | b >_r a\}|$. In particular, $r_C(a) = r(a)$. Using these notations amounts to see r as the bijection that maps a candidate $a \in C$ to its rank $r(a) \in [\![1; m]\!]$ in the whole set of candidates[1] and to see r_S as the bijection that maps a candidate $a \in S$ to its rank $r_S(a) \in [\![1; |S|]\!]$ in the subset S. Given a non-empty subset S of C and a ranking $r \in R$, $\mathrm{top}_r(S)$ is the candidate which is the most preferred by r among the candidates of S. In other words, $\mathrm{top}_r(S) = r_S^{-1}(1)$ if we identify r_S with the bijection from S to $[\![1; |S|]\!]$ as explained above. By convention, $\mathrm{top}_r(\emptyset) = a_\emptyset$ where $a_\emptyset$ is a default alternative when none of the candidates are available (for instance, postpone the election). Another convention is that, if r is a ranking, $r_\emptyset(a_\emptyset) = r(a_\emptyset) = 1$[2].

Under the assumption of anonymity, we will consider voting situations [4] that we here model as multisets of rankings (since the same ranking may occur several times). Considering $n \in \mathbb{N}$ fixed throughout the paper, we use $\mathcal{V}$ to denote the collection of the multisets of n rankings (representing voters).

Given a ranking r and $i \in [\![1; m]\!]$, we introduce $S_i(r) = \{a \in C | a \leq_r r_i\} = \{r_j | j \in [\![i; m]\!]\}$, the set of candidates that, in the ranking r, are in position i or worse and $\mathscr{S}_i(r) = \mathscr{P}(S_i(r))$ the powerset of $S_i(r)$. In addition, $\mathscr{T}_i(r) = \{S \cup \{r_i\} | S \in \mathscr{S}_i(r)\} = \{S \in \mathscr{S}_i(r) | r_i \in S\}$ is the collection of all sets of candidates that contain r_i, and no candidate better than r_i in the ranking r. Note that $\mathscr{T}_i(r)$ is also equal to $\{S \subseteq C | \mathrm{top}_r(S) = r_i\}$ i.e. the collection of all sets of candidates whose top-element according to r is r_i.

---

[1] With this point of view, the candidate in the i-th position can be seen as the preimage of i by r, i.e. $r_i = r^{-1}(i)$.

[2] These conventions are aimed at simplifying the proofs and do not interfere with the search of optimal rankings since $a_\emptyset$ is not in C.

2.2   Unavailable Candidate Model

We consider a setting where the availability of the candidates is uncertain. We assume a probability distribution P on $\mathscr{P}(C)$ that we will refer to as the *unavailability distribution* (not necessarily known by the voters). For every $S \subseteq C$, $P(S)$ denotes the probability that the set of available candidates is exactly S[3]. We use $\mathcal{P}_C$ to denote the set of probability distributions on $\mathscr{P}(C)$. Given $P \in \mathcal{P}_C$ and an application g defined on $\mathscr{P}(C)$, the expectation of g under P is $\underset{S \sim P}{\mathbb{E}}[g(S)] := \sum_{S \subseteq C} P(S)g(S)$.

Definition 1. $P \in \mathcal{P}_C$ *is* pair-sensitive *if* $\forall S \subseteq C, |S| = 2 \Rightarrow P(S) > 0$.

Definition 2. $P \in \mathcal{P}_C$ *is said* impartial *if it satisfies* $\forall S \subseteq C, \forall S' \subseteq C, |S| = |S'| \Rightarrow P(S) = P(S')$. $\tilde{\mathscr{P}}_C$ *denotes the set of impartial probability distributions of* $\mathcal{P}_C$. *We then define, for* $P \in \tilde{\mathscr{P}}_C$, $\tilde{P}$ *as follows: for* $k \in [\![1; m]\!]$, $\tilde{P}(k)$ *is the probability* $P(S)$ *for any* $S \subseteq C$ *of cardinality* k[4]. P *is impartial means that* P *treats all the candidates equally.*

Lu and Boutilier [15] propose to evaluate a ranking with respect to the expected disagreement that measures, for each voter, the binary disagreement that evaluates to one if the winner (the elected candidate) is not the same as the voter's preferred choice among the available choices and zero otherwise. Assuming that the probability of a candidate to be unavailable is $p \in]0; 1[$, the expected number of disagreements between two rankings r and r' is defined as

$$D_p(r, r') := \sum_{S \subseteq C} p^{m-|S|}(1-p)^{|S|} \mathbb{1}[\text{top}_r(S) \neq \text{top}_{r'}(S)] \qquad (1)$$

Observation 1. D_p *is a metric.*

An optimal ranking is a ranking that minimises $\sum_{v \in V} D_p(r, v)$. Note that D_p implicitly assumes that the best decision would be to follow plurality when the set of available candidates is revealed. Since a ranking has to be produced before the set S is revealed, the optimal ranking is the one that best approximates "plurality a posteriori". Plurality is a rule that is not perceived satisfactory from the point of view of social theory [14] because it only takes into account the first candidate of the voter's preference and thus loses a lot of information. Precisely, a major advantage of plurality is its simplicity and the small quantity of information needed and, thus, the cognitive load for the voters is reduced. This is not the case in the UCM where the complete ranking of preferences is anyway needed for each voter. Generalising scoring rules via ex post dissatisfaction and not only plurality provides a richer model and does not require more information, except the disutility functions which may be, in a lot of scenarios, the same for all voters.

---

[3] The atomic elements being the subsets of C, P should actually be defined on $\mathscr{P}(\mathscr{P}(C))$ and the probability that the set of available candidates is equal to S would be $P(\{S\})$. We nevertheless write $P(S)$ for the sake of readability.

[4] The notation $\tilde{P}(k)$ must not be confused with the probability that the set of available candidates is of cardinality k, which is actually equal to $\tilde{P}(k) \times \binom{m}{k}$.

3   Generalised Unavailable Candidate Model

In this work we provide a generalisation of the UCM by evaluating a ranking with respect to the expected dissatisfaction that it imposes on the voters. A ranking defines a policy for making a choice under uncertain availability of candidates. The final choice is the top ranked candidate among the available ones. The voters suffer a degree of dissatisfaction that depends on the position of the final choice in their own ranking. In order to introduce our model, as the first thing, we need to define a function that maps rank positions to "disutility" values.

Definition 3. *A* disutility function *(DF)* ρ *is an increasing mapping from* $[\![1; m]\!]$ *to* $\mathbb{R}$. ρ *may be represented as a sequence i.e.* $\rho = (\rho(i))_{i \in [\![1;m]\!]}$. $\rho(i)$ *measures how much a voter is unsatisfied by the item at the i-th position in its ranking.*

The overall idea of our generalised model is the following. Given a ranking r, when the set S of available candidates is observed, the candidate $\text{top}_r(S)$ is declared the winner; in other words the winner is the highest ranked candidate in S with respect to r. As far as the voter v is concerned, its most desired candidate is $\text{top}_v(S)$. We allow the disutility functions to depend on the voter and note ρ_v the disutility function associated to voter v. This implies that, in the generalised UCM, considering a voter v is actually considering the ranking of the preference of v *and* the DF ρ_v, the definition of $\mathcal{V}$ being consequently adapted. Nevertheless, in the absence of ambiguity and to lighten the notations, v may refer to either the voter or its preference order in the remainder of the paper. ρ_v and the position in v of a candidate a determine the disutility that voter v suffers from the election of a. The dissatisfaction, with respect to S, is then the difference between the disutility of $\text{top}_r(S)$ and that of $\text{top}_v(S)$. Finally, the total dissatisfaction is the sum of the dissatisfactions of all the voters in V. When we produce a ranking, the set of available candidates is not known but we know the distribution P. We thus aim at providing the ranking that minimises the total dissatisfaction in expectation over P.

In the following we make this reasoning more concrete and discuss two different methods to compute dissatisfaction that differ on whether full or restricted rankings are considered.

3.1   Ex ante Dissatisfaction

In this approach, the dissatisfaction is computed using the positions of the candidates in the whole set of candidates C. When a candidate $a \in S$ is chosen as winner, with S being the set of available candidates, voter v suffers disutility $\rho_v(v(a))$, i.e. *the disutility associated with the position of a in the full ranking v*, while its most preferred candidate in S would have given him disutility $\rho_v(v(\text{top}_v(S)))$; this gives $\rho_v(v(a)) - \rho_v(v(\text{top}_v(S)))$ as value of dissatisfaction. We then take the expectation of such a value under P, since the set S is not known beforehand. Finally, we obtain $\hat{\Delta}_P(v, r)$, that we formally define now.

Definition 4. *Let $P \in \mathcal{P}_C$ be a probability distribution, v be a voter (or ranking), ρ_v be the DF associated to v, and r a ranking. We define*

$$\hat{\Delta}_P(v,r) := \mathop{\mathbb{E}}_{S \sim P}[\rho_v(v(\mathrm{top}_r(S))) - \rho_v(v(\mathrm{top}_v(S)))]$$

and, if $V \in \mathcal{V}$, $\hat{\Delta}_P(V,r) := \sum_{v' \in V} \hat{\Delta}_P(v',r)$.

Proposition 1. *For any $P \in \mathcal{P}_C$, $\hat{\Delta}_P \geq 0$ and, for any voter v, $\hat{\Delta}_P(v,v) = 0$[5].*

We call $\hat{\Delta}_P$ the *ex ante dissatisfaction measure* associated to P and $\rho_v(v(\mathrm{top}_r(S))) - \rho_v(v(\mathrm{top}_v(S)))$ is the *ex ante dissatisfaction* induced by $r \in R$ in $S \subseteq C$ to the voter v. The ex ante dissatisfaction induced by r in the empty set to v is $\rho(v(a_\emptyset)) - \rho(v(a_\emptyset)) = 0$ so the empty set does not contribute to the ex ante dissatisfaction measure. Note that the singletons do not contribute either since they cannot generate dissatisfaction. We also call the application $\hat{\Delta}_P$ defined on $\mathcal{V} \times R$, the ex ante dissatisfaction measure associated to P.

Definition 5. *The* optimal rankings *for $V \in \mathcal{V}$ are defined as the elements of $\arg\min_{r \in R}(\hat{\Delta}_P(V,r))$. We also write this set $\hat{R}_P^*(V)$.*

Example 1. Let $m = 3$, $C = \{a,b,c\}$. Let $P \in \mathcal{P}_C$ be the uniform probability distribution: for all $S \subseteq C$, $P(S) = \frac{1}{8}$. Let $n = 11$. Let $V \in \mathcal{V}$ be a set of 11 voters, 4 of them voting according to the ranking $r' = (a,b,c)$ and 7 of them voting according to the ranking $r'' = (c,a,b)$. We assume that all voters have the same DF $\rho = (0,1,2)$. Note that, for every $r \in R$, $\hat{\Delta}_P(V,r) = \mathop{\mathbb{E}}_{S \sim P}[4\rho(r'(\mathrm{top}_r(S))) + 7\rho(r''(\mathrm{top}_r(S)))] - \chi(V)$ where $\chi(V) := \mathop{\mathbb{E}}_{S \sim P}[4\rho(r'(\mathrm{top}_{r'}(S))) + 7\rho(r''(\mathrm{top}_{r''}(S)))]$ does not depend on r. The following array displays, for all $r \in R$, from left to right, the value of $\mathrm{top}_r(S)$ for every non-empty and non-singleton $S \subseteq C$, $\sum_{S \subseteq C} \rho(r'(\mathrm{top}_r(S)))$, $\sum_{S \subseteq C} \rho(r''(\mathrm{top}_r(S)))$ and $\Sigma = 4\sum_{S \subseteq C} \rho(r'(\mathrm{top}_r(S))) + 7\sum_{S \subseteq C} \rho(r''(\mathrm{top}_r(S))) = 8[\hat{\Delta}_P(V,r) + \chi(V)]$.

r	S						
	abc	ab	ac	bc	$\rho(r'(\mathrm{top}_r(S)))$	$\rho(r''(\mathrm{top}_r(S)))$	Σ
(a,b,c)	a	a	a	b	$0+0+0+1$	$1+1+1+2$	39
(a,c,b)	a	a	a	c	$0+0+0+2$	$1+1+1+0$	**29**
(b,a,c)	b	b	a	b	$1+1+0+1$	$2+2+1+2$	61
(b,c,a)	b	b	c	b	$1+1+2+1$	$2+2+0+2$	62
(c,a,b)	c	a	c	c	$2+0+2+2$	$0+1+0+0$	31
(c,b,a)	c	b	c	c	$2+1+2+2$	$0+2+0+0$	42

We deduce that the only optimal ranking is (a,c,b).

---

[5] Referring to the definition of $\hat{\Delta}_P$, one can note that the first argument of $\hat{\Delta}_P$ here is voter v itself while the second argument is its preference order.

3.2   Ex post Dissatisfaction

In this model, we assume that the disutility felt by a voter when a candidate is elected depends on its position *within* the set S of the actually available candidates. More precisely, when a candidate $a \in S$ is chosen as winner, voter v suffers disutility $\rho_v(v_S(a))$, that is the disutility value associated to $v_S(a)$, *the position of a in the ranking obtained by restricting v to the set S.* Voter v's most preferred candidate in S would have given him disutility $\rho_v(v_S(\text{top}_v(S))) = \rho_v(1)$, yielding dissatisfaction $\rho_v(v_S(a)) - \rho_v(1)$. The expectation of such value under P gives $\Delta_P(v, r)$, that we formally define now.

Definition 6. *Let $P \in \mathcal{P}_C$, v be a voter (or ranking) with DF ρ_v and r be a ranking.*

$$\Delta_P(v, r) := \mathop{\mathbb{E}}_{S \sim P}[\rho_v(v_S(\text{top}_r(S))) - \rho_v(1)].$$

and, if $V \in \mathcal{V}$, $\Delta_P(V, r) := \sum_{v' \in V} \Delta_P(v', r)$.

Proposition 2. *For any $P \in \mathcal{P}_C$, $\Delta_P \geq 0$ and, for any voter v, $\Delta_P(v, v) = 0$.*

We call Δ_P the *ex post dissatisfaction measure* associated to P whether it is defined on R^2 or $\mathcal{V} \times R$. For a voter v, a ranking r, $S \subseteq C$, $\rho_v(v_S(\text{top}_r(S))) - \rho_v(1)$ is the *ex post dissatisfaction* induced by r in S to v. The ex post dissatisfaction induced by r in the empty set to v is $\rho_v(v_\emptyset(a_\emptyset)) - \rho_v(1) = \rho_v(1) - \rho_v(1) = 0$ so the empty set does not contribute to the ex post dissatisfaction measure. As in the ex ante approach, the singletons do not contribute either.

Definition 7. *The optimal rankings for $V \in \mathcal{V}$ are defined as the elements of $\arg\min_{r \in R}(\Delta_P(V, r))$. We also write this set $R_P^*(V)$.*

Example 2. We will take the situation of Example 1 and study it with the ex post approach. We keep the same notations as in Example 1. Let $r \in R$. Since $\rho(1) = 0$, we directly have $\Delta_P(V, r) = \mathop{\mathbb{E}}_{S \sim P}[4\rho(r_S'(\text{top}_r(S))) + 7\rho(r_S''(\text{top}_r(S)))]$. Let us summarise the computations in the following array, as in Example 1:

r	S				$\rho(r_S'(\text{top}_r(S))$	$\rho(r_S''(\text{top}_r(S))$	Σ
	abc	ab	ac	bc			
(a, b, c)	a	a	a	b	$0 + 0 + 0 + 0$	$1 + 0 + 1 + 1$	21
(a, c, b)	a	a	a	c	$0 + 0 + 0 + 1$	$1 + 0 + 1 + 0$	18
(b, a, c)	b	b	a	b	$1 + 1 + 0 + 0$	$2 + 1 + 1 + 1$	43
(b, c, a)	b	b	c	b	$1 + 1 + 1 + 0$	$2 + 1 + 0 + 1$	40
(c, a, b)	c	a	c	c	$2 + 0 + 1 + 1$	$0 + 0 + 0 + 0$	**16**
(c, b, a)	c	b	c	c	$2 + 1 + 1 + 1$	$0 + 1 + 0 + 0$	27

Hence, the optimal ranking is (c, a, b). Note that this is a different ranking from the optimal ranking obtained in the ex ante approach, namely (a, c, b).

The question whether the ex ante or the ex post approach is more relevant is open. On the one hand, the ex ante model captures the idea that the "utility" perceived by a voter when a candidate is elected should be independent of whether another candidate is available or not. On the other hand, the ex post approach is more relevant if we want to focus on the regret of the voters: if the candidate a is elected, the more available candidates that voter v preferred to a have been excluded, the more regret v will experiment. We also believe that the ex post approach reduces the incentive of manipulating with regards to the unavailability distribution. Indeed, a voter who knows the unavailability distribution (but not the preferences of the other voters) may use this knowledge to manipulate the vote in both approaches. Nevertheless, in a situation where two candidates are much more likely to be available than the others for instance (case of a partial unavailability distribution), the benefit of manipulating must be higher in the ex ante approach since ranking these two candidates at the extreme positions (first and last) will increase the influence of the voter in the result of the aggregation. Quantitative results confirming this intuition would require further work. We will see in the following that both ex ante and ex post models are interestingly linked to several well known voting rules. Note that the ex post approach is also considered in [2, 11] and, in particular, by Lu and Boutilier in [15] as the following proposition shows.

Proposition 3. *Let $p \in]0; 1[$. D_p is the ex post dissatisfaction measure Δ_P whose probability distribution is $P: S \subseteq C \mapsto p^{m-|S|}(1-p)^{|S|}$ (a Bernoulli distribution as defined in Sect. 5) and with the DF $1 - \mathbb{1}_{\{1\}} = (0, 1, ..., 1)$ for all voters. We call $1 - \mathbb{1}_{\{1\}}$ the binary DF.*

Note that Lumet et al. [16] also proposed a double ex ante/ex post approach in a problem of fair allocation of indivisible goods with the assumption that some goods may turn out to be in bad condition and thus unusable. Choosing between their two approaches consists of choosing whether the aggregation over the agents is performed *before* or *after* the expectation over the conditions of the objects in the computation of what the authors call the *collective utility*. The utility of an object in good condition is well defined and fixed. By contrast, in our work in which the aggregator over the voters is simply the sum[6] and commutes with the expectation, the ex ante/ex post distinction is based on the definition of the dissatisfaction and on the fact that the disutility can be computed either *before* or *after* knowing the actual set of available candidates.

4   Ex ante Dissatisfaction

4.1   Link with Scoring Rules

We now develop the theory of ranking with respect to ex ante dissatisfaction measure and show the connection with positional scoring rules.

---

[6] Studying other aggregators is a perspective that would allow us to give more focus on fairness in the consensus production.

Definition 8. *Let $V \in \mathcal{V}$. We define $s_V : a \in C \mapsto \sum_{v \in V} \rho_v(v(a))$.*

Definition 9. *Let $V \in \mathcal{V}$. We define $R^{\uparrow}(V)$ the set of rankings where the candidates are ranked in the increasing order of s_V (there can be several such rankings if some candidates have equal scores).*

The following theorem characterises the ex ante dissatisfaction rule. It shows that, regardless of the unavailability distribution P, an optimal ranking can be found by sorting alternatives with respect to their score. Moreover, if P is pair-sensitive, the set of all optimal rankings is the exact output of the scoring rule with scoring function $-\rho_v$ for voter v. This result echoes the one from [20] which characterises scoring rule via minimisation of the positional Spearman semi-metric, with the restrictive assumption of strictly increasing scores.

Theorem 1. *Let $P \in \mathcal{P}_C$ and $V \in \mathcal{V}$. Then, $R^{\uparrow}(V) \subseteq \hat{R}_P^*(V)$.*

If, besides, P is pair-sensitive, $R^{\uparrow}(V) = \hat{R}_P^*(V)$.

Sketch of Proof. We show that minimising $\Delta_P(V, r)$ amounts to minimising $\psi(V, r) = \mathop{\mathbb{E}}_{S \sim P}[s_V(\text{top}_r(S))]$ and that, for any $S \subseteq C$, $s_V(\text{top}_r(S))$ is minimal if $r \in R^{\uparrow}(V)$. In the case where P is pair-sensitive, if $r \notin R^{\uparrow}(V)$ we can find $i \in [\![1; m-1]\!]$ such that $s_V(r_i) > s_V(r_{i+1})$. Exchanging the positions of r_i and r_{i+1} in r strictly reduces the value of $\psi(V, r)$ so r is not optimal. Note that, if P is not pair-sensitive and $P(\{a, b\}) = 0$, in the case where $r^* \in R^{\uparrow}(V)$ with $r_{m-1}^* = a$ and $r_m^* = b$, then r deduced from r^* by swapping a and b is also optimal, but not in $R^{\uparrow}(V)$ if $s_V(a) < s_V(b)$.

Corollary 1. *An optimal ranking for $V \in \mathcal{V}$ in the ex ante dissatisfaction model can be found in polynomial time in m and n.*

Corollary 2. *Let $V \in \mathcal{V}$. Let $P \in \mathcal{P}_C$ and $P' \in \mathcal{P}_C$ be two pair-sensitive probability distributions. Then $\hat{R}_P^*(V) = \hat{R}_{P'}^*(V)$.*

The previous corollary expresses that the set of optimal rankings for V does not depend on the unavailability distribution as long as the latter is pair-sensitive.

Example 3. Let $C = \{a, b, c, d\}$. Let V be the multiset of three voters among which one votes according to (d, b, a, c), one according to (d, a, b, c) and one according to (b, c, a, d). We suppose that all voters have the DF $\rho = (0, 1, 2, 3)$. We have: $s_V(a) = 2 + 1 + 2 = 5$, $s_V(b) = 1 + 2 + 0 = 3$, $s_V(c) = 3 + 3 + 1 = 7$, $s_V(d) = 0 + 0 + 3 = 3$. The rankings that rank the candidates in the increasing order of s_V thus are (b, d, a, c) and (d, b, a, c). We deduce that these two rankings are the optimal rankings for V whatever is the pair-sensitive unavailability distribution.

5 Ex post Dissatisfaction

In this section, we study ex post dissatisfaction rule from an algorithmic point of view and under natural assumptions on the unavailability distribution and the

DF. It is obvious that the ex post dissatisfaction is unchanged by a translation on any of the voters' DF. Hence, without loss of generality, we will consider in the following that all the DF are null in 1 and therefore non-negative on $[\![1; m]\!]$.

Remark 1. Let $P \in \mathcal{P}_C$. If all the voters' DF are null in 1 then, for all $(V, r) \in \mathcal{V} \times R$, $\Delta_P(V, r) = \sum_{v \in V} \mathop{\mathbb{E}}_{S \sim P} \rho_v(v_S(\text{top}_r(S)))$.[7]

Definition 10. *A DF ρ is discriminating at the top (DT) if $\rho(1) < \rho(2)$.*

Proposition 4. *Let $P \in \mathcal{P}_C$ pair-sensitive. Let v be a voter with DF ρ_v. We have $(\forall r \in R, \Delta_P(v, r) = 0 \Longleftrightarrow r = v)$ if, and only if, ρ_v is DT.*
Conversely, let $P \in \mathcal{P}_C$ not pair-sensitive. For any voter v, there exists a ranking $r \neq v$ such that $\Delta_P(v, r) = 0$ (ρ_v being either DT or not).

Observation 2. *Multiplying all voters' DF by $\alpha \in \mathbb{R}_+^*$ does not change $R_P^*(V)$.*

Proof. Multiplying all the voters' DF by the same α multiplies Δ_P by α.

Definition 11. *A DF ρ is overnormalised if it is DT, $\rho(1) = 0$ and $\rho(2) \geq 1$.*

Proposition 4 incentives us to study only DF that are DT. In this context, Observation 2 shows that we can restrict our study to overnormalised DF without loss of generality - it suffices to multiply all the DF by $\frac{1}{\min_{v \in V} \rho_v(2)}$.

As in the work of Lu and Boutilier [15], we introduce a specific kind of probability distribution of $\mathcal{P}_C$. Assuming that the probability of a candidate to be unavailable is independent of the presence of the other candidates and that this probability is equal to $p \in]0; 1[$ for all candidates, we get the probability distribution $S \subseteq C \mapsto p^{m-|S|}(1-p)^{|S|}$. We call it the *Bernoulli distribution of parameter p*. Note that a Bernoulli distribution is impartial and pair-sensitive.

5.1   Connections with Kendall's Tau Metric

In the following, for any $(r, r') \in R^2$, $\kappa(r, r') := \sum_{i < j} \mathbb{1}[r_i <_{r'} r_j]$ is the Kendall's tau metric between r and r' and, for any $(V, r) \in \mathcal{V} \times R$, $\kappa(V, r) := \sum_{v \in V} \kappa(v, r)$.

Remark 2. Let us suppose that all voters have the DF $\rho = (0, 1, ..., 1)$, i.e. the binary DF. Let P be the impartial probability distribution of $\mathcal{P}_C$ that is non-null only on the pairs; in other words, for $S \subseteq C$, if $|S| = 2$, $P(S) = \frac{2}{n(n-1)}$, otherwise $P(S) = 0$. For any $(V, r) \in \mathcal{V} \times R$, $\Delta_P(V, r) = \frac{2}{n(n-1)} \kappa(V, r)$. Hence, Δ_P is the Kendall's tau metric within a strictly positive factor.

A result of Baldiga and Green in [2] shows that the Kemeny rule and the ex post dissatisfaction rule with binary DF for all voters and general impartial unavailability distribution may produce different rankings and highlights in that way the role of the unavailability distribution.

---

[7] This remark motivates our choice of considering disutilities instead of utilities because, when $\rho_v(1) = 0$ for all $v \in V$, the ex post dissatisfaction measure can be seen as a simple sum of expectations of $\rho_v(v_S(\text{top}_r(S)))$.

Definition 12. *For $(r, r') \in R^2$, $i \in [\![1; m]\!]$, we define $x_{r,r'}(i) := |\{a \in C | a >_r r'_i \wedge a <_{r'} r'_i\}| = r_{S_i(r')}(r'_i) - 1$ the number of candidates ranked after r'_i in r' but before r'_i in r.*

We now recall a proposition established with other notations in the proof of Theorem 11 of [15] that will be useful in the following.

Proposition 5. *[15] Let $(r, r') \in R^2$. $\kappa(r, r') = \sum_{i=1}^{m} x_{r,r'}(i) = \sum_{i=1}^{m-1} x_{r,r'}(i)$.*

Definition 13. *The* Borda DF *maps $i \in [\![1; m]\!]$ to $i - 1$.*

The Borda DF is overnormalised. The name Borda DF is justified by the fact that, if all voters have the Borda DF and $P \in \mathcal{P}_C$ is an arbitrary pair-sensitive probability distribution, Theorem 1 shows that the ex ante dissatisfaction rule based on the minimisation of $\hat{\Delta}_P$ is equivalent to the Borda rule.

Lemma 1. *If ρ is a DF, $P \in \mathcal{P}_C$ is a Bernoulli distribution of parameter $p \in]0; 1[$, $(r, r') \in R^2$ and $i \in [\![1; m]\!]$,*

$$\sum_{S \in \mathcal{F}_i(r')} P(S)\rho(r_S(\text{top}_{r'}(S))) = \sum_{j=0}^{x_{r,r'}(i)} \rho(j+1)\binom{x_{r,r'}(i)}{j}(1-p)^{j+1}p^{x_{r,r'}(i)+i-j-1}.$$

The next proposition establishes the link with Kendall's tau.

Proposition 6. *Let $p \in]0; 1[$ and $P \in \mathcal{P}_C$ be the Bernoulli distribution of parameter p. Let $r \in R$ and v be a voter with the Borda DF.*
It holds: $\Delta_P(v, r) = (1 - p)^2 \sum_{i=1}^{m-1} x_{v,r}(i)p^{i-1}$.

Proposition 6 shows that the ex post dissatisfaction measure associated to the Borda DF and a Bernoulli distribution can be seen as a weighted version of the Kendall's tau metric but in a different sense than in [19] or [10] since here the weight associated to an inversion in the computation of $\Delta_P(v, r)$ does not depend on the ranks of both candidates but only on the rank of the candidate ranked before in r and after in v.

5.2   Solving the Ex post Dissatisfaction Problem

Complexity of the Ex post Dissatisfaction Rule. In the sequel, given $q \in [1; +\infty[$, a DF ρ is said to be *q-sub-geometrical* if ρ is overnormalised and, for any $k \in [\![2; m]\!]$, $\rho(k) \leq q^{k-2}\rho(2)$. Theorem 2 shows that optimising ex post dissatisfaction measure is NP-hard.

Theorem 2. *We suppose that all the voters have the same DF ρ and that ρ is q-sub-geometrical, for a $q \in [1; +\infty[$. Let $\epsilon \in]0, \frac{1}{nm(m-1)+1}[$. Let $p \in [\max((1 - \epsilon)^{\frac{1}{m-1}}, \frac{q-(1+\epsilon)^{\frac{1}{m-1}}}{q-1}); 1[$[8]. Let $P \in \mathcal{P}_C$ be the Bernoulli distribution of parameter p, and $V \in \mathcal{V}$. Any ranking in $R_P^*(V)$ is also a Kemeny consensus.*

---

[8] $\epsilon > 0$ so both $(1 - \epsilon)^{\frac{1}{m-1}}$ and $\frac{q-(1+\epsilon)^{\frac{1}{m-1}}}{q-1}$ are strictly lower than 1.

Sketch of Proof. Using the assumptions on p and ρ, we prove, for all $(V, r) \in \mathcal{V} \times R$: $\rho(2)(1-p)^2(1-\epsilon)\kappa(V, r) \leq \Delta_P(V, r) \leq \rho(2)(1-p)^2(1+\epsilon)\kappa(V, r)$. The bounds on ϵ and these inequalities enable us to show a contradiction if we suppose the existence of an optimal ranking which is not a Kemeny consensus.

Corollary 3. *For any number of candidates $m \in \mathbb{N}^*$, we suppose that the voters' DF are all q_m-sub-geometrical, with $q_m \in [1; +\infty[$. For $m \in \mathbb{N}^*$, we consider $\epsilon_m \in]0, \frac{1}{nm(m-1)+1}[$, $p_m \in [\max((1-\epsilon_m)^{\frac{1}{m-1}}, \frac{q_m-(1+\epsilon_m)^{\frac{1}{m-1}}}{q_m-1}); 1[$, $P_m \in \mathcal{P}_C$ the Bernoulli distribution of parameter p_m, and V_m a voting situation for m candidates. The problem of size m of finding $r \in R$ minimising $\Delta_{P_m}(V_m, r)$ is NP-hard.*

Proof. Since finding a Kemeny consensus is NP-hard [3], Theorem 2 gives us the NP-hardness of the problem where all the voters have the same DF. Therefore, the more general problem where DF depend on the voters is also NP-hard.

A Polynomial-Time Approximation Scheme. We here exhibit an approximation algorithm for the ex post dissatisfaction rule and present some intermediary results that allow us to prove this is a PTAS under natural assumptions. In this subsection, P denotes the Bernoulli distribution of parameter $p \in]0; 1[$. Firstly, let us introduce some definitions necessary to the construction of the algorithm.

Definition 14. *If there is no ambiguity on P and V, we define $f \colon (\mathcal{S}, a) \in \mathcal{P}(\mathcal{P}(C)) \times C \mapsto \sum_{v \in V} \sum_{\substack{S \in \mathcal{S} \\ a \in S}} P(S)\rho_v(v_S(a))$. For $a \in C$, let $f(a) := f(\mathcal{P}(C), a)$.*

Observation 3. *Let $V \in \mathcal{V}$, $r \in R$. $\Delta_P(V, r) = \sum_{i=1}^m f(\mathcal{S}_i(r), r_i)$.*

Definition 15. *For $V \in \mathcal{V}$, $r \in R$, $k \in [\![1; m]\!]$, $l \in [\![k; m]\!]$, we denote $\Delta_P^{k,l}(V, r) := \sum_{v \in V} \sum_{i=k}^l \sum_{S \in \mathcal{T}_i(r)} P(S)\rho_v(v_S(r_i)) = \sum_{i=k}^l f(\mathcal{S}_i(r), r_i)$ the contribution to $\Delta_P(V, r)$ of the subsets of C for which one of the candidates of $\{r_i | i \in [\![k; l]\!]\}$ is winning according to r.*

We now define our *MyopicTop* algorithm (Algorithm 1). It is conceptually similar to the one of Lu and Boutilier but makes use of a new notion of dominance (Corollary 4) which encapsulates the complexity of ex post dissatisfaction rule and applies to any DF.

Example 4. Let $C = \{a, b, c, d, e\}$, $p = \frac{1}{4}$, $K = 2$, V constituted of three voters (a, b, c, d, e), one voter (a, c, b, d, e) and one voter (c, a, d, e, b), all voters with Borda DF. a is dominant in C. In $C \setminus \{a\}$, there is no dominant candidate and the rankings of $C \setminus \{a\}$ minimising $\Delta_P^{2,3}(V, r)$ are the rankings starting with (c, b). Then, MyopicTop algorithm outputs (a, c, b, d, e) or (a, c, b, e, d).

Lemma 2. *Let $V \in \mathcal{V}$ and $a \in C$. If there exists an optimal ranking that ranks a in first position, then, for every candidate $b \in C$, $(1 + p)f(b) \geq (1 - p)f(a)$.*

Corollary 4. *Let $V \in \mathcal{V}$ be a voting situation and $a \in C$. If, for all $b \in C \setminus \{a\}$, $(1+p)f(a) < (1-p)f(b)$, then, a is the first candidate of all optimal rankings. In this case, we call a the dominant candidate.*

Algorithm 1: MyopicTop

Input: C, p, V, K
Output: $r = (r_1, ..., r_m)$
$C' \leftarrow C$; $i \leftarrow 1$;
while $C' \neq \emptyset$ *and* C' *has a dominant candidate* **do**
    $r_i \leftarrow$ the dominant candidate in C';
    $C' \leftarrow C' \setminus \{r_i\}$;
    $i \leftarrow i + 1$;
end
Determine $(r_i, ..., r_{i+K-1}) \in C^K$ such that, for any $r' \in R_{C \setminus \{r_1, ..., r_{i+K-1}\}}$,
$(r_1, ..., r_{i+K-1}, r'_1, ..., r'_{m+1-i-K}) \in \underset{\substack{r'' \in R \\ r''_j = r_j, j < i}}{\arg\min} \, \Delta_P^{i, i+K-1}(V, r'')$;
Arbitrarily order the remaining $m + 1 - i - K$ candidates;

Proposition 7. *The* MyopicTop *algorithm runs in $O(nm^{\max(3, K+2)})$.*

Sketch of Proof. Lemma 1 shows that the while loop can be performed in $O(nm^3)$. The second part of the algorithm needs to test only $\frac{m!}{(m-K)!} = O(m^K)$ rankings since, once the $i - 1$ first candidates are fixed, $\Delta_P^{i, i+K-1}(V, r)$ only depends on the candidates between the ith and the $i + K - 1^{th}$ positions in r. Besides, we can reuse Lemma 1 to show that $\Delta_P^{i, i+K-1}$ is computed in $O(nm^2)$. Hence, the second part is performed in $O(nm^{K+2})$. The last part is done in $O(m)$.

Lemma 3 allows us to apply Corollary 4 to subsets of C. We can then show Theorem 3 which proves that *MyopicTop* algorithm is a PTAS for the ex post dissatisfaction rule when the normalised DF are bounded[9]. Note that here parameter p is assumed to be fixed and independent on n and m. This differs from the assumptions made to establish the NP-hardness result in Theorem 2. The question whether the problem with p fixed is NP-hard or not remains open.

Lemma 3. *Let $V \in \mathcal{V}$, $r^* \in R_P^*(V)$ be an optimal ranking. For all $k \in [\![1; m]\!]$, $(r_k^*, ..., r_m^*) \in \underset{r \in R_{S_k(r^*)}}{\arg\min} \left(\sum_{v \in V} \sum_{S \in \mathscr{S}_k(r^*)} P(S)[\rho_v(v_S(\text{top}_r(S)))] \right)$ i.e. $(r_k^*, ..., r_m^*)$ is an optimal ranking for the reduced set of candidates $S_k(r^*) = \{r_k^*, ..., r_m^*\}$.*

---

[9] It is unclear whether there exists a PTAS if the normalised DF are not bounded.

Theorem 3. *We consider voting situations where voters' DF are overnormalised and bounded by a fixed $M \in \mathbb{R}_+^*$: for any voter v, for all $i \in [\![1;m]\!]$, $\rho_v(i) \leq M$. Let $\epsilon > 0$, $K = \lceil \log_{\frac{1}{p}}(\frac{2M}{(1-p)^3\epsilon}) \rceil$ and $V \in \mathcal{V}$. Let $r^* \in R_P^*(V)$ be an optimal ranking and r be the ranking obtained via the MyopicTop algorithm with inputs C, p, V, K. If $\Delta_P(V, r^*) = 0$ then $\Delta_P(V, r) = 0$. Otherwise, $\frac{\Delta_P(V,r)}{\Delta_P(V,r^*)} \leq 1 + \epsilon$.*

Sketch of Proof. First of all, we show that, if the while loop runs until there is no remaining candidates, then $r = r^*$. This happens, for instance, when $\Delta_P(V, r^*) = 0$ and then, in this case, we also have $\Delta_P(V, r) = 0$. If the while loop ends before, at an index $i - 1$, where $i \in [\![1;m]\!]$, r coincides with r^* for the first $i - 1$ candidates and $\Delta_P^{i,i+K-1}(V, r) \leq \Delta_P^{i,i+K-1}(V, r^*)$ by construction of r. Besides, we show that $\Delta_P^{i+K,m}(V, r) \leq nMp^{i+K-1}$ and, the fact that r_i^* is not dominant in $\mathscr{S}_i(r^*)$ enables us to show that $f(\mathscr{S}_i(r^*), r_i^*) \geq \frac{n}{2}p^{i-1}(1 - p)^3$. Finally, by construction of K, we get $\frac{\Delta_P(V,r)}{\Delta_P(V,r^*)} \leq 1 + \frac{\Delta_P^{i+K,m}(V,r)}{f(\mathscr{S}_i(r^*),r_i^*)} \leq 1 + \frac{2Mp^K}{(1-p)^3} \leq 1 + \epsilon$.

Note that, for the algorithm to be polynomial, the bound M *must not* depend on the number m of candidates. The assumption whereby the DF are bounded may seem restrictive but is actually quite reasonable if we suppose that a voter cannot cognitively conceive an unbounded dissatisfaction. Indeed, one can naturally consider that, after a fixed rank, all alternatives are equally disliked by the voter, the disutility value M would then mean "I completely dislike this alternative". DF can even be strictly increasing, but converging towards M. Hence, this upperbound assumption models the increasing difficulty for the voters to discriminate between alternatives as they go further in their preference rankings.

6   Conclusion

We provided an extension of the UCM accounting for rank-based dissatisfaction. The voters' preferences are aggregated in a ranking that minimise the overall expected dissatisfaction and is used to select the winner once the available candidates are known. We considered two different settings, ex ante and ex post, corresponding to different ways of defining dissatisfaction, provided a theoretical analysis of both cases, and gave algorithms for finding or approximating an optimal ranking. This analysis showed that the assumption used to define dissatisfaction has a crucial impact on the complexity of the voting rule. Interestingly, we showed that this two-sided model provides a unified representation for very different voting rules, spanning from positional scoring rules to Kemeny rule.

Future works may include analysis of practical performance with simulations. We now provide some theoretical directions of research. First of all, we are interested in studying other probability distributions (including non impartial ones) than Bernoulli and thereby emphasising the fact that ex post model generalises voting rules studied in [2, 11, 15]. Other aggregators than the sum could enable us to include fairness considerations [16]. A quite different idea, inspired from [15], would be to analyse the link between optimal rankings and optimal policies - i.e.

choice functions - in the ex post approach. We believe that there is a connection between the comparison ex ante/ex post optimal rankings and the rationalisability of optimal policies; this would echo Theorem 1 from [2] that links the rationalisability of optimal policies to the influence of the unavailability distribution. More quantitavely, we could study the role of the inconsistency (in the sense of [1]) of the optimal policies. Comparing ex ante and ex post approaches in terms of manipulation could also be fruitful as mentioned in Sect. 3.2. Finally, we are also interested in studying elicitation of preferences in a context of uncertain availability.

References

1. Apesteguia, J., Ballester, M.A.: A measure of rationality and welfare. J. Polit. Econ. **123**(6), 1278–1310 (2015)
2. Baldiga, K.A., Green, J.R.: Assent-maximizing social choice. Soc. Choice Welfare **40**(2), 439–460 (2013)
3. Bartholdi, J., Tovey, C.A., Trick, M.A.: Voting schemes for which it can be difficult to tell who won the election. Soc. Choice welfare **6**(2), 157–165 (1989)
4. Berg, S., Lepelley, D.: On probability models in voting theory. Statistica Neerlandica **48**(2), 133–146 (1994)
5. Boutilier, C., Lang, J., Oren, J., Palacios, H.: Robust winners and winner determination policies under candidate uncertainty. In: Twenty-Eighth AAAI Conference on Artificial Intelligence. AAAI, Québec (2014)
6. Brandt, F., Conitzer, V., Endriss, U., Lang, J., Procaccia, A.D.: Handbook of Computational Social Choice. Cambridge University Press, Cambridge (2016)
7. Chevaleyre, Y., Lang, J., Maudet, N., Monnot, J., Xia, L.: New candidates welcome! possible winners with respect to the addition of new candidates. Math. Soc. Sci. **64**(1), 74–88 (2012)
8. Dutta, B., Jackson, M.O., Le Breton, M.: Strategic candidacy and voting procedures. Econometrica **69**(4), 1013–1037 (2001)
9. Dutta, B., Jackson, M.O., Le Breton, M.: Voting by successive elimination and strategic candidacy. J. Econ. Theory **103**(1), 190–218 (2002)
10. García-Lapresta, J.L., Pérez-Román, D.: Consensus measures generated by weighted kemeny distances on weak orders. In: 2010 10th International Conference on Intelligent Systems Design and Applications, pp. 463–468. IEEE Computer Society, Cairo (2010)
11. Gilbert, H., Portoleau, T., Spanjaard, O.: Beyond pairwise comparisons in social choice: a setwise kemeny aggregation problem. In: Thirty-Fourth AAAI Conference on Artificial Intelligence, pp. 1982–1989 (2020)
12. Klamler, C.: A distance measure for choice functions. Soc. Choice Welfare **30**(3), 419–425 (2008)
13. Lang, J., Maudet, N., Polukarov, M.: New results on equilibria in strategic candidacy. In: Vö, B. (ed.) SAGT 2013. LNCS, vol. 8146cience. Springer, Heidelberg (2013). https://doi.org/10.1007/978-3-642-41392-6_2
14. Laslier, J.F.: And the loser is... plurality voting. In: Felsenthal, D., Machover, M. (eds.) Electoral Systems. SCW, pp. 327–351. Springer, Heidelberg (2012). https://doi.org/10.1007/978-3-642-20441-8_13

15. Lu, T., Boutilier, C.E.: The unavailable candidate model: a decision-theoretic view of social choice. In: Proceedings of the 11th ACM Conference on Electronic Commerce, pp. 263–274 (2010)
16. Lumet, C., Bouveret, S., Lemaître, M.: Fair division of indivisible goods under risk (2012)
17. Naamani-Dery, L., Kalech, M., Rokach, L., Shapira, B.: Reducing preference elicitation in group decision making. Expert Syst. Appl. **61**, 246–261 (2016)
18. Oren, J., Filmus, Y., Boutilier, C.: Efficient vote elicitation under candidate uncertainty. In: IJCAI, pp. 309–316 (2013)
19. Shieh, G.S.: A weighted kendall's tau statistic. Stat. Prob. Lett. **39**(1), 17–24 (1998)
20. Viappiani, P.: Characterization of scoring rules with distances: application to the clustering of rankings. In: Twenty-Fourth International Joint Conference on Artificial Intelligence (2015)

Simultaneous Elicitation of Scoring Rule and Agent Preferences for Robust Winner Determination

Beatrice Napolitano[1]([✉])[iD], Olivier Cailloux[1], and Paolo Viappiani[2]

[1] Université Paris-Dauphine, Université PSL, CNRS, LAMSADE, Paris, France
beatrice.napolitano@dauphine.fr
[2] LIP6, UMR 7606, CNRS and Sorbonne Université, Paris, France

Abstract. Social choice deals with the problem of determining a consensus choice from the preferences of different agents. In the classical setting, the voting rule is fixed beforehand and full information concerning the preferences of the agents is provided. This assumption of full preference information has recently been questioned by a number of researchers and several methods for eliciting the preferences of the agents have been proposed. In this paper we argue that in many situations one should consider as well the voting rule to be partially specified. Focusing on positional scoring rules, we assume that the chair, while not able to give a precise definition of the rule, is capable of answering simple questions requiring to pick a winner from a concrete profile. In addition, we assume that the agent preferences also have to be elicited. We propose a method for robust approximate winner determination and interactive elicitation based on minimax regret; we develop several strategies for choosing the questions to ask to the chair and the agents in order to converge quickly to a near-optimal alternative. Finally, we analyze these strategies in experiments where the rule and the preferences are simultaneously elicited.

Keywords: Uncertainty in AI · Computational social choice · Preference elicitation

1 Introduction

Aggregation of preference information is a central task in many computer systems (recommender systems, search engines, etc.). In many situations, such as in group recommender systems, the goal is to find a consensus choice; social choice theory can provide foundations for such applications. The traditional approach to social choice assumes that 1) the full preference orderings of the agents and 2) the social choice function are expressed beforehand. These represent two strong hypotheses. Requiring agents to express full preference orderings can be prohibitively costly (in terms of cognitive and communication cost). This observation has motivated several works assuming partial preference orders: one

© Springer Nature Switzerland AG 2021
D. Fotakis and D. Ríos Insua (Eds.): ADT 2021, LNAI 13023, pp. 51–67, 2021.
https://doi.org/10.1007/978-3-030-87756-9_4

early work is by Conitzer and Sandholm [7] who studied the complexity of communication when using different voting rules; Konczak and Lang [15] studied the computation of possible and necessary winners for various voting rules; Xia and Conitzer [34] then showed that, while the identification of a necessary co-winner in scoring rules is polynomial, the determination of possible co-winners is NP-hard; additional complexity results were given by Walsh [32] and Pini et al. [25].

Since in many practical situations there would be too many possible winners but no necessary winners, several works addressed the problem of agent preferences elicitation using a variety of approaches (minimax regret, Bayesian methods, etc.) with the goal of converging to a necessary winner [2,9,14,21,24,26]. Among those, Walsh [33] and Conitzer [6] analyzed when to stop the elicitation process.

A second concern is the ability of the chair (the person or organization supervising the voting process) to provide a precise definition of the voting rule, suggesting the relaxation of the second hypothesis. Indeed, it is often difficult for non-experts to formalize a voting rule on the basis of some generic preferences over a desired aggregation method. Here we provide two examples of such situations.

Consider, as a first example, a chair that is about to hire a new employee whose performances are evaluated by several experts. The members of the chair may not have a voting rule in mind at the start of the process, and might not wish to agree on a specific voting rule. However, they might be willing to answer a few questions requiring to select who should be the winner out of specific profiles.

Consider, as a second example, the reviewing process of a conference where the best paper must be elected. The agents express their preferences on the papers they reviewed, but they are not aware of the voting rule the Program Chair will apply when aggregating them. Nonetheless, reviewers are still willing to participate in the process. Also, the PC may not have a specific voting rule in mind, and she will find it hard to provide a precise scoring vector if asked. Maybe she strongly believes that being ranked once in the first position is "much more" valuable than being ranked two times second, but does not know exactly how much more (though she can judge example cases).

In this paper, we focus on positional scoring rules with convex weights, that are a particularly common method used to aggregate rankings. We develop methods, based on the notion of minimax regret, for determining a robust "winner" under uncertainty of both the voting rule and the agent preferences. We provide incremental elicitation methods that at each step of the elicitation question either one of the agents or the chair, and we discuss several heuristics to choose questions that quickly reduce the regret. Answers to questions are encoded as constraints; questions to the agents are comparisons between pairs of alternatives while questions to the chair ask to select a winner out of a synthetic profile.

While some previous works have considered partially specified aggregation methods [20,30,31], we do not know of any work considering both sources of uncertainty at the same time. Actually, very few works altogether have considered the problem of eliciting a voting rule by asking questions to the chair. We mention the work of Cailloux and Endriss [5] that assumes a different representation for the rule. Additionally, some works address the manipulability of voting rules [1,8,10,11] and strategic behaviors [12,17,27].

Our approach is evaluated on simulations with synthetic and real datasets where both the voting rule and the agent preferences are initially unknown to the system and incrementally revealed through questioning. We assume the chair to be human, thus able to answer questions about a limited number of alternatives, so we focus on small scale social choice situations. We compare the effectiveness of several questioning strategies based on the current knowledge of the rule and preferences. To summarize our contributions: 1) we provide a novel mechanism for eliciting a voting rule by translating abstract questions about weights to a choice of an alternative given a concrete profile; 2) we show that with our elicitation method it is possible to reach low regret with a reasonable number of questions; 3) we present elicitation strategies that achieve good results within reasonable computation time; 4) we show that for the class of rules considered, asking a few questions to the chair suffice to reach low regret; 5) our experiments suggest that low degree of similarity among preferences (as in impartial culture) is a more challenging setting than less varied profiles.

2   Social Choice with Partial Information

We now introduce some basic concepts. We consider a set A of m alternatives (products, restaurants, public projects, job candidates, etc.) and an infinite set $\mathbb{N}$ of potential agents.

A *profile* $(\succ_j, j \in N)$ considers a finite subset of agents $N \subset \mathbb{N}$ and associates to each agent a preference order $\succ_j \in \mathcal{L}(A)$, a linear order over the alternatives. A profile is equivalently represented by $\boldsymbol{v} = (v_j, j \in N)$ where $v_j(x) \in \{1, \ldots, m\}$ denotes the rank of alternative x in the preference order $\succ_j$. A social choice function $f : \cup_{\emptyset \neq N \subset \mathbb{N}, N \text{finite}} \mathcal{L}(A)^N \to \mathscr{P}^*(A)$ associates to each profile a set of (tied) winners, where $\mathscr{P}^*(A)$ is the powerset of A excluding the empty set. Among the many possible social choice functions, we consider convex *positional scoring rules (PSRs)*. A PSR $f^{\boldsymbol{w}}$ is parameterized by a *scoring vector* $\boldsymbol{w}$ associating weights $w_r \in [0, 1]$ to positions, with $1 = w_1 \geq w_2 \geq \ldots \geq w_m = 0$. Let α_r^x be the number of times that alternative x was ranked in the r-th position. Given $\boldsymbol{v}$ and $\boldsymbol{w}$, an alternative $x \in A$ obtains the score

$$s(x; \boldsymbol{v}, \boldsymbol{w}) = \sum_{j \in N} w_{v_j(x)} = \sum_{r=1}^{m} \alpha_r^x w_r. \tag{1}$$

The winners $f^{\boldsymbol{w}}(\boldsymbol{v})$ are the alternatives with highest score.

An important class of PSRs is the one using convex weights [19,30], meaning that the difference between the weight of the first position and the weight of the second position is at least as large as the difference between the weights of the second and third positions, etc.

$$\forall r \in \{1, \ldots, m-2\} : w_r - w_{r+1} \geq w_{r+1} - w_{r+2}. \qquad (2)$$

The constraint above is a natural and common assumption, often used when aggregating rankings in sport competitions (such as F1 racing, alpine skiing world cup): losing ranks at the top is more damaging than losing ranks at the bottom. Let $\mathcal{W}$ denote the set of such convex weight vectors.

We consider a specific finite set of agents $N^* \subset \mathbb{N}$ and let $\boldsymbol{v}^* = (\succ_j^*, j \in N^*)$ and $\boldsymbol{w}^*$ denote the profile and weight vector, unknown to us, that represent the preferences of the agents in N^* and of the chair.

At a given time, our knowledge of agent j's preference is encoded by a partial order $\succ_j^{\mathrm{p}} \subseteq \succ_j^*$ over the alternatives, a transitive and asymmetric relation (we assume that preference information is truthful). An incomplete profile $\boldsymbol{p} = (\succ_j^{\mathrm{p}}, j \in N^*)$ maps each agent to a partial preference. Let $C(\succ_j^{\mathrm{p}}) = \{\succ \in \mathcal{L}(A) \mid \succ_j^{\mathrm{p}} \subseteq \succ\}$ denote the set of possible completions of $\succ_i^{\mathrm{p}}$ and $C(\boldsymbol{p}) = \prod_{j \in N} C(\succ_j^{\mathrm{p}})$ the set of complete profiles extending $\boldsymbol{p}$. Note that $\boldsymbol{v}^* \in C(\boldsymbol{p})$.

The vector $\boldsymbol{w}^*$ is also unknown but we assume that the chair is able to specify additional preference information taking the form of linear constraints about $\boldsymbol{w}^*$. Let $W \subseteq \mathcal{W}$ denote the set of weight vectors compatible with the preferences expressed by the chair about the scoring vector. We will show in Sect. 4 that the additional preferences we use can be elicited by showing a complete profile of a synthetic election and asking who should be elected in this case.

3   Robust Winner Determination

It is desirable in an elicitation protocol such as ours to be able to stop before reaching full knowledge of the agent preferences or of the preferences of the chair about the voting rule. As, often, there are no necessary winners and too many possible winners, it is useful to declare a winner given partial information. As a decision criterion to determine a winner, we propose to use minimax regret [29]. This decision criterion has been used for robust optimization under data uncertainty [16] as well as in decision-making with uncertain utility values [3,28]. In particular, Lu and Boutilier [21] have adopted minimax regret for winner determination in social choice where the preferences of agents are partially known, while the social choice function is known.

We consider the simultaneous presence of incomplete knowledge in agent preferences and in the weights of the PSR. We use *maximum regret* to quantify the worst-case error, and let the alternatives that minimize this quantity win, giving some robustness in face of ignorance. Intuitively, the quality of a proposed alternative a is how far a is from the optimal one in the worst case, given the current knowledge.

Given $\boldsymbol{p}$ and W (that represent the current knowledge about agent preferences and the PSR), the maximum regret is considered by assuming that an adversary can both 1) extend the partial profile $\boldsymbol{p}$ into a complete profile, and 2) instantiate the weights choosing among any weight vector in W. We formalize the notion of minimax regret in multiple steps. First of all, $\text{Regret}(x, \boldsymbol{v}, \boldsymbol{w})$ is the "regret" of selecting x as a winner instead of the optimal alternative under $\boldsymbol{v}$ and $\boldsymbol{w}$:

$$\text{Regret}(x, \boldsymbol{v}, \boldsymbol{w}) = \max_{y \in A} s(y; \boldsymbol{v}, \boldsymbol{w}) - s(x; \boldsymbol{v}, \boldsymbol{w}).$$

The *pairwise maximum regret* of x relative to y given the partial profile $\boldsymbol{p}$ and the set of weights W is the worst-case loss of choosing x instead of y under all possible realizations of the full profile *and* all possible instantiations of the weights:

$$\text{PMR}(x, y; \boldsymbol{p}, W) = \max_{\boldsymbol{w} \in W} \max_{\boldsymbol{v} \in C(\boldsymbol{p})} s(y; \boldsymbol{v}, \boldsymbol{w}) - s(x; \boldsymbol{v}, \boldsymbol{w}).$$

The maximum regret is the worst-case loss of x:

$$\text{MR}(x; \boldsymbol{p}, W) = \max_{y \in A} \text{PMR}(x, y; \boldsymbol{p}, W) = \max_{\boldsymbol{w} \in W} \max_{\boldsymbol{v} \in C(\boldsymbol{v})} \text{Regret}(x; \boldsymbol{v}, \boldsymbol{w}). \quad (3)$$

$\text{MR}(x; \boldsymbol{p}, W)$ is the result of an adversarial selection of the complete profile $\boldsymbol{v} \in C(\boldsymbol{p})$ and of the scoring vector $\boldsymbol{w} \in W$ that jointly maximize the loss between x and the true winner under $\boldsymbol{v}$ and $\boldsymbol{w}$. Finally, $\text{MMR}(\boldsymbol{p}, W) = \min_{x \in A} \text{MR}(x; \boldsymbol{p}, W)$ is the value of *minimax regret* under $\boldsymbol{p}$ and W, obtained when recommending a *minimax optimal* alternative $x^*_{\boldsymbol{p}, W} \in A^*_{\boldsymbol{p}, W} = \text{argmin}_{x \in A} \text{MR}(x; \boldsymbol{p}, W)$. Picking as consensus choice an alternative associated with minimax regret provides a recommendation that gives worst-case guarantees. In cases of ties, we can return all minimax alternatives $A^*_{\boldsymbol{p}, W}$ as winners or pick one of them using some tie-breaking strategy.

Observe that if $\text{MMR}(\boldsymbol{p}, W) = 0$, then any $x^*_{\boldsymbol{p}, W} \in A^*_{\boldsymbol{p}, W}$ is a necessary winner: any valid completion of the profile and choice of $\boldsymbol{w} \in W$ gives to $x^*_{\boldsymbol{p}, W}$ the highest score.

We note that our notion of regret gives some cardinal meaning to the scores: instead of just being used to select winners under the corresponding PSR, their differences are considered as representing the regret of the chair.

Computation of Minimax Regret. Given a voting rule and a partially specified profile, Xia and Conitzer [34] determine necessary winners by showing constructions that attempt to maximize the score difference between a proposed winner and a chosen alternative. This reasoning was later adopted by Lu and Boutilier [21] who used the considerations on the worst-case completions for computing the minimax regret.

In order to compute pairwise maximum regret, and therefore minimax regret, we decompose the PMR into the contributions associated to each agent by adapting this same reasoning to our setting. The context is however more challenging due to the presence of uncertainty in the weights.

Recall that, in the computation of $s(x; \boldsymbol{v}, \boldsymbol{w})$, $w_{v_j(x)}$ represents the score that x obtains in the ranking v_j (see Eq. (1)). Since scoring rules are additively decomposable, we can consider separately the contribution of each agent to the total score. Thus, we can write the actual regret of choosing x instead of y as $s(y; \boldsymbol{v}, \boldsymbol{w}) - s(x; \boldsymbol{v}, \boldsymbol{w}) = \sum_{j \in N} w_{v_j(y)} - w_{v_j(x)}$, and we obtain

$$\text{PMR}(x, y; \boldsymbol{p}, W) = \max_{w \in W} \sum_{j \in N} \max_{v_j \in C(\succ_j^p)} [w_{v_j(y)} - w_{v_j(x)}].$$

The following propositions show that the procedure for completing a partial profile, proposed by Lu and Boutilier [21] when considering a fixed weight vector, also applies in our setting. We write $a \succeq_j^p b$ iff $a \succ_j^p b \vee a = b$ and adopt the canonical notation when considering a relation as a function, writing $\succeq_j^p(x)$ for $\{y \mid x \succeq_j^p y\}$.

Proposition 1. *There exists a completion $\hat{\boldsymbol{v}} \in C(\boldsymbol{p})$ of the partial profile $\boldsymbol{p}$ such that $\text{PMR}(x, y; \boldsymbol{p}, W) = \max_{w \in W}[s(y; \hat{\boldsymbol{v}}, \boldsymbol{w}) - s(x; \hat{\boldsymbol{v}}, \boldsymbol{w})]$ and such that the linear order $\hat{v}_j$ of each agent j satisfies:*

$$a \succ_j x \Leftrightarrow \neg(x \succeq_j^p a); \tag{4}$$

$$y \succ_j a \Leftrightarrow \neg(a \succeq_j^p y) \wedge \neg((x \succeq_j^p y) \wedge \neg(x \succeq_j^p a)). \tag{5}$$

Proof Sketch. Consider our knowledge $\succeq_j^p$ about the preference of the agent j. The adversary's goal is to make the score of y as high as possible and the score of x as low as possible. To do this, he should complete $\succ_j^p$ to $\succ_j$ by placing above x as many alternatives as possible; that is, all the alternatives except those that are known to be worse than x (those a such that $x \succeq_j^p a$); and similarly, he should put below y all the alternatives he can. Two conditions must be excluded for a to go below y. The alternatives such that $a \succeq_j^p y$ can't be put below y. Furthermore, the first objective must take priority over the second one: when an alternative should go above x according to the first objective (because $\neg(x \succeq_j^p a)$), and x is known to be better than y (thus $x \succeq_j^p y$), then a should be put above x (irrespective of whether $a \succeq_j^p y$), which will move both x and y one rank lower than if a had been put below y. This maximizes the adversary's interests: because the weight vector is convex, the score difference will be lower when both alternatives are ranked lower (Eq. 2), and that difference of scores is in favor of x when $x \succ_j^p y$, thus to be minimized from the adversary's point of view. $\square$

Proposition 2. *The rank of x in the PMR-maximizing linear orders of agent j is $\hat{v}_j(x) = 1 + |A| - |\succeq_j^p(x)|$, and the rank of y is $\hat{v}_j(y) = 1 + |\prec_j^p(y)| + |\beta|$, where $|\beta| = |A \setminus (\succeq_j^p(x) \cup \prec_j^p(y))|$ if $(x \succeq_j^p y)$ and $|\beta| = 0$ otherwise.*

Proof. The rank of x is directly obtained from Eq. (4). The rank of y is obtained by complementing Eq. (5), obtaining $a \succeq_j y \Leftrightarrow (a \succeq_j^p y) \vee ((x \succeq_j^p y) \wedge \neg(x \succeq_j^p a))$, and, observing that $a \succ_j y \Leftrightarrow a \neq y \wedge a \succeq_j y$, obtaining that $a \succ_j y$ if and only if

$$(a \neq y) \wedge [(a \succeq_j^p y) \vee ((x \succeq_j^p y) \wedge \neg(x \succeq_j^p a))], \tag{6}$$

or equivalently, if and only if

$$(a \succ_j^p y) \vee ((x \succeq_j^p y) \wedge \neg(x \succeq_j^p a)). \tag{7}$$

Indeed, (6) $\Rightarrow$ (7), and (7) $\Rightarrow$ (6) because $(x \succeq_j^P y) \land \neg(x \succeq_j^P a) \Rightarrow a \neq y$ (as when $a = y$, $(x \succeq_j^P y)$ and $\neg(x \succeq_j^P a)$ are opposite claims). Suffices now to rewrite Eq. (7) to let the two disjuncts designate disjoint sets:

$$a \succ_j y \Leftrightarrow (a \succ_j^P y) \lor ((x \succeq_j^P y) \land \neg(x \succeq_j^P a) \land \neg(a \succ_j^P y)). \tag{8}$$

$\square$

Note that in Proposition 2, in the case $(x \succeq_j^P y)$, β is the number of alternatives incomparable with both x and y.

Proposition 3. *The* PMR *can be written as:*

$$\text{PMR}(x, y; \boldsymbol{p}, W) = \max_{w \in W} \sum_{j \in N} w_{\hat{v}_j(y)} - w_{\hat{v}_j(x)} = \max_{w \in W} \sum_{r=1}^{m} (\hat{\alpha}_r^y - \hat{\alpha}_r^x) w_i, \tag{9}$$

where $\hat{\alpha}_r^y$ (resp. $\hat{\alpha}_r^x$) is the number of times y (resp. x) has rank r in the complete profile $\hat{v}$ defined in Proposition 2.

Proposition 3 shows that PMR is linear in the weights. The pairwise max regret $\text{PMR}(x, y; \boldsymbol{p}, W)$ can thus be obtained by solving the following linear program defined on the variables $w_1, ..., w_m$:

$$\max_{\boldsymbol{w}} \sum_{r=1}^{m} (\hat{\alpha}_r^y - \hat{\alpha}_r^x) w_r \quad \text{s.t. } w_1 = 1 \geq ... \geq w_m = 0, \text{Eq. (2) and } \boldsymbol{w} \in W. \tag{10}$$

The max regret $\text{MR}(x; \boldsymbol{p}, W)$ is determined by computing the pairwise regret of x with all other alternatives in A, and the recommended alternatives are the ones with least max regret. Observe that when the PMR of an alternative x (against some other alternative y) exceeds the best MR value found so far, we do not need to further evaluate x. This idea can be exploited using a minimax-search tree [4].

4   Interactive Elicitation

We propose an incremental elicitation method based on minimax regret. At each step, the system may ask a question either to one of the agents about her preferences or to the chair about the voting rule. The goal is to obtain relevant information to reduce minimax regret as quickly as possible. The elicitation can be terminated either after a given number of questions, or when the minimax regret is lower than a threshold (or when it drops to zero if we wish optimality).

Question Types. We distinguish between questions asked to the agents and questions asked to the chair. As *questions asked to the agents* we consider comparison queries relating two alternatives. The effect of a response to a question asked to an agent is the increase in our knowledge about the agent rankings, thus augmenting the partial profile $\boldsymbol{p}$. If agent j answers a comparison query stating that alternative a is preferred to b, then the partial order $\succ_j^P$ is augmented with $a \succ_j^P b$ and by transitive closure.

A bit more discussion is needed about *questions asked to the chair*. Such questions aim at refining our knowledge about the scoring rule; a response gives us a constraint on the weight vector $\boldsymbol{w}$. In particular, we want to obtain constraints of the type $w_r - w_{r+1} \geq \lambda(w_{r+1} - w_{r+2})$ for $r \in \{1, \ldots, m-2\}$, relating the difference between the importance of ranks r and $r+1$ with the difference between ranks $r+1$ and $r+2$.

Building Concrete Questions for the Chair. Even if the chair might be considered able to answer directly such abstract questions, we want to ensure that these questions can also, in principle, be asked in a more concrete way: in terms of winners of example profiles. Such questions have clear semantics whose understanding can be assumed to be shared by the chair, contrary to abstract questions about weights. Moreover, this way of questioning the chair is independent of the voting rule that is being elicited; whereas questions about weights only make sense when considering PSRs. Asking who should win in specific profiles has been used in experimental settings investigating the feeling of justice of individuals [13], but, to the best of our knowledge, the use of such questions to systematically guide an elicitation process about voting rules is novel. This is similar to favor, in decision theory, direct choice questions ("please choose either a or b") compared to, say, questioning the decision maker about the shape of her utility function. The former are considered "observable": acts of choice are translated to preference statements [22, Ch. 1].

Although questioning in terms of profiles and in terms of weights is logically equivalent in our setting, there is no a priori certainty that questioning the chair using different phrasing would yield logically equivalent answers: research in experimental psychology shows that participants' answers differ widely when changing the phrasing of preference-related questions [18]. To get out of such conundrums, we need a language considered "fundamental". Questions of the form "In this profile, who should win?" arguably provides such a natural language.

Thus, our task is to build a profile, given λ and $r \leq m - 2$, in such a way that the set of (tied) winners picked by the chair reveals whether $w_r - w_{r+1} \geq \lambda(w_{r+1} - w_{r+2})$.

Proposition 4. *Given a rational $\lambda = p/q > 1$ and a rank r between 1 and $m-2$, there exists a profile P such that, for any weight vector $\boldsymbol{w} \in \mathcal{W}$, $a \in f(P)$ iff $w_r - w_{r+1} \geq \lambda(w_{r+1} - w_{r+2})$ and $b \in f(P)$ iff $w_r - w_{r+1} \leq \lambda(w_{r+1} - w_{r+2})$, where f is the PSR parameterized with $\boldsymbol{w}$.*

Proof. Define a linear order $>_1$ over A as placing a at rank r, b at rank $r+1$, and the remaining alternatives arbitrarily. Define $>_2$ over A as placing a at rank $r+2$, b at rank $r+1$, and the remaining alternatives arbitrarily. Define an arbitrary linear ordering $>$ over $A \setminus \{a, b\}$. Define a linear order $>_3$ as placing a first, b second, and following the order of $>$ for the remaining positions. Finally, define a linear order $>_4$ as placing b first, a second, and following the *inverse* order of $>$ for the remaining positions.

Define P as the profile of $3(p+q)$ agents containing q times $>_1$, p times $>_2$, and $>_3$ and $>_4$ each $p+q$ times. As a result, a obtains the following ranks: q times r, p times $r+2$, $p+q$ times first, and $p+q$ times second. The alternative b obtains the ranks $r+1$, 2 and 1, each $p+q$ times. Consider any alternative $c \in A \setminus \{a, b\}$. Its score is maximal when it comes first in $>_1$, first in $>_2$ and first in $>$, by convexity of the weights. In that case, c is positioned at the ranks 1, 3 and m, each $p+q$ times.

Letting $s(x)$ denote the score of x at P, we obtain $s(a) = qw_r + pw_{r+2} + (p+q)w_1 + (p+q)w_2$, thus, $s(a) \geq (p+q)w_m + (p+q)w_1 + (p+q)w_2$; $s(b) = (p+q)w_{r+1} + (p+q)w_2 + (p+q)w_1$; and, $\forall c \in A \setminus \{a,b\}$, $s(c) \leq (p+q)w_1 + (p+q)w_3 + (p+q)w_m$. It follows that a or b maximize s (as $s(a) \geq s(c)$). We conclude by observing that $a \in f(P) \Leftrightarrow s(a) \geq s(b) \Leftrightarrow qw_r + pw_{r+2} \geq (p+q)w_{r+1} \Leftrightarrow w_r - w_{r+1} \geq (p/q)(w_{r+1} - w_{r+2})$, and similarly for $b \in f(P)$. $\qquad\square$

Example. Suppose we want to ask the following question to the chair: $w_2 - w_3 \geq 2(w_3 - w_4)$. We show the profile in Fig. 1a to the chair and ask who should win (each column is the preference of one agent). Both a and b have scores higher than c and d for all convex weights, thus either a or b will be picked under our hypothesis; and $s(a) \geq s(b) \Leftrightarrow w_2 + 2w_4 \geq 3w_3$. Figure 1b represents the same profile using a compressed view, the numbers in bold indicating the number of agents having the preference in the corresponding column. As the proof shows, constructed profiles require only four different linear orders.

c d d a a a b b b	**1 2 3 3**
a c c b b b a a a	c d a b
b b b c c c d d d	a c b a
d a a d d d c c c	b b c d
	d a d c
(a)	(b)

Fig. 1. Profile representing a question to the chair in extended (a) and compact (b) form.

Elicitation Strategies. We develop several strategies for simultaneous elicitation of agent preferences and of the PSR. While it is of course possible to first fully elicit the agent preferences and afterwards elicit weights, we want to investigate approaches that are able to recommend winning alternatives before obtaining complete knowledge of the profile or the rule. We define here various strategies; a strategy tells us, given the current partial knowledge (p, W), which question to ask next.

The *Random* strategy is used as a baseline. It first chooses equiprobably whether to question the chair or the agents. In the first case, it draws one rank in $1 \leq r \leq m - 2$ equiprobably, takes the middle of the interval of values for λ that are still possible considering our knowledge so far, and asks whether $w_r - w_{r+1} \geq \lambda(w_{r+1} - w_{r+2})$. In the second case, it draws equiprobably among the agents whose preference is not known entirely; it then draws an alternative a among those involved in some incomparabilities in $\succ_j^P$ and an alternative b among those incomparable with a in $\succ_j^P$.

Let $(x^*, \bar{y}, \bar{v}, \bar{w})$ be the current solution of the minimax regret, where x^* is the minimax optimal alternative and $\bar{y}, \bar{v}, \bar{w}$ the corresponding adversarial choices. The *Pessimistic* strategy considers a set of $n + (m - 2)$ candidate questions: one per agent, and one per rank (excluding the first and the last one which are known).

The candidate questions to the agents are chosen by extending the idea of Lu and Boutilier [21], that privilege learning about the relationship of x^* and $\bar{y}$ to the other alternatives if possible. Given $j \in N^*$, if x^* and $\bar{y}$ are incomparable in $\succ_j^P$, the candidate question concerns the pair $(x^*, \bar{y})$, otherwise, it concerns the pair (x^*, z) for some z incomparable to x^* (randomly chosen), or if none such z exist, the pair $(\bar{y}, z)$ for some z incomparable to $\bar{y}$, or, if both x^* and $\bar{y}$ are comparable to every alternatives in $\succ_j^P$, any incomparable pair is picked at random.

The candidate questions to the chair are determined as in the Random strategy.

Once having selected $n + m - 2$ candidate questions, the Pessimistic strategy selects the one that leads to minimal regret in the worst case. Assume that a question q_1 has type t_1 (being "chair" or "agent"), and leads to the new knowledge states (p_1, W_1) if answered positively and (p_1', W_1') if the answer is negative. Define

$$R_1^{\max} = \max\{\mathrm{MMR}(p_1, W_1), \mathrm{MMR}(p_1', W_1')\}$$

and

$$R_1^{\min} = \min\{\mathrm{MMR}(p_1, W_1), \mathrm{MMR}(p_1', W_1')\}\epsilon_t + \epsilon_t'.$$

The terms ϵ_t and ϵ_t' are real numbers associated to the type t of question; these parameters are used to fine tune the choice of the question type. Define similarly t_2, $R_2^{\max}$ and $R_2^{\min}$ for q_2. Pessimistic considers question q_1 to be better than q_2 iff $R_1^{\max} < R_2^{\max}$ or $[R_1^{\max} = R_2^{\max}$ and $R_1^{\min} < R_2^{\min}]$. In other words if the maximal *a posteriori* MMR of two questions are (approximately) equal, then it considers the (penalized) minimal MMR values.

The *Extended pessimistic* strategy uses the same criterion as the pessimistic strategy, but extending it to a bigger set of candidate questions, the same as those considered by the Random strategy. These candidate questions are then evaluated using the same operator as for the Pessimistic strategy. Extended pessimistic is applicable only to very small problem instances: its complexity is in $O(n^2 m^5)$, because we consider $O(m^2)$ questions for each agent and need for each question to compute MMR twice, whose complexity is $O(nm^3)$.

The *Two phases* strategy is developed in order to investigate the effect of varying the proportion of questions of the two types, when asking all questions to the chair at the beginning or at the end. It is parameterized by q_c, the number of questions to be asked to the chair. The *Two phases-ca* variant first asks q_c questions to the chair, then $k - q_c$ questions to the agents, using in both cases Pessimistic to select the specific questions; whereas the *Two phases-ac* variant starts with $k - q_c$ questions to the agents, then questions the chair.

Finally, the *Elitist* strategy aims at uncovering as quickly as possible the top alternatives of all agents. For any agent j, it asks to compare an alternative currently undominated in $\succ_j^P$ with one that is currently incomparable. Thus, the top alternative for j will be known after having asked exactly $m - 1$ questions to j. After having asked $n(m-1)$ questions to the agents, it questions the chair only, using the same approach as Pessimistic. This strategy can be expected to perform well when the chair assigns a large weight to the first rank, as compared to the other ranks. It is used to further challenge Pessimistic, which is not specifically tailored to such a situation.

5 Empirical Evaluation

We performed several numerical experiments using both real data and randomly generated profiles in order to validate our approach and test the performance of our elicitation strategies.

Given a problem size (m, n), a number of questions k and a strategy to test, we first create an "oracle", representing the true preferences of the agents (randomly generated or coming from real data) and the weights associated with the chair's scoring rule (randomly generated). We start with empty knowledge ($\boldsymbol{p} = \emptyset, W = \mathcal{W}$) about the preference orderings of the agents and the weights of the chair. We obtain the first question to be asked using the strategy under test. We then use the oracle to answer the question and update the system's knowledge, which is thus used to obtain the next question. This is repeated until k answers have been obtained, computing the resulting MMR values along the way for various values of k. We repeat this whole experiment a variable number of times, for a given (m, n, k), and report the average resulting MMR and standard deviation sd. The sizes of the considered scenarios are comparable to the ones used by Cailloux and Endriss [5].

The oracle is built as follows. For the real preferences, we used three datasets from PrefLib [23]: *T Shirt* (researchers voted on tee shirt designs; $m = 11, n = 30$), *Courses* (students voted on courses; $m = 9, n = 146$; referred to as AGH on PrefLib) and *Skate* (judges voted on skaters at the Euros Pairs Short Program; $m = 14, n = 9$). For the synthetic datasets, we follow an Impartial Culture (IC) assumption: the linear order of each agent is drawn i.i.d. uniformly. We believe IC to be a challenging situation and expect the number of questions to ask, in order to reach a certain level of regret, to decrease with less varied profiles. To generate the scoring rule weights, we first draw $m - 1$ numbers uniformly at random (in the interval $[\![0, 1]\!]$ representing weight "differences"), normalize and sort them; a sequence of convex decreasing weights is then obtained by a decumulative sum. The penalty parameters for the Pessimistic and Extended pessimistic strategies are $\epsilon_{\text{chair}} = 1.1$, $\epsilon'_{\text{chair}} = 10^{-6}$, $\epsilon_{\text{agent}} = 1.0$ and $\epsilon'_{\text{agent}} = 0$.

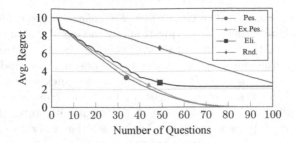

Fig. 2. Average MMR in problems of size $(5, 10)$ after k questions.

Table 1. Average MMR in problems of size $(10, 20)$ after k questions assuming geometric weights.

k	Pes. ± sd	Eli. ± sd
0	20.0 ± 0.0	20.0 ± 0.0
50	16.0 ± 0.5	17.3 ± 0.4
100	12.5 ± 0.9	15.6 ± 0.4
150	9.6 ± 1.4	13.9 ± 0.8
200	7.4 ± 1.3	11.0 ± 1.1
250	5.3 ± 1.5	6.6 ± 0.8
300	3.5 ± 1.3	6.6 ± 0.8

Comparison of Strategies. Our first experiment concerns small size situations. Figure 2 compares some of our strategies in the case $m = 5, n = 10$ (variations around this size yield similar conclusions), where the results are averaged over 200 runs. We see that asking random questions does not allow to reach a low regret level even after having asked 100 questions, whereas a low regret level (MMR $= 1$) is reached by Pessimistic before having asked 60 questions. This also holds for other problem sizes. For instance, for $m = 10$, $n = 20$ and 500 questions, Random strategy reaches an average regret (over 20 runs) of 9.3 (± 0.7) and Pessimistic 0.5 (± 0.5). We notice that Pessimistic performs slightly better than Extended pessimistic, showing that Pessimistic chooses candidate questions wisely; this is good news since Pessimistic is much faster: it takes on average only 16 s for a complete elicitation session (for $m = 5$, $n = 10$ and 100 questions), while Extended pessimistic takes 50 s. Although their performance is close, Pessimistic performs systematically better in multiple runs of the experiment.

We also compared the Pessimistic strategy against Elitist in a situation specifically tailored to advantage Elitist. For that experiment specifically, instead of drawing the weights of the oracle randomly, we fix it to a "geometric" weight vector, such that $w_r - w_{r+1} = 2(w_{r+1} - w_{r+2})$, for all $r \leq m - 2$, so as to dramatically increase the importance of the weights associated to the top ranks. Even in that case, we see in Table 1 that Pessimistic performs better than Elitist.

Evaluation of Pessimistic Strategy. Our next set of experiments evaluate the Pessimistic strategy in absolute terms. We first wonder how many questions should be asked in order to achieve low regret, fixed at $n/10$: this is equivalent to the difference of score of an alternative x that results from switching from a profile P to a profile P' where a tenth of the agents rank x last instead of first. Table 2, first five columns, contains the result: it displays, for each dataset, the number of questions asked to the chair ($q_c^{\text{MMR} \leq n/10}$), and the quartiles of the number of questions asked to the agents ($q_a^{\text{MMR} \leq n/10}$), averaged over 20 runs. It is interesting to note that about twenty or thirty questions per agent on average suffice to reach a low regret in those instances. We find also noteworthy that the Pessimistic strategy chooses to ask zero questions to the chair but still achieves low regret, in most of those instances.

Another interesting measure is the average number of questions asked to the chair ($q_c^{\text{MMR}=0}$) and to the agents ($q_a^{\text{MMR}=0}$) before reaching zero regret. The results for various sizes are displayed in the last two columns of Table 2. Here, we see that the Pessimistic strategy does choose to question the chair when reaching low enough regret values. The m15n30 dataset did not reach zero regret in 1000 questions.

Table 2. Questions asked by Pessimistic strategy on several datasets to reach $\frac{n}{10}$ regret, columns 4 and 5, and zero regret, last two columns.

dataset	m	n	$q_c^{\text{MMR} \leq n/10}$	$q_a^{\text{MMR} \leq n/10}$	$q_c^{\text{MMR}=0}$	$q_a^{\text{MMR}=0}$
m5n20	5	20	0.0	[4.3	5.0	5.8]	5.3	[5.4	6.2	7.2]
m10n20	10	20	0.0	[13.9	16.1	18.4]	32.0	[19.7	21.8	24.7]
m11n30	11	30	0.0	[16.6	19.0	22.3]	45.2	[23.1	25.7	28.9]
tshirts	11	30	0.0	[13.1	16.6	19.6]	43.2	[28.2	32.0	35.6]
courses	9	146	0.0	[6.0	7.0	7.0]	0.0	[6.8	7.0	7.0]
m14n9	14	9	5.4	[30.3	33.5	36.7]	64.1	[37.6	40.5	44.3]
skate	14	9	0.0	[11.4	11.6	12.3]	0.0	[11.5	11.8	12.8]
m15n30	15	30	0.0	[25.0	29.5	33.7]		

Figure 3 shows the decrease in MMR according to the number of questions asked for various problem sizes. In particular, this shows important differences between some real datasets and the problems generated using IC. In the *Skate* problem, the value MMR $= 1$ is reached after less than 100 questions, while the IC case of the same size ($m = 14, n = 9$) requires more than 200 questions to reach that value. This reasoning also applies to the *Courses* dataset but not to the *T Shirt* dataset. This can be explained by the high degree of similarity in the preference rankings of the *Skate* and the *Courses* problems, which helps reducing the regret faster. For example, in *Skate* the top-2 alternatives are the same for all agents, and 8 out of 9 agents rank the same alternative at position 3. By contrast, in *T Shirt*, the alternatives are evenly distributed in the preference rankings.

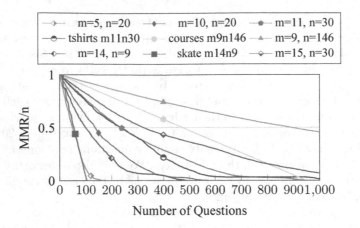

Fig. 3. Average MMR (normalized by n) after k questions with pessimistic strategy for different datasets.

Table 3. Average MMR in problems of size $(10, 20)$ after 500 questions, among which q_c to the chair.

q_c	2 ph. ca $\pm$ sd	2 ph. ac $\pm$ sd
0	0.6 ± 0.6	0.6 ± 0.6
15	0.5 ± 0.5	0.6 ± 0.6
30	0.3 ± 0.5	0.4 ± 0.5
50	0.1 ± 0.2	0.1 ± 0.3
100	0.4 ± 0.6	0.2 ± 0.5
200	2.1 ± 1.6	2.4 ± 1.2
300	5.8 ± 1.7	6.7 ± 1.5
400	11.3 ± 1.1	11.9 ± 1.2
500	20.0 ± 0.0	20.0 ± 0.0

Comparison with Two Phases. The experiments so far let the strategy free to question either the chair or an agent at each step. One may wonder what is lost in terms of regret by asking different proportions of questions to the chair and the agents. Such restrictions may be useful because of (partial) unavailability of the chair, or because the estimated cognitive costs may differ sensibly.

Table 3 shows the MMR value reached in problems of size $m = 10, n = 20$ after 500 questions, using the Two phases strategy, in the "ca" (chair then agents) and in the "ac" (agents then chair) variants. These numbers are to be compared with the MMR value reached after 500 questions with the Pessimistic strategy (displayed in Fig. 3), which is 0.7; the Pessimistic strategy asks on average 13 (± 13) questions to the chair in this setting. The line $q_c = 0$, where no question is asked to the chair, suggest that it is possible to obtain a good-quality recommendation while knowing only that the voting rule is a scoring rule with convex weights, which is our basic hypothesis. However, we observe that asking no questions to the chair does not permit to reach MMR = 0. The strategy, indeed, obtains full knowledge of the profile after an average of 500 questions to the agents but never reaches zero.

6   Conclusions

In this paper we have considered a social choice setting with partial information about the agent preferences and voting rule. We have proposed the use of minimax regret both as a means of robust winner determination and as a guide to the process of simultaneous elicitation of preferences and voting rule. Our experimental results suggest that regret-based elicitation is effective and allows to quickly reduce worst-case regret significantly. They also show that, in our setting, good quality (low regret) recommendations can be achieved short of having full knowledge of weights or profile.

As part of our contribution, we provide an open-source library that can be found at https://github.com/oliviercailloux/minimax, to reproduce our experiments and perform many more.

Some directions for future works include developing new elicitation strategies, considering alternative heuristics; extending the elicitation to voting rules beyond scoring rules; eliciting preferences while restraining to concrete and easy questions.

References

1. Baumeister, D., Hogrebe, T.: Manipulative design of scoring systems. In: Proceedings of AAMAS 2019 (2019). https://dl.acm.org/doi/10.5555/3306127.3331928
2. Benabbou, N., Di Sabatino Di Diodoro, S., Perny, P., Viappiani, P.: Incremental preference elicitation in multi-attribute domains for choice and ranking with the borda count. In: Proceedings of SUM 2016 (2016)
3. Boutilier, C., Patrascu, R., Poupart, P., Schuurmans, D.: Constraint-based optimization and utility elicitation using the minimax decision criterion. Artif. Intell. (2006). https://doi.org/10.1016/j.artint.2006.02.003
4. Braziunas, D.: Decision-theoretic elicitation of generalized additive utilities. Ph.D. thesis (2012). https://dl.acm.org/doi/book/10.5555/2519022
5. Cailloux, O., Endriss, U.: Eliciting a suitable voting rule via examples. In: Proceedings of ECAI 2014 (2014). https://doi.org/10.3233/978-1-61499-419-0-183
6. Conitzer, V.: Eliciting single-peaked preferences using comparison queries. J. Artif. Intell. Res. (2009). https://doi.org/10.1613/jair.2606
7. Conitzer, V., Sandholm, T.: Communication complexity of common voting rules. In: Proceedings of EC 2005 (2005). https://doi.org/10.1145/1064009.1064018
8. Conitzer, V., Walsh, T., Xia, L.: Dominating manipulations in voting with partial information. In: AAAI 2011 (2011). https://dl.acm.org/doi/10.5555/2900423.2900525
9. Dey, P., Misra, N.: Preference elicitation for single crossing domain. In: Proceedings of IJCAI 2016 (2016). https://dl.acm.org/doi/10.5555/3060621.3060653

10. Dey, P., Misra, N., Narahari, Y.: Complexity of manipulation with partial information in voting. Theor. Comput. Sci. (2018). https://doi.org/10.1016/j.tcs.2018.03.012
11. Elkind, E., Erdélyi, G.: Manipulation under voting rule uncertainty. In: Proceedings of AAMAS 2012 (2012). https://dl.acm.org/doi/10.5555/2343776.2343786
12. Endriss, U., Obraztsova, S., Polukarov, M., Rosenschein, J.S.: Strategic voting with incomplete information. In: IJCAI 2016 (2016). https://dl.acm.org/doi/10.5555/3060621.3060655
13. Giritligil, A., Sertel, M., Kara, A.: Does majoritarian approval matter in selecting a social choice rule? an exploratory panel study. Soc. Choice Welfare (2005). https://doi.org/10.1007/s00355-005-0024-8
14. Kalech, M., Kraus, S., Kaminka, G.A., Goldman, C.V.: Practical voting rules with partial information. Auton. Agent. Multi-Agent Syst. (2011). https://doi.org/10.1007/s10458-010-9133-6
15. Konczak, K., Lang, J.: Voting procedures with incomplete preferences. In: IJCAI 2005 (2005)
16. Kouvelis, P., Yu, G.: Robust Discrete Optimization and Its Applications. Kluwer, Alphen aan den Rijn (1997)
17. Lev, O., Meir, R., Obraztsova, S., Polukarov, M.: Heuristic voting as ordinal dominance strategies. In: Proceedings of the AAAI 2019 (2019). https://doi.org/10.1609/aaai.v33i01.33012077
18. Lichtenstein, S., Slovic, P.: The Construction of Preference. Cambridge University Press, Cambridge (2006).https://doi.org/10.1017/CBO9780511618031
19. Llamazares, B.: Ranking candidates through convex sequences of variable weights. Group Decis. Negot. (2016). https://doi.org/10.1007/s10726-015-9452-8
20. Llamazares, B., Peña, T.: Aggregating preferences rankings with variable weights. Eur. J. Oper. Res. (2013). https://doi.org/10.1016/j.ejor.2013.04.013
21. Lu, T., Boutilier, C.: Robust approximation and incremental elicitation in voting protocols. In: Proceedings of IJCAI 2011 (2011). https://doi.org/10.5591/978-1-57735-516-8/IJCAI11-058
22. Mas-Colell, A., Whinston, M.D., Green, J.R.: Microeconomic Theory (1995)
23. Mattei, N., Walsh, T.: Preflib: a library of preference data preflib.org. In: Proceedings of ADT 2013 (2013). https://www.preflib.org/
24. Naamani-Dery, L., Golan, I., Kalech, M., Rokach, L.: Preference elicitation for group decisions using the borda voting rule. Group Decis. Negot. (2015). https://doi.org/10.1007/s10726-015-9427-9
25. Pini, M.S., Rossi, F., Venable, K.B., Walsh, T.: Incompleteness and incomparability in preference aggregation. In: Proceedings of IJCAI 2007 (2007). https://doi.org/10.1016/j.artint.2010.11.009
26. Pini, M.S., Rossi, F., Venable, K.B., Walsh, T.: Aggregating partially ordered preferences. J. Log. Comput. (2009). https://doi.org/10.1093/logcom/exn012
27. Reijngoud, A., Endriss, U.: Voter response to iterated poll information. In: Proceedings of AAMAS 2012 (2012). https://dl.acm.org/doi/10.5555/2343776.2343787
28. Salo, A.A., Hämäläinen, R.P.: Preference ratios in multiattribute evaluation (PRIME)-elicitation and decision procedures under incomplete information. IEEE Trans. Syst. Man Cybern. (2001). https://doi.org/10.1109/3468.983411
29. Savage, L.J.: The Foundations of Statistics. Wiley, Hoboken (1954)
30. Stein, W.E., Mizzi, P.J., Pfaffenberger, R.C.: A stochastic dominance analysis of ranked voting systems with scoring. EJOR (1994). https://doi.org/10.1016/0377-2217(94)90205-4

31. Viappiani, P.: Positional scoring rules with uncertain weights. In: Ciucci, D., Pasi, G., Vantaggi, B. (eds.) SUM 2018. LNCS, vol. 11142, pp. 306–320. Springer, Cham (2018). https://doi.org/10.1007/978-3-030-00461-3_21
32. Walsh, T.: Uncertainty in preference elicitation and aggregation. In: Proceedings of AAAI 2007 (2007). https://dl.acm.org/doi/10.5555/1619645.1619648
33. Walsh, T.: Complexity of terminating preference elicitation. In: Proceedings of AAMAS 2008 (2008). https://dl.acm.org/doi/abs/10.5555/1402298.1402357
34. Xia, L., Conitzer, V.: Determining possible and necessary winners under common voting rules given partial orders. In: Proceedings of AAAI 2008 (2008). https://doi.org/10.1613/jair.3186

Preference Elicitation

Incremental Elicitation of Preferences: Optimist or Pessimist?

Loïc Adam and Sébastien Destercke[⊠]

UMR CNRS 7253 Heudiasyc, Sorbonne Université, Université de Technologie de
Compièegne, 60319 - 60203 Compiègne Cedex, France
{loic.adam,sebastien.destercke}@hds.utc.fr

Abstract. In robust incremental elicitation, it is quite common to make
recommendations and to select queries by using a minimax regret criterion, which corresponds to a pessimistic attitude. In this paper, we
explore its optimistic counterpart, showing this new approach enjoys
the same convergence properties. While this optimistic approach does
not offer the same kind of guarantees than minimax approaches, it still
offers some other interesting properties. Finally, we illustrate with some
experiments that the best approach amongst the two approaches heavily
depends on the underlying setting.

Keywords: Preferences elicitation · Incremental · Minimax regret ·
Maximax utility

1 Introduction

Preference elicitation by interacting with an agent or a user is a crucial step to
identify and formalise her preferences. While there are different ways to interact
with a user, incremental elicitation [2] is a very interesting approach since each
new question takes into account the preferential information provided previously.
In the literature, one of the main approaches of incremental elicitation is the
robust approach, based on a Minimax regret optimisation [3,4]. Provided their
underlying hypotheses[1] are satisfied, the interest of using such approaches is that,
due to their pessimistic stance (minimising the regret in the worst situation),
they come with strong guarantees about the recommended alternative. They
also converge in a reasonable number of steps to a good recommendation, as
the space of possible models is guaranteed to shrink after each question. In this
paper, we will work under the same hypotheses as the robust approaches to
simplify our exposure, as we could easily adapt our proposal to extensions of the
robust approach [10] which are able to deal with errors.

Minimax regret approach [14] is a popular choice for making decisions under
uncertainty, as it minimises the worst-case regret. Such a decision rule provides

---

[1] The user is an oracle, and the chosen family of preference model includes the right
model.

D. Fotakis and D. Ríos Insua (Eds.): ADT 2021, LNAI 13023, pp. 71–85, 2021.
https://doi.org/10.1007/978-3-030-87756-9_5

rather safe recommendations, corresponding to a situation where the agent is rather pessimistic on the outcomes of a decision, fearing a possibly rare but disastrous worst-case scenario. However, Minimax regret is only one amongst many other decision rules under uncertainty (see, e.g., [16] for an account of those), and as all other rules, it has drawbacks one may not appreciate. For instance, it is sensitive to the addition of irrelevant alternatives, and does not guarantee potentially optimal recommendations, i.e. recommendations that are the best for at least one particular model within the set of possible models.

In this paper, we consider a somewhat opposed view, using a Maximax optimist approach to make recommendations. There are multiple reasons to investigate such an alternative: one is that such optimistic approaches in presence of uncertainty are often used in learning under uncertainty, for example to deal with missing data [6] or to identify optimal models [13], hence adopting such an optimistic view in preference elicitation that shares many similarities with the aforementioned learning setting seems relevant; another is that optimistic recommendations do not suffer the same drawbacks we have mentioned for the Minimax regret[2], hence may be acceptable in situations where the Minimax regret is not. Last but not least, such an approach can be computationally more efficient than regret based ones, since regret typically involves comparing pairs of alternatives, whereas Maximax decision typically involves alternative-wise computations.

We introduce in Sect. 2 all the necessary elements in preference elicitation to understand our work. We then present in Sect. 3 an optimist approach that we call Maximax gain, adapting the Current Solution Strategy (CSS) heuristic. We also discuss some of its interesting properties. Lastly, Sect. 4 shows some simulated experiments whose goal are to investigate whether there are situations in which an optimist approach also increases recommendation performances.

2   Preliminaries

In this section, we introduce the various elements necessary to understand the rest of our work. We also introduce a running example that we will repeatedly use to illustrate the introduced notions.

2.1   Notations

Alternatives. We define $\mathbb{X}$ as the finite set of available alternatives. Alternatives within $\mathbb{X}$ are denoted $x_1, x_2, ..., x_k$. We assume that alternatives are summarised by q real values, the criteria, such that $x \in \mathbb{R}^q$. The *ith* criterion value of an alternative $x \in \mathbb{X}$ is denoted x^i. Given two alternatives $x, y \in \mathbb{X}$, we denote:

- $x \succ_p y$ if and only if x is strictly preferred to y,
- $x \succeq_p y$ if and only if x is preferred or equally preferred to y.

---

[2] It does not, however, offer the same robust guarantees.

Example 1 (Choosing the Best Sandwich (Running Example)). *We imagine that a user wants to choose the best possible sandwich among multiple ones (the alternatives). Each sandwich is characterised by two criteria: flavour and price. Each criterion is valued between 0 (worst) and 10 (best). Table 1 lists the available sandwiches. We can see for instance that cheese sandwich is very cheap but with a mediocre flavour, while duck sandwich is full of flavour yet overpriced.*

Table 1. Grades of sandwiches

	Flavour	1/price
Cheese	5	9
Duck	10	0
Fish	8	4
Ham	7	7

Aggregation Models. We consider that each alternative x is valuated by its utility, and that the utility depends on the preferences of the agent. We also suppose that such an utility is modelled by a function $f_\omega(x)$, parameterised by $\omega \in \Omega$, aggregating the different criteria. ω is also known as the preferential model. Given this evaluation function f_ω, it is possible to compare two alternatives $x, y \in \mathbb{X}$:

$$x \succeq_\omega y \Longleftrightarrow f_\omega(x) \geq f_\omega(y). \tag{1}$$

A first model we consider is the weighted sum (WS) model. This is a very simple model, which can be considered as the basic building block of decision theory and is still widely used in multi-criteria decision-making. Given a vector of weights $\omega = \{\omega^1, ..., \omega^q\} \in \mathbb{R}^q$, we have:

$$f_\omega(x) = \sum_{i=1}^{q} \omega^i x^i, \tag{2}$$

with $\omega^i \geq 0$ and $\sum_i \omega^i = 1$.

We also consider the Ordered weighted averaging (OWA) model [17]. This model generalises aggregation operators such as the arithmetic mean, median, min or max. With an OWA model, criteria values are ordered increasingly. Given a vector of weights $\omega \in \mathbb{R}^q$ and the ordered criteria values $x^{(1)} \leq ... \leq x^{(q)}$, we have:

$$g_\omega(x) = \sum_{i=1}^{q} \omega^i x^{(i)}, \tag{3}$$

with $\omega^i \geq 0$ and $\sum_i \omega^i = 1$.

These two models are simple and can be used in most situations. More complex models like Choquet integrals [7,9] exist that can be used, e.g., to model interactions between criteria. Provided the models are linear in ω once x is fixed, Equation (1) is equivalent to a linear constraint, meaning that we can use them within a linear program [2]. All mentioned models so far are linear in ω.

Example 2 (Application of Models). *Given the sandwiches we presented in Table 1, we assume the user evaluates each sandwich with a WS model such that* $\omega = (0.8, 0.2)$*. This means she values the flavour over the price. She then prefers the duck sandwich, as it scores 8, over the fish sandwich, with a score of 7.2.*

If she evaluates with an OWA model such that $\omega = (0.8, 0.2)$*, meaning she penalises a sandwich that is bad on at least one criterion, she will prefer the more balanced fish sandwich:* $f_\omega(fish) = 4 \times 0.8 + 8 \times 0.2 = 4.8$ *while* $f_\omega(duck) = 0 \times 0.8 + 10 \times 0.2 = 2$.

2.2   Robust Elicitation with Minmax Regret

Motivation. Finding a unique model ω from pairwise comparisons is difficult. However, it is often possible to draw reasonable inferences without complete information. Robust recommendation approaches aim at identifying a subset Ω' of possible models ω from preferential information. We then identify the preferences that hold for every model $\omega \in \Omega'$. This results in a partial preorder over $\mathbb{X}$ where:

$$x \succeq^{\Omega'} y \Longleftrightarrow \forall \omega \in \Omega' \; f_\omega(x) \geq f_\omega(y). \tag{4}$$

A good elicitation strategy needs to reduce Ω' as quickly as possible to make good recommendations without exhausting the budget of questions. Such a strategy should also make good recommendations even if $\succeq^{\Omega'}$ does not have a single maximal element, as in practice information collection may end before that.

Regret Based Elicitation. Minmax regret is a well-known notion for decision problems under uncertainty and set-valued information [14]. It still provides strong guarantees on the recommendation quality, while being less conservative than standard Minmax.

We are now introducing the different measures to compute the Minmax regret. The regret of choosing an alternative x over the alternative y given a model ω is defined by:

$$R_\omega(x, y) = f_\omega(y) - f_\omega(x). \tag{5}$$

Given a set Ω' of models, the pairwise max regret is:

$$PMR(x, y, \Omega') = \max_{\omega \in \Omega'} R_\omega(x, y), \tag{6}$$

which is the maximum regret of choosing x over y given any model $\omega \in \Omega'$.
The max regret of choosing x is:

$$MR(x, \Omega') = \max_{y \in \mathbb{X}} PMR(x, y, \Omega'), \tag{7}$$

which is the regret of choosing x in the worst case scenario, i.e., considering the worst model for its strongest opponent.

Finally, the min max regret of a set $\mathbb{X}$ of alternatives given a set Ω' of possible models is:

$$\mathrm{mMR}(\Omega') = \min_{x \in \mathbb{X}} \mathrm{MR}(x, \Omega'), \qquad (8)$$

In this approach $x^* = \arg \mathrm{mMR}(\Omega')$ is the alternative giving the minimal regret in a worst-case scenario, and is the current recommendation if no further information can be collected.

Example 3 (Initial Choice with a Minmax Regret). *We want to pick the alternative which minimises the maximum regret in the worst-case scenario, the preferences being evaluated with a WS model. The evolution of the score of each alternative depending on the parameter $w^{1/price}$ is depicted on Fig. 1. The alternative which minimises the maximum regret is the ham sandwich, with MR(ham) = 3 when we pick it instead of the duck sandwich for $w^{1/price} = 0$ (meaning only the flavour is considered).*

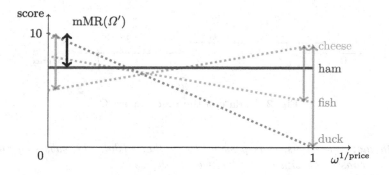

Fig. 1. Choice of the best alternative given a Minmax approach

2.3 Elicitation Sequence and Regret CSS

Preferential information is often collected through pairwise comparison: we present a pair (x, y) to the user, and she tells which one she prefers. We will denote by

$$\omega^{x \succeq y} = \{\omega \in \Omega : f_\omega(x) \geq f_\omega(y)\}, \qquad (9)$$

the subset of models consistent with the assessment $x \succeq y$, and $\omega^{y \succeq x}$ the subset for $y \succeq x$. In an elicitation sequence, we alternatively present a pair to the user, and update the information with the answer. In the robust approach, if Ω^k is the possible subset of models at the kth step, the next step is to present a couple (x, y) to the user, and then compute $\Omega^{k+1} = \Omega^k \cap \omega^{x \succeq y}$ if the user prefers x, $\Omega^{k+1} = \Omega^k \cap \omega^{x \preceq y}$ otherwise.

Choosing a good pair (x, y) is therefore a critical step. We consider for our work the well-known CSS strategy [5], where given a subset Ω', the user compares the current regret-based recommendation $x^* = \arg \text{mMR}(\Omega')$ (so our best option w.r.t this criterion) to its worst opponent:

$$y^* = \arg \max_{y \in \mathcal{X}} \text{PMR}(x^*, y, \Omega'). \tag{10}$$

This heuristic strategy provides good results in general, and guarantees that the updated set will be non-empty.

Example 4 (Updating the Model Space). *We assume the user decides with a WS model, where $\omega^* = (0.8, 0.2)$. In a first question q_1, she has to choose her favourite sandwich among the pair (x_D, x_F). We have already shown in Example 2 that she prefers x_D $(f_\omega(x_D) = 8, f_\omega(x_F) = 7.2)$. We assume she answers correctly that $x_D \succeq x_F$. We then have Ω' the set of models consistent with her known preferences, such that $\Omega' = \omega^{x_D \succeq x_F} = \{\omega \in \Omega : \sum_{i=1}^{2} \omega^i.(x_D^i - x_F^i) \geq 0\} = \{\omega \in \Omega : \omega^1 \geq 2\omega^2\}$, where ω^1 corresponds to the flavour, and ω^2 to 1/price. The updated model space Ω^1 is shown on Fig. 2.*

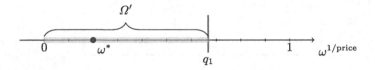

Fig. 2. Update of the model space Ω

With more questions, it is possible to further update the subset of possible models Ω' consistent with the preferences of the user.

An important property of robust approaches combined with CSS is that, by construction, they guarantee that the elicitation sequence will converge, as we remind here:

Proposition 1 *[1,5]. Given $\Omega^{k+1} \subseteq \Omega^k$, the sets of possible model at steps k and $k + 1$, we have that:*

$$PMR(x, y, \Omega^k) \geq PMR(x, y, \Omega^{k+1}), \tag{11}$$

$$MR(x, \Omega^k) \geq MR(x, \Omega^{k+1}), \tag{12}$$

$$mMR(\Omega^k) \geq mMR(\Omega^{k+1}). \tag{13}$$

Proof. PMR. Suppose we have a function f and two sets Ω, Ω' such that $\Omega' \subseteq \Omega$. We have $\max_{x \in \Omega} f(x) \geq \max_{x \in \Omega'} f(x)$, the maximum of Ω being either in Ω' or in $\Omega \setminus \Omega'$. We can replace f by the PMR, Ω by Ω^k and Ω' by Ω^{k+1} since $\Omega^{k+1} \subseteq \Omega^k$. (11) is then proved. Proof for MR and mMR directly follows, as they are maximum and minimum taken over decreasing values.

3 Optimist Approach

Minmax regret is based on a pessimist decision rule: the user wants the alternative that minimises the maximum loss, i.e., in the worst-case scenario. The user is risk-averse and does not mind if the gain is lower on average. However, it is unclear in a preference framework that the user will always be risk-averse, rather than opportunity-seeking. This is why we now consider the Maximax gain approach and its direct CSS adaptation, that considers recommendations based on another decision rule, where the user wants to maximise their gain in the best-case scenario. The choice of the Maximax gain approach can be justified in various ways: we show later in this section that an alternative suggested by a Maximax gain approach is the best possible for at least one situation (one model ω). This is not necessary the case with a Minmax regret, meaning that a risk-adverse user may actually select an option known to be necessarily sub-optimal. It does not mean that one strategy is better than the other, just that they have different properties, as the choice can depend on the willingness of the user for taking risks to maximise their possible gain. In this section, we discuss why such an approach may be an interesting alternative to Minmax regret approaches.

3.1 Robust Elicitation with Maximax Gain

Given an alternative x, the maximal gain over a set Ω' of possible models is:

$$MG(x, \Omega') = \max_{\omega \in \Omega'} f_\omega(x), \tag{14}$$

which corresponds to the gain in the best-case scenario. Given a set of alternatives $\mathbb{X}$ and the set Ω', the max maximal gain is:

$$MMG(\Omega') = \max_{x \in \mathbb{X}} MG(x, \Omega'), \tag{15}$$

$x^* = \arg MMG(\Omega')$ is the alternative giving the maximal possible gain in the corresponding best-case scenario. As with Minmax regret, x^* is the current recommendation if no additional information can be collected. When it comes to choosing the question, we still retain the CSS heuristic approach that chooses $y^* = \arg\max_{y \in \mathbb{X}} PMR(x^*, y, \Omega')$ as an adversary. For convenience, we will refer to the corresponding elicitation as *gain CSS*.

Example 5 (Initial Choice with an Optimist Approach). *In Example 3 we have shown how to pick the best alternative based on a Minmax regret, when we have no information on the preferences of a user. We will now find the best alternative based on a Maximax gain, and show it can be different from the one proposed with the Minmax regret, and the preferences are still evaluated with a WS model.*

As shown on Fig. 3, the maximum gain obtainable for the duck sandwich is 10 when $\omega = (1, 0)$. We also deduce that $MG(x_F, \Omega) = 8$ for $\omega = (1, 0)$, and $MG(x_C, \Omega) = 9$ for $\omega = (0, 1)$. The maximum gain of the ham sandwich is a particular case, since $MG(x_H, \Omega) = 7 \ \forall \omega \in \Omega$.

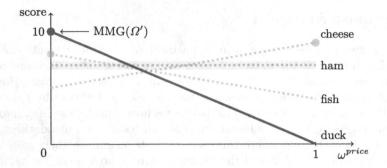

Fig. 3. Choice of the best alternative given a Maximax approach

We then conclude that $MMG(\Omega) = 10$ and that $x^* = x_D$. Pessimist and optimist approaches give different current solutions. The duck sandwich is a great candidate for maximising the gain in the best case scenario $\omega = (0,1)$, but this alternative is the worst if $\omega^{price} \gtrsim 0.36$.

3.2 Optimality

We first introduce the notion of possibly Ω'–optimal solutions [3]. Given the set $\mathbb{X}$ and a subset $\Omega' \subseteq \Omega$ of possible models, the set $PO_{\Omega'}$ is defined by:

$$PO_{\Omega'} = \{x : \exists \omega \in \Omega', x \in \arg\max_{\mathbb{X}} f_w(y)\}. \tag{16}$$

In other words, an alternative $x \in \mathbb{X}$ is possibly Ω'–optimal if x is the best alternative for at least one model $\omega \in \Omega'$. An optimist robust elicitation, based on a Maximax gain, is interesting for the following property, that shows that the recommended item could be the best (not guaranteed by a Minmax approach):

Proposition 2. *The Maximax gain alternative $x^* = \arg MMG(\Omega')$ is possibly Ω'–optimal.*

Proof. Consider the model ω for which is obtained $x^* = \max_{x \in \mathbb{X}} [\max_{\omega \in \Omega'} f_\omega(x)]$. It is clear that for this model which is within Ω', x^* is the best alternative, hence it is possibly Ω'–optimal.

In Fig. 4, x_D, x_C, and a modified version of x_H equals to $(0.6, 0.6)$ noted x_{H^*} are displayed; x_{H^*} being the Minmax regret recommendation. As we can see, such approaches cannot be expected to satisfy Proposition 2. On the other side, one can see in Fig. 4 that the Maximax gain recommendation can be a very bad choice in some situations, meaning that checking whether we are in such situations may be of importance.

3.3 Convergence

In the case of regret CSS, the elicitation provably converges to the optimal model as long as no errors are made. It is guaranteed that after the kth update, $\Omega^k \subseteq \Omega^{k-1}$ will never be empty, and that the inclusion will be strict under mild assumptions. In the case of gain CSS, we have the same property:

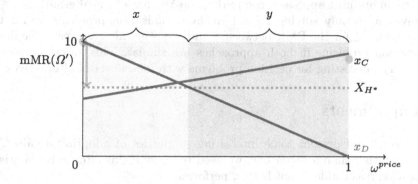

Fig. 4. Visualisation of optimality of 3 alternatives

Proposition 3. *Consider the pair* $x^k = \arg MMG(\Omega^k)$, $y^k = \arg\max_X PMR(x^k, y, \Omega^k)$ *chosen at the kth step of gain CSS. Then we have:*

$$\Omega^k \cap \omega^{x^k \succeq y^k} \neq \emptyset,$$

$$\Omega^k \cap \omega^{x^k \preceq y^k} \neq \emptyset.$$

Guaranteeing that $\Omega^k \subseteq \Omega^{k-1}$.

Proof. We prove that whatever the answer, the intersection with Ω^k is non-empty:

- $\Omega^k \cap \omega^{x^k \succeq y^k} \neq \emptyset$: immediate since $x^k \in PO_{\Omega^k}$, by Proposition 2, indicating that there is a model $\omega \in \Omega^k$ such that $f_\omega(x^k) \succeq f_\omega(y^k)$.
- $\Omega^k \cap \omega^{x^k \preceq y^k} \neq \emptyset$: since we picked y^k in accordance with the CSS, Eq. (10) tells us that $y^k = \arg\max_{y \in \mathcal{X}} PMR(x^k, y, \Omega^k)$. Since $PMR(x^k, y^k, \Omega^k) = \max_{\omega \in \Omega^k} [f_\omega(y^k) - f_\omega(x^k)] \geq 0$, if follows that there is a $\omega \in \Omega^k$ (e.g., the one for which the PMR is reached) with $f_\omega(y^k) \geq f_\omega(x^k)$, ending the proof.

This proposition tells us that our space of possible models will shrink after each question in a non-degenerate way, guaranteeing us to converge to the true model. Note that we have a strict inclusion, i.e., $\Omega^k \subset \Omega^{k-1}$ relation if the $\arg MMG$ is unique, and if the corresponding PMR is strictly positive.

3.4 Computation Complexity

Another advantage of an optimist approach based on a Maximax gain is its lower computational complexity. With a pessimist approach based on a Minmax regret, computing the PMR for all the possible pairs (x, y) such that $x, y \in \mathbb{X}$ and $x \neq y$ is equivalent to solving $n^2 - n$ linear optimisation problems, where $n = |\mathbb{X}|$.

With an optimist approach, computing the MG for all the alternatives $x \in \mathbb{X}$ is equivalent to only solving $2n - 1$ linear optimisation problems (n for the MMG and $n - 1$ for the PMR between x and the other alternatives). The linear optimisation problems in both approaches are similar. This lower complexity cost is very interesting for decision problems with a large set of alternatives.

4 Experiments

While Sect. 3 did present some interesting properties of adopting a gain CSS approach rather than a regret one, we need to check if this alternative heuristic can provide reasonable, if not better, performances.

This section brings some elements of answer, by comparing regret and gain CSS strategies in different situations.

4.1 Experimental Protocol

We performed numerical simulations to compare the performances of regret and gain CSS approaches in different contexts, in which we change the kinds of alternatives we consider, as well as the kind of models.

The first element of comparison we consider is the choice of the alternatives:

- In a first setting, we generated randomly multiple alternatives x_i with 8 criteria from a uniform distribution, such that $x_i \in [0, 1]^8$, until we obtained a Pareto front of 100 alternatives. In this case, most alternatives have quite high values on the different criteria, and we consider them to be *good alternatives*
- In a second setting, we generated randomly 100 alternatives x_i with 8 criteria from a Dirichlet distribution, such that $x_i \in [0, 1]^8$ and $\sum_{j=1}^{8} x_i^j = 1$. We then have alternatives whose average utility is the same, and on which trade-offs have to be made. Since such alternatives are poorly noted (average utility of 1 when 8 is the best), we consider them to be *bad alternatives*

The second element of comparison is the choice of a function f_ω to estimate the utility of an alternative. We compared both approaches with 4 different functions, all generated randomly from different Dirichlet distributions:

- WSB: a "balanced" weighed sum (WS), where all criteria values are close. Parameters: $\alpha = 1000.(1/8, 1/8, ..., 1/8)$.
- WSU: an "unbalanced" weighed sum (WS), where some criteria can have significantly higher values than the others. Parameters: $\alpha = (1/8, 1/8, ..., 1/8)$.

- OWAU: an "unfair" OWA, which favours the criteria with the higher values. Parameters: $\alpha = 50.(1/36, 2/36, ..., 8/36)$.
- OWAR: a "redistributive" OWA, which favours the criteria with the lowest values. Parameters: $\alpha = 50.(8/36, 7/36, ..., 1/36)$.

We propose two measures for evaluating the prediction quality of each approach. A first measure is the real score of the current recommendation, computed from the hypothetical true preference model of the user. A second measure is the position of the current recommendation compared to the other alternatives, given the real score of each alternative. 0 means we have the best alternative and 99 the worst one.

We also reduce the variability of the two measures by averaging them on 200 simulations, and by computing a confidence interval of 95%.

4.2 Results

This section discusses the results of our experiments. Since displaying all graphs renders the reading difficult, we only display some of them, picturing different behaviours. Some synthetic statistics on all cases are given in Tables 2 and 3.

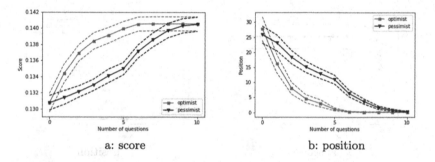

a: score   b: position

Fig. 5. Score and position with poor alternatives on a balanced WS model

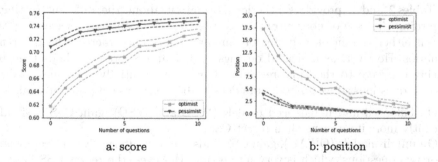

a: score   b: position

Fig. 6. Score and position with good alternatives on a balanced WS model

On one extreme of the spectrum, we can see on Figs. 5 and 6 the score and position for the balanced WS. In this case, the superiority of a method highly depends on the kind of available alternatives. On the other end of the spectrum, we can see on Figs. 7 and 8 the results for the fair OWA model. While there is a slight advantage for the regret CSS strategy, it is not remarkable, and even not significant in the case of poor alternatives.

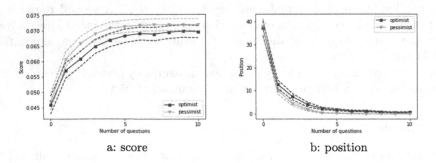

a: score   b: position

Fig. 7. Score and position with poor alternatives on a fair OWA model

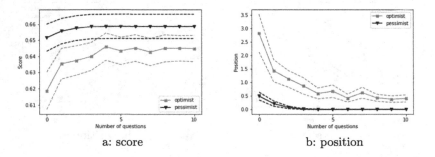

a: score   b: position

Fig. 8. Score and position with good alternatives on a fair OWA model

Tables 2 and 3 provide synthetic information about the different settings. Regarding the case of poor alternatives, the Gain CSS approach seems to give overall either significantly better or similar performances across the different scenarios. However, as indicated on Figs. 5, 6, 7 and 8, both methods tend to quickly converge to the same result, and provide essentially the same quality after 15 questions. If we go into more details about the case of poor alternatives:

- On balanced and unbalanced models (WSB and WSU), gain CSS is significantly more interesting than regret CSS.
- On unfair models (OWAU), gain CSS finds the best alternative only after one or two questions, which is very interesting. However, the regret CSS finds it after around 5 questions, and the score difference is non-significant whatever the number of questions is.

– On redistributive models (OWAR), gain CSS is usually a bit less effective than regret CSS. However, the differences are very small.

Regarding the case of good alternatives, the gain CSS appears overall less effective than regret CSS. On all the experiments but one, our optimist approach is slower to find a good solution. Again, both approaches converge to the best alternative after some questions. Let us now give a bit more details about the two approaches in the case of good alternatives:

– On balanced models, gain CSS is significantly worse than regret CSS.
– On unbalanced models, gain CSS is slightly worse than regret CSS, but they quickly converge to the same recommendation (after 7 or 8 questions).
– On unfair and redistributive models, gain CSS is slightly slower to find the best solution. The difference in score and position is small, yet significant.

Table 2. Current solution score after 5 questions on different contexts

Method	Poor alternatives				Good alternatives			
	WSB	WSU	OWAU	OWAR	WSB	WSU	OWAU	OWAR
Optimist/gain	**0.140**	**0.607**	**0.231**	0.068	0.693	0.881	0.846	0.643
Pessimist/regret	0.135	0.451	**0.231**	**0.071**	**0.740**	**0.887**	**0.853**	**0.659**

Table 3. Current solution position after 5 questions on different contexts

Method	Poor alternatives				Good alternatives			
	WSB	WSU	OWAU	OWAR	WSB	WSU	OWAU	OWAR
Optimist/gain	**1.075**	**2.165**	**0.81**	1.86	5.27	2.91	0.475	0.675
Pessimist/regret	10.91	7.505	1.335	**0.34**	**0.795**	**2.065**	0	0

4.3 Summary

The performances of gain CSS compared to the performances of regret CSS are qualitatively summarised on Table 4.

The main conclusion to draw from this table is that both the nature of the true underlying model, and the values of the alternatives, may have a huge impact on the results.

In our opinion, this observation has two important impacts: the first is that how simulations are carried out in the validation of preference elicitation methods can have a huge impact on the results of these simulations, calling both for deeper theoretical studies about the situations in which a given heuristic has chances to work better, and for simulations considering large spectrum of situations.

Those results also show that in absence of strong inductive bias or refined knowledge about the alternatives, choosing one elicitation technique can hardly be based on performance requirements, and should therefore focus on which axioms should be satisfied in a given problem.

Table 4. Performance interest of gain CSS compared to regret CSS (++: quite inter-esting, −−: quite uninteresting)

	Poor alternatives	Good alternatives
Balanced	++	−−
Unbalanced	++	−
Unfair	∼	−
Redistributive	∼	−

5   Conclusion

We studied the use of an optimist approach using a Maximax gain criterion for recommending an alternative in a robust preference elicitation, instead of a pessimist approach using a Minimax regret. We demonstrated that such an optimist approach possesses the same convergence properties as the classical regret-based one, and has interesting optimality and computational properties. Experiments on simulated data have shown that an optimist approach can be more effective in some contexts.

Our work has shown that the choice of the alternatives has some impacts on the performances of both approaches. We believe that it could be interesting to study more precisely the influence of the alternatives on the computation of the regret. This could be useful for determining the best strategy to choose alternatives in future works.

We cannot therefore give a definite answer to the question we asked in the title. Section 3 gives some pros and cons in terms of properties and axioms that are similar to the pros and cons of optimist and pessimist approaches one can find in other settings [12]. However, selecting the strategy using a performance requirement clearly calls for more theoretical studies regarding the situations in which different heuristics will perform better. A more empirical way to solve this issue could be to characterise elicitation problems through various quality measures (see, e.g., [8]), and see if we can predict the optimal/winning strategy from that, taking inspiration from machine learning methods [15]. Finally, let us note that while we looked at the problem with a greedy approach, whose interest is its efficiency and its agnosticity w.r.t. to the remaining number of questions, it may also be interesting (but also far more difficult) to consider the sequential version of our decision problem (see, e.g., [11]).

References

1. Benabbou, N., Gonzales, C., Perny, P., Viappiani, P.: Minimax regret approaches for preference elicitation with rank-dependent aggregators. EURO J. Decis. Processes, 29–64 (2015). https://doi.org/10.1007/s40070-015-0040-6
2. Benabbou, N., Perny, P., Viappiani, P.: Incremental elicitation of choquet capacities for multicriteria choice, ranking and sorting problems. Artif. Intell. **246**, 152–180 (2017)

3. Bourdache, N., Perny, P.: Anytime algorithms for adaptive robust optimization with OWA and WOWA. In: Rothe, J. (ed.) ADT 2017. LNCS, vol. 10576, pp. 93–107. Springer, Cham (2017). https://doi.org/10.1007/978-3-319-67504-6_7
4. Boutilier, C.: Computational decision support: regret-based models for optimization and preference elicitation. In: Comparative Decision Making: Analysis and Support Across Disciplines and Applications, pp. 423–453 (2013)
5. Boutilier, C., Patrascu, R., Poupart, P., Schuurmans, D.: Constraint-based optimization and utility elicitation using the minimax decision criterion. Artif. Intell. **170**(8–9), 686–713 (2006)
6. Cabannnes, V., Rudi, A., Bach, F.: Structured prediction with partial labelling through the infimum loss. In: International Conference on Machine Learning, pp. 1230–1239. PMLR (2020)
7. Choquet, G.: Theory of capacities. In: Annales de l'institut Fourier, vol. 5, pp. 131–295 (1954)
8. Ciomek, K., Kadziński, M., Tervonen, T.: Heuristics for prioritizing pair-wise elicitation questions with additive multi-attribute value models. Omega **71**, 27–45 (2017)
9. Grabisch, M.: The application of fuzzy integrals in multicriteria decision making. Eur. J. Oper. Res. **89**(3), 445–456 (1996)
10. Guillot, P.L., Destercke, S.: Preference elicitation with uncertainty: extending regret based methods with belief functions. In: Ben Amor, N., Quost, B., Theobald, M. (eds.) SUM 2019. LNCS, vol. 11940, pp. 289–309. Springer, Cham (2019). https://doi.org/10.1007/978-3-030-35514-2_22
11. Holloway, H.A., White Iii, C.C.: Question selection for multi-attribute decision-aiding. Eur. J. Oper. Res. **148**(3), 525–533 (2003)
12. Hüllermeier, E., Destercke, S., Couso, I.: Learning from imprecise data: adjustments of optimistic and pessimistic variants. In: Ben Amor, N., , B., Theobald, M. (eds.) SUM 2019. LNCS, vol. 11940, pp. 266–279. Springer, Cham (2019). https://doi.org/10.1007/978-3-030-35514-2_20
13. Maron, O., Moore, A.W.: The racing algorithm: model selection for lazy learners. Artif. Intell. Rev. **11**(1), 193–225 (1997)
14. Savage, L.J.: The theory of statistical decision. J. Am. Stat. Assoc. **46**(253), 55–67 (1951)
15. Tornede, A., Wever, M., Hüllermeier, E.: Extreme algorithm selection with dyadic feature representation. In: Appice, A., Tsoumakas, G., Manolopoulos, Y., Matwin, S. (eds.) DS 2020. LNCS, vol. 12323, pp. 309–324. Springer, Cham (2020). https://doi.org/10.1007/978-3-030-61527-7_21
16. Troffaes, M.C.: Decision making under uncertainty using imprecise probabilities. Int. J. Approx. Reason. **45**(1), 17–29 (2007)
17. Yager, R.R.: On ordered weighted averaging aggregation operators in multicriteria decisionmaking. IEEE Trans. Syst. Man Cybern. **18**(1), 183–190 (1988)

Probabilistic Lexicographic Preference Trees

Xudong Liu[1(✉)] and Miroslaw Truszczynski[2]

[1] School of Computing, University of North Florida, Jacksonville, USA
xudong.liu@unf.edu
[2] Department of Computer Science, University of Kentucky, Lexington, USA
mirek@cs.uky.edu

Abstract. We introduce *probabilistic lexicographic preference trees* (or PrLPTs for short). We show that they offer intuitive and often *compact* representations of non-deterministic qualitative preferences over alternatives in multi-attribute (or, combinatorial) binary domains. We specify how a PrLPT defines the probability that a given outcome has a given rank, and the probability that a given outcome is preferred to another one, and show how to compute these probabilities in polynomial time. We also show that computing outcomes that are optimal with the probability equal to or exceeding a given threshold for some classes of PrLP-trees is in P, but for some other classes the problem is NP-hard.

Keywords: Preference representation and reasoning · Lexicographic preference trees · Probabilistic preference models

1 Introduction

Preferences play a key role in fields such as multi-agent systems, recommender systems, marketing, and decision theory. To model and reason about preferences, researchers proposed several deterministic preference formalisms including quantitative models such as penalty logic [7] and possibilistic logic [8], and qualitative models such as lexicographic preference trees and forests [2,9,13,14], conditional preference networks [3], and conditional importance networks [4].

However, preferences are often non-deterministic. For example, a restaurant customer may have a preference *"If having steak for the entree, with 80% probability the user prefers to drink beer and with 20% wine."* Given such non-deterministic statements collected or learned by a learning algorithm, *probabilistic* queries, arise such as *probabilistic dominance testing* (computing the probability of one alternative is preferred over another), and *probabilistic optimality testing* (computing the probability of some alternative being optimal). To address such scenarios, formalisms for non-deterministic preference are needed.

Research of qualitative preference reasoning with uncertainty has resulted in several formal preference models. Most notable are probabilistic conditional preference networks [1,6], and probabilistic preference logic networks [15]. Although

D. Fotakis and D. Ríos Insua (Eds.): ADT 2021, LNAI 13023, pp. 86–100, 2021.
https://doi.org/10.1007/978-3-030-87756-9_6

both models represent uncertainty in an intuitive way, answering some decision questions is computationally intractable. For instance, probabilistic dominance testing for both formalisms is #P-hard.

Experimental studies on human decision making in economics and psychology [5,10,17] show that humans often make decisions in a lexicographic manner, and that human decision makers who use other decision models often translate them to lexicographic models for a more compelling verbal communication [16]. Motivated by this research and building on the concept of lexicographic preference trees [2,14], we propose *probabilistic lexicographic preference trees*, or *PrLP-trees*, as another model of qualitative preferences with uncertainty. We also identify four classes of PrLP-trees that allow compact representations.

We show that PrLP-trees define the probability that a given outcome has a given rank, and the probability that an outcome is preferred to another one. We show that the probabilities mentioned above can be computed in polynomial time, both for "full" PrLP-trees and for all classes of compact PrLP-trees. On the other hand, we show that finding the alternative with the *highest* probability of being optimal is NP-hard.

The paper is organized as follows. Section 2 formally introduces PrLP-trees and their probabilistic semantics, illustrates the concept, and gives some key properties. Section 3 presents the collapsing algorithm and introduces classes of compact PrLP-trees. It is followed by a discussion of complexity results and then by conclusions.

2   Probabilistic Lexicographic Preference Trees

We consider preference relations over combinatorial objects. Formally, let $\mathcal{A}$ be a finite set of n binary *attributes*, each attribute $X \in \mathcal{A}$ having the domain $\{x, \bar{x}\}$. A *combinatorial domain* over $\mathcal{A}$, $CD(\mathcal{A})$, is the Cartesian product of the domains of all attributes in $\mathcal{A}$: $CD(\mathcal{A}) = \times_{X \in \mathcal{A}}\{x, \bar{x}\}$. We call elements of $CD(\mathcal{A})$ *alternatives* or *outcomes*.

A *probabilistic lexicographic preference tree* (PrLP-tree, for short) over a combinatorial domain $CD(\mathcal{A})$ is a tree satisfying the following conditions:

1. Each node in the tree is a *box* node or a *circle* node.
2. The root of the tree, denoted by *root*, and the leaves are box nodes; box and circle nodes alternate on each path from the root to a leaf.
3. Edges from a box node v to its children $c \in C(v)$ are labeled by positive reals p_c^v so that $\sum_{c \in C(v)} p_c^v = 1$, where $C(v)$ is the set of children of node v.
4. Every circle node v is labeled by an attribute, say $X \in \mathcal{A}$, and by a real $p_{v,x}$ satisfying $0 \leq p_{v,x} \leq 1$. The latter represents the probability that x is preferred to $\bar{x}$ at v. We set $p_{v,\bar{x}} = 1 - p_{v,x}$ to represent the probability that $\bar{x}$ is preferred to x at v.
5. Every circle node v has two children, *left* and *right*. Assuming that v is labeled with an attribute $X \in \mathcal{A}$, the edge to the left child is labeled with x and to the right child with $\bar{x}$. (With this convention, we do not need to make these edge labels explicit.)

6. Every attribute in $\mathcal{A}$ appears exactly once on every path from the root to a leaf.

To evaluate outcomes in a PrLP-tree, we start at the root and move down the tree. Along the way, the role of a box node is to identify the next attribute to consider, and the role of a circle node labeled with an attribute X is to specify which of the two values x and $\bar{x}$ is preferred at that node. Deterministic LP-trees [2] are special PrLP-trees, where each non-leaf box node has exactly one child (determinism for the choice of the next attribute to consider), and the reals $p_{v,x}$ are either 0 or 1 (determinism for the preference between x and $\bar{x}$.

To illustrate these intuitions (cf. Fig. 1), let us consider the domain of dinners with three attributes: *Appetizer* (A) with values *salad* (a) and *soup* ($\bar{a}$), *Entree* (E) with values *beef* (e) and *fish* ($\bar{e}$), and *Drink* (D) with values *beer* (d) and *wine* ($\bar{d}$).

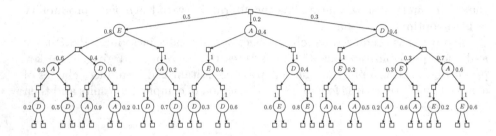

Fig. 1. A PrLP-tree over the dinner domain

To determine which of the two outcomes, say $\alpha = ae\bar{d}$ and $\beta = a\bar{e}d$, is preferred we proceed from the root to a leaf. First, at the root, we select the attribute which (at the time of the decision) is regarded as the most important one. The selection is random and follows the probability distribution given by the labels of the edges starting at the root. Here *Entree* has probability 0.5, *Appetizer*—0.2, and *Drink*—0.3. Say we selected *Entree* (the most likely choice). We then move down to the corresponding circle node (labeled with E for *Entree* in Fig. 1). Since the two outcomes differ on *Entree*, we can determine right at that node whether α is better than β or the other way around. To do so, we randomly select one of the values of the selected attribute (here, e or $\bar{e}$) based on the probability of e being preferred, which is given as the label of that node. Say, we chose that e as preferred (this happens 80% of the time). Thus, under these random choices, α is preferred to β.

But, it could turn out differently. At the root, we might have selected *Appetizer* as the most important attribute (this happens 20% of the time) and move down to the circle node labeled A (for *Appetizer*). However, α and β have the same value a for this attribute. So, unlike in the earlier scenario, we are unable to decide at this node which of α and β is preferred. Thus, we move down the left edge (it corresponds to the value a shared by both outcomes) to reach the next box node. In this box node, there is no uncertainty about the next most important attribute. It is *Entree* and we move down to the corresponding circle node. Since the two outcomes differ on this attribute, we will now be able to order α and β. We use the probability information there (e is preferred to $\bar{e}$ with the probability 0.4) to select randomly the preferred value. Say, we selected $\bar{e}$ (happens with the probability 0.6). Then this time around, β is preferred to α.

Thus, each time we ask whether α is preferred to β and run this process, we may get a different answer based on random choices made at box and circle nodes according the probability distributions defined there. It is of interest then to establish the probability that α is preferred to β. The formulas we present in the paper provide the answer, formally specifying the random variable *dominance* that returns for each pair (α, β) of distinct outcomes either $\alpha \succ \beta$ or $\beta \succ \alpha$.

A similar process defines the random variable *rank* of an outcome α, which counts how many other outcomes are preferred to α. To illustrate, let us consider the outcome $\alpha = ae\bar{d}$. As above, we start at the root and select the attribute to take as the most important one. Say, we generated *Entree*. We descend to the corresponding circle node and select one of the values e and $\bar{e}$ as preferred. We do it at random following the distribution specified in that node. Say, we selected e as preferred (it happens 80% of the time). This means that the rank of α is at most 3 (because every outcome with $\bar{e}$ as its value for *Entree* is less preferred than α and so, at most three outcomes are more preferred than α).

We can refine this estimate by continuing down the tree following the left edge (according to our convention, this edge corresponds to the value α has for *Entree*. In that box node we randomly select the next most important attribute (using the probability distribution the box node defines). Say we selected *Drink* (which happens 40% of the time). We descend to the corresponding circle node. We randomly select d or $\bar{d}$ as preferred (according to the probability labeling this node). Say d is selected as preferred. This selection makes the outcomes ade and $ad\bar{e}$ preferred to α. Thus, we now have that the rank of α is 2 or 3. Finally, we descend to the right box node (because α has $\bar{d}$ on *Drink*). There is only one attribute left, *Appetizer*, so there is no choice there and we descend to the corresponding circle node. In that node, we randomly select the preferred value for *Appetizer* (here, a is preferred with the probability 0.2). If we select a, the rank of α is 2. If we select $\bar{a}$, the rank of α is 3, as $\bar{a}e\bar{d}$ is preferred to α.

As before, each time we run this process to establish the rank of an outcome, we may get a different result. It is then important to know what is the probability (given a PrLP-tree) that an outcome has a specified rank, in particular, the probability that an outcome is optimal (has rank 0). These probabilities formally define the random variable *rank*.

We now present the formulas for the probabilities discussed above to define the random variables rank and dominance.

Let T be a PrLP-tree over a combinatorial domain $\mathcal{A}$. We write $Pr_T(r(\alpha) = k)$ for the probability that an outcome α has rank k in T. Similarly, we write $Pr_T(\alpha \succ \beta)$ for the probability that an outcome α is (strictly) preferred to an outcome β in T. We define both probabilities by induction.

In the base case we have $\mathcal{A} = \emptyset$. In that case, $CD(\mathcal{A})$ consists of just one outcome, the empty outcome, which we denote by ϵ. Further, every PrLP-tree T over $\mathcal{A}$ consists of a single node (which must be a box node). We set $Pr_T(r(\epsilon) = k) = 1$, if $k = 0$, and $Pr_T(r(\epsilon) = k) = 0$, otherwise. Similarly, for every outcomes $\alpha, \beta \in CD(\mathcal{A})$, since they necessarily are equal (each being equal to ϵ), we set $Pr_T(\alpha \succ \beta) = 0$.

Let us now assume that $|\mathcal{A}| = n > 0$ and let $A \in \mathcal{A}$ be an attribute labeling one of the children, say v, of the root. We first define the probability

$$Pr_T(r(\alpha) = k|A)$$

that an outcome α has rank k in T given that A is chosen as the most important attribute (this happens with the probability given by p_v^{root}).

Let T_a and $T_{\bar{a}}$ be the two PrLP-trees rooted in the two children of v. Clearly, T_a and $T_{\bar{a}}$ are PrLP-trees over the set of attributes $\mathcal{A} \setminus \{A\}$. Let us consider an outcome α. We set $\alpha' = \alpha[\mathcal{A} \setminus \{A\}]$ and write x for $\alpha[A]$ (clearly, $x = a$ or $\bar{a}$). We also adopt the convention that $\bar{\bar{a}} = a$. If $k < 2^{n-1}$, then for $r(\alpha) = k$ to hold, we must have that x is preferred to $\bar{x}$ at v, which happens with the probability $p_{v,x}$, and that α' has rank k in the tree T_x. If $k \geq 2^{n-1}$, then for $r(\alpha) = k$ to hold, we must have that $\bar{x}$ is preferred to x at v, which happens with the probability $p_{v,\bar{x}}$, and that α' has rank $k - 2^{n-1}$ in the tree T_x. Therefore, we define,

$$Pr_T(r(\alpha) = k|A) =$$
$$\begin{cases} p_{v,x} Pr_{T_x}(r(\alpha') = k) & \text{if } k < 2^{n-1} \\ p_{v,\bar{x}} Pr_{T_x}(r(\alpha') = k - 2^{n-1}) & \text{if } k \geq 2^{n-1}. \end{cases} \tag{1}$$

Next, we define $Pr_T(r(\alpha) = k)$ by setting

$$Pr_T(r(\alpha) = k) = \sum_{v \in C(root)} p_v^{root} Pr_T(r(\alpha) = k|A_v), \tag{2}$$

where A_v denotes the attribute labeling the node v.

A similar reasoning leads to a recursive definition of $Pr_T(\alpha \succ \beta)$. Using the notation introduced above and writing $\beta' = \beta[\mathcal{A} \setminus \{A\}]$, we define the probability $Pr_T(\alpha \succ \beta|A)$ that α is preferred to β, assuming that A is the first (most important attribute) in the evaluation of the dominance, by setting

$$Pr_T(\alpha \succ \beta|A) =$$
$$\begin{cases} p_{v,\alpha[A]} & \text{if } \alpha[A] \neq \beta[A] \\ Pr_{T_{\alpha[A]}}(\alpha' \succ \beta') & \text{if } \alpha[A] = \beta[A]. \end{cases} \tag{3}$$

We then define $Pr_T(\alpha \succ \beta)$ by the formula

$$Pr_T(\alpha \succ \beta) = \sum_{v \in C(root)} p_v^{root} Pr_T(\alpha \succ \beta|A_v), \tag{4}$$

where A_v denotes the attribute labeling the node v.

It is clear from the recursive definitions that the probabilities $Pr_T(r(\alpha) = k)$ and $Pr_T(\alpha \succ \beta)$ can be computed efficiently.

Theorem 1. *The probabilities $Pr_T(r(\alpha) = k)$ and $Pr_T(\alpha \succ \beta)$ can be computed in linear time in the size of T.*

The following three theorems capture natural properties to be expected of the probabilities we defined above. That they hold provides an additional support for the soundness of our definitions.[1]

Theorem 2. *For every PrLP-tree T over a set of attributes $\mathcal{A}$ and an integer k, $0 \leq k \leq 2^{|\mathcal{A}|} - 1$, $\sum_\alpha Pr_T(r(\alpha) = k) = 1$.*

Proof. We prove the theorem by induction. If $\mathcal{A} = \emptyset$, the only outcome in the corresponding combinatorial domain is the empty outcome ϵ. Moreover, $k = 0$. By definition, $Pr(r(\epsilon) = 0) = 1$ and the assertion holds.

Let us now consider a tree T over a set of attributes $\mathcal{A}$, where $|\mathcal{A}| > 0$ and let us assume that the claim holds for all trees over a smaller set of attributes. Let us denote $n = |\mathcal{A}|$. We will present the argument for the case when $k < 2^{n-1}$. The other case is similar. By the formulas (1) and (2) in the main paper,

$$\sum_\alpha Pr_T(r(\alpha) = k)$$

$$= \sum_\alpha \sum_{v \in C(root)} p_v^{root} p_{v,\alpha[A_v]} Pr_{T_{\alpha[A_v]}}(r(\alpha') = k)$$

$$= \sum_{v \in C(root)} \sum_\alpha p_v^{root} p_{v,\alpha[A_v]} Pr_{T_{\alpha[A_v]}}(r(\alpha') = k)$$

$$= \sum_{v \in C(root)} p_v^{root} \sum_\alpha p_{v,\alpha[A_v]} Pr_{T_{\alpha[A_v]}}(r(\alpha') = k).$$

The set of outcomes over $\mathcal{A}$ splits into two disjoint sets: one with all outcomes that have a as the value on A and the other one with all outcomes that have $\bar{a}$ as the value of A. Thus,

$$\sum_\alpha p_{v,\alpha[A_v]} Pr_{T_{\alpha[A_v]}}(r(\alpha') = k)$$

$$= \sum_\beta p_{v,a} Pr_{T_a}(r(\beta) = k) + \sum_\beta p_{v,\bar{a}} Pr_{T_{\bar{a}}}(r(\beta) = k)$$

$$= p_{v,a} + p_{v,\bar{a}} = 1.$$

The sums on the right hand side of the equality in the second line are over all outcomes β over $\mathcal{A} \setminus \{A\}$ and the first equality in the last line follows by the induction hypothesis applied to the trees T_a and $T_{\bar{a}}$.

It follows that

$$\sum_\alpha Pr_T(r(\alpha) = k) = \sum_{v \in C(root)} p_v^{root} = 1,$$

which completes the inductive step argument. □

---

[1] Due to space restrictions, we provide a proof to the first of these results only.

Theorem 3. *For every PrLP-tree T over $\mathcal{A}$ and an outcome $\alpha \in CD(\mathcal{A})$,*
$$\sum_{k \in [0, 2^{|\mathcal{A}|} - 1]} Pr_T(r(\alpha) = k) = 1.$$

Theorem 4. *For every PrLP-tree T and two outcomes α and β such that $\alpha \neq \beta$, $Pr_T(\alpha \succ \beta) + Pr_T(\beta \succ \alpha) = 1$.*

Our PrLP-tree model can be viewed as a *concise* representation of certain probability distributions over the space of standard deterministic LP-trees [2]. Let T be a PrLP-tree. Selecting for each box node exactly one of its circle children (according to the probability distribution at that node), and selecting at each circles node one of the two possible preference orders on the corresponding attributes domain (again, according to the probability distribution at that node) yields an LP-tree and its probability (the product of the probabilities of each random selection made).

Let $LP(T)$ be the probability space described above: the set of LP-trees derivable from T and their probabilities. Each such space supports reasoning about uncertain preferences. For instance, given an outcome α, to find its rank, we select an LP-tree R from the family $LP(T)$ at random according to the probabilities of trees in $LP(T)$, and then use R to compute the rank of α. In this approach, the probability that α has rank k is equal to the sum of probabilities of all trees R in $LP(T)$ in which α has rank k. The dominance for two outcomes α and β could be handled similarly.[2] We conjecture that this approach yields probabilities for the rank and dominance that are identical to those proposed originally for PrLP-trees. Reasoning with probability spaces of LP-trees is more direct, while PrLP-trees provide a much more concise representation. An in-depth study of this relation is the subject of future work.

3   Compact Representation

The size of a PrLP-tree is exponential in the size of the attribute domains. However, PrLP-trees can be collapsed to a more compact representation. Let us consider a circle node v labeled by attribute X. If the two subtrees T_x and $T_{\bar{x}}$ of v are the same except possibly for the probability labels on the corresponding circle nodes, then these two subtrees can be merged into one.

To merge T_x and $T_{\bar{x}}$, we replace them with a single subtree of v, say R, that has the same structure as T_x (or, equivalently, as $T_{\bar{x}}$), and we modify the probability labels on the circle nodes in R. Let us consider a circle node w in R, let w' and w'' be the corresponding nodes in T_x and $T_{\bar{x}}$, respectively, and let p' and p'' be the probability labels of w' and w''. If $p' = p''$ then we label w in R with p', the common probability label of w' and w''. If $p' \neq p''$, the probability label of w is a conditional probability table (CPT) that consists of two rows: $x : p'$ and $\bar{x} : p''$. The table gives the *conditional* probabilities that a

---

[2] We refer to the work by Lang et al. [11] for the definitions of ranks and dominance for LP-trees.

is preferred to $\bar{a}$ (where we write A for the attribute labeling w' and w''). The condition is the value of the attribute X and the probabilities reflect those in the corresponding nodes in the subtrees T_x and $T_{\bar{x}}$.

We note, that when we merge the subtrees of v, the probability labels of circle nodes in T_x and $T_{\bar{x}}$ may be CPTs and not reals (since we may already have applied merging at nodes other than v). Let us now consider this more general case. That is, let us assume that the labels of w' and w'' are CPTs P' and P''. To produce a CPT P to label w in R, we rewrite each row $u' : p'$ in P' with $x, u' : p'$, each row $u'' : p''$ in P'' with $\bar{x}, u'' : p''$, and we put all rewritten rows together. In this way, the merged CPTs in R contain conditional probabilities that a is preferred to $\bar{a}$, with the conditions representing values on the attributes labeling the ancestor nodes of w at which we applied the merging.

To collapse the tree we apply merging at all circle nodes at which merging is possible. One can show that the order in which we select nodes for merging is immaterial. Once this phase is complete, we may have box nodes with just one child. These nodes carry no information (because there is no uncertainty about the next circle node). Therefore, we remove them. We also remove the box nodes representing leaves, for they also do not convey any specific meaning.

This process is formalized in Algorithm 2, which performs individual merging steps bottom-up by calling a recursive Algorithm 1, and then applies post-processing that removes leaves and all box nodes that only have one child. It is illustrated with several examples of collapsible PrLP-trees and the corresponding compacted trees in Fig. 2.

When a PrLP-tree is collapsed into one path of circle nodes labeled by probability values (not CPTs), we call such a tree an *unconditional importance and unconditional preference* (UI-UP) tree. If a collapsed tree only has one path of circle nodes some of which are labeled by CPTs, we call it an *unconditional importance and conditional preference* (UI-CP) tree. We show such types of trees in Fig. 2b and d. UI trees indeed are compact with linear size in the attribute domains, if the number of conditioning attributes in trees is bounded by a small constant.

In general, even after collapsing, the tree structure can still have branching points and so multiple paths from the root to leaves. We refer to such trees as *conditional importance* (CI) trees. If, in a CI tree, nodes with the same attribute unanimously are labeled by the same probability label, we call the tree a *conditional importance and unconditional preference* (CI-UP) tree. All other CI trees are called *conditional importance and conditional preference* (CI-CP) trees. One example of a CI-CP tree is given in Fig. 2f. Note that, should the numerals for all appearances of each attribute be the same in Fig. 2e, we would obtain a CI-UP tree with the same tree structure as in Fig. 2f but with all nodes labeled by probability numerals, not CPTs. We see that CI trees are of polynomial size when both the number of box nodes and the number of conditioning attributes in trees are bounded by a small constant.

PrLP-trees can be used to simulate agent's reasoning about the quality of outcomes. We already discussed the use of a PrLP-tree to find a rank of an outcome

---

Algorithm 1: The recursive procedure *collapseRec* in method *collapse*

Input: A full PrLP-tree T

Output: T collapsed

1 **if** T *is a leaf* **then**

2 | **return**;

3 **else**

4 | **for** *Each child c of T* **do**

5 | | collapseRec(c);

6 | **end**

7 | **if** $T \neq \square$ **then**

8 | | **if** T_x *and* $T_{\bar{x}}$ *are equivalent* **then**

9 | | | Merge T_x and $T_{\bar{x}}$ to R;

10 | | | Disconnect T_x and $T_{\bar{x}}$ from T;

11 | | | Connect R straight down as a child to T;

12 **end**

Algorithm 2: The procedure *collapse* to collapse a full PrLP-tree

---

Input: A full PrLP-tree T

Output: T collapsed

1 collaspseRec(T);

2 Eliminate all leaf nodes and all box nodes in T that only has one outgoing edge;

---

or to compare two outcomes. The corresponding probabilities $Pr_T(r(\alpha) = k)$ and $Pr_T(\alpha \succ \beta)$ can also be evaluated efficiently based on a compact representation of a PrLP-tree T.

Theorem 5. *(1) There is a linear time algorithm to solve the* PROBRANK *problem: given a collapsed PrLP-tree T in any class of $\{UI, CI\} \times \{UP, CP\}$, an outcome α and an integer $0 \leq k \leq 2^{n-1}$, compute $Pr_T(r(\alpha) = k)$.*

(2) There is a linear time algorithm to solve the PROBDOM *problem: given a collapsed PrLP-tree T in any class of $\{UI, CI\} \times \{UP, CP\}$, and two different outcomes α and β, compute $Pr_T(\alpha \succ \beta)$.*

Proof (Sketch). One can show that Algorithms 3 and 4 presented below have the claimed properties. $\qquad\square$

We illustrate these two algorithms with the two examples below.

Example 1. Let us consider the CI-CP tree in Fig. 2f. Let outcome $\alpha = ade$ and a rank of 5. Algorithm 3 starts by calling $probRank(t_1, \alpha, 5, 0, 1, 0)$. Since $r < 2^2$ is false, P is updated to $1 * 0.6 = 0.6$, and $probRank(t_2, \alpha, 1, 1, 0.6, 0)$ is called. Because t_2 is a box node, the algorithm goes through its both children nodes $t3$ and t_5 to call $probRank$ for t_3 and $probRank$ for t_5, respectively.

First, for $probRank(t_3, \alpha, 1, 1, 0.36, 0)$, since $r < 2^1$ is true, P is updated to $0.36 * 0.3 = 0.108$, and $probRank(t_4, \alpha, 1, 2, 0.108, 0)$ is called. For this call,

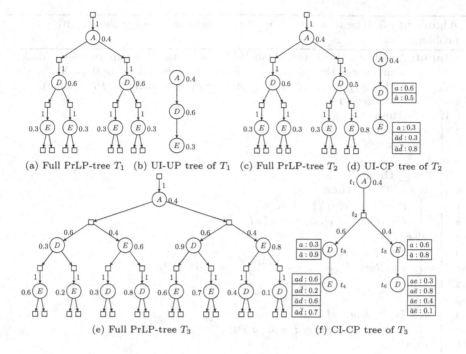

Fig. 2. Collapsing to compact PrLP-trees

because $r < 2^0$ is false, P is updated to $0.108 * 0.4 = 0.0432$ and $probRank(\emptyset, \alpha, 0, 3, 0.0432, 0)$ is called. Therein, PP is updated to 0.0432.

Secondly, for $probRank(t_5, \alpha, 1, 1, 0.24, 0.0432)$, since $r < 2^1$ is true, P is updated to $0.24 * 0.6 = 0.144$, and $probRank(t_6, \alpha, 1, 2, 0.144, 0.0432)$ is called. For this call, because $r < 2^0$ is false, P is updated to $0.144 * 0.7 = 0.1008$ and $probRank(\emptyset, \alpha, 0, 3, 0.1008, 0.0432)$ is called. Therein, PP finally is updated to 0.144.

Example 2. Let us again consider the CI-CP tree in Fig. 2f. Let outcomes $\alpha = ad\bar{e}$ and $\beta = a\bar{d}\bar{e}$. Algorithm 4 starts by calling $probDom(t_1, \alpha, \beta, 1, 0)$. Since $\alpha[A] = \beta[A]$, $probDom(t_2, \alpha, \beta, 1, 0)$ is called. Because t_2 is a box node, the algorithm goes through its both children nodes $t3$ and t_5 to call $probDom$ for t_3 and $probDom$ for t_5, respectively.

First, for $probDom(t_3, \alpha, \beta, 0.6, 0)$, since $\alpha[D] = \beta[D]$, $probDom(t_4, \alpha, \beta, 0.6, 0)$ is called. For this call, because $\alpha[E] \neq \beta[E]$, PP is updated to $0 + 0.6 * 0.6 = 0.36$ and this call returns.

Second, for $probDom(t_5, \alpha, \beta, 0.4, 0.36)$, since $\alpha[E] \neq \beta[E]$, PP is updated to $0.36 + 0.4 * 0.6 = 0.6$, the final result.

Algorithm 3: The recursive procedure *probRank* to solve the PROBRANK problem

---

Input: A collapsed PrLP-tree T labeled by attribute X, an alternative α, rank r, number of circle ancestors c preceding T initialized to 0, path probability P initialized to 1, and resulting probability PP (global) initialized to 0

1   **if** $c = n$ **then**
2     $PP \leftarrow PP + P$;
3   **return**;
4   **end**
5   **if** $T \neq \square$ **then**
6   **if** $r < 2^{n-c-1}$ **then**
7       $P \leftarrow P * Pr(\alpha[X] \succ \overline{\alpha[X]})$;
8       $probRank(T_{\alpha[X]}, \alpha, r, c+1, P, PP)$;
9   **else**
10       $P \leftarrow P * Pr(\overline{\alpha[X]} \succ \alpha[X])$;
11       $probRank(T_{\alpha[X]}, \alpha, r - 2^{n-c-1}, c+1, P, PP)$;
12   **end**
13   **else**
14   **for** *Each child i of T* **do**
15       $probRank(i, \alpha, r, c, P * p_c^T, PP)$;
16   **end**
17   **end**

Algorithm 4: The recursive procedure *probDom* to solve the PROBDOM problem

---

Input: A collapsed PrLP-tree T labeled by attribute X, two distinct alternatives α and β, path probability P initialized to 1, and resulting probability PP (global) initialized to 0

1   **if** $T \neq \square$ **then**
2   **if** $\alpha[X] \neq \beta[X]$ **then**
3       $PP \leftarrow PP + P * Pr(\alpha[X] \succ \beta[X])$;
4   **else**
5       $probDom(T_{\alpha[X]}, \alpha, \beta, P, PP)$;
6   **end**
7   **else**
8   **for** *Each child i of T* **do**
9       $probDom(i, \alpha, \beta, P * p_c^T, PP)$;
10   **end**
11   **end**

---

4   Complexity Results

Reasoning about optimality of outcomes or, in the case of uncertainty, about the probability of an outcome to be optimal, is an essential component of preference reasoning. In this section we study the complexity of the optimality problem for PrLP-trees. Specifically, we consider the following two problems.

1. (OPTIMALITY): Given a collapsed PrLP-tree T in any class of $\{UI, CI\} \times \{UP, CP\}$, find an alternative with the highest probability of being optimal, in symbols, find $\text{argmax}\, Pr_T(r(\alpha) = 0)$.
   $$\alpha$$
2. (OPTIMALITY-D) (the decision version of the previous problem): Given a collapsed PrLP-tree T in any class of $\{UI, CI\} \times \{UP, CP\}$, and positive integers $a \leq b$, decide whether there is an alternative α such that $Pr_T(r(\alpha) = 0) \geq a/b$.

It turns out that the complexity of these problems depends on whether probabilities assigned to attributes in an input PrLP-tree are conditional or not.

Theorem 6. *For UP PrLP-trees, the problems* OPTIMALITY *and* OPTIMALITY-D *can be solved in linear time.*

Proof (Sketch). Let T be an input tree. By the assumption, for every attribute A, the probability label of any node in T labeled with A is the same. Let us denote this common value by p_A. To define an optimal outcome, say α, we set $\alpha[A] = a$, if $p_A \geq 1/2$, otherwise, we set $\alpha[A] = \bar{a}$. One can prove that for that outcome α, $Pr(r(\alpha) = 0)$ is maximized. This solves the OPTIMALITY problem. Next, one can show that this maximum probability is given by the products q of the quantities $\max\{p_A, 1 - p_A\}$ taken over all attributes A. Now, the problem OPTIMALITY-D has an answer YES for a threshold a/b if and only if $q \geq a/b$. $\square$

When the input trees allow for conditional probabilities, both problems are intractable.

Theorem 7. *The problem* OPTIMALITY-D *is NP-complete for the class of CP PrLP-trees.*

Proof. Due to space restrictions, we only show proof for UI-CP trees. The membership in NP is clear. To prove hardness, we construct a reduction from 3-SAT. Let Φ be a CNF formula over propositional variables $X_1, \ldots, X_n$ and with clauses $C_1, \ldots, C_n$. We construct an instance of the problem OPTIMALITY-D, restricted to UI-CP trees as follows. We consider a combinatorial domain $\mathcal{C}$ in which the variables X_i and the clauses C_j serve as attributes. When we view X_i as an attribute, X_i has values x_i and $\bar{x}_i$. With some abuse of notation, we will take the values x_i and $\bar{x}_i$ as the logical values *true* and *false*, respectively. Thus, outcomes in $\mathcal{C}$ restricted to attributes $X_1, \ldots, X_n$ can be identified with truth assignments to propositional variables $X_1, \ldots, X_n$.

We construct a UI-CP tree T_Φ, its structure is given by a path of $n + m$ nodes, by labeling the nodes of the path with first $X_1, \ldots X_n$ (in any order) and then with $C_1, \ldots, C_m$ (also in any order). We assign a probability $1/2$ to every node labeled with X_i (these probabilities are unconditional). We label each node labeled with a clause C_p with a conditional probability table. To describe it, let us assume that $C_p = \ell(X_i) \vee \ell(X_j) \vee \ell(X_k)$, where $\ell(X_i), \ell(X_j)$, and $\ell(X_k)$ are the three literals of C_p, and X_i, X_j, and X_k are the atoms in these literals.

We make the probabilities that c_p is preferred to $\bar{c}_p$ at that node depend on the attributes, X_i, X_j and X_k. For each combination of values for which C_p is satisfiable, we set the probability that c_p is preferred to $\bar{c}_p$ to 1. For the only combination of values that falsifies the clause, we set that probability to $1/2$.

To complete the construction of an instance to the OPTIMALITY-D problem, we set $a = 1$ and $b = 2^n$. It is clear the entire construction can be done in time linear in the size of Φ. We now observe that if a truth assignment α to $X_1, \ldots, X_m$ satisfies every clause in Φ, then the outcome consisting of α appended with $c_1, c_2, \ldots, c_m$ is optimal with the probability $1/2^n$. That is, there is an outcome α' such that $Pr_T(r(\alpha') = 0) \geq a/b$.

Conversely, let α' be an outcome of the combinatorial domain $\mathcal{C}$ such that $Pr_T(r(\alpha') = 0) \geq a/b$. It follows that every attribute C_j contributes a factor of 1 to the product defining $Pr_T(r(\alpha') = 0)$ (if any of these attributes contributed $1/2$, we would have $Pr_T(r(\alpha') = 0) \leq 1/2^{n+1} < a/b$). This implies each clause C_j is satisfied by the assignment given by $\alpha'[X_1, \ldots, X_n]$, that is, Φ is satisfiable.

Consequently, our construction is indeed a reduction from 3-SAT and so, the hardness part for the problem OPTIMALITY-D follows.   □

This result has an immediate consequence for the OPTIMALITY problem.

Corollary 1. OPTIMALITY *is NP-hard for CP PrLP-trees.*

Indeed, if one could compute in polynomial time an outcome α with the highest probability of being optimal, one could next compute that probability, say p, also in polynomial time. Therefore, the answer to the problem whether there is an outcome α such that $Pr_T(r(\alpha) = 0) \geq a/b$ is YES if an only if $p \geq a/b$.

We conclude this section with a summary of our complexity analysis in Fig. 3, omitted from which due to space limit is that OPTIMALITY is in P for UP trees and is NP-hard for CI trees.

	UP	CP
UI	P (Thm 5)	P (Thm 5)
CI	P (Thm 5)	P (Thm 5)

(a) PROBRANK & PROBDOM

	UP	CP
UI	P (Thm 6)	NPC (Thm 7)
CI	P (Thm 6)	NPC (Thm 7)

(b) OPTIMALITY-D

Fig. 3. A summary of complexity results

5   Related Work

Probabilistic Conditional Preference Networks. Generalizing the deterministic CP-net models, PCP-net introduces uncertainty to the conditional preference tables, where each local preference is annotated with a probabilistic distribution [1]. The dependency graph in the PCP-net still is deterministic, rendering it unable to express uncertainty over the dependency relationships between nodes.

A PCP-net can be seen as a compact representation of deterministic CP-nets coming from multiple agents. However, these agents must all share the same dependency graph and the relationship between the PCP-net and its CP-nets is not clear. For acyclic PCP-nets, the PROBDOM problem is #P-hard, and the probabilistic optimality problem, i.e., a special case of the PROBRANK problem with the rank being 0, is in P. However, the OPTIMALITY problem for acyclic PCP-nets still remains open.

Probabilistic Preference Logic Networks. Unlike PrLP-trees and PCP-nets that are structured preference relations over combinatorial domains, PPL-nets [15] express *non-structured* preferences over *explicitly given* alternatives. In a PPL-net, a set of pairwise comparisons with their importance degree is given, and we are to compute a normalized weight for every world represented by a permutation of the alternatives. The probability of a preference query is the sum of all the normalized weights of the worlds where the query holds true. As for the complexity of reasoning, the PROBDOM problem for PPL-nets is #P-hard [15]. However, other problems we considered has yet to be formalized and studied.

6   Conclusion and Future Work

We introduced *probabilistic lexicographic preference trees*, an intuitive and often compact representation to reason about qualitative preferences with uncertainty over alternatives of binary attributes. We defined the semantics of PrLP-trees, proposed for them a compact representation, and identified four important classes of compact PrLP-trees. We studied reasoning with PrLP-trees. We showed that the PROBDOM and PROBRANK problems can be solved in polynomial time for "full" PrLP-trees and for all classes of compact PrLP-trees. We proved that the OPTIMALITY problem is in P for UP trees but is NP-hard for CP trees. Future work includes extending these results to partial lexicographic preference trees [12], algorithms for learning compact PrLP-trees from datasets, and a study of the relationship between PrLP-trees and probability spaces consisting of LP-trees.

Acknowledgements. This work was partially supported by the NSF grant IIS-1618783.

References

1. Bigot, D., Zanuttini, B., Fargier, H., Mengin, J.: Probabilistic conditional preference networks. In: Proceedings of the Conference on Uncertainty in Artificial Intelligence (2013)

2. Booth, R., Chevaleyre, Y., Lang, J., Mengin, J., Sombattheera, C.: Learning conditionally lexicographic preference relations. In: ECAI, pp. 269–274 (2010)
3. Boutilier, C., Brafman, R., Domshlak, C., Hoos, H., Poole, D.: CP-nets: a tool for representing and reasoning with conditional ceteris paribus preference statements. J. Artif. Intell. Res. **21**, 135–191 (2004)
4. Bouveret, S., Endriss, U., Lang, J.: Conditional importance networks: a graphical language for representing ordinal, monotonic preferences over sets of goods. In: IJCAI (2009)
5. Colman, A.M., Stirk, J.A.: Singleton bias and lexicographic preferences among equally valued alternatives. J. Econ. Behav. Org. **40**(4), 337–351 (1999)
6. Cornelio, C., Grandi, U., Goldsmith, J., Mattei, N., Rossi, F., Venable, K.B.: Reasoning with PCP-nets in a multi-agent context. In: AAMAS, pp. 969–977 (2015)
7. De Saint-Cyr, F.D., Lang, J., Schiex, T.: Penalty logic and its link with Dempster-Shafer theory. In: Uncertainty Proceedings 1994, pp. 204–211. Elsevier (1994)
8. Dubois, D., Lang, J., Prade, H.: Possibilistic logic. In: Handbook of Logic in Artificial Intelligence and Logic Programming, vol. 3, pp. 439–513. Oxford University Press (1994)
9. Fishburn, P.C.: Utilities and decision rules: a survey', lexicographic orders. Manag. Sci. **20**, 1442–1471 (1974)
10. Ford, J.K., Schmitt, N., Schechtman, S.L., Hults, B.M., Doherty, M.L.: Process tracing methods: contributions, problems, and neglected research questions. Organ. Behav. Hum. Decis. Process. **43**(1), 75–117 (1989)
11. Lang, J., Mengin, J., Xia, L.: Voting on multi-issue domains with conditionally lexicographic preferences. Artif. Intell. **265**, 18–44 (2018)
12. Liu, X., Truszczynski, M.: Learning partial lexicographic preference trees over combinatorial domains. In: Proceedings of the 29th AAAI Conference on Artificial Intelligence (AAAI), pp. 1539–1545. AAAI Press (2015)
13. Liu, X., Truszczynski, M.: Reasoning with preference trees over combinatorial domains. In: Walsh, T. (ed.) ADT 2015. LNCS, vol. 9346, pp. 19–34. Springer, Cham (2015). https://doi.org/10.1007/978-3-319-23114-3_2
14. Liu, X., Truszczynski, M.: Voting-based ensemble learning for partial lexicographic preference forests over combinatorial domains. Ann. Math. Artif. Intell. **87**(1–2), 137–155 (2019)
15. Lukasiewicz, T., Martinez, M.V., Simari, G.I.: Probabilistic preference logic networks. In: ECAI, pp. 561–566 (2014)
16. Moshkovich, H., Mechitov, A., Olson, D.: Verbal decision analysis. In: Multiple Criteria Decision Analysis: State of the Art Surveys. ISOR, vol. 78, pp. 609–633. Springer, New York (2005). https://doi.org/10.1007/0-387-23081-5_15
17. Westenberg, M.R.M., Koele, P.: Multi-attribute evaluation processes: methodological and conceptual issues. Acta Psychol. **87**(2–3), 65–84 (1994)

Incremental Preference Elicitation with Bipolar Choquet Integrals

Hugo Martin[✉] and Patrice Perny

Sorbonne Université, CNRS, Laboratoire d'Informatique de Paris 6, LIP6,
Paris, France
{hugo.martin,patrice.perny}@lip6.fr

Abstract. Preference modeling and preference learning are crucial issues in multicriteria decision-making to formulate recommendations that are tailored to the Decision Maker. In the field of multicriteria analysis, various aggregation functions have been studied to scalarize performance vectors and compare solutions. Nonetheless, most of these models do not take into account the presence of reference points in the criteria scales. Since it has been observed that decision makers may exhibit different attitudes towards aggregation depending on whether evaluations are above or below reference values, we consider here bipolar extensions of well-known aggregation models and propose incremental preference elicitation methods based on these models. In particular, we consider the elicitation of a 2-additive bipolar Choquet Integral, of a bipolar Weighted Ordered Weighted Average (WOWA), and of a non-weighted bipolar OWA. We propose a general approach that is implemented in all these cases and provide numerical tests showing its practical efficiency.

Keywords: Bipolar Choquet Integral · biWOWA · biOWA · Capacity elicitation

1 Introduction

In the field of multicriteria decision support, various decision models have been proposed to determine the optimal choice within the set of Pareto optimal solutions. The most common approach is to define an overall utility value from any performance vector using an aggregation function synthesizing the advantages and weaknesses of the solution considered. This aggregation function is often parameterized by weighting coefficients allowing to control the relative importance of criteria and possibly their interaction in the aggregation. These parameters must be taylored to the value system of the Decision Maker (DM) to make personalized recommendations.

There are many contributions on preference elicitation in the recent literature, proposing to assess the parameters of a decision model. A first stream of research concerns complete elicitation methods aiming to the determination of precise weighting parameters. This approach requires much preference information from

© Springer Nature Switzerland AG 2021
D. Fotakis and D. Ríos Insua (Eds.): ADT 2021, LNAI 13023, pp. 101–116, 2021.
https://doi.org/10.1007/978-3-030-87756-9_7

the DM but has the advantage to allow the construction of a fully specified decision model that can be used to derive personalized recommendations or predicting choices of the DM in any set of alternatives. Another approach, quite popular in the last decade, is to perform an incremental and adaptive elicitation of preferences. The goal is a bit less ambitious here, it is to obtain a sufficient amount of preference information to be able find the preferred option in a given set of alternatives. This second approach significantly reduces the elicitation burden.

Various incremental elicitation methods have been proposed in the literature. While some of them manage a probability distribution over the set of parameters and use Bayesian revisions to progressively pass from a prior distribution representing ignorance to a more specific distribution concentrated on a subregion of the parameter space (see e.g., [6]), some others proceed by a progressive reduction of the parameter space until the optimal choice can be identified without ambiguity [5,14,26,28]. Both approaches are interesting but we will focus in this paper on the latter approach. In the field of multicriteria decision making, this approach is generally implemented by maintaining a set of possible parameter vectors (named the *uncertainty set* hereafter) defined as a convex polyedron that is progressively reduced as new (linear) constraints appear from new preference statements. This approach obviously applies to weighted sums, but more generally to any scalarizing function linear in its parameter. Assume indeed that the overall utility of any performance vector $x = (x_1, \ldots, x_n)$ is defined by $f_w(x_1, \ldots, x_n)$ where $f_w : \mathbb{R}^n \to \mathbb{R}$ is a scalarizing function parameterized by w and *linear* in w, then any preference statement of type "x is as least as good as y" for any two vectors x and y translates into the constraint $f_w(x) \geq f_w(y)$ which is linear in w.

This approach based on uncertainty sets defined as convex polyedra is not restricted to weighted sums. For instance it can also be applied to rank-dependent aggregation functions such as Ordered Weighted Averages (OWA) [29] and Choquet integrals [13] as shown in [2,3]. In this paper we consider more general rank-dependent decision models recently introduced in multicriteria analysis for preference aggregation with bipolar preferences, in particular biOWA [16] and biChoquet integrals [13,17]. The motivation for this is twofold:

- it has been observed in different contexts that DMs tend to think of outcomes relative to a certain reference point and may exhibit different attitudes towards positive evaluations (i.e. , evaluations above the reference point) and negative evaluations (i.e., evaluations below the reference point) see, e.g., [25]). The biOWA aggregator and more generally any biChoquet integral allow to model such decision behaviors and their parameters must be elicited.
- the descriptive power of bipolar models comes at a cost: bipolarity requires using more weighting parameters to keep the possibility to model different attitudes in the aggregation, depending on whether we are in the positive side or in the negative side of the evaluation space. Therefore the elicitation process is more demanding in terms of preference information and there is a

need of testing the practical feasibility of incremental elicitation methods on such models.

The aim of this paper is to propose a preference elicitation method based on the biChoquet integral for multicriteria evaluation with bipolar scales. Our approach is to progressively specify the two capacities used in the model by an iterative reduction of their uncertainty sets using preference queries. We will implement this approach on the general biChoquet model and also on specific subclasses, for interactive decision support on explicit sets. The paper is organized as follows: Sect. 2 introduces some background on biChoquet integrals and on regret-based incremental elicitation methods. Section 3 introduces an incremental elicitation algorithm for the case of 2-additive biChoquet integrals. We then propose an adaptation of our elicitation algorithm to the case of bipolar weighted ordered weighted averages (biWOWA, Sect. 4), which is then further specialized in the case of bipolar weighted ordered averages (biOWA, Sect. 5). In all cases, we provide the results of numerical tests to show their performance both in terms of computation time and number of generated preference queries.

2 Background

2.1 Choquet and BiChoquet Integrals

Let $N = \{1, \ldots, n\}$ denote the set of criteria under consideration to assess the performance of a solution in the decision problem. We assume that any feasible solution is characterized by a performance vector $x = (x_1, \ldots, x_n)$ where x_i represents the value of x w.r.t. the ith criterion. In order to model the preferences of the DM we consider here the Choquet integral which is a widely-used model in decision theory [9,21] with various applications in multicriteria decision making [7,10,15,23] and AI [1,3,8,22].

The Choquet integral is a kind of weighted aggregation operator where weights are not only assigned to every criteria but also to groups of criteria. This enables to model positive or negative interactions among criteria, giving enhanced descriptive possibilities compared to linear models. The weights are defined using a set function named *capacity* and defined as follows;

Definition 1. *A capacity on N is a set function $v : 2^N \rightarrow [0, 1]$ such that $v(\emptyset) = 0$ and for all $A, B \subseteq N, A \subseteq B \Rightarrow v(A) \leq v(B)$.*

Throughout the paper we will always assume that the capacities under consideration are normalized, i.e., $v(N) = 1$. Then the Choquet integral can be defined from any capacity as follows:

Definition 2. *For any vector $x = (x_1, \ldots, x_n) \in \mathbb{R}^n$, the Choquet integral w.r.t. capacity v is a scalarizing function $C_v : \mathbb{R}^n \to \mathbb{R}$ defined by:*

$$C_v(x) = \sum_{i=1}^{n} [v(X_{(i)}) - v(X_{(i+1)})] x_{(i)} \tag{1}$$

$$= \sum_{i=1}^{n} [x_{(i)} - x_{(i-1)}] v(X_{(i)}) \tag{2}$$

where (.) is any permutation such that $x_{(1)} \leq \ldots \leq x_{(n)}$, and $X_{(i)} = \{x_{(i)}, \ldots, x_{(n)}\}$ is the set of objectives where the performance is at least as good as $x_{(i)}$, for $i = 1, \ldots, n$. Furthermore we assume that $x_{(0)} = 0$ and $X_{(n+1)} = \emptyset$.

Given a capacity v we can define the *dual* capacity by $\bar{v}(A) = 1 - v(N \backslash A)$. For any vector $x \in \mathbb{R}^n$, we have: $C_v(x) = -C_{\bar{v}}(-x)$. Let us give an example of the use of the Choquet integral in preference modelling.

Example 1. *Let $X = \{a, b, c\}$ be a set of alternatives evaluated according to 3 criteria as follows:*

	a	b	c
criterion 1	-2	-1	0
criterion 2	5	2	-2
criterion 3	0	1	5

Assume that the DM prefers b to a and c. One can easily check that such a preference is not representable by a weighted linear aggregator. However, it is easily representable by a Choquet integral. For instance, let us consider the following capacity (Table 1):

Table 1. Capacity of the decision maker in Example 1

	$\emptyset$	$\{1\}$	$\{2\}$	$\{3\}$	$\{1,2\}$	$\{1,3\}$	$\{2,3\}$	$\{1,2,3\}$
v	0	0.1	0.2	0.1	0.4	0.4	0.4	1

We have then $C_v(a) = (-2) \; v(N) + [0 - (-2)] \; v(\{2,3\}) + [5 - 0] \; v(\{2\}) = -2 + 1 + 1 = 0$, $C_v(b) = (-1) \; v(N) + [1 - (-1)] \; v(\{2,3\}) + [2 - 1] \; v(\{2\}) = -1 + 1 + 0.3 = 0.3$ and $C_v(c) = -2 + 0.8 + 0.5 = -0.7$, which implies that b is preferred to a and c.

Despite its descriptive appeal, the Choquet integral has itself some descriptive limits, especially when the DM uses different aggregation logics in the positive and in the negative part of the utility scale, as illustrated in Example 2.

	a	b	c	d
criterion 1	2	3	−3	−4
criterion 2	5	3	−3	−1

Example 2. *Let $X = \{a, b, c, d\}$ be a set of alternatives evaluated according to 2 criteria as follows:*

Assume that the value system of the DM is as follows: when performances are positive she wants to maximize the average performance. However, when some performances are negative, she adopts a more cautious behavior towards losses and favors a solution having a more balanced profile. Hence her preference order could be a, b, c, d. Any representation of this preference order by the Choquet integral should satisfy: $x \succ y \Leftrightarrow C_v(x) > C_v(y)$ for all $x, y \in X$.

We have $a \succ b$, therefore:

$$2 + 3\, v(\{2\}) > 3$$
$$\Leftrightarrow 3\, v(\{2\}) > 1$$
$$\Leftrightarrow v(\{2\}) > 1/3$$

We have $c \succ d$, therefore:

$$-4 + 3\, v(\{2\}) < -3$$
$$\Leftrightarrow 3\, v(\{2\}) < 1$$
$$\Leftrightarrow v(\{2\}) < 1/3$$

The obtained contradiction demonstrates that no capacity v exists to represent the prescribed ranking. Therefore, the Choquet integral cannot model these preferences.

The observation of such behaviors motivated the development of models able to capture preferences that may vary depending on the position of the performances relatively to some reference values. In this paper we will assume that a vector of reference values $p = (p_1, \ldots, p_n)$ is known where p_j is a neutral evaluation on criterion j, separating the good and the bad part of the scale. For the rest of this paper, when referring to a solution $x = (x_1, \ldots, x_n)$, we will consider that it has already been centered on p, (i.e., $x = x'$ - p where x' is the original solution vector). Hence, 0 becomes the neutral value on all criteria scales. The existence of such bipolar scales has motivated the introduction of the following extension of the Choquet integral, defined for criterion values expressed on a bipolar scale [11, 12, 17].

Definition 3. *Let $x \in \mathbb{R}^n$ and u and v be two capacities. The bipolar extension of the Choquet (biChoquet integral for short) is defined as follows:*

$$C_{u,v}(x) = C_u(x^+) - C_v(x^-) \tag{3}$$

where $x^+ = \max(x, 0)$ and $x^- = \max(-x, 0)$.

If we reinterpret Example 2 with this model, We obtain $u(\{2\}) > 2/6$ and $v(2) < 2/6$ which is no longer contradictory. Actually this model can easily describe the preference order given in Example 2 due to the combination of two capacities, one for the positive side and the other for negative side. This general model includes several interesting subclasses. For example, when $u(A) = \varphi(\sum_{i \in A} p_i)$ and $v(A) = \psi(\sum_{i \in A} p_i)$ for some functions φ and ψ strictly increasing on the unit interval and such that $\varphi(0) = \psi(0) = 0$ and $\varphi(1) = \psi(1) = 1$, then the biChoquet integral is nothing else but the model proposed by Kahneman and Tversky in their Cumulative Prospect Theory (CPT) [25]. In the context of CPT, p_i represents the probability of a state i. If we import the CPT model in the context of multicriteria aggregation, p_i must be interpreted as the weight of criterion i and we obtain a bipolar version of the WOWA operator introduced by Torra [24] (when all criterion values are positive, we exactly obtain a WOWA). If we further specify the model by setting $p_i = 1/n$ then capacities u and v are symmetric (their values only depend on the cardinality of the set) and the resulting model is known as biOWA [16]. We will come back to these models in the following sections.

Until now, we have made the assumption that preferences were known. This is a strong hypothesis and, in practice, the parameters of these models must be elicited. We now recall some background on incremental elicitation methods based on the minimisation of regrets.

2.2   Elicitation Based on Regret Minimization

We consider an aggregation function f_w where w is the unknown weighting vector used in the model used to represent DM's preferences. When no preference information is available the uncertainty set defined as the set of all admissible weighting vectors is defined by $\Omega = \{w \in \mathbb{R}_+^\kappa, \text{such that } \sum_{i=1}^n w_i = 1\}$. When a set P of preference statements is eventually observed (under the form of a list of ordered pairs of alternatives where the first is preferred to the second), the initial set Ω can be reduced to a subset denoted Ω_P using the linear constraints induced by the preferences in P. Hence Ω_P is a convex polyhedron, at any step of the elicitation process.

Given an uncertainty set Ω_P, an alternative is said to be necessarily optimal in X, if $f_w(x) \geq f_w(y)$ for all $y \in X$ and all $w \in \Omega_P$. In this context, the goal of an incremental preference elicitation method is to iteratively generate preference queries to collect preference statements and further restrict Ω_P until a necessarily optimal element can be identified in X. In order to generate informative preference queries and to identify a necessarily optimal element as soon as possible, we can use the notion of max-regret as suggested in [27]. Let us recall the definition of regrets used in the elicitation process:

For two alternatives x and y, the Pairwise Max Regret (which quantifies the regret of choosing x instead of y) is defined by:

$$\text{PMR}(x, y, \Omega_P) = \max_{w \in \Omega_P} (f_w(y) - f_w(x)) \tag{4}$$

Then, the Max Regret attached to a solution $x \in X$ is defined by:

$$\text{MR}(x, \Omega_P) = \max_{y \in \chi} \text{PMR}(x, y, \Omega_P) \tag{5}$$

The MinMaxRegret (MMR) is then the minimal value of the MaxRegret for all elements in X.

$$\text{MMR}(x, \Omega_P) = \min_{x \in \chi} \text{MR}(x, \Omega_P) \tag{6}$$

A necessarily optimal solution has a Max Regret value of 0. Hence, we have to collect preference statements until the MMR drops to 0. Preference queries are generated using the current solution strategy that consists in comparing a solution x^* having a minimal MR value to its strongest challenger, i.e., any solution y^* maximizing $PMR(x^*, y^*, \Omega_P)$. In practice, to save a significant number of queries, we can stop the process when MMR drops below a given $\epsilon > 0$ without loosing much in the quality of the returned solution. When the set of alternatives X is finite and defined explicitly, this general elicitation process interleaving preference queries and the exploration of the set of alternatives is formalized in Algorithm 1. The computation of the PMR is specific to each aggregation function and will be discussed later in the paper.

Result: x^*: a necessarily optimal alternative
initialization;
$X = \{x_1, \dots, x_m\}$, $\Omega_P = \Omega$;
do
    for $x \in X$ **do**
        for $y \in X$ **do**
            | Compute PMR(x, y, Ω_P);
        end
        Update MR(x, Ω_P);
    end
    x^*, mmr = Update MMR;
    $(a, b) :=$ Select a query for the DM in X ;
    preference$(a, b) :=$ Ask (a, b) to the DM ;
    Update$(\Omega_P, \text{preference}(a, b))$;
while $mmr \geq \epsilon$;
Return x^*

Algorithm 1: Elicitation of preferences

This algorithm can be used to incrementally elicit the capacities u and v in $C_{u,v}$ (Eq. 3). In this case, the aggregator f_w is the biChoquet integral and its parameter w is defined by the pair of capacities u, v. In the following sections, we introduce some computational models based on linear programming to efficiently obtain the PMR values in the simultaneous elicitation of u and v. We successively consider three different families of instances of biChoquet integrals.

3 Elicitation of a 2-Additive BiChoquet Integral

Capacities u and v are useful mathematical functions to model the interactions among criteria but their definition or approximation would require to work with

$2(2^n - 2)$ weighting coefficients where n is the number of criteria. In order to introduce more compact representations of interactions while keeping some flexibility in the model, we use the Möbius inverse of the capacities. Given a capacity v, the Möbius inverse of v is defined by $m_v(A) = \sum_{B \subseteq A} (-1)^{|A \setminus B|} v(B)$ for all $A \subseteq N$. Then, $C_{u,v}$ can be rewritten from m_u and m_v as follows:

Proposition 1. *Let u and v be two capacities and x a performance vector, we have:*

$$C_{u,v}(x) = \sum_{A \subseteq N} m_u(A) \min_{i \in A} x_i^+ - \sum_{A \subseteq N} m_{\bar{v}}(A) \max_{i \in A} x_i^- \qquad (7)$$

with m_u the Möbius inverse of u and $m_{\bar{v}}$ the Möbius inverse of $\bar{v}$ the dual capacity of v, $x^+ = \max(x, 0)$ and $x^- = \max(-x, 0)$.

Proof. We recall that $C_v(x) = -C_{\bar{v}}(-x)$ and that $C_v(x) = \sum_{A \subseteq N} m(A) \min_{i \in A} x_i$. We have then $C_{u,v}(x) = C_u(x^+) - C_v(x^-) = C_u(x^+) - (-C_{\bar{v}}(-x^-))$ and therefore $C_{u,v}(x) = C_u(x^+) - (-\sum_{A \subseteq N} m_{\bar{v}}(A) \min_{i \in A}(-x_i^-)) = \sum_{A \subseteq N} m_u(A) \min_{i \in A} x_i^+ - \sum_{A \subseteq N} m_{\bar{v}}(A) \max_{i \in A} x_i^+$.

This formulation suggests that more compact representations of subclasses of biChoquet integrals can easily be obtained. We can indeed restrict u and $\bar{v}$ to k-additive capacities. A capacity u is said to be k-additive if and only if its Möbius inverse m_u verifies for all $A \subseteq N, |A| > k, m_u(A) = 0$ and it exists at least $B \subseteq N, |B| = k$ such as $m_u(B) \neq 0$. For example, a 2-additive capacity is completely characterized by $n(n+1)/2$ Möbius masses (one for every singleton and every pair). Such a capacity makes it possible to model non-linearities due to pairwise interactions between pairs of criteria while involving only a polynomial number of parameters. Moreover, by restricting u and v to 2-additive capacities, we can exploit the following result [18]:

Proposition 2. *We set $Q = \{A \subseteq N, 1 \leq |A| \leq 2\}$ and $Q' = \{B \subseteq N, |B| = 2\}$. The class of 2-additive capacities forms a convex polytope whose extreme points are of two types:*

- *For all $A \in Q$, we define the extreme point M_A as, for all $X \subseteq N$, $M_A(X) = 1$ if $X = A$, 0 otherwise.*
- *For all $B \in Q'$, we define the extreme point M'_B as, for all $X \subseteq N$, $M'_B(X) = -1$ if $X = B$, 1 if $\emptyset \neq X \subset B$, 0 otherwise.*

Every 2-additive capacity has then its Möbius inverse m defined as a convex combination of those extreme points:

$$m = \sum_{A \in Q} \alpha_A \cdot M_A + \sum_{B \in Q'} \alpha'_B \cdot M'_B$$

with $\forall A \in Q, \alpha_A \geq 0, \forall B \in Q', \alpha'_B \geq 0$ and $\sum_{A \in Q} \alpha_A + \sum_{B \in Q'} \alpha'_B = 1$. Therefore, every 2-additive capacity is defined by an unique positive vector of size $2 \times \binom{n}{2} + n$, formed by the concatenation of α and α'. In our context, we consider two 2-additive capacities u and $\bar{v}$ and their Möbius inverse m_u and $m_{\bar{v}}$. Their coefficients in the combination of extreme points of the polytope will be denoted $(\alpha_A^u, \alpha_B'^u)$ and $(\alpha_A^{\bar{v}}, \alpha_B'^{\bar{v}})$ in the sequel. Using the previous notions and definitions, we present the following linear program to compute the PMR between x and y, with a set of possible parameters Ω_P, defined as the set of all possible 2-additives capacities u and $\bar{v}$ characterized by their Möbius masses m_u and $m_{\bar{v}}$ and described by variables $\alpha_A^u, \alpha_B'^u, \alpha_A^{\bar{v}}, \alpha_B'^{\bar{v}}$.

$$\max \sum_{A \subseteq Q} m_u(A)(\min_{i \in A} y_i^+ - \min_{i \in A} x_i^+) - \sum_{A \subseteq Q} m_{\bar{v}}(A)(\max_{i \in A} y_i^- - \max_{i \in A} x_i^-)$$

$$(\mathcal{P}_1) \begin{cases} m_u(X) = \sum_{A \in Q} \alpha_A^u M_A(X) + \sum_{B \in Q'} \alpha_B'^u M_B'(X), \; \forall X \subseteq Q \\ m_{\bar{v}}(X) = \sum_{A \in Q} \alpha_A^{\bar{v}} M_A(X) + \sum_{B \in Q'} \alpha_B'^{\bar{v}} M_B'(X), \; \forall X \subseteq Q \\ \sum_{X \in Q} \alpha_X^u + \sum_{X \in Q'} \alpha_X'^u = 1 \\ \sum_{X \in Q} \alpha_X^{\bar{v}} + \sum_{X \in Q'} \alpha_A'^{\bar{v}} = 1 \\ \sum_{A \subseteq Q} m_u(A) \min_{a_i^+ \in A} a_i^+ - m_{\bar{v}}(A) \max_{a_i^- \in A} a_i^- \geq \\ \quad \sum_{A \subseteq Q} m_u(A) \min_{b_i^+ \in A} b_i^+ - m_{\bar{v}}(A) \max_{b_i^- \in A} b_i^-, \; \forall (a,b) \in \Omega_P \\ m_u(A), \; m_{\bar{v}}(A) \geq 0, \; \forall A \subseteq N \\ \alpha_A^u, \alpha_A^{\bar{v}} \geq 0, \; \forall A \subseteq Q, \;\; \alpha_B'^u, \alpha_B'^{\bar{v}} \geq 0, \; \forall B \subseteq Q' \end{cases}$$

For any two 2-additive capacities u and $\bar{v}$, this linear program has $6\binom{n}{2}+3n$ continuous variables and $2\binom{n}{2} + 2n + |P|$ constraints, with $|P|$ the number of added constraints induced by preferences statements. We implemented Algorithm 1 to elicit u and v in $C_{u,v}$ using program $\mathcal{P}_1$ for the computation of PMR values. To run our tests, we used Gurobi 8.1.1 solver, a cluster of computers with 252 GB of RAM and 32 Intel(R) Xeon(R) CPU E5-2630 v3 @ 2.40 GHz processors. Our elicitation algorithm has been tested on randomly generated instances with capacities randomly drawn using Proposition 2. Alternatives are randomly sampled to

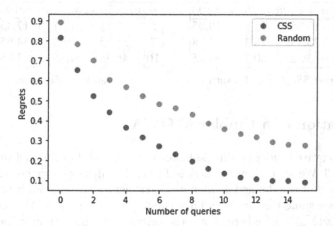

Fig. 1. Comparison of CSS and Random strategies - Average regrets evolution - 5 criteria and 25 alternatives

form a Pareto-front with a significant part of unsupported solutions (which do not belong to the frontier of the convex envelope and thus that cannot be considered optimal by a linear model). For each instance, the elicitation sequence is iterated until we observe a decrease of the MMR value of at least 90% ($\epsilon = 0.1$). The following Figure compares the evolution of the regret (i.e., the MMR value) throughout the elicitation process, for strategies CSS and Random (in the latter, the preference query is randomly selected at each step) (Fig. 1).

The curves reflecting the decay of MMR values show that when $\epsilon = 0.1$, the elicitation algorithm stops after about 15 preference queries with the CSS whereas much more queries would be necessarily for the Random strategy. As observed for other models, CSS appears to be effective to find informative queries reducing regrets. Through the rest of this paper, we will focus on this query strategy to perform tests with different families of bipolar models.

Now, we report the average computation times and the average number of preference queries used to solve instances of different sizes (n the number of criteria varies from 5 to 10 and m, the size of the set of alternatives, varies from 50 to 100).

Even though the number of parameters is two times greater than in the monopolar case (standard 2-additive Choquet integrals), we observe that the solution times of Algorithm 1 remain reasonably low. Actually, they are in the same order of magnitude that the computation times we use to obtain for standard 2-additive Choquet integrals. Moreover, we observe that the elicitation cost in terms of number of preference queries asked to the DM does not increase drastically when considering the bipolar extension of the Choquet Integral.

n	$m = 50$	$m = 75$	$m = 100$
5	2.85	7.1	9.77
7	4.6	10.63	19.43
10	7.42	21	40.07

(a) CSS-times (s) - Choquet

n	$m = 50$	$m = 75$	$m = 100$
5	5.12	9.37	17.27
7	8.93	18.56	33.64
10	14.63	35.98	70.48

(b) CSS-times (s) - BiChoquet

n	$m = 50$	$m = 75$	$m = 100$
5	13.95	15.8	19.05
7	23.35	28.3	27.8
10	39.2	50.7	54.35

(a) CSS-queries - Choquet

n	$m = 50$	$m = 75$	$m = 100$
5	16.15	17.1	17.35
7	31.45	35.5	34.95
10	46.35	64.85	78.5

(b) CSS-queries - BiChoquet

4   Elicitation of a Bipolar WOWA

In this Section we consider another subclass of the general biChoquet model introduced in Eq. 3. We are no longer restricted to two additive capacities but consider all capacities that are defined as monotonic transformed of an additive measure. Formally, we assume that u and v have the following forms: $u(A) = \varphi(\sum_{i \in A} p_i)$ and $v(A) = \psi(\sum_{i \in A} p_i)$ where p_i are the criteria weights. As mentioned at the end of Subsect. 2.1, the resulting subclass of biChoquet functions is a counter-part of CPT in the setting of multicriteria aggregation. The aggregators in this family can also be seen as bipolar extensions of WOWA (the weighted extension of OWA proposed in [24]). For this reason, these operators are named biWOWA hereafter. More formally they are defined as follows:

Definition 4. *Let $x \in \mathbb{R}^n$ be a performance vector, $p \in \mathbb{R}^n$ an importance vector over the set of criteria, φ and ψ two increasing functions with $\varphi(0) = \psi(0) = 0$ and $\varphi(1) = \psi(1) = 1$. The Bipolar Ordered Weighted Averaging operator (biWOWA) is defined as the aggregation function $f_{\varphi,\psi} : \mathbb{R}^n \to \mathbb{R}$ such that:*

$$f_{\varphi,\psi}(x) = \sum_{i=1}^{n} \varphi(\sum_{k=i}^{n} p_{(k)})[x_{(i)}^+ - x_{(i-1)}^+] - \sum_{i=1}^{n} \psi(\sum_{k=i}^{n} p_{(k)})[x_{(i)}^- - x_{(i-1)}^-]$$

with $x^+ = \max\{x,0\}$, $x^- = \max\{-x,0\}$ and $(.)$ the permutation of criteria which sorts x in the increasing order.

We assume here that the weighting vector p is known and we focus on the elicitation of φ and ψ. This is a challenging problem because we have to consider a continuous set of non-linear increasing functions. To overcome this difficulty, we use a spline representation of φ and ψ. Spline functions are piecewise polynomials whose elements connect with a high level of smoothness. Further details on spline functions can be found in [19,20], but an interesting property of these functions is that they can be generated with a linear combination of basis monotonic spline functions. This allows to reduce the elicitation of φ and ψ to their corresponding weighting vectors b^φ and b^ψ in the spline basis. More precisely, we have $\varphi(x) = \sum_{j=1}^{r} b_j^\varphi I_j(x)$ and $\psi(x) = \sum_{j=1}^{r} b_j^\psi I_j(x)$, where $I_j(x), j = \{1,\ldots,l\}$ are the basic monotonic spline functions (see Fig. 2). A similar approach, based on the use of spline functions, has been proposed for the WOWA model in [4] on the robust assignment problem.

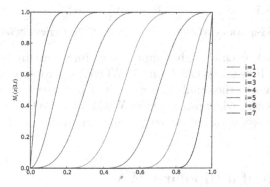

Fig. 2. I-spline basis of order 3

In order to compute PMR(x, y) for two alternatives x and y when preferences are represented with the biWOWA model, we propose the linear program $\mathcal{P}_2$.

For any two increasing functions φ and ψ, this linear program has $2r$ continuous variables and $2l + |P|$ constraints, with $|P|$ the number of added constraints induced by preferences statements and l the size of the spline functions basis.

$$\max \; [\sum_{i=0}^{n}[\sum_{j=1}^{l} b^{\varphi}(j)I_j(\sum_{k=i}^{n} p_{(k)}))](y_{(i)}^+ - y_{(i-1)}^+) - [\sum_{j=1}^{l} b^{\varphi}(j)I_j(\sum_{k=i}^{n} p_{(k)}))](x_{(i)}^+ - x_{(i-1)}^+)]$$

$$-[\sum_{i=0}^{n}[\sum_{j=1}^{l} b^{\psi}(j)I_j(\sum_{k=i}^{n} p_{(k)}))](y_{(i)}^- - y_{(i-1)}^-) - [\sum_{j=1}^{l} b^{\psi}(j)I_j(\sum_{k=i}^{n} p_{(k)}))](x_{(i)}^- - x_{(i-1)}^-))]$$

$$(\mathcal{P}_2) \begin{cases} \sum_{j=1}^{l} b^{\varphi}(j) = 1 \\ \sum_{j=1}^{l} b^{\psi}(j) = 1 \\ f_{\varphi,\psi}(a) \geq f_{\varphi,\psi}(b), \; \forall(a,b) \in \Omega_P \\ b^{\varphi}(j), b^{\psi}(j) \geq 0, \; j = 1, \dots, l \end{cases}$$

We implemented Algorithm 1 using program $\mathcal{P}_2$ to compute the PMR values to elicit φ and ψ in $f_{\varphi,\psi}$. In our tests, we used the experimental setting of Sect. 3. We simulated the DM's answers to preferences queries using $f_{\varphi',\psi'}$ where φ' and ψ' were randomly drawn using the basis of spline functions. The execution times and the number of queries asked to the DM are given in the tables below;

n	$m = 50$	$m = 75$	$m = 100$
5	1.51	2.75	3.71
7	2.05	3.55	5.61
10	3.6	5.15	6.34

(a) CSS-times (s) - WOWA

n	$m = 50$	$m = 75$	$m = 100$
5	3.43	5.89	9.89
7	6.42	8.08	18.66
10	12.45	21.40	37.25

(b) CSS-times (s) - biWOWA

n	$m = 50$	$m = 75$	$m = 100$
5	4.75	4.65	4.8
7	4.7	5.6	4.75
10	5.5	5.25	5.2

(a) CSS-queries - WOWA

n	$m = 50$	$m = 75$	$m = 100$
5	5.35	4.8	5.4
7	5.8	5.2	5.8
10	6.5	6.2	6.55

(b) CSS-queries - biWOWA

We observe that, even though computation times remain low in the bipolar case, the increase between WOWA and BiWOWA operator is significant. However, when it comes to the number of generated preference queries, we observe only a slight increment when passing from WOWA to biWOWA. The increase of computation time seems to be related to the computation of PMR using linear program $\mathcal{P}_2$.

5 Elicitation of a Bipolar OWA

In this section we consider the bipolar Ordered Weighted Averaging (biOWA), introduced in [16] to generalize OWA to the bipolar case. As mentioned before, this is an instance of the biWOWA obtained when the criteria weights are equal ($p_i = 1/n$). This instance can be elicited by the general method proposed for biWOWA but we are going to introduce a more specific and direct method, taking advantage of the fact that biOWA admits a much simpler formulation when all criteria weights have the same value.

Definition 5. *Let $x \in \mathbb{R}^n$ be a performance vector and $\alpha, \beta \in \mathbb{R}^n_+$ two weighting vectors, the* bipolar ordered weighted averaging *(biOWA) is the aggregation function $g_{\alpha,\beta} : \mathbb{R}^n \to \mathbb{R}$ defined by:*

$$g_{\alpha,\beta}(x) = \alpha \cdot x^+_\uparrow - \beta \cdot x^-_\downarrow \tag{8}$$

with $x^+ = \max\{x, 0\}$, $x^- = \max\{-x, 0\}$ and $x_\uparrow$ (resp. $x_\downarrow$) is the vector obtained from x by rearranging its components in the increasing (resp. decreasing) order.

In this case, the parameter of the model is defined by the pair α, β of weighting vectors defining how the DM focuses on good and bad positive (resp. negative) evaluations. As in the OWA operator, these weights are not attached to criterion values but to their rank in the ordered list of positive (resp. negative) criterion values. In order to compute the PMR values for this model, we use program $(\mathcal{P}_3)$ given below:

$$\max \sum_{i=1}^n \alpha_i \, (y^+_{(i)} - x^+_{(i)}) - \beta_i \, (y^-_{(i)} - x^-_{(i)})$$

$$(\mathcal{P}_3) \begin{cases} \sum_{i=1}^n \alpha_i = 1 \\ \sum_{i=1}^n \beta_i = 1 \\ \sum_{i=1}^n \alpha_i \, a^+_{(i)} - \beta_i \, a^-_{(i)} \geq \sum_{i=1}^n \alpha_i \, b^+_{(i)} - \beta_i \, b^-_{(i)}, \ \forall (a,b) \in \Omega_P \\ \alpha_i, \beta_i \geq 0, \ \forall i \in \{1, .., n\} \end{cases}$$

For any two weighting vectors α and β, this linear program has $2n$ continuous variables and $2 + |P|$ constraints, with $|P|$ the number of added constraints induced by preferences statements. We implemented Algorithm 1 using program $\mathcal{P}_3$ to compute PMR values and elicit a biOWA. We also implemented the method for the elicitation of a standard OWA in order to compare execution times and the number of asked queries. To run our tests, we used the same experimental setting as before and the DM's answers were simulated using $g_{\alpha',\beta'}$ where α' and β' were randomly drawn weighting vectors. The results of the tests are given in the tables hereafter.

n	$m = 50$	$m = 75$	$m = 100$
5	0.27	0.41	0.57
7	0.28	0.48	0.87
10	0.31	0.66	0.82

(a) CSS-times (s) - OWA

n	$m = 50$	$m = 75$	$m = 100$
5	0.45	0.61	1.42
7	0.55	0.96	1.45
10	0.61	1.27	1.65

(b) CSS-times (s) - biOWA

n	$m = 50$	$m = 75$	$m = 100$
5	5.65	5.8	5.65
7	7.25	7.45	8.05
10	8.55	9.75	10.0

(a) CSS-queries - OWA

n	$m = 50$	$m = 75$	$m = 100$
5	16.1	6.0	6.3
7	9.05	8.9	8.45
10	11.95	13.15	11.7

(b) CSS-queries - biOWA

As in Sect. 3, we observe that computation times are at most two times more important when passing from OWA to biOWA. This good result can be explained by the partial elicitation of preferences and the efficient computation of PMR values due to program $\mathcal{P}_3$. The number of preference queries is similar for OWA and biOWA, which is an encouraging result for the use of biWOWA in other incremental elicitation contexts (e.g., when the alternatives are defined implicitly).

6 Conclusion

Preferences modeling and learning in multicriteria decision-making problems are crucial issues. Moreover, bipolar models are gaining importance in the field of decision theory to overcome descriptive limitations of usual aggregation functions when a reference point must be considered. For these reasons, we have proposed new computational models to perform an incremental elicitation of preferences based on a bipolar rank-dependent model. We applied our approach to biOWA, biWOWA and 2-additive biChoquet integrals, which extends our previous contributions on the elicitation of monopolar models (OWA, WOWA and Choquet).

The elicitation methods proposed here and the tests performed concern the case where the set of alternatives is defined explicitly. A natural continuation of this work would be to extend the approach to sets of alternatives that are implicitly defined (e.g., for preference-based combinatorial optimization). Another extension could be to consider the elicitation of the biChoquet model for more general classes of capacities, and the elicitation of bi-capacities for the Choquet model. The major challenge will be the increased number of parameters to be learned in the model and the efficient computation of PMR values.

Acknowledgement. We wish to thank anonymous reviewers for their useful comments on a preliminary version.

References

1. Beliakov, G., Calvo, T., James, S.: Aggregation functions for recommender systems. In: Ricci, F., Rokach, L., Shapira, B. (eds.) Recommender Systems Handbook, pp. 777–808. Springer, Boston (2015). https://doi.org/10.1007/978-1-4899-7637-6_23
2. Benabbou, N., Gonzales, C., Perny, P., Viappiani, P.: Minimax regret approaches for preference elicitation with rank-dependent aggregators. EURO J. Decis. Process. **3**(1), 29–64 (2015)
3. Benabbou, N., Perny, P., Viappiani, P.: Incremental elicitation of Choquet capacities for multicriteria choice, ranking and sorting problems. Artif. Intell. **246**, 152–180 (2017)
4. Bourdache, N., Perny, P.: Anytime algorithms for adaptive robust optimization with OWA and WOWA. In: Rothe, J. (ed.) Algorithmic Decision Theory. ADT 2017. LNCS, vol. 10576, pp. 93–107. Springer, Cham (2017). https://doi.org/10.1007/978-3-319-67504-6_7
5. Braziunas, D., Boutilier, C.: Minimax regret based elicitation of generalized additive utilities. In: UAI, pp. 25–32 (2007)
6. Chajewska, U., Koller, D., Parr, R.: Making rational decisions using adaptive utility elicitation. In: AAAI, pp. 363–369 (2000)
7. Galand, L., Perny, P., Spanjaard, O.: Choquet-based optimisation in multiobjective shortest path and spanning tree problems. Eur. J. Oper. Res. **204**(2), 303–315 (2010)
8. Gonzales, C., Perny, P., Dubus, J.P.: Decision making with multiple objectives using gai networks. Artif. Intell. **175**(7–8), 1153–1179 (2011)
9. Grabisch, M.: The application of fuzzy integrals in multicriteria decision making. Eur. J. Oper. Res. **89**(3), 445–456 (1996)
10. Grabisch, M., Labreuche, C.: A decade of application of the Choquet and Sugeno integrals in multi-criteria decision aid. Ann. Oper. Res. **175**(1), 247–286 (2010)
11. Grabisch, M., Labreuche, C.: Bi-capacities - I: definition, Möbius transform and interaction. Fuzzy Sets Syst. **151**(2), 211–236 (2005)
12. Grabisch, M., Labreuche, C.: Bi-capacities - II: the Choquet integral. Fuzzy Sets Syst. **151**(2), 237–259 (2005)
13. Grabisch, M., Marichal, J.L., Mesiar, R., Pap, E.: Aggregation Functions, vol. 127. Cambridge University Press, Cambridge (2009)
14. Ha, V., Haddawy, P.: Problem-focused incremental elicitation of multi-attribute utility models. In: UAI, pp. 215–222 (1997)
15. Labreuche, C., Grabisch, M.: The Choquet integral for the aggregation of interval scales in multicriteria decision making. Fuzzy Sets Syst. **137**(1), 11–26 (2003)
16. Martin, H., Perny, P.: BiOWA for preference aggregation with bipolar scales: application to fair optimization in combinatorial domains. In: Kraus, S. (ed.) Proceedings of the Twenty-Eighth International Joint Conference on Artificial Intelligence, IJCAI 2019, Macao, China, 10–16 August 2019, pp. 1822–1828 (2019)
17. Martin, H., Perny, P.: New computational models for the Choquet integral. In: ECAI 2020 - 24th European Conference on Artificial Intelligence. Frontiers in Artificial Intelligence and Applications, vol. 325, pp. 147–154 (2020)
18. Miranda, P., Combarro, E.F., Gil, P.: Extreme points of some families of nonadditive measures. Eur. J. Oper. Res. **174**(3), 1865–1884 (2006)
19. Perny, P., Viappiani, P., Boukhatem, A.: Incremental preference elicitation for decision making under risk with the rank-dependent utility model. In: Uncertainty in Artificial Intelligence (2016)

20. Ramsay, J.O., et al.: Monotone regression splines in action. Stat. Sci. **3**(4), 425–441 (1988)
21. Schmeidler, D.: Integral representation without additivity. Proc. Am. Math. Soc. **97**(2), 255–261 (1986)
22. Tehrani, A.F., Cheng, W., Dembczynski, K., Hüllermeier, E.: Learning monotone nonlinear models using the Choquet integral. Mach. Learn. **89**(1–2), 183–211 (2012)
23. Timonin, M.: Maximization of the Choquet integral over a convex set and its application to resource allocation problems. Ann. Oper. Res. **196**(1), 543–579 (2012)
24. Torra, V.: The weighted OWA operator. Int. J. Intell. Syst. **12**(2), 153–166 (1997)
25. Tversky, A., Kahneman, D.: Advances in prospect theory: cumulative representation of uncertainty. J. Risk Uncertain. **5**(4), 297–323 (1992)
26. Wang, T., Boutilier, C.: Incremental utility elicitation with the minimax regret decision criterion, pp. 309–316 (2003)
27. Wang, T., Boutilier, C.: Incremental utility elicitation with the minimax regret decision criterion. In: IJCAI, vol. 3, pp. 309–316 (2003)
28. White III, C.C., Sage, A.P., Dozono, S.: A model of multiattribute decisionmaking and trade-off weight determination under uncertainty. IEEE Trans. Syst. Man Cybern. **14**(2), 223–229 (1984)
29. Yager, R.R.: On ordered weighted averaging aggregation operators in multicriteria decisionmaking. IEEE Trans. Syst. Man Cybern. **18**(1), 183–190 (1988)

Preference Aggregation and Voting

In the Beginning There Were n Agents: Founding and Amending a Constitution

Ben Abramowitz[1,2]($\boxtimes$), Ehud Shapiro[2], and Nimrod Talmon[3]

[1] Rensselaer Polytechnic Institute, Troy, NY 12180, USA
abramb@rpi.edu
[2] Weizmann Institute of Science, Rehovot, Israel
ehud.shaprio@weizmann.ac.il
[3] Ben-Gurion University, Be'er Sheva, Israel
talmonn@bgu.ac.il

Abstract. Consider n agents forming an egalitarian, self-governed community. Their first task is to decide on a decision rule to make further decisions. We start from a rather general initial agreement on the decision-making process based upon a set of intuitive and self-evident axioms, as well as simplifying assumptions about the preferences of the agents. From these humble beginnings we derive a decision rule. Crucially, the decision rule also specifies how it can be changed, or amended, and thus acts as a *de facto* constitution. Our main contribution is in providing an example of an initial agreement that is simple and intuitive, and a constitution that logically follows from it. The naive agreement is on the basic process of decision making – that agents approve or disapprove proposals; that their vote determines either the acceptance or rejection of each proposal; and on the axioms, which are requirements regarding a constitution that engenders a self-updating decision making process.

Keywords: Voting · Constitutions · Self-governance · Democracy

1 Introduction

Consider a group of n agents that gather to form a self-governed community. To make collective decisions, they first need to agree on a rule by which to do so; but offering to vote on the voting rule leads to infinite regress.

Thus, in this paper we offer a different approach. In particular, we assume that the agents first agree on a common set of axioms, and aim to infer an initial decision rule from these axioms. Ideally, the axioms will logically imply an identifiable set of permissible decision rules, and a unique decision rule from this set to serve as the initial rule, so that an agent who agrees to the axioms cannot object to the initial rule they imply, nor to the set of permissible rules.

Furthermore, it is crucial that the decision rule be amendable. Once the agents have passed the first hurdle and founded their rudimentary constitution,

© Springer Nature Switzerland AG 2021
D. Fotakis and D. Ríos Insua (Eds.): ADT 2021, LNAI 13023, pp. 119–131, 2021.
https://doi.org/10.1007/978-3-030-87756-9_8

agent preferences over the various permissible decision rules become material. Thus, we assume that the agents also agree on a common set of axioms that relate to the amendment procedure to change the decision rule itself. Again, ideally, the axioms will logically imply a unique amendment procedure such that, taking agent preferences over possible decision rules into account, agents who agree to the axioms would have no objections to the amendment procedure.

We view the initial decision rule as the foundation of the *constitution* of the community and the amendment procedure as the process by which the constitution changes itself. We look at our work as a step towards understanding how a group of agents can congeal into a self-governed, rule-based, egalitarian community with a decision making mechanism that is not dictated by an external agent.

1.1 Paper Structure

We discuss some related work in Sect. 2. In Sect. 3 we discuss how to found and amend a constitution; we do so in an informal way, to mimic the way by which the agents may freely discuss the principles by which they wish to be self-governed, and to introduce the intuitions behind the formal treatment that follows. In Sect. 4 we formally define the set of axioms and their logical consequences – a unique decision rule and amendment procedure. In Sect. 5 we present an alternative axiomatic basis, explore its consequences, and compare it to the initial axiomatic basis proposed. We conclude, in Sect. 6, with a discussion on future research directions.

2 Related Work

Kenneth May's seminal paper [21] demonstrates that majority rule is the only voting rule that satisfies a compact, intuitive set of axioms. We have taken May's Theorem as inspiration to answer the question of how a group of agents could establish a set of rules for themselves where none existed before. While our axioms lead the agents to use majority rule initially when founding their constitution, they also enable the agents to change this rule according to their collective preferences. While there is a large literature on characterizing social choice rules axiomatically (see, e.g., [7,26]) and implementation theory (see, e.g., [17]), and a small body of work on voting on criteria [10,23,31], we believe that our work is the first to introduce a set of axioms that simultaneously lead to the formulation of an initial rule and the process by which it can amend itself. Perhaps the closest work to this comes from the study of legislative procedures and procedural choice [11,12] We consider two alternative axioms for how the agents can determine what proposal should win if multiple incompatible proposals are viable, namely Condorcet-consistency and Conservatism. Extensions of the Condorcet principle have a long history in social choice [5,15], and the principle of Conservatism comes from reality-aware parameter update rules [27]. Our axiom of Minimalism and assumptions about agent preference structure overcome the general impossibility result of [16] without concern of infinite regress [30].

One line of research on constitutional amendments assumes that voter preferences over possible rules are derived from their preferences over the outcomes that each rule produces [19]. This is the "consequentialist approach" used in game theoretic analysis [20]. Perhaps the work here most similar to ours is that of Bhattacharya [4] as they consider single-peaked preferences. We take a different approach to modeling preferences over constitutional amendments, more similar to that of Barbera and Jackson [3]. In their model, constitutions are just a set of voting rules. A constitution may be a single rule used to update itself, or it can consist of a fixed amendment rule used to update the decision rule. In contrast to Barbera and Jackson, who conclude that a rule used to update itself is too simple for their setting, we see this simplicity as a virtue in bootstrapping a constitution.

3   Founding and Amending a Constitution, Informally

We begin by informally discussing the processes of founding and amending a constitution; later, we delve into a rigorous mathematical treatment.

3.1   Founding a Constitution, Informally

We assume that all n agents agree that they need a procedure for accepting or rejecting proposals. Proposals are considered individually, and for each proposal, each agent either approves it or not. The agents wish to establish a decision rule that determines whether the group as a whole accepts or rejects a given proposal, based on the votes of its members. To achieve that, they first agree on the following axioms that a decision rule must fulfill.

We first introduce the axioms without mathematical precision, to demonstrate how unassuming they are, and show that the claims that follow from them make intuitive sense. In the next Sect. 4 we rephrase the axioms mathematically, rewrite the claims as propositions, and prove the propositions formally.

Axioms for a Decision Rule

1. **Decisiveness:** For a given proposal, the decision is either to accept or to reject it, as determined by the votes.
2. **Monotonicity:** A decision to accept a proposal will remain so if some agents that do not approve the proposal change their vote to approve.
3. **Anonymity:** A decision to accept a proposal shall not depend on the identities of the agents approving it.
4. **Concordance:** If two proposals are incompatible, in the sense that no agent may approve both, then not both can be accepted.
5. **Minimality:** Among the decision rules consistent with the axioms above, choose an initial rule that requires a minimal number of approvals.

Axioms 1–3 imply that the decision rule is based on the size of the fraction of the agents approving the proposal.

Informal Claim 1 (Fractional Approval Voting). *Axioms 1–3 imply that the decision rule shall accept a proposal if approved by at least some fraction of the agents.*

Adding Axiom 4 implies that the fraction shall be at least a half.

Informal Claim 2 (Supermajority Approval). *Axioms 1–4 imply that the fraction of approvals needed to accept a proposal is greater than $\frac{1}{2}$.*

We refer to a fraction of approvals greater than $\delta \geq \frac{1}{2}$ as a δ-*supermajority*, with the case of $\delta = \frac{1}{2}$ being *simple majority* and $\delta = 1 - \epsilon$, where $0 < \epsilon \leq \frac{1}{n}$, being *unanimous consent*.

Note that, unlike Axioms 1–4, which effectively restrict the set of possible decision rules, Axiom 5 compares and prioritizes some decision rules over others.

Informal Claim 3 (Initially, Simple Majority Approval). *Axioms 1–5 imply that the initial decision rule is approval via a simple majority.*

To summarize, we claim that, according to Axioms 1–5, the decision rule must always be a δ-supermajority approval rule, $\frac{1}{2} \leq \delta < 1$, with the initial supermajority being a simple majority $\delta = \frac{1}{2}$.

3.2   Amending a Constitution, Informally

The initial decision rule of the community is simple majority according to Axiom 5, but the axioms are consistent with amending it to any δ-supermajority, with $\frac{1}{2} \leq \delta < 1$. So, once the initial rule is established as simple majority, some agents may wish to amend it to a higher supermajority rule. Some agents may aspire for a decision by unanimous consent, while others may prefer to require more than simple majority but less than unanimous consent, for example to protect minorities. Our assumption that agents may prefer greater supermajorities for decisions is consistent with analytical and empirical research that suggests this often holds even for agents in the majority [11]. Agents may also wish to employ different rules for different decisions in the future. Furthermore, agents' preferences may change over time. Thus, below we focus on the fundamental question of how the decision rule can be amended.

Naturally, the axioms agreed upon by the community apply to the decision of whether to accept a newly-proposed decision rule just like they would for any other decision. Hence, the decision to change the present δ-supermajority decision rule must be approved by a δ-supermajority. However, proposals to amend the decision rule itself require additional considerations, reflected in the consistency requirements introduced below. But first we provide some background in order to express them.

We assume that agents have preferences over decision rules, and that an agent approves a proposal to amend the decision rule if and only if the agent prefers the newly-proposed rule over the decision rule in force. Concretely, we identify agents with their index $i \in [n]$, and make the simplifying assumption that agent preferences are single-peaked in the following sense: Every agent $i \in [n]$ has an *ideal point* δ_i, $\frac{1}{2} \leq \delta_i < 1$, such that: (i) agent i strictly prefers δ_i over all other proposed δ's; (ii) among any two proposed values larger than δ_i, the agent prefers the smaller of the two; (iii) among any two proposed values smaller than δ_i, the agent prefers the larger of the two; and (iv) the agent i has no other pairwise preferences.

Secondly, we say that a proposal p is *preferred* over another proposal (aka *dominates*) p' if the set of agents that prefer p over p' are a majority. A proposal p is *most-preferred* if no other proposal p' is preferred over p. Note that, if p is the only most-preferred proposal, then it is a Condorcet winner [6]. Also, as preference is transitive and there are at most $\frac{n}{2}$ significant values of δ ($\frac{1}{2}, \frac{1}{2} + \frac{1}{n}, \ldots, \frac{n-1}{n}$), a knock-out tournament among these values for δ, choosing the preferred value in each match, may quickly result in a most-preferred value for δ.

Using these terms, we express two amendment axioms. The first axiom, named Posterior Consistency, says that, to change the current decision rule to a new decision rule, this amendment decision must be justifiable in retrospect according to the new rule, as opposed to being consistent only with current rule in force. This is in essence an "anti-hypocrisy" axiom. The second axiom requires that among all alternative decision rules consistent with the other axioms, we choose a most-preferred decision rule.

Axioms for a Decision Rule Applied to Amend Itself

6. **Posterior Consistency:** A proposal to amend the decision rule is accepted only if accepted according to the newly accepted decision rule as well.
7. **Condorcet Consistency:** Among the decision rules that satisfy all axioms above, choose a most-preferred one.

Given our assumptions on agent preferences and Axioms 1–5, we claim that there is a unique amendment process that is also consistent with Axioms 1–7.

Informal Claim 4 (Condorcet Amendment Rule). *Axioms 1–7 together with our assumptions on agent preferences imply that the process to amend the current δ-supermajority approval decision rule must be:*

1. *Increase δ to $\delta' > \delta$, if and only if there is a δ'-supermajority that prefers δ' over δ and δ' is the **maximal** supermajority with this property;*
2. *Decrease δ to $\delta' < \delta$, if and only if there is a δ-supermajority that prefers δ' over δ and δ' is the **minimal** supermajority with this property; and*
3. *Retain the present δ-supermajority rule otherwise.*

This completes the informal presentation.

4   Founding and Amending a Constitution, Formally

In this section we rephrase the assumptions, axioms, and claims formally.

4.1   Founding a Constitution, Formally

Let $B = \{0, 1\}$; these values represent approving or disapproving a given proposal by each agent, as well as accepting or rejecting the proposal by the group of agents as a whole. For the set of n agents, we refer to $V \in B^n$ as a voter *profile*. For a profile V, we define $[\![V]\!] := |\{i \mid V_i = 1\}|$ to be the number of approvals in V. For two profiles $V, V' \in B^n$, define $V \leq V'$ if $V_i \leq V_i'$ for all $1 \leq i \leq n$.

Indeed, we assume a set of n agents that wish to agree on a decision rule d that takes a proposal and a profile V, where each voter specifies either *approve* (1) or *not approve* (0); and produces (i.e., outputs) a decision of either to *accept* (1) or to *reject* (0). In the axioms below, $V, V' \in B^n$ and the given proposal is assumed and not specified as an explicit parameter.

Formal Axioms for a Decision Rule

9. **Decisiveness:** The decision rule d is a function $d : B^n \to B$.
10. **Monotonicity:** If $d(V) = 1$ and $V \leq V'$ then $d(V') = 1$.
11. **Anonymity:** If V' is a permutation of V then $d(V) = d(V')$.
12. **Concordance:** If $V_i + V_i' \leq 1$ for all $1 \leq i \leq n$, then $d(V) + d(V') \leq 1$.
13. **Minimality:** If d and d' are consistent with the axioms above, and $d'(V') \leq d(V)$ implies $[\![V']\!] \leq [\![V]\!]$ for all $V, V' \in B^n$, but not vice versa (exchanging d and d'), then prefer d' over d.

We now rephrase Informal Claims 1–3 into propositions, and prove them.

Proposition 1 (Fractional Approval Voting). *Axioms 9–11 imply that there is some fraction $0 \leq f < 1$ such that $d(V) = 1$ if and only if $[\![V]\!]/n > f$.*

Proof. Let $V^* \in B^n$ be a voter profile $V^* = \arg\min_{\substack{V \in B^n \\ d(V)=1}} [\![V]\!]$, and let $f = \frac{[\![V^*]\!]-1}{n}$. Naturally, for any profile $[\![V']\!] < [\![V^*]\!]$, $d(V') = 0$ by the definition of V^*. Let V' be a profile for which $[\![V^*]\!] \leq [\![V']\!]$ and hence $f < [\![V^*]\!]/n \leq [\![V']\!]/n$. We wish to show that $d(V') = 1$. Consider a voter profile $V'' \in B^n$ such that $V'' \leq V'$ and $[\![V'']\!] = [\![V^*]\!]$. By anonymity, $d(V'') = d(V^*)$ and by monotonicity, $d(V'') \geq d(V') = 1$, assuming of course that d is a function $d : B^n \to B$ (Axiom 9). $\square$

With n voters, a decision rule d using fraction $f = \frac{k}{n}$ for $k \in \mathbb{N}$ will return the same outcome as any decision rule d' using fraction $\frac{k}{n} + \epsilon$ for $\epsilon < \frac{1}{n}$, for all possible profiles. We therefore do not differentiate between these functionally equivalent rules. For example, the decision rule using fraction $f = \frac{1}{2}$ is majority rule whether the number of voters is even or odd.

Proposition 2 (Supermajority Approval). *If $d(V) = 1$ then $[\![V]\!] > \frac{n}{2}$.*

Proof. By way of contradiction, assume $d(V) = 1$ and $[\![V]\!] \leq \frac{n}{2}$. Then V has a non-overlapping permutation V' such that $V_i + V_i' \leq 1$ for all $i \in [n]$. By Axiom 11, $d(V') = d(V)$, implying via the assumption that $d(V) + d(V') = 2$, contradicting Axiom 12. $\square$

Proposition 3 (Initially, Simple Majority Approval). *Axioms 1–5 imply that the initial decision rule is $d(V) = 1$ if $[\![V]\!]/n > \frac{1}{2}$.*

Proof. Let d be the initial decision rule. By Proposition 1 it is fractional, with some fraction $0 \leq f < 1$. By Proposition 2, $f \geq \frac{1}{2}$. We claim that $f \leq \frac{1}{2}$. By way of contradiction, assume $f = \frac{1}{2} + \frac{1}{n}$, and consider a decision rule d' with a fraction $f' = \frac{1}{2}$. Just like d, the rule d' is consistent with all the Axioms 9–13. Furthermore, it satisfies that $d'(V') \leq d(V)$ implies $[\![V']\!] \leq [\![V]\!]$, as $f < f'$. However, the converse is not true. Assume $[\![V]\!] = \frac{1}{2} + \frac{1}{n}$ and $[\![V']\!] = \frac{1}{2}$. Then $d(V) \leq d'(V')$ as both equal 0, but $[\![V]\!] > [\![V']\!]$, contradicting Axiom 13. Hence the initial decision rule is fractional with $f = \frac{1}{2}$, namely approval by a simple majority. $\square$

Let $\Delta := [\frac{1}{2}, 1)$. To summarize, Propositions 1–3 prove that, according to Axioms 9–13, the decision rule must always be a δ-supermajority approval rule, $\delta \in \Delta$, with the initial supermajority being a simple majority $\delta = \frac{1}{2}$. Next, we rewrite Axioms 6–7(a, b) formally, recast Claims 4–6 as proper theorems, and prove them.

4.2 Amending a Constitution, Formally

We have established that all decision rules are δ-supermajority rules for some $\delta \in \Delta$. We therefore identify each decision rule d with its δ, and denote it by d^δ.

Agent Preferences. Recall that each agent i has a preference $\preceq_i \subseteq \Delta \times \Delta$ over the decision rules, where $x \prec_i y$ if $x \preceq_i y$ but not vice versa. Furthermore, agent preferences are single-peaked in that every agent $i \in [n]$ has an *ideal point* $\delta_i \in \Delta$, such that: (i) $\delta \prec_i \delta_i$ for all $\delta \neq \delta_i \in \Delta$; (ii) if $\delta_i < \delta < \delta'$ then $\delta' \prec_i \delta$; (iii) if $\delta_i > \delta > \delta'$ then $\delta' \prec_i \delta$; and (iv) $\prec_i$ is the smallest relation satisfying (i)–(iii).

Hence, a voter approves a proposals if and only if it is in between the status quo and her ideal point, and the voter profile $V^{\delta,\delta'}$ on the amendment decision $d^\delta(\delta')$ for $\delta \neq \delta'$ is defined accordingly by $V_i^{\delta,\delta'} = 1$ iff $\delta_i \leq \delta' < \delta$ or $\delta_i \geq \delta' > \delta$.

We can now specify formally Posterior Consistency:

14. **Posterior Consistency:** Given an ideal points profile δ_i, $i \in [n]$, and $\delta, \delta' \in \Delta$, $d^\delta(V^{\delta,\delta'}) = 1$ only if $d^{\delta'}(V^{\delta,\delta'}) = 1$.

According to Axiom 15, the amendment process is not a simple approval vote on a proposed amendment, but a selection of an most-preferred decision rule based on the reported ideal points of the agents. Hence, the decision on an amendment is of a different type, $\Delta \times \Delta^n \to \Delta$, which takes the δ in force and the ideal points profile δ_i, $i \in [n]$, as input, and produces the amended δ' as output.

Recall that a proposal p is *preferred* over (aka *dominates*) another proposal p' if $|\{i \in [n] : p' \prec_i p\}| > \frac{n}{2}$. A proposal p is *most-preferred* (aka *undominated*) if no other proposal p' is preferred over p. With this we can specify Condorcet Consistency:

15. **Condorcet Consistency:** Given an ideal points profile δ_i, $i \in [n]$, choose a most-preferred[1] $\delta' \in \Delta$ for which $d^\delta(V^{\delta,\delta'}) = 1$.

In other words, given the current δ-supermajority approval rule, Condorcet Consistency requires the choice of an amendment δ' that maximizes $[\![V^{\delta,\delta'}]\!]$.

We can now state our main theorem.

Theorem 1 (Condorcet Amendment). *Given ideal points profile δ_i for $i \in [n]$, Axioms 9–15 and our assumptions on agent preferences imply that d^δ should be amended to $d^{\delta'}$ for $\delta' \neq \delta$ if either:*

1. *$\delta' = \arg\max_{\delta < x < 1} |\{i \in [n] : \delta_i \geq x\}| > xn$ exists, or*

2. *$\delta' = \arg\min_{1/2 \leq x < \delta} |\{i \in [n] : \delta_i \leq x\}| > \delta n$ exists*

Otherwise, no proposal to amend the decision rule is accepted.

Remark 1. Note that the two definitions of δ' could be made more similar textually by replacing the right-hand side of the equations with $> max(x, \delta) \cdot n$.

Proof. First, recall that, in all cases, the result is a δ-supermajority rule, which is consistent with Axioms 9–13. Next we argue that the δ' computed by the rules above is uniquely consistent with Axioms 9–15:

1. If such a $\delta' > \delta$ exists, then d^δ would accept a proposal to amend δ to δ' since δ' is preferable to δ by a δ'-supermajority, which is more than a δ-supermajority. Furthermore, it is Posterior Consistent as the proposal to amend δ to δ' will be approved by $d^{\delta'}$ for the same reason. Now we argue that δ' uniquely satisfies the axioms. A $\delta'' > \delta'$ cannot be chosen without violating Posterior Consistency, since there is no δ''-supermajority approval for that amendment by the maximality of δ'. A $\delta'' < \delta'$ cannot be chosen without violating Condorcet consistency, since δ' is preferred over δ'' by a δ'-supermajority and hence by a majority.
2. If such a $\delta' < \delta$ exists, note that accepting $d^{\delta'}$ as the new decision rule satisfies d^δ by the definition of δ', namely the use of δn on the right hand side. Furthermore, since δ' is preferable to δ by a δ-supermajority, which is greater than a δ'-supermajority, it is also Posterior Consistent. We now argue that δ uniquely satisfies the axioms. A $\delta'' < \delta'$ cannot be accepted by d^δ, since there is no δ-supermajority approval for amending δ to δ'', by the minimality of δ'. And a $\delta'' > \delta'$ cannot be chosen without violating Condorcet consistency, since δ' is preferred over δ'' by a δ'-supermajority and hence by a majority.
3. If there is no δ-supermajority to increase or decrease δ then of course δ should not change. No amendment proposal could be accepted according to δ itself, so remaining at δ in the face of any proposal is Posterior Consistent and trivially Condorcet-consistent.

---

[1] The standard definition of Condorcet Consistency is the selection of the unique Condorcet winner (namely undominated alternative) when it exists. Our definition is a bit more general.

This finishes the proof of Theorem 1.   □

We have shown that a group of n agents that agree on a simple and self-evident axioms and a minimal set of assumption, can obtain from these a definite decision rule that also determines its own amendment process, via formal logical deduction.

5   Alternative Approaches: Conservative Amendment

Naturally, there are many possible alternatives to the axiomatic basis proposed herein, as well as to the assumptions made. Here we explore an alternative to Axiom 7.

7. **a. Conservatism:** Among the decision rules that satisfy all axioms above, choose the one closest to the current decision rule.

Given our assumptions on agent preferences and Axioms 1–5, we also claim that there is a unique amendment process that is consistent with Axioms 1–7(a).

In practice, Condorcet Amendment happens to select the furthest δ-supermajority rule from the current rule that satisfies Axioms 1–6. By contrast, Conservative Amendment requires that the rule selected be as close to the current rule as possible.

Informal Claim 5 (Conservative Amendment). *Axioms 1–7(a) together with our assumptions on agent preferences imply that the process to amend the current δ-supermajority approval decision rule must be:*

1. *Increase δ to $\delta' > \delta$, if and only if there is a δ'-supermajority that prefers δ' over δ and δ' is the **minimal** supermajority with this property;*
2. *Decrease δ to $\delta' < \delta$, if and only if there is a δ-supermajority that prefers δ' over δ and δ' is the **maximal** supermajority with this property; and*
3. *Retain the present δ-supermajority rule otherwise.*

While the Condorcet and Conservative Amendment rules seem rather different, they are related in the sense that the second leads to the same result of the first if applied iteratively, in an exhaustive way – in case the rule in force is simple majority.

Informal Claim 6 (Iterate from Simple Majority). *For any given ideal points profile, if the rule in force is $\delta = \frac{1}{2}$ then the iterative application of the Conservative Amendment rule until no further amendments occur halts at the decision rule characterized by Informal Claim 4.*

To recap the informal presentation, if agents found their constitution according to Axioms 1–5, their initial decision rule will be simple majority. If they amend their initial rule by applying iteratively either of two amendment rules that satisfy posterior consistency, with a given set of ideal points, the result would be the δ characterized by Informal Claim 4.

We now proceed with the formal analysis, considering Conservatism instead of Condorcet Consistency. First, we offer a formal statement of Conservatism:

15. **a. Conservatism:** Given an ideal points profile δ_i, $i \in [n]$, choose a $\delta' \neq \delta \in \Delta$ closest to δ for which $d^\delta(V^{\delta,\delta'}) = 1$.

The resulting Theorem 2 looks very similar to Theorem 1, but the two cases swap the minimization and maximization over values of δ. The proof is similar as well.

Theorem 2 (Conservative Amendment). *Given ideal points δ_i for $i \in [n]$, Axioms 9–15(a) imply that d^δ should be amended to $d^{\delta'}$ for $\delta' \neq \delta$ if either:*

1. $\delta' = \arg\min_{\delta < x < 1} |\{i \in [n] : \delta_i \geq x\}| > xn$ exists, or

2. $\delta' = \arg\max_{1/2 \leq x < \delta} |\{i \in [n] : \delta_i \leq x\}| > \delta n$ exists

Otherwise, no proposal to amend the decision rule is accepted.

Proof. First, recall that in all cases the result is a δ-supermajority rule, which is consistent with Axioms 9–13. Next we argue that the δ' computed by the rules above is uniquely consistent with Axioms 9–15(a):

1. If such a $\delta' > \delta$ exists, then d^δ would accept a proposal to amend δ to δ' since δ' is preferable to δ by a δ'-supermajority, which is more than a δ-supermajority. Furthermore, it is Posterior Consistent as the proposal to amend δ to δ' will be approved by $d^{\delta'}$ for the same reason. We now show that δ' uniquely satisfies the axioms. Any $\delta'' > \delta'$ cannot be chosen without violating Conservatism because δ' would be accepted. No $\delta'' < \delta'$ could be accepted, because this would imply that δ'' is preferred by a δ''-supermajority according to Posterior Consistency, which violates the definition of δ'.
2. If such a $\delta' < \delta$ exists, note that accepting $d^{\delta'}$ as the new decision rule satisfies d^δ from the definition of δ', namely the use of δn on the right hand side. Furthermore, since δ' is preferable to δ by a δ-supermajority, which is greater than a δ'-supermajority, it is also Posterior Consistent. We now argue that δ' uniquely satisfies the axioms. A $\delta'' > \delta'$ cannot be chosen because according to Posterior Consistency it must be preferred by a δ''-supermajority, which violates the definition of δ'. A $\delta'' < \delta'$ cannot be accepted without violating Conservatism because δ' would be accepted.
3. If there is no δ-supermajority to increase or decrease δ then of course δ should not change. No amendment proposal could be accepted according to δ itself, so remaining at δ is Posterior Consistent and trivially Conservative.

This finishes the proof of Theorem 2.   □

Proposition 4 (Increasing Condorcet Amendment is Idempotent). *Given the initial decision rule $\delta = \frac{1}{2}$ and any ideal points profile, the application Condorcet Consistent Amendment is idempotent.*

Proof. When the current rule is $\delta = \frac{1}{2}$, any change to the decision rule must increase δ. From Theorem 1, under Condorcet Amendment, δ' will be the largest such that the number of agents with ideal points at least δ' is strictly greater than $\delta'n$. Let us refer to this decision rule as $\hat{\delta} \in \Delta$. Since a δ'-supermajority prefer $\hat{\delta}$ over δ, we know the majority cannot prefer any $\delta'' < \hat{\delta}$ over $\hat{\delta}$, so once $\hat{\delta}$ takes over as the current decision rule, no proposal to decrease it will be accepted. For an amendment $\delta'' > \hat{\delta}$ to occur, there would have to be a δ''-supermajority who prefer δ'' over $\hat{\delta}$, but this violates the definition of $\hat{\delta}$. Thus, Condorcet Amendment is idempotent and chooses $\hat{\delta}$.

Proposition 5 (Iterate from Simple Majority). *Given the initial decision rule $\delta = \frac{1}{2}$ and any ideal points profile, the result of iterative application to completion of Conservative Amendment is the same as a single application of Condorcet Amendment.*

Proof. Suppose that the iterative application of Conservative Amendment were to select some $\delta' \neq \hat{\delta}$. First, suppose that $\delta' < \hat{\delta}$. A $\hat{\delta}$-supermajority prefers $\hat{\delta}$ to δ', which is larger than a δ-supermajority, so Conservative Amendment, given δ', would approve the proposal $\hat{\delta}$. This contradicts the fact that the iterative application of Conservative Amendment reached completion. Second, suppose $\delta' > \hat{\delta}$. From Theorem 2, this implies that more than $\delta'n$ agents have ideal points of at least δ', because it is accepted when proposed against a smaller δ. This contradicts the definition of $\hat{\delta}$.   □

Here, we have offered an alternative axiomatic basis, replacing Condorcet Consistency (Axioms 7, 15) with Conservatism (Axioms 7(a) and 15(a)), discussed its consequences, and compared it to the initial axiomatic basis proposed.

6   Discussion

We considered a set of agents forming an egalitarian, self-governed community needing to establish by what process they will make decisions.

We started by assuming that the agents agree on a small set of intuitive axioms, and showed that from these axioms alone arises a simple constitution – an initial decision rule for making decisions on whether to accept or reject proposals, which can be applied to itself if the agents wish to change the rule. One of the axioms, Minimality, was unique among the axioms because it compared possible rules and prioritized one over the rest rather than restricting the set of possible rules. There are possible replacements to consider, including Unanimity.

We have argued that a rule that amends itself requires additional considerations, and offered additional axioms for that case. We have shown that basic assumptions on the structure of agent preferences over possible rules and these axioms result in a unique amendment process. We have considered two alternatives for the final axiom - Condorcet Consistency and Conservatism. While Condorcet Consistency results in a different amendment process from Conservatism, we have shown that repeated application of the Conservatism process, starting from the initial simple majority rule, results in the same rule produced by the Condorcet Consistent amendment process.

One natural next step following our work is to consider different structures of agent preferences over rules (e.g., metric preferences) with the same axioms.

A second is to determine what alternative sets of axioms might lead to different sets of rules that can also be used to amend themselves.

Lastly, a key assumption we have made here is that the set of agents is fixed. However, any self-governed community must determine who its members are when founding (i.e. whose votes count), and must decide for itself how to add and remove members in the future [1,2,8,9,13,14,18,22,24,25,28,29,32–34].

Acknowledgements. Ehud Shapiro is the Incumbent of The Harry Weinrebe Professorial Chair of Computer Science and Biology. We thank the generous support of the Braginsky Center for the Interface between Science and the Humanities. Nimrod Talmon was supported by the Israel Science Foundation (ISF; Grant No. 630/19). Ben Abramowitz was supported in part by NSF award CCF-1527497.

References

1. Alcantud, J.C.R., Laruelle, A.: Collective identity functions with status quo. Math. Soc. Sci. **93**, 159–166 (2018)
2. Alcantud, J.C.R., Laruelle, A.: Independent collective identity functions as voting rules. Theory Decis. **89**, 1–13 (2020)
3. Barbera, S., Jackson, M.O.: Choosing how to choose: self-stable majority rules and constitutions. Q. J. Econ. **119**(3), 1011–1048 (2004)
4. Bhattacharya, M.: Constitutionally consistent voting rules over single-peaked domains. Soc. Choice Welf. **52**(2), 225–246 (2019)
5. Brandt, F., Brill, M., Harrenstein, P.: Extending tournament solutions. Soc. Choice Welf. **51**(2), 193–222 (2018)
6. Brandt, F., Conitzer, V., Endriss, U., Lang, J., Procaccia, A.D.: Handbook of Computational Social Choice. Cambridge University Press, Cambridge (2016)
7. Campbell, D.E., Kelly, J.S.: Impossibility theorems in the Arrovian framework. Handb. Soc. Choice Welf. **1**, 35–94 (2002)
8. Cho, W.J., Saporiti, A.: Group identification with (incomplete) preferences. J. Public Econ. Theory **22**(1), 170–189 (2020)
9. Danezis, G., Mittal, P.: Sybilinfer: detecting sybil nodes using social networks. In: NDSS, San Diego, CA, pp. 1–15 (2009)
10. de Almeida, A.T., Morais, D.C., Nurmi, H.: Criterion based choice of rules. In: Systems, Procedures and Voting Rules in Context. AGDN, vol. 9, pp. 57–66. Springer, Cham (2019). https://doi.org/10.1007/978-3-030-30955-8_7
11. Diermeier, D., Prato, C., Vlaicu, R.: Procedural choice in majoritarian organizations. Am. J. Polit. Sci. **59**(4), 866–879 (2015)
12. Diermeier, D., Prato, C., Vlaicu, R.: Self-enforcing partisan procedures. J. Polit. **82**(3), 937–954 (2020)
13. Dimitrov, D., Sung, S.C., Xu, Y.: Procedural group identification. Math. Soc. Sci. **54**(2), 137–146 (2007)
14. Fioravanti, F., Tohmé, F.: Asking infinite voters 'Who is a J?': group identification problems in N. J. Classif. **37**(1), 58–65 (2020)
15. Fishburn, P.C.: Condorcet social choice functions. SIAM J. Appl. Math. **33**(3), 469–489 (1977)
16. Houy, N., et al.: A note on the impossibility of a set of constitutions stable at different levels. Technical report, Université Panthéon-Sorbonne (Paris 1) (2004)

17. Jackson, M.O.: A crash course in implementation theory. Soc. Choice Welf. **18**(4), 655–708 (2001)
18. Kasher, A., Rubinstein, A.: On the question "Who is a J?" a social choice approach. Logique et Analyse **160**, 385–395 (1997)
19. Koray, S.: Self-selective social choice functions verify Arrow and Gibbard-Satterthwaite theorems. Econometrica **68**(4), 981–996 (2000)
20. Lagunoff, R.: Dynamic stability and reform of political institutions. Games Econ. Behav. **67**(2), 569–583 (2009)
21. May, K.O.: A set of independent necessary and sufficient conditions for simple majority decision. Econometrica: J. Econom. Soc. **20**, 680–684 (1952)
22. Miller, A.D.: Group identification. Games Econ. Behav. **63**(1), 188–202 (2008)
23. Nurmi, H.: The choice of voting rules based on preferences over criteria. In: Kamiński, B., Kersten, G., Szapiro, T. (eds.) Outlooks and Insights on Group Decision and Negotiation. GDN 2015. LNBIP, vol. 218, pp. 241–252. Springer, Cham (2015). https://doi.org/10.1007/978-3-319-19515-5_19
24. Poupko, O., Shahaf, G., Shapiro, E., Talmon, N.: Sybil-resilient conductance-based community growth. In: van Bevern, R., Kucherov, G. (eds.) Computer Science – Theory and Applications. CSR 2019. LNCS, vol. 11532, pp. 359–371. Springer, Cham (2019). https://doi.org/10.1007/978-3-030-19955-5_31
25. Poupko, O., Shahaf, G., Shapiro, E., Talmon, N.: Building a sybil-resilient digital community utilizing trust-graph connectivity. IEEE/ACM Trans. Network. **PP**, 1–13 (2021)
26. Sen, A.: Collective Choice and Social Welfare. Harvard University Press, Cambridge (2017)
27. Shahaf, G., Shapiro, E., Talmon, N.: Sybil-resilient reality-aware social choice. arXiv preprint arXiv:1807.11105 (2018)
28. Shahaf, G., Shapiro, E., Talmon, N.: Genuine personal identifiers and mutual sureties for sybil-resilient community formation. arXiv preprint arXiv:1904.09630 (2019)
29. Sung, S.C., Dimitrov, D.: On the axiomatic characterization of "Who is a J?". Logique et Analyse **48**, 101–112 (2005)
30. Suzuki, T., Horita, M.: How to order the alternatives, rules, and the rules to choose rules: when the endogenous procedural choice regresses. In: Kamiński, B., Kersten, G., Szapiro, T. (eds.) Outlooks and Insights on Group Decision and Negotiation. GDN 2015. LNBIP, vol. 218, pp. 47–59. Springer, Cham (2015). https://doi.org/10.1007/978-3-319-19515-5_4
31. Suzuki, T., Horita, M.: A characterization for procedural choice based on dichotomous preferences over criteria. Group Decis. Negot.: Multidisc. Perspect. **388**, 91 (2020)
32. Wei, W., Xu, F., Tan, C.C., Li, Q.: SybilDefender: defend against sybil attacks in large social networks. In: 2012 Proceedings IEEE INFOCOM, pp. 1951–1959. IEEE (2012)
33. Yu, H., Gibbons, P.B., Kaminsky, M., Xiao, F.: SybilLimit: a near-optimal social network defense against sybil attacks. In: 2008 IEEE Symposium on Security and Privacy (SP 2008), pp. 3–17. IEEE (2008)
34. Yu, H., Kaminsky, M., Gibbons, P.B., Flaxman, A.D.: SybilGuard: defending against sybil attacks via social networks. IEEE/ACM Trans. Network. **16**(3), 576–589 (2008)

Unveiling the Truth in Liquid Democracy with Misinformed Voters

Ruben Becker[1], Gianlorenzo D'Angelo[1], Esmaeil Delfaraz[1(✉)], and Hugo Gilbert[2]

[1] Gran Sasso Science Institute, L'Aquila, Italy
{ruben.becker,gianlorenzo.dangelo,esmaiel.delfaraz}@gssi.it
[2] Université Paris-Dauphine, Université PSL, CNRS, LAMSADE, 75016 Paris, France
hugo.gilbert@dauphine.psl.eu

Abstract. This paper investigates the so-called *ODP*-problem that has been formulated by Caragiannis and Micha [8]. This problem considers a setting with two election alternatives out of which one is assumed to be correct. In *ODP*, the goal is to organise the delegations in a social network in order to maximize the probability that the correct alternative is elected. While the problem is known to be computationally hard, we strengthen existing hardness results and show that the approximation hardness of ODP highly depends on the connectivity of the social network and the individual accuracies. Interestingly, under some assumptions, on either the accuracies of voters or the connectivity of the network, we obtain a polynomial-time 1/2-approximation algorithm. Lastly, we run extensive simulations and observe that simple algorithms relying on the abilities of liquid democracy outperform direct democracy on a large class of instances.

Keywords: Liquid democracy · Truth revelation · Approximation algorithms

1 Introduction

Liquid Democracy (LD) is a recent voting paradigm which aims to modernize the way we make collective decisions to make it more flexible, interactive and accurate [3,6]. In a nutshell, LD allows voters to delegate transitively along a Social Network (SN). Indeed, each voter may decide to vote directly or to delegate her vote to one of her neighbors. This neighbor can in turn delegate her vote and the ones that have been delegated to her to someone else. As a result, these delegations will flow until they reach a voter who decides to vote. This voter is called the *guru* of the people she represents. LD is implemented in several online tools [3,21,22] and has been used by several political parties (e.g., the German Pirate party) for inner-decision making. The framework is praised for its flexibility, as it enables voters to vote directly for issues on which they feel both concerned and expert and to delegate for others.

D. Fotakis and D. Ríos Insua (Eds.): ADT 2021, LNAI 13023, pp. 132–146, 2021.
https://doi.org/10.1007/978-3-030-87756-9_9

Recently, several questions related to LD have been investigated. This questions are for instance related to (1) the propensity to which LD leads to an unequal voting power distribution [12,18,24,26]; (2) inconsistencies that could occur if voters would vote on different but related issues [7,9]; (3) the ability that LD has to incentivise participation in the delegation/voting process [10,27]; or (4) the stability of LD's delegation process [5,16,17,25,32].

Another line of research examines the accuracy of LD as a collective decision paradigm [8,24,33]. Indeed, as delegations can be motivated by the will to find a more expert representative than oneself, LD should concentrate the voting power in the hands of the most expert voters. This seems as a desirable feature in particular when the election aims to discover a *ground truth* (i.e., one of the alternative is the correct one to elect). This claim was previously investigated by Kahng, Mackenzie, and Procaccia [24] and Caragiannis and Micha [8]. Their works follow the long-standing line of research in social choice studying the accuracy of group-judgmental processes using the uncertain dichotomous choice model [4,11,19,30] and is closely related to recent works analysing the accuracy of other voting frameworks allowing delegations [28,29]. Both of these two works considered the following simple model. The election has two alternatives: a correct alternative and an incorrect one. Voters are nodes in a SN, modeled by a directed graph and each voter has an accuracy associated to her that indicates how well-informed she is. The authors investigated the accuracy of the majority of voters in LD and mostly provided negative results. In particular, Caragiannis and Micha [8] showed that the *Optimal Delegation Problem* (ODP for short) in which one aims to set the delegations of the voters so as to maximize the probability of selecting the ground truth is an NP-hard problem.

This work further investigates the accuracy of the LD paradigm. While some of our results are also negative, our gaze is not particularly severe on LD. Indeed, we believe that a loss of accuracy resulting from concentrating the voting power in too few hands is a pitfall which has been well understood since the Condorcet jury theorem [11]. Moreover, we believe that the hardness of ODP is the rule more than the exception for non-trivial graph problems involving probabilities. Our aim here is to provide indications through approximation results and simulations on the type of elections on which LD can be beneficial or problematic.

Our Contribution. We prove that, for any constant $C > 0$, unless $P = NP$, there is no polynomial-time algorithm for *ODP* that achieves an approximation guarantee of $\alpha \geq (\ln n)^{-C}$, where n is the number of voters. The reduction designed for this result uses poorly connected SNs in which some voters suffer from misinformation. Interestingly, under some assumptions on either the voters' accuracies or the SN's connectivity, we instead obtain 1/2-approximation algorithms. Lastly, we run extensive simulations and observe that simple algorithms outperform direct democracy on a large class of instances. The simulations also show that increasing the SN's connectivity increases the accuracy of all heuristics confirming that it is a key feature for LD's accuracy. Omitted proofs and material is available in a long version of the paper [2].

2   Preliminaries

We consider a binary election involving a set $\{T, F\}$ of two alternatives and a set $V = \{1, \ldots, n\}$ of n voters. For the sake of simplicity, we assume that the number of voters n is odd. The alternative T denotes the ground truth, i.e., T is more desirable than F. However, voters do not have a direct access to which alternative is the ground truth. Indeed, we consider a simple model in which each voter v_i has a probability p_i of voting for T if she votes directly. Value p_i is called the accuracy of voter i and measures her expertise level. Note that we assume that voters vote independently from one another. We denote by $\boldsymbol{p}$ the accuracy vector of size n defined by $\boldsymbol{p}[i] = p_i$. One might expect that values p_i should be greater than or equal to 0.5. Indeed, even if a voter is ignorant of a topic she cannot do worse than a random choice. However, similarly as Caragiannis and Micha [8], we allow probabilities to be lower than 0.5 modeling the fact that some voters may suffer from misinformation on a sensitive topic. Hence, even if such a voter could make a better choice by flipping a coin, this argument would necessitate that the voter actually knows she is misinformed.

Moreover, we assume the voters to be nodes V of a SN modeled as a digraph $G = (V, E)$, where a (directed) edge $(i, j) \in E$ corresponds to a social relation between voter i and voter j. The set of out-neighbors of voter i is denoted by $\mathsf{Nb}_{out}(i) = \{j \in V | (i, j) \in E\}$.

Each voter i has two possible choices: either she can vote directly, or she can delegate her vote and all the votes she has received through delegations to one of her neighbors in $\mathsf{Nb}_{out}(i)$. In the first (resp. second) case, she is called a *guru* (resp. *follower*). These different choices are formalized by a *delegation function* $d : V \to V$ such that $d(i) = j$ if i delegates to $j \in \mathsf{Nb}_{out}(i)$, and $d(i) = i$ if i votes directly. A delegation function d implies a *delegation graph* H_d which is the subgraph of G where there is an edge (i, j) iff $d(i) = j$ and $i \neq j$. Hence, the set of gurus denoted by $Gu(d)$ corresponds to the set of nodes with outdegree 0.

Example 1. Let us illustrate our notations with an example. We consider an instance with 7 voters involved in the SN displayed by Fig. 1 and an accuracy vector $\boldsymbol{p} = [0.9, 0.65, 0.45, 1, 0.5, 0.35, 0.8]$. Consider the following delegation function d defined by $d(1) = d(2) = 1$, $d(3) = d(4) = 4$, $d(5) = 6$, and $d(6) = d(7) = 7$. Hence voters $2, 3, 5$ and 6 are followers, and voters $1, 4$ and 7 are gurus. The graph H_d is represented in Fig. 1.

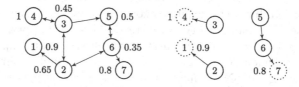

Fig. 1. The SN and voters' accuracy levels (left side of the figure) and the delegation graph with gurus being dotted (right side of the figure).

We will consider delegation functions d such that H_d is acyclic and we denote by $\Delta(G)$ the set of *delegation graphs* that can be obtained from G. Under this assumption, the graph H_d is a forest of directed in-trees $\{t_1, \ldots, t_\ell\}$ such that t_i is rooted in some guru g_i. Then, each guru g_i receives a voting weight $w(g_i) = |t_i|$, where $|t|$ denotes the number of nodes in tree t. We also extend this notation to subsets S of $Gu(d)$ by letting $w(S) = \sum_{g \in S} w(g)$.

Given a delegation graph, the election uses a weighted majority rule where guru g has weight $w(g)$. Given d and $\boldsymbol{p}$, the probability $P_{d,\boldsymbol{p}}[T]$ of electing T can be computed as follows:

$$P_{d,\boldsymbol{p}}[T] = \sum_{S \subseteq Gu(d)} \prod_{i \in S} p_i \prod_{j \in Gu(d) \setminus S} (1 - p_j) \mathbb{1}_{w(S) > n/2}. \tag{1}$$

Indeed, the probability that a set $S \subseteq Gu(d)$ corresponds to the gurus voting correctly is $\prod_{i \in S} p_i \prod_{j \in Gu(d) \setminus S} (1 - p_j)$ and this set will then be successful in electing T iff $w(S) > n/2$. When clear from the context, we will write $P_{d,\boldsymbol{p}}$ in place of $P_{d,\boldsymbol{p}}[T]$. In Example 1, as $p_4 = 1$, T wins when voter 1 or 7 (or both) votes correctly. Hence, we get that $P_{d,\boldsymbol{p}} = 0.98$.

We now make two remarks. First, note that looking for an optimal delegation function d to maximize $P_{d,\boldsymbol{p}}$ is related to looking for an optimal weighting function. By observing that $w(S) > n/2$ if and only if $w(Gu(d) \setminus S) < n/2$, we can conclude that an upper bound to $P_{d,\boldsymbol{p}}$ can be obtained if weights are set such that $w(S) > n/2$ when $\prod_{i \in S} p_i \prod_{j \in Gu(d) \setminus S}(1 - p_j) > \prod_{j \in Gu(d) \setminus S} p_j \prod_{i \in S}(1 - p_i)$. These conditions can be obtained easily if for each guru $g = v_i$, $w(g)$ is proportional to $\log(p_i/(1 - p_i))$ [30]. Of course, these weights may not be compatible with any delegation function as they may be non-integral or even negative.

Second, note that using Eq. 1 does not make it possible to compute $P_{d,\boldsymbol{p}}$ in polynomial time. However, we propose a simple recursive formula (which is novel to the best of our knowledge) to compute $P_{d,\boldsymbol{p}}$ in a tractable way. Start by ordering the gurus in $Gu(d)$ from g_1 to $g_{|Gu(d)|}$. Then, $P_{d,\boldsymbol{p}}$ can be computed by using the following recursive formula where $F(\tau, i)$ denotes the probability that the set of gurus voting for T in $\{g_i, \ldots, g_{|Gu(d)|}\}$ has weight at least τ:

$$F(\tau, i) = \begin{cases} 1 & \text{if } \tau \leq 0 \\ p_{g_i} \mathbb{1}_{w(g_i) \geq \tau} & \text{if } i = |Gu(d)| \\ p_{g_i} F(\tau - w(g_i), i+1) + (1 - p_{g_i}) F(\tau, i+1) & \text{otherwise.} \end{cases} \tag{2}$$

Equation 2 is composed of three sub-cases. The first one states that this probability is trivially one if $\tau \leq 0$. If it does not apply, the third sub-case states that if g_i votes correctly (which occurs with probability p_{g_i}) then the threshold τ will be met if gurus voting for T in $\{g_{i+1}, \ldots, g_{|Gu(d)|}\}$ have weight at least $\tau - w(g_i)$, while if g_i votes incorrectly (which occurs with probability $1 - p_{g_i}$) then the threshold will be met if they have weight at least τ. This reasoning can only be applied if g_i is not the last guru. This is sub-case 2 stating that this last guru can meet the threshold τ if she votes correctly and has weight at least τ.

Obviously, $P_{d,p} = F(\lceil n/2 \rceil, 1)$. To compute $F(\lceil n/2 \rceil, 1)$, it suffices to compute values $F(\tau, i)$ for $\tau \in \{0, \ldots, \lceil n/2 \rceil\}$ and $i \in \{1, \ldots, |Gu(d)|\}$ (where $|Gu(d)| \leq n$). Hence, using memoization, we get:

Proposition 1. *Given a delegation function d, the probability $P_{d,p}$ of electing T can be computed using $O(n^2)$ operations.*

Following the works of Caragiannis and Micha [8] and Kahng, Mackenzie, and Procaccia [24], we investigate the *Optimal Delegation Problem* (ODP) which aims to coordinate the delegations to maximize $P_{d,p}$.

ODP

Input: A social network G and an accuracy vector $\boldsymbol{p}$.

Feasible Solution: A delegation function d such that H_d is acyclic.

Measure: $P_{d,p}$ to maximize.

Remarks. Several remarks should be made on the ODP problem. First, the problem suggests that a central authority knows voters' accuracy values. Note that they may be approximated from past elections and that it is not because the central authority knows the accuracy values that she knows which alternative is the ground truth. Second, one could argue that if a central authority can have access to the accuracy values of the voters, then she could reverse the votes of the misinformed voters to make a more correct decision or even use the optimal weights from [30]. We do not consider such actions as they could be considered undemocratic. Third, the ODP problem suggests that a central authority may select the delegations of the voters which may not seem acceptable. In fact, we assume when studying ODP that G is more specific than just a SN. We indeed assume that there is an arc between i and j if i knows of j and agrees to delegate to her. Put another way, all voters specify a subset of neighbors they would accept to delegate to and ask for a central authority to guide their choice. A similar approach has been taken by Gölz et al. [18]. Lastly, note that another motivation to investigate the ODP problem is to investigate an upper bound on the performance achievable by voters when acting in a decentralized way.

Caragiannis and Micha [8] have shown that ODP is hard to approximate within an additive term of $1/16$. The next section provides a complementary approximation hardness result. Notably, this result will show that from an approximation viewpoint, the complexity of the problem is sensitive to the connectivity of the SN as well as the presence of misinformation (i.e., accuracies below 0.5).

3   Hardness of ODP

For $r \in (0, 1)$, let ODP_r be the restriction of problem ODP to instances in which all voters have accuracy greater than r. In this section, we show that for any $r \in (0, 0.5)$ and any constant $C > 0$, ODP_r cannot be approximated within a factor of $\alpha \geq (\ln n)^{-C}$ unless $P = NP$. This provides a strong approximation

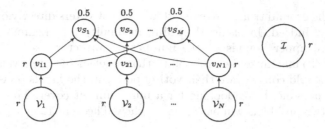

Fig. 2. An example of graph G resulting from the reduction. Each set $\mathcal{V}_i$ includes $L-1$ nodes and there is an arc from each node in $\mathcal{V}_i$ to the element voter v_{i1}. The set $\mathcal{I}$ includes K isolated nodes with accuracy r.

hardness result for ODP whenever some voters suffer from misinformation. Two interesting observations can be made. First, this result does not require very small accuracy values as it holds for any value $r = 0.5 - \epsilon$ with $\epsilon > 0$. Second, this result is in strong contrasts to ODP_r with $r \geq 0.5$, for which the direct voting strategy (and in fact any strategy) provides a $1/2$-approximation:

Theorem 1. *For any $r \in (0, 0.5)$ and constant $C > 0$, there is no polynomial-time algorithm for ODP_r that achieves an approximation guarantee of $\alpha \geq (\ln n)^{-C}$, unless $P = NP$.*

Our result is obtained through a reduction from Minimum Set Cover (MSC) which cannot be approximated better than within a factor of $(1 - o(1)) \ln N$ (N is the number of elements in the MSC instance) unless $P = NP$ [13]. In the MSC problem, we are given a universe $U = \{x_1, \ldots, x_N\}$ of N elements, and a collection $S = \{S_1, S_2, \ldots, S_M\}$ of subsets of U. The goal is to find the minimum number of sets from S, denoted by OPT_{SC}, covering all elements of U.

The Reduction. We set $\beta \in (0, 0.5)$ and $r = 0.5 - \beta$. From an instance $I = (U, S)$ of MSC, we create an instance I' of ODP_r as follows. As Fig. 2 illustrates, the graph $G = (V, E)$ is compounded of the following elements:

- A set $\mathcal{I}$ with $K = 8N^2M/\beta^2$ isolated nodes of accuracy r.
- For each element $x_i \in U$, we create L voters $v_{i1}, v_{i2}, \ldots, v_{iL}$. Let $\mathcal{V}_i = \{v_{i2}, v_{i3}, \ldots, v_{iL}\}$. We create an arc from each node in $\mathcal{V}_i$ to voter v_{i1}. Each voter v_{ij} has an accuracy of r and L is set to $\left\lceil \frac{\beta(4N-1)}{N(2N-1)}K + \frac{M}{N} \right\rceil + 1$. These voters will be called element voters in the following.
- One node v_{S_i} is created for each $S_i \in S$ with an accuracy of 0.5. These voters will be called set voters in the following. For every $i \in \{1, \ldots, N\}$ and $j \in \{1, \ldots, M\}$, we create a directed edge from v_{i1} to v_{S_j} if $x_i \in S_j$.

Note that, interestingly, each voter can only delegate to voters which are at least as accurate as them in this reduction. We denote by n the number of voters obtained, i.e., $n = |V| = K + NL + M$.

Idea of the Reduction. The reduction is built with the following idea: the value of K, i.e., the number of nodes in $\mathcal{I}$, is chosen carefully so that with large probability, the ground truth is elected iff all element voters directly or indirectly vote correctly. Indeed, by using Hoeffding inequality (see Lemma 1), we can give lower and upper bounds on the number of correct votes in $\mathcal{I}$ which are likely to hold. To maximize the probability that all element voters vote correctly, these voters should concentrate their voting power in the hands of as few many set voters as possible, hence looking for a minimum set cover. This connection between the two problems will enable us to show Theorem 1.

Lemma 1. *(Hoeffding inequality [23]) Let S be the number of successes in K trials of a Bernoulli random variable, which takes value 1 with probability r. Then, for every $\epsilon > 0$:*

$$Pr[S \geq (r + \epsilon)K] \leq \exp(-2\epsilon^2 K), \quad Pr[S \leq (r - \epsilon)K] \leq \exp(-2\epsilon^2 K).$$

We start formalizing the ideas expressed in the previous paragraph by giving a sequence of lemmas. Thereafter, we set $\epsilon = \frac{\beta}{2(2N-1)}$. Lemma 1 shows that it is likely that the number of voters voting correctly in $\mathcal{I}$ belongs to $((r - \epsilon)K, (r + \epsilon)K)$. The next two lemmas will be used to argue that in that case the ground truth will be elected if all element voters vote (directly or indirectly) correctly.

Lemma 2. *If at least $(r - \epsilon)K$ voters in $\mathcal{I}$ vote correctly, then it is enough that all NL element voters vote correctly to elect the ground truth.*

Lemma 3. *Let K' be a number such that $n \geq K' \geq K$. If at most $(r + \epsilon)K'$ voters out of K' voters vote correctly, then it is not enough that $\min\{(N-1)L + M, n - K'\}$ other voters vote correctly to elect the ground truth.*

We now create a connection between a delegation function d in I' and a set cover $X \subseteq S$ in I. For this purpose, we introduce a transformation on delegation functions. Given a delegation function d, we define $\tilde{d}$ as the delegation function obtained from d by making all element voters which are not delegating (directly or indirectly) to a set voter do so.[1] For a delegation function d, let us denote by X_d the subset of gurus in $\{v_{S_1}, v_{S_2}, ..., v_{S_M}\}$ that receive some delegations according to $\tilde{d}$. Importantly, note that X_d corresponds to a set cover in I.

Let $\mathcal{J}_d$ be the set of voters in $\mathcal{I} \cup \{v_{i,j} | 1 \leq i \leq N$ and $2 \leq j \leq L\}$ which vote directly according to d and $K_d = |\mathcal{J}_d| \geq K$. Note that these voters have necessarily a weight of 1 as they may not receive any delegation. Let S (resp. S_d) be a random variable representing the number of voters voting correctly in $\mathcal{I}$ (resp. $\mathcal{J}_d$). Let $\mathcal{X} = (S \leq (r - \epsilon)K)$, $\mathcal{Y} = (S \geq (r + \epsilon)K)$, and $\mathcal{Y}_d = (S_d \geq (r + \epsilon)K_d)$. Importantly, note that due to Lemmas 2 and 3, we have that $P_{\tilde{d},p}[T | \overline{\mathcal{X}} \cap \overline{\mathcal{Y}}] = 2^{-|X_d|}$ (where $\overline{\mathcal{Z}}$ denotes the complement of $\mathcal{Z}$). Hence, if $\mathcal{X} \cup \mathcal{Y}$ is a rare event, then $P_{\tilde{d},p}$ will be highly dependent on the size of X_d. Interestingly, Lemma 1 allows us to show that events $\mathcal{X} \cup \mathcal{Y}$ and $\mathcal{X} \cup \mathcal{Y}_d$ are indeed rare.

Lemma 4. *We have the following inequalities:*

$$P[\mathcal{X} \cup \mathcal{Y}] \leq 2\exp(-M), \quad P[\mathcal{X} \cup \mathcal{Y}_d] \leq 2\exp(-M).$$

---

[1] The choice of a set voter can be done arbitrarily when several choices are possible.

Let η be a constant in $(0,1)$. We can assume M large enough such that $2\exp(-M) \leq \eta$ and $2\exp(M\ln(2))/(\exp(M) - 2) \leq \eta$. Using Lemma 4, we prove the following relations between $P_{\tilde{d},p}$ and $|X_d|$.

Lemma 5. *The following inequalities hold,*

$$P_{\tilde{d},p} \geq (1 - 2\exp(-M))2^{-|X_d|} \geq 2\exp(-M)/\eta \text{ and } P_{\tilde{d},p} \leq 3 \times 2^{-|X_d|}.$$

Lastly, we provide an inequality between $P_{d,p}$ and $P_{\tilde{d},p}$:

Lemma 6. *The following inequality holds between $P_{d,p}$ and $P_{\tilde{d},p}$.*

$$P_{d,p} \leq (1 + \eta)P_{\tilde{d},p}.$$

We are now ready to prove Theorem 1.

Proof of Theorem 1. Let OPT_{SC} and $P_{d^*,p}$ be the optimal values in I and I' respectively. Moreover, Let d_{SC} be a strategy such that $X_{d_{SC}} = OPT_{SC}$. Then

$$P_{d^*,p} \geq P_{\tilde{d}_{SC},p} \geq (1 - 2\exp(-M))2^{-|X_{d_{SC}}|} \geq (1 - \eta)\left(\frac{1}{2}\right)^{OPT_{SC}} \tag{3}$$

using Lemma 5 and the fact that $2\exp(-M) \leq \eta$.

Let us assume that there exists a polynomial-time approximation algorithm A for ODP with approximation factor $\alpha \geq (\ln n)^{-C}$ for some constant $C > 0$. We obtain that $\alpha \geq (\ln(4)N)^{-C}$.[2] Let c be a constant such that $\eta < c < 1$, and $D = \frac{3(1+\eta)}{1-\eta}$, then we can assume that $(\ln(4)N)^{-C}/D \geq N^{-(C+\eta)}$, $OPT_{SC} \geq \frac{C+\eta}{(c-\eta)\ln(2)}$ and $\ln(N)\eta \geq 1$, otherwise, OPT_{SC} is bounded by some constant and we can solve I in polynomial time. Hence, from $\alpha \geq (\ln n)^{-C}$ we obtain that:

$$\frac{\ln(\alpha/D)}{\ln(0.5)} \leq \frac{C+\eta}{\ln(2)}\ln(N) \leq OPT_{SC}(c-\eta)\ln(N) \leq OPT_{SC}(c\ln(N) - 1)$$

Let d be the solution returned by algorithm A, we deduce that

$$(0.5)^{|X_d|} \geq \frac{P_{\tilde{d}p}}{3} \geq \frac{P_{dp}}{3(1+\eta)} \geq \frac{\alpha}{3(1+\eta)}P_{d^*,p} \geq \frac{\alpha(1-\eta)}{3(1+\eta)} \times (0.5)^{OPT_{SC}}$$

using Lemma 5 and Eq. 3. We conclude that

$$(0.5)^{|X_d|} \geq \alpha/D(0.5)^{OPT_{SC}} \Rightarrow |X_d| \leq \frac{\ln(\alpha/D)}{\ln(0.5)} + OPT_{SC} \leq c\ln(N)OPT_{SC}.$$

Hence, A would provide a $c\ln(N)$ approximation with $c < 1$ for minimum set cover which is not possible unless $P = NP$.   □

---

[2] We use that $n \leq 4^N$, whose proof is available in a long version of the paper [2].

Note that if all voters have an accuracy greater than or equal to 0.5, then direct voting or any other delegation strategy would yield a 0.5-approximation. Moreover, there exists a straightforward polynomial-time approximation scheme for ODP_r, with $r > 0.5$. Indeed, in this case, for any ϵ, the Condorcet jury theorem states that there exists a constant dependent on ϵ and r such that if the number of voters exceeds this constant then direct voting will have an accuracy greater than $1 - \epsilon$ (this can easily be turned into an $1 - \epsilon'$ multiplicative approximation factor). Below this constant, one can simply use brute-force.

We now show that if the graph is strongly connected, the simple strategy in which all voters delegate (directly or indirectly) to the same voter $v^* \in \arg\max\{p_i | i \in V\}$ among the most competent voters yields a $1/2$-approximation algorithm for ODP. We call this strategy the *Best Guru Strategy* (BGS).

Theorem 2. *When the SN is strongly connected, the best guru strategy leads to a $1/2$-approximation algorithm for ODP.*

As this result is straightforward when $p_{\max} = \max\{p_i | i \in V\}$ is greater than or equal to 0.5, we focus on the case where $p_{\max} < 0.5$. We show that in this case, BGS is in fact optimal. For this purpose, we will require the following lemma.

Lemma 7. *Given an accuracy vector p, let $p_{\max}$ be the vector obtained from p by raising all entries to $p_{\max}$, then for any delegation function d, $P_{d,p} \leq P_{d,p_{\max}}$.*

Once all the entries of p have been raised to $p_{\max}$, we can use a result by Berend and Chernyavsky (Theorem 3 in [4]). This result states that the expert rule (where one voter has all the voting power) is the less effective rule to elect the ground truth when $p \geq 0.5$. We equivalently use it to state that it is the most effective rule to elect the ground truth when $p \leq 0.5$. Moreover, note that in their setting, an important difference is that weights are not attached to voters but rather distributed uniformly at random before voting. However, note that their setup is equivalent to ours when all voters have the same accuracy as with $p_{\max}$. Indeed, in this case, the way in which the weights are allocated to voters do not impact $P_{d,p}$. We may then state the following lemma to prove Theorem 2.

Lemma 8. *When the SN is strongly connected, BGS is an optimal solution for ODP when $p_{\max} = \max\{p_i | i \in V\} \leq 0.5$.*

The conclusion of this section is that educating the members of the SN and making them more connected are two levers to address the limits pointed out by Theorem 1 as they can lead to easier instances from an approximation viewpoint. In the next section, we will provide exact and heuristic approaches for ODP.

4   Exact and Heuristic Methods

To solve ODP optimally, we provide a Mixed Integer Linear Program (MILP). This MILP (omitted due to lack of space) uses two batches of constraints, one to ensure that the variables encode a valid delegation function d without any cycle, and one to ensure that the objective function corresponds to $P_{d,p}$. While this MILP is not efficient, it provides an exact method to solve small instances. In the rest of this section, we propose some heuristic methods.

Centralized Heuristic Methods for ODP. We first design some centralized heuristic methods. The methods maintain a set S of mandatory gurus and iteratively modify S following either a greedy or a local search strategy. Given S, we provide two centralized methods to organize the delegations. The first one, greedy_delegation, works in the following way. It considers each guru $g \in S$ in descending order of accuracy value and allocates to g all remaining non-gurus $v \in V \setminus S$ that can reach g in $G[(V \setminus S) \cup \{g\}]$. Voters that cannot reach any guru in S vote for themselves. The second one, voronoi_delegation, works by making each non-guru $v \in V \setminus S$ delegate to the "closest" guru in S. To take into account accuracies, the "distance" between a voter v and a guru g is defined as the length of the shortest path from v to g divided by the accuracy of g. This yields a kind of weighted Voronoï graph structure [15]. Once again, voters who cannot reach any guru in S vote for themselves. In what follows, the delegation function d_S corresponding to a set S is obtained using one of the two procedures: greedy_delegation or voronoi_delegation.

Our greedy heuristics start from the delegation function $d_1 = d$ corresponding to the direct voting strategy, and an empty set $S = \{\}$. At each iteration i, we determine the node v whose addition to S provides the largest $P_{d_{S \cup \{v\}},p}$ value. If adding v to S results in a positive increment larger than some small ϵ value, we set S to $S \cup \{v\}$ and update $d_i = d_{S \cup \{v\}}$ accordingly. Otherwise the method returns the current delegation function.

Our local search heuristics start from an initial delegation function $d_1 = d$, and a set S initialized as $Gu(d)$. Then, at each iteration i, we determine the single-node addition or removal operation on S which leads to the largest increment value. If Z is the set resulting from S by this optimal modification, then d_i is set to d_Z. The local search stops when no add or removal operation can result in a positive increment larger than some small ϵ value.

Decentralized Heuristics. Another method, called emerging, simulates what could happen in an LD election without any central authority organizing the delegations. This method assumes that voters have some intuition about the ranking of the accuracy values and the difference between them, which we model in a probabilistic manner. More formally, it assumes that each voter i approves the set $A_i := \{j \in \mathrm{Nb}_{out}(i) : p_j > p_i\} \cup \{i\}$ as possible delegates. Then, each voter i delegates to a voter $j \in A_i$ with probability $p_j / \sum_{k \in A_i} p_k$ and votes with probability $p_i / \sum_{k \in A_i} p_k$. The resulting delegation graph is necessarily acyclic.

5 Numerical Tests

This section performs simulations to evaluate the performance of the heuristics presented in Sects. 4. We notably estimate how a decentralized LD approach would perform compared to a more centralized one by comparing the performance of emerging w.r.t. the more centralized approaches. The performance of the heuristics is confronted to the ones of the GreedyCap algorithm [24] and direct democracy. Our simulations were executed on a computer server running Ubuntu 16.04.5LTS with 24 Intel(R) Xeon(R) CPU E5-2643 3.40 GHz cores and a total of 128 GB RAM. Our algorithms are implemented in python using networkx [20] and our code was executed with python version 3.7.6. We used gurobi version 9.0.2 for solving the MILPs in order to obtain the exact solutions to ODP.

Experimental Setting. We tested our algorithms on randomly generated networks built using the following different models: the $G_{n,m}$ model [14], i.e., graphs are chosen uniformly at random from the set of graphs with n nodes and m edges; the Barabási-Albert preferential attachment model [1]; and the Newman- Watts-Strogatz small-world model [31]. Let us motivate the use of these networks. The empirical analysis by Kling et al. [26] suggests that delegative networks have some voters receiving most delegations while most voters receive very little. This suggests that delegative networks can share several observed features of SNs as scale-freeness (the Barabási-Albert model respects this property). The use of the $G_{n,m}$ model easily allows to explore how the density of the graph impacts the results. Lastly, the use of the Watts- Strogatz model allows to study small world networks which is a property observed in many SNs and hence may remain to some extent in delegation networks.

To be close to a real world setting, voters' accuracies are generated as a mixture of Gaussians, where there is one Gaussian for experts $\mathcal{N}(0.7, 0.1)$ (10% of the voters), one for misinformed voters $\mathcal{N}(0.3, 0.1)$ (20% of the voters) and one for average voters $\mathcal{N}(0.5, 0.1)$ (70% of the voters). These values are sampled until they are in $(0, 1)$. We suppose that each method (except the MILP) does not have access to the exact accuracy values but only approximations. For this purpose, we partition the interval $(0, 1)$ in cells of the shape $(i\texttt{prec}, \min((i + 1)\texttt{prec}, 1)]$ with $i \in \{0, 1, \ldots, \lfloor 1/\texttt{prec} \rfloor\}$ and where prec is a parameter indicating the precision with which the accuracies can be approximated (when not specified prec $= 0.1$). The approximation of an accuracy value p is set to the arithmetic mean of the interval $I = [i\texttt{prec}, \min((i + 1)\texttt{prec}, 1)]$ for which $p \in I$.

Errorbars in our plots denote 95%-confidence intervals. The measurement points in our plots are averages over 50 experiments, 5 generations of random accuracies on each of 10 random graphs generated according to the respective graph model. For the experiments involving the MILP and for testing the impact of the parameter prec, in order to further reduce variance, we generate accuracies 10 times on each of the graphs.

The seven heuristic algorithms evaluated are: greedy_cap [24];[3] ls_gr, ls_vo, greedy_gr and greedy_vo our local search and greedy strategies organizing the delegations using either greedy_delegation or voronoi_delegation; for these last four methods the parameter ϵ is set to 0.05; emerging and direct_demo.

---

[3] This method uses a parameter α set to 1 and a cap function $C : x \to 10 \log(x)^{1/3}$.

We evaluate the solutions returned by the different methods on the probability of electing the ground truth, denoted by `score` in the figures below.

Research Questions. We investigate the three following questions. (i) How well do the heuristics (and notably the decentralized one) perform w.r.t. direct democracy and w.r.t. the best possible delegation function obtained by the MILP? (ii) How much does parameter `prec` impact these results? Stated differently, how well do we need to evaluate voters' accuracies to have efficient heuristics? (iii) How much does $m = |E|$ impact these results? Indeed, the theoretical results of Sect. 3 suggest that connectivity is key for ODP.

Results. The probabilities of finding the ground truth resulting from applying our heuristics to random networks with increasing values of $n = |V|$ are plotted in Fig. 3. We observe that, for all three types of random networks, all heuristics achieve high scores with the local search methods performing best. Interestingly, the `greedyCap` and `emerging` methods which are less centralized methods also perform well, electing the ground truth with large probability. In particular, the performance of the `emerging` method suggests that an LD election would lead to a highly accurate decision even without the help of a centralized entity. As illustrated in the first plot of Fig. 3, all these methods outperform by far direct voting. Indeed, as in our setting, the average accuracy is slightly below 0.5, direct voting will perform poorly and its accuracy will not increase in n. Conversely, as the LD heuristics make it possible to concentrate the voting power in the hands of the most expert voters, we observe that the probability of electing the ground truth increases with the number of such voters and hence in n.

To complement these results on the accuracy of our heuristics, we compare the probabilities of finding the ground truth that they yield with the one of the optimal delegation function computed using the MILP on small $G_{n,m}$ graphs. In the first plot of Fig. 4, we observe that local search strategies seem to provide solutions almost as accurate as the optimal ones.

Lastly, we evaluate the impact of parameters m and `prec`. The probabilities of finding the ground truth resulting from applying our heuristics to $G_{n,m}$ networks with increasing values of m (resp. `prec`) are shown in the second (resp. third) plot of Fig. 4. We observe that, for all heuristics, their accuracy increases with the connectivity of the network confirming the importance of this parameter. Indeed, the more connected the network is, the easier it is for voters to find a suitable guru. Conversely, increasing `prec` decreases the accuracies of all methods as it becomes increasingly difficult to estimate voters' accuracies. However, even for a large value of `prec` as 0.3, the methods remain quite efficient. This suggests that LD can be an efficient collective decision framework, even if voters cannot perfectly evaluate the accuracy of their neighbors.

Evaluation on Other Measures. We also evaluated the solutions on the number of gurus; the average distance from voters to their guru; and the average accuracy of gurus. Related plots are provided in a long version of the paper [2]. We observe that greedy strategies have much fewer gurus, much larger average accuracy values and average distance values than other heuristics. Interestingly, all other

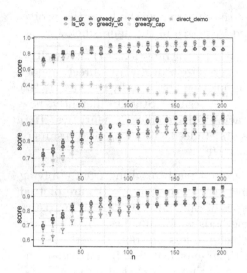

Fig. 3. Results for `score` using random graphs, n increasing from 11 to 201 in steps of 10: (1) $G_{n,m}$ graphs with $m = 4n$; (2) Barabási- Albert graphs ($m = 2$); (3) Watts-Strogatz graphs ($k = 2$, $p = 0.1$). The scores of `direct_demo` are only displayed in the first plot as they are the same for these three plots.

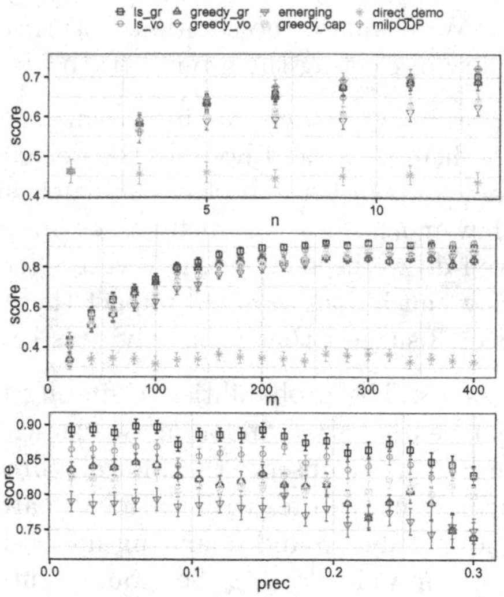

Fig. 4. Results for `score` using random $G_{n,m}$ graphs with (1) $m = 2n$ edges, n increasing from 1 to 13 in steps of 2; (2) $n = 101$ nodes, m increasing from 20 to 400 in steps of 20; (3) $n = 101$ nodes, $m = 2n$ and `prec` increasing from 0.015 to 0.3 in steps of 0.015.

heuristics yield low average distance values and a higher number of gurus which are desirable features for LD's acceptability.

6   Conclusion

Following recent works by Caragiannis and Micha [8] and Kahng, Mackenzie, and Procaccia [24], we have provided new results on the accuracy of the LD paradigm. We have showed that the complexity of finding a delegation graph that maximizes the accuracy of voters as a whole is hard from an approximation viewpoint. Moreover, we have stressed that this hardness result depends on the connectivity of the social network and on the voters' accuracies. Lastly, we have provided an exact and several heuristic methods and we have argued through simulations that some simple strategies in the LD framework could yield accurate decisions even when decentralized and relying on approximate accuracy values.

As a future work, designing other algorithms that would provide interesting approximation guarantees under some conditions on the social network would be a worthwhile contribution. Moreover, it would be interesting to study the accuracy of the LD framework by using alternative and maybe more complex models than the one of the uncertain dichotomous choice model.

Acknowledgments. This work is partially supported by the Italian MIUR PRIN 2017 Project ALGADIMAR "Algorithms, Games, and Digital Markets".

References

1. Barabási, A.L., Albert, R.: Emergence of scaling in random networks. Science **286**(5439), 509–512 (1999)
2. Becker, R., D'Angelo, G., Delfaraz, E., Gilbert, H.: When can liquid democracy unveil the truth? arXiv preprint arXiv:2104.01828 (2021)
3. Behrens, J., Kistner, A., Nitsche, A., Swierczek, B.: The principles of LiquidFeedback. Interaktive Demokratie (2014)
4. Berend, D., Chernyavsky, Y.: Effectiveness of weighted majority rules with random decision power distribution. J. Publ. Econ. Theory **10**(3), 423–439 (2008)
5. Bloembergen, D., Grossi, D., Lackner, M.: On rational delegations in liquid democracy. In: Proceedings of the 33rd AAAI Conference on Artificial Intelligence, AAAI 2019, Honolulu, Hawaii, USA, 27 January– 1 February 2019, pp. 1796–1803 (2019)
6. Brill, M.: Interactive democracy. In: Proceedings of the 17th International Conference on Autonomous Agents and MultiAgent Systems, AAMAS 2018, Stockholm, Sweden, 10–15 July 2018, pp. 1183–1187 (2018)
7. Brill, M., Talmon, N.: Pairwise liquid democracy. In: Proceedings of the 27th International Joint Conference on Artificial Intelligence, IJCAI 2018, Stockholm, Sweden, 13–19 July 2018, pp. 137–143 (2018)
8. Caragiannis, I., Micha, E.: A contribution to the critique of liquid democracy. In: Proceedings of the 28th International Joint Conference on Artificial Intelligence, IJCAI 2019, Macao, China, 10–16 August 2019, pp. 116–122 (2019)
9. Christoff, Z., Grossi, D.: Binary voting with delegable proxy: an analysis of liquid democracy. In: Proceedings of the 16th Conference on Theoretical Aspects of Rationality and Knowledge, TARK 2017, Liverpool, UK, 24–26 July 2017, pp. 134–150 (2017)
10. Colley, R., Grandi, U., Novaro, A.: Smart voting. In: Proceedings of the 29th International Joint Conference on Artificial Intelligence, IJCAI 2020, pp. 1734–1740 (2020)
11. De Condorcet, N., et al.: Essai sur l'application de l'analyse à la probabilité des décisions rendues à la pluralité des voix. Cambridge University Press (1785)
12. Dey, P., Maiti, A., Sharma, A.: On parameterized complexity of liquid democracy. In: Mudgal, A., Subramanian, C.R. (eds.) CALDAM 2021. LNCS, vol. 12601, pp. 83–94. Springer, Cham (2021). https://doi.org/10.1007/978-3-030-67899-9_7
13. Dinur, I., Steurer, D.: Analytical approach to parallel repetition. In: Proceedings of the 46th ACM Symposium on Theory of Computing, STOC 2014, New York, NY, USA, 31 May– 03 June 2014, pp. 624–633 (2014)
14. Erdös, P., Rényi, A.: On random graphs I. Publ. Math. Debrecen **6**(290–297), 18 (1959)
15. Erwig, M.: The graph Voronoi diagram with applications. Networks **36**(3), 156–163 (2000)
16. Escoffier, B., Gilbert, H., Pass-Lanneau, A.: The convergence of iterative delegations in liquid democracy in a social network. In: Fotakis, D., Markakis, E. (eds.) SAGT 2019. LNCS, vol. 11801, pp. 284–297. Springer, Cham (2019). https://doi.org/10.1007/978-3-030-30473-7_19

17. Escoffier, B., Gilbert, H., Pass-Lanneau, A.: Iterative delegations in liquid democracy with restricted preferences. In: Proceedings of the 34th AAAI Conference on Artificial Intelligence, AAAI 2020, New York, NY, USA, 7–12 February 2020, pp. 1926–1933 (2020)
18. Gölz, P., Kahng, A., Mackenzie, S., Procaccia, A.D.: The fluid mechanics of liquid democracy. In: Christodoulou, G., Harks, T. (eds.) WINE 2018. LNCS, vol. 11316, pp. 188–202. Springer, Cham (2018). https://doi.org/10.1007/978-3-030-04612-5_13
19. Grofman, B., Owen, G., Feld, S.L.: Thirteen theorems in search of the truth. Theory Decis. **15**(3), 261–278 (1983)
20. Hagberg, A., Swart, P., S Chult, D.: Exploring network structure, dynamics, and function using networkx. Technical report, Los Alamos National Lab, US (2008)
21. Hainisch, R., Paulin, A.: Civicracy: establishing a competent and responsible council of representatives based on liquid democracy. In: Proceedings of the 2016 Conference for E-Democracy and Open Government, CeDEM 2016, Krems, Austria, 18–20 May, 2016, pp. 10–16 (2016)
22. Hardt, S., Lopes, L.C.: Google votes: a liquid democracy experiment on a corporate social network. Technical report (2015)
23. Hoeffding, W.: Probability inequalities for sums of bounded random variables. In: Fisher, N.I., Sen, P.K. (eds.) The Collected Works of Wassily Hoeffding. Springer Series in Statistics (Perspectives in Statistics), pp. 409–426. Springer, New York (1994). https://doi.org/10.1007/978-1-4612-0865-5_26
24. Kahng, A., Mackenzie, S., Procaccia, A.D.: Liquid democracy: an algorithmic perspective. J. Artif. Intell. Res. **70**, 1223–1252 (2021)
25. Kavitha, T., Király, T., Matuschke, J., Schlotter, I., Schmidt-Kraepelin, U.: Popular branchings and their dual certificates. In: Bienstock, D., Zambelli, G. (eds.) IPCO 2020. LNCS, vol. 12125, pp. 223–237. Springer, Cham (2020). https://doi.org/10.1007/978-3-030-45771-6_18
26. Kling, C.C., Kunegis, J., Hartmann, H., Strohmaier, M., Staab, S.: Voting behaviour and power in online democracy: a study of liquidfeedback in germany's pirate party. In: Proceedings of the 9th International Conference on Web and Social Media, ICWSM 2015, Oxford, UK, 26–29 May 2015, pp. 208–217 (2015)
27. Kotsialou, G., Riley, L.: Incentivising participation in liquid democracy with breadth-first delegation. In: Proceedings of the 19th International Conference on Autonomous Agents and Multiagent Systems, AAMAS 2020, Auckland, New Zealand, 9–13 May, 2020, pp. 638–644 (2020)
28. Magdon-Ismail, M., Xia, L.: A mathematical model for optimal decisions in a representative democracy. In: Proceedings of the 32nd Annual Conference on Neural Information Processing Systems, NeurIPS 2018, Montréal, Canada, 3–8 December, 2018, pp. 4707–4716 (2018)
29. Meir, R., Amir, O., Cohensius, G., Ben-Porat, O., Xia, L.: Truth discovery via proxy voting. arXiv preprint arXiv:1905.00629 (2019)
30. Nitzan, S., Paroush, J.: Optimal decision rules in uncertain dichotomous choice situations. Int. Econ. Rev., 289–297 (1982)
31. Watts, D.J., Strogatz, S.H.: Collective dynamics of 'small-world' networks. Nature **393**(6684), 440–442 (1998)
32. Zhang, Y., Grossi, D.: Power in liquid democracy. In: Proceedings of the 35th AAAI Conference on Artificial Intelligence, AAAI 2021, Virtual Event, 2–9 February 2021, pp. 5822–5830 (2021)
33. Zhang, Y., Grossi, D.: Tracking truth by weighting proxies in liquid democracy. arXiv preprint arXiv:2103.09081 (2021)

Computing Kemeny Rankings
from d-Euclidean Preferences

Thekla Hamm[✉], Martin Lackner, and Anna Rapberger

TU Wien, Vienna, Austria
{thekla.hamm,martin.lackner,anna.rapberger}@tuwien.ac.at

Abstract. Kemeny's voting rule is a well-known and computation-ally intractable rank aggregation method. In this work, we propose an algorithm that finds an embeddable Kemeny ranking in d-Euclidean elections. This algorithm achieves a polynomial runtime (for a fixed dimension d) and thus demonstrates the algorithmic usefulness of the d-Euclidean restriction. We further investigate how well embeddable Kemeny rankings approximate optimal (unrestricted) Kemeny rankings.

Keywords: Euclidean preferences · Kemeny's voting rule · Rank aggregation algorithms · Computational complexity

1 Introduction

Rank aggregation is the problem of combining a collection of rankings into a social "consensus" ranking, with applications ranging from multi-agent planning [22] and collaborative filtering [32] to internet search [5,17]. The classic application of rank aggregation is voting and thus rank aggregation methods are extensively studied in social choice theory, where rankings correspond to voters' preferences. A prominent rank aggregation method is *Kemeny's voting rule*, also known as Kemeny-Young method. This method is based on the Kendall-tau distance between rankings and outputs a *consensus ranking (or Kemeny ranking)* that minimizes the sum of distances to the input rankings.

Kemeny's voting rule is of particular importance for two reasons: First, it is the only rank aggregation method satisfying three desirable properties (neutrality, consistency, and being a Condorcet method) [39]. Second, it is the maximum likelihood estimator for the "correct" ranking if the input is viewed as noisy perceptions of a ground truth (assuming a very natural noise model) [40]. However, Kemeny's rule has a main disadvantage: its computational complexity [7,27]. In particular, computing the Kemeny score is NP-hard even for four voters [17].

Due to the importance of Kemeny's rule, much algorithmic research has been conducted with the goal to overcome this computational barrier. The majority of this work has focused on approximation algorithms, parameterized algorithms and heuristical methods (see related work below). In this paper, we take an approach that is widely used in computational social choice: to restrict the input to a smaller preference domain [20]. If the input rankings possess a favorable

© Springer Nature Switzerland AG 2021
D. Fotakis and D. Ríos Insua (Eds.): ADT 2021, LNAI 13023, pp. 147–161, 2021.
https://doi.org/10.1007/978-3-030-87756-9_10

structure, it may be possible to circumvent hardness results that hold in the general case. For Kemeny's rule, this is the case if the input has a certain 1-dimensional structure; more specifically, Kemeny's rule is polynomial-time computable for single-peaked rankings [11] and for rankings with bounded single-peaked or single-crossing width [14]. In contrast, Kemeny's rule remains NP-hard for preferences that are single-peaked on a circle [34] and, as very recently shown in [24], for d-Euclidean preferences with $d \geq 2$. In fact, both preference domains admit an interesting connection: In [38] it has been shown that preferences that are single-peaked on a circle can capture specific 2-Euclidean preferences.

The d-Euclidean preference domain [10, 21] is a d-dimensional spatial model based on the assumption that voters and candidates can be placed in $\mathbb{R}^d$ and a voter's preference ranking is derived from the Euclidean distance between her coordinates and the candidates—closer candidates being more preferable. This model captures situations where voters' preferences are mainly determined by real-valued attributes of candidates (e.g., a political candidate may be placed in a two-dimensional space with axes corresponding to her position on economic and social issues, or a textbook might be judged on its focus on theory/applications and on its complexity level). It is intuitively clear that a one-dimensional model is too simplistic to capture most real-world situations, and more dimensions greatly increase the applicability of this domain. However, as mentioned before, it is not the case that simply restricting the input to d-Euclidean preferences yields a computational advantage as the problem remains NP-hard [24].

The goal of our paper is to find an efficient algorithm for Kemeny's voting rule given d-Euclidean preferences (for $d \geq 2$) by additionally imposing reasonable restrictions on the output. We work under the assumption that an embedding witnessing the d-Euclidean property is known and that the consensus ranking (i.e., the output) has to be embeddable via the same embedding. The embeddability of the consensus ranking is a sensible assumption as it extends the explanation of the preference structure to the consensus ranking, i.e., if voters' preferences can be understood as points in a d-dimensional space, then also the output should be explainable via this space. Our main result is that this problem can be solved in time in $\mathcal{O}(|\mathcal{C}|^{4d})$ for strict orders and $\tilde{\mathcal{O}}(|\mathcal{C}|^{4.746 \cdot d + 2})$ for weak orders (with ties), i.e., it is solvable in polynomial time for a fixed dimension d. This algorithm makes use of a correspondence between embeddable rankings and faces of a hyperplane arrangement in which each hyperplane is equidistant to two embedded candidates. The determination of an embeddable consensus ranking is then performed on an appropriately constructed vertex- and edge-weighted graph, which is extracted from the arrangement.

We further show that this algorithm can be adapted to an egalitarian variant of the Kemeny problem, which minimizes the maximum Kendall-tau distance. Finally, we study the restriction of requiring an embeddable consensus ranking in more detail. We prove that an embeddable consensus ranking has at most twice the Kemeny score of the optimal, unrestricted Kemeny ranking. In numerical experiments, we show that the embeddable Kemeny ranking and the optimal Kemeny ranking coincide in most small instances.

Related Work. In addition to the results by Escoffier et al. [24] who showed NP-hardness of Kemeny's voting rule given d-Euclidean preferences for $d \geq 2$, the work of Peters [33] on the recognition of d-Euclidean elections is of particular importance to our problem. Peters shows that this problem is NP-hard for $d \geq 2$ [33] (it is even $\exists \mathbb{R}$-complete). Thus, one cannot hope for a polynomial-time algorithm for our problem if the embedding is removed from the input. Instead, we assume that the embedding is either found in a preprocessing stage (with sufficient time available) or is known due to understanding the origin of preferences (which adhere to a d-dimensional geometry). In contrast, recognizing 1-Euclidean elections is possible in polynomial time [16,29].

As mentioned before, Kemeny's rule has attracted much attention from an algorithmic perspective: exponential-time search-based techniques [6,13,15], approximation algorithms [1,28], parameterized algorithms [8,14], and heuristical algorithm [2,36]. As Kemeny's voting rule is of practical importance, much work has also been invested in runtime benchmarks [3].

2 Preliminaries

A weak order $\succeq$ over a set X is a complete ($x \succeq y$ or $y \succeq x$ for all $x, y \in X$) and transitive binary relation. We write $x \succ y$ if $x \succeq y$ but not $y \succeq x$. Further, we write $x \sim y$ if $x \succeq y$ and $y \succeq x$. A weak order $\succeq$ is a *strict order* if it has no ties, i.e., if $x \neq y$ then either $x \succ y$ or $y \succ x$.

We define an *election* $(\mathcal{C}, \mathcal{V}, (\succeq_v)_{v \in \mathcal{V}})$ as a set of *candidates* $\mathcal{C}$, a set of *voters* $\mathcal{V}$, and for each $v \in \mathcal{V}$, a weak order $\succeq_v$ over the candidates called the *preference* (order) of v. Whenever $c \succeq_v c'$, we say that v *prefers* c over c'

Let d be positive integer and let $p : \mathcal{C} \cup \mathcal{V} \to \mathbb{R}^d$ be an *embedding* in the d-dimensional space. Further, let $\| \cdot \|_d$ denote the Euclidean norm in $\mathbb{R}^d$. We say that a voter's preference order $\succeq_v$ for $v \in \mathcal{V}$ on $\mathcal{C}$ is *p-embeddable* if for all $c, c' \in \mathcal{C}$, $c \succeq c'$ if and only if $\|p(v) - p(c)\|_d \leq \|p(v) - p(c')\|_d$. Generally for a weak order $\succeq$ on $\mathcal{C}$ that do not coincide with a voter's preference order, we say $\succeq$ is *p-embeddable* if there is some $x \in \mathbb{R}^d$ such that for all $c, c' \in \mathcal{C}$, $c \succeq c'$ if and only if $\|x - p(c)\|_d \leq \|x - p(c')\|_d$. An election $(\mathcal{C}, \mathcal{V}, (\succeq_v)_{v \in \mathcal{V}})$ is said to be *p-embeddable* if $\succeq_v$ for all $v \in \mathcal{V}$ are p-embeddable. Finally, an election is *d-Euclidean* if it it is p-embeddable for some p.

We define the *Kendall-tau distance* of two weak orders $\succeq, \succeq'$ over $\mathcal{C}$ as

$$K(\succeq, \succeq') = \sum_{\{x,y\} \subseteq \mathcal{C}} d_{\succeq, \succeq'}(x, y), \quad \text{where}$$

$$d_{\succeq, \succeq'}(x, y) = \begin{cases} 2 & \text{if } (x \succ y \text{ and } y \succ' x) \text{ or } (y \succ x \text{ and } x \succ' y) \\ 1 & \text{if } (x \sim y \text{ and } x \not\sim' y) \text{ or } (x \not\sim y \text{ and } x \sim' y) \\ 0 & \text{otherwise (i.e., } \succ \text{ and } \succ' \text{ agree on the order of } x \text{ and } y). \end{cases}$$

Equivalently,

$$K(\succeq, \succeq') = |\{\{x, y\} \subseteq \mathcal{C} \mid (x \succeq y \wedge y \succ' x) \vee (y \succeq x \wedge x \succ' y)\}|$$
$$+ |\{(x, y\} \subseteq \mathcal{C} \mid (x \succeq' y \wedge y \succ x) \vee (y \succeq' x \wedge x \succ y)\}|.$$

For strict orders $\succ$ and $\succ'$, this definition simplifies to $K(\succ, \succ') = |\{(x, y) \in \mathcal{C}^2 \mid (x \succ y \wedge y \succ' x) \vee (y \succ x \wedge x \succ' y)\}|$.

We can now define Kemeny's voting rule and the corresponding consensus rankings, which we refer to as *optimal Kemeny rankings* in the following.

Definition 1. Given an election $(\mathcal{C}, \mathcal{V}, (\succeq_v)_{v \in \mathcal{V}})$, a strict order $\succ$ on $\mathcal{C}$ is an *optimal Kemeny ranking* if there is no other strict order $\succ'$ on $\mathcal{C}$ with

$$\sum_{v \in \mathcal{V}} K(\succ', \succeq_v) < \sum_{v \in \mathcal{V}} K(\succ, \succeq_v),$$

i.e., an optimal Kemeny ranking minimizes the sum of Kendall-tau distances to the preference orders. We refer to $\sum_{v \in \mathcal{V}} K(\succ, \succeq_v)$ as the Kemeny score of $\succ$.

We note that Definition 1 could be adapted to define Kemeny rankings as weak orders; this would not change our results.

From a computational viewpoint, Kemeny's voting rule is captured by the following NP-hard decision problem [7,17,27]:

KEMENY SCORE
Instance: An election $(\mathcal{C}, \mathcal{V}, (\succeq_v)_{v \in \mathcal{V}})$ and an objective value $z \in \mathbb{N}$.
Question: Is there a strict order $\succ$ on $\mathcal{C}$ such that $\sum_{v \in \mathcal{V}} K(\succ, \succeq_v) \leq z$?

We furthermore consider an *egalitarian variant* which minimizes the maximal dissatisfaction of each voter.

Definition 2. Given an election $(\mathcal{C}, \mathcal{V}, (\succeq_v)_{v \in \mathcal{V}})$, we say that a strict order $\succ$ on $\mathcal{C}$ is an *egalitarian Kemeny ranking* if there is no other strict order $\succ' \neq \succ$ on $\mathcal{C}$ with $\max_{v \in \mathcal{V}} K(\succ', \succeq_v) < \max_{v \in \mathcal{V}} K(\succ, \succeq_v)$.

Like for KEMENY SCORE, the corresponding decision problem EGALITARIAN KEMENY SCORE, i.e., given $(\mathcal{C}, \mathcal{V}, (\succeq_v)_{v \in \mathcal{V}})$, $z \in \mathbb{N}$, decide whether there is a strict order $\succ$ on $\mathcal{C}$ such that $\max_{v \in \mathcal{V}} K(\succ, \succeq_v) \leq z$, is NP-hard even for four voters which was independently proved by Biedl et al. [9] and Popov [35].

3   Embeddable Kemeny Rankings

The main focus of this paper is on the constrained setting of d-Euclidean elections, that is, we assume that the input is an embedding p as well as a p-embeddable election. In addition, we require that the output (i.e., the Kemeny ranking) is also p-embeddable.

Definition 3. Given an embedding $p : \mathcal{C} \cup \mathcal{V} \to \mathbb{R}^d$ and a p-embeddable election $(\mathcal{C}, \mathcal{V}, (\succeq_v)_{v \in \mathcal{V}})$, a strict order $\succ$ on $\mathcal{C}$ is a *p-embeddable Kemeny ranking* if $\succ$ is p-embeddable and there is no other p-embeddable strict order $\succ'$ on $\mathcal{C}$ such that $\sum_{v \in \mathcal{V}} K(\succ', \succeq_v) < \sum_{v \in \mathcal{V}} K(\succ, \succeq_v)$.

A *p-embeddable egalitarian Kemeny ranking* is defined analogously.

First we observe that a p-embeddable Kemeny ranking does not need to coincide with any optimal Kemeny rankings for a given p-embeddable election.

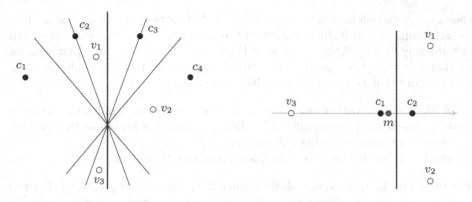

Fig. 1. Election from Example 1.   **Fig. 2.** Election from Example 2

Example 1. Consider the voting setting depicted in Fig. 1. The preferences of voter v_1 are given by $c_2 \succ_1 c_3 \succ_1 c_1 \succ_1 c_4$, the preferences of voter v_2 are $c_4 \succ_2 c_3 \succ_2 c_2 \succ_2 c_1$ and v_3 prefers $c_1 \succ_3 c_4 \succ_3 c_2 \succ_3 c_3$. The unique Kemeny ranking is $c_4 \succ c_2 \succ c_3 \succ c_1$ (with a Kemeny score of 14) since $K(\succ, \succ_1) = 6$, $K(\succ, \succ_2) = 2$, $K(\succ, \succ_3) = 6$, and $\sum_{i \leq 3} K(\succ', \succ_{v_i}) > 14$ for all $\succ' \neq \succ$. Now observe that $\succ$ is not embeddable in Fig. 1. Among embeddable rankings, the Kemeny score is minimized by $\succ_1$, $\succ_2$, and $\succ_3$, all of which achieve a Kemeny score of 16. These are the embeddable Kemeny rankings.

One may ask whether it is sensible to use an ordinal voting rule such as Kemeny's rule in our setting where voters and candidates can be represented in a coordinate space. It is important to note that we do *not* assume that a voter's position in $\mathbb{R}^d$, given by an embedding, is actually a correct representation of this voter's preferences. In particular, we do not assume that distances between voters and candidates is an accurate measure of *intensities*. That is, a voter prefers a candidate with distance 1 over a candidate with distance 2, but not necessarily twice as much. Hence, our assumption of embeddability in d-Euclidean space is significantly weaker than assuming a model where distances correspond to voters' utilities. In such a model, ordinal voting rules indeed are less useful and choosing the geometric median of the set of voter points[1] is more natural than computing a Kemeny ranking (in contrast to Kemeny's rule, the geometric median can be computed efficiently [12]). The next example shows that these two concepts differ.

Example 2. Consider a 2-Euclidean election with two candidates $\mathcal{C} = \{c_1, c_2\}$ and three voters $\mathcal{V} = \{v_1, v_2, v_3\}$ (cf. Fig. 2). The embedding p is given by $p(c_2) = -p(c_1) = (1, 0)$; $p(v_1) = (3, 6)$, $p(v_2) = (3, -6)$, $p(v_3) = (-10, 0)$. Voters v_1, v_2 prefer c_2 over c_1 while voter v_3 prefers c_1 over c_2. The optimal Kemeny ranking is

---

[1] The geometric median of a set of points S is a point that minimizes the sum of distances to points in S (as does the Kemeny ranking albeit for a different metric).

thus $c_2 \succ c_1$ (which is clearly p-embeddable). In contrast, the geometric median m is the point $\approx (-0.46, 0)$ which lies on the side of $p(c_1)$ and thus corresponds to the ordering $c_1 \succ c_2$. The crucial point here is that if we changed the embedding so that $p(v_1) = (4, 6)$, $p(v_2) = (4, -6)$, the geometric median would lie at $\approx (0.54, 0)$ and thus correspond to the Kemeny ranking.

A similar observation can be made in the case of the egalitarian Kemeny ranking; minimizing the maximum Euclidean distance is known as the 1-center problem or smallest enclosing ball problem.

For the 1-dimensional case, the question is easy to answer.

Proposition 1. *In a p-embeddable 1-Euclidean election, any optimal Kemeny ranking is also p-embeddable and coincides with the geometric median.*

As we have seen before, Proposition 1 does not extend to higher dimensions: Examples 1 and 2 are counter-examples for $d = 2$.

4   Computing Embeddable Kemeny Rankings

In this section, we give a brute-force algorithm to determine all p-embeddable Kemeny rankings of a given p-embeddable election. In order to traverse all strict p-embeddable orders, we observe their correspondence to faces of the hyperplane arrangement that contains all hyperplanes consisting of points equidistant to any two embedded candidates. This correspondence is also important for our main algorithm (Sect. 5), which drastically improves the asymptotic runtime.

Consider a d-Euclidean election $(\mathcal{C}, \mathcal{V}, (\succeq_v)_{v \in \mathcal{V}})$ embedded via $p : \mathcal{C} \cup \mathcal{V} \to \mathbb{R}^d$. For any pair $c, c' \in \mathcal{C}$ of candidates we consider the hyperplane $S_{c,c'} = \{x \in \mathbb{R}^d \mid \|x - p(c)\|_d = \|x - p(c')\|_d\}$. Each $S_{c,c'}$ divides $\mathbb{R}^d$ into two halfspaces—one containing $p(c)$, we also say this halfspace *lies on the same side* of $S_{c,c'}$ as c; and one containing $p(c')$. Each halfspace is assumed to be closed, that is, it contains its bounding hyperplane. A *face* of the hyperplane arrangement $\{S_{c,c'} \mid c, c' \in \mathcal{C}\}$ is a connected non-empty subspace of $\mathbb{R}^d$ obtained by intersecting halfspaces of the arrangement with at least one halfspace chosen for each hyperplane $S_{c,c'}$. We write $\mathcal{P}$ to denote the set of all faces of the arrangement.

Let $f \in \mathcal{P}$ be a face. For any pair of candidates $c, c' \in \mathcal{C}$, we say that f *lies on the same side* of $S_{c,c'}$ as c, if it is a subset of the halfspace that lies on the same side of $S_{c,c'}$ as c. This allows us to identify f by the set $X = \{(c, c') \in \mathcal{C}^2 \mid c$ and the subspace lie on the same side of $S_{c,c'}\}$; we write f_X to denote the face identified by X, i.e., $f_X = f$. A face f is called k-face if it has dimension k. Observe that for every face f_X, either $(c, c') \in X$ or $(c', c) \in X$ for every pair $c, c' \in \mathcal{C}$. Further note that X can also contain both tuples (c, c'), (c', c)—in that case, $f_X \subseteq S_{c,c'}$. For a face f_X, if $(c, c') \in X$ then $f_X \subseteq \{x \in \mathbb{R}^d \mid \|x - p(c)\|_d \leq \|x - p(c')\|_d\}$. Additionally we denote the set of d-dimensional faces as $\mathcal{R}$ and refer to them as *regions*. In the following, we use the standard notation $f°$ for the interior of a set f.

Intuitively, each face f_X corresponds to a weak p-embeddable order for the given d-Euclidean election and embedding p. This correspondence is formally captured by the following result.

Lemma 1. *Let $\Phi \colon \mathcal{P} \to \{\succeq\ \subseteq C^2 \mid\ \succeq$ is a p-embeddable weak order$\}$ be a function defined by $\Phi(f_X) =\ \succeq$ where $c \succeq c' \Leftrightarrow (c, c') \in X$. Then Φ is a bijection.*

Since we require that Kemeny rankings are strict, the following observation showing that each region corresponds to a strict p-embeddable ordering for the given embedded d-Euclidean election is also useful.

Lemma 2. *Let $\Phi' \colon \mathcal{R} \to \{\succeq\ \subseteq C^2 \mid\ \succeq$ is a p-embeddable strict order$\}$ be the restriction of Φ (from Lemma 1) to regions. Also Φ' is a bijection.*

For a face $f \in \mathcal{P}$, we write $\succeq_f$ instead of $\Phi(f)$ (this is a weak order). Further, for a region R, we write $\succ_R$ instead of $\Phi'(R)$ (this is a strict order).

We can now use the preceding correspondences to give a straightforward polynomial time algorithm that enumerates all p-embeddable strict orders.

Theorem 1. *Determining all p-embeddable Kemeny rankings for a d-Euclidean election $(\mathcal{C}, \mathcal{V}, (\succeq_v)_{v \in \mathcal{V}})$ given by $p : \mathcal{C} \cup \mathcal{V} \to \mathbb{R}^d$ is possible in time polynomial in $|\mathcal{C}|$, more specifically in time in $\mathcal{O}(|\mathcal{C}|^{6d})$.*

Proof. Consider the d-Euclidean preference profile given by the function $p : \mathcal{C} \cup \mathcal{V} \to \mathbb{R}^d$. For every $f \in \mathcal{P}$, let $\#(f)$ denote the number of voters in f, i.e. $\#(f) = |\{v \in \mathcal{V} \mid p(v) \in f\}|$. By comparing the corresponding values for each $R \in \mathcal{R}$, we can determine $R \in \mathcal{R}$ which minimizes $\sum_{f' \in \mathcal{P}} \#(f') \cdot \mathrm{K}(\succeq_{f'}, \succ_R)$, and denote such an R by $R_{\min}$. We return $\succ_{R_{\min}}$ as p-embeddable Kemeny ranking.

Correctness. For $R \in \mathcal{R}$ and $f' \in \mathcal{P}$,

$$\sum_{f' \in \mathcal{P}} \#(f') \cdot \mathrm{K}(\succeq_{f'}, \succ_R) = \sum_{f' \in \mathcal{P}} \sum_{\substack{v \in \mathcal{V} \\ p(v) \in f'}} \mathrm{K}(\succeq_{f'}, \succ_R)$$

$$= \sum_{f' \in \mathcal{P}} \sum_{\substack{v \in \mathcal{V} \\ p(v) \in f'}} \mathrm{K}(\succeq_v, \succ_R)$$

$$= \sum_{v \in \mathcal{V}} \mathrm{K}(\succeq_v, \succ_R)$$

Since we are looking for a p-embeddable Kemeny ranking, it has to have the form $\succ_R$ for some $R \in \mathcal{R}$ by Lemma 2, which implies correctness.

Runtime. The hyperplane arrangement induces $\mathcal{O}(|\mathcal{C}|^{2d})$ faces (by [26, Corollary 28.1.2] as we consider at most $\binom{|\mathcal{C}|}{2}$ distinct hyperplanes) and can be computed in time in $\mathcal{O}(|\mathcal{C}|^{2d})$ [18, Theorem 7.6]. For each face $R \in \mathcal{R}$, the computation and comparison of the objective function naively requires time in $\mathcal{O}(|\mathcal{P}|^2) \subseteq \mathcal{O}(|\mathcal{C}|^{4d})$. Thus the overall complexity of the procedure lies in $\mathcal{O}(|\mathcal{C}|^{6d})$. $\qquad\square$

An analogous procedure works for the egalitarian variant.

5 Increasing Efficiency

To achieve a better runtime—in particular for large d—we conduct a more in-depth graphical analysis of the relation of p-embeddable orders to each other. Specifically, this section is dedicated to proving our following main result.

Theorem 2 (Main Theorem). *Determining all p-embeddable Kemeny rankings for a d-Euclidean election $(\mathcal{C}, \mathcal{V}, (\succeq_v)_{v \in \mathcal{V}})$ given by $p : \mathcal{C} \cup \mathcal{V} \to \mathbb{R}^d$ is possible in time in $\tilde{\mathcal{O}}(|\mathcal{C}|^{2(d \cdot \omega + 1)})$, where $\omega < 2.373$ [4] is the exponent of matrix multiplication.*

5.1 Preference Graph

We define the preference graph G_{pref} as the edge-weighted graph given by setting

- $V(G_{\mathrm{pref}}) = \{v_f \mid f \in \mathcal{P}\}$;
- $E(G_{\mathrm{pref}}) = \{\{v_f, v_{f'}\} \mid (\dim(f) = \dim(f') - 1 \wedge f \subset f') \vee (\dim(f') = \dim(f) - 1 \wedge f' \subset f)\}$; and
- $w : E(G_{\mathrm{pref}}) \to \mathbb{N}$, $\{v_f, v_{f'}\} \mapsto |\{\{c, c'\} \subseteq \mathcal{C} \mid (\dim(f' \cap S_{c,c'}) = \dim(f') \wedge \dim(f \cap S_{c,c'}) \neq \dim(f)) \vee (\dim(f \cap S_{c,c'}) = \dim(f) \wedge \dim(f' \cap S_{c,c'}) \neq \dim(f'))\}|$.

In other words, vertices corresponding to faces one of which is contained in the other are connected to each other by edges in G_{pref} whenever the dimension of one face differs from the other by exactly one. The edge weights correspond to the number of pairs (c, c') of candidates inducing this respective hyperplane. An example is given in Fig. 3. By a bound on the number of faces [26, Corollary 28.1.2] and since we consider at most $\binom{|\mathcal{C}|}{2}$ different hyperplanes, we can bound the number of vertices by $|V(G_{\mathrm{pref}})| \in \mathcal{O}(|\mathcal{C}|^{2d})$.

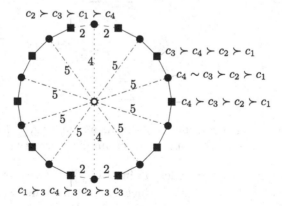

Fig. 3. G_{pref} for candidates as given in Example 1. Vertex shapes encode the dimensions of the corresponding faces, and dash-styles encode weights where edges without weight labels have unit-weight. Exemplary vertices are annotated with the corresponding p-embeddable orders.

G_{pref} without weights coincides with the *incidence graph* of a hyperplane arrangement as defined in [18] which is constructed in $\mathcal{O}(|\mathcal{C}|^{2d})$-time [18, Theorem 7.6]. We modify this procedure to include appropriate edge weights for G_{pref}.

Lemma 3. G_{pref} *can be constructed in time in* $\mathcal{O}(|\mathcal{C}|^{2d})$.

Note that at this point, we have set up or shown natural bijective correspondences between: the vertices of G_{pref}, the faces in $\mathcal{P}$, all p-embeddable orders of $\mathcal{C}$ and sets of pairs of candidates in $\mathcal{C}$ which explicitly encode the pairwise comparisons according to such p-embeddable orders. In this way, it will be natural to write any $v \in V(G_{\text{pref}})$ as v_f for some $f \in \mathcal{P}$, any p-embeddable order of $\mathcal{C}$ as $\succeq_f$ for some $f \in \mathcal{P}$, and any $f \in \mathcal{P}$ as f_X for some $X \subseteq \mathcal{C}^2$.

5.2 Shortest Paths in the Preference Graph

The crucial property of the preference graph, apart from capturing p-embeddable orders through its vertices, is that the chosen edge weights reflect the Kendall-tau distance between embeddable orders. We first can show this for single edges.

Lemma 4. *For* $\{v_f, v_{f'}\} \in E(G_{\text{pref}})$, $w(\{v_f, v_{f'}\}) = \mathrm{K}(\succeq_f, \succeq_{f'})$.

This previous lemma acts as the base case for the general correspondence of distances in G_{pref} and the Kendall-tau distance between the orders associated to the vertices of G_{pref} (i.e. the p-embeddable orders). We denote by $\mathrm{dist}_{G_{\text{pref}}}(v, w)$ the length of a shortest (in terms of summed edge weights) v-w-path in G_{pref}.

Lemma 5. *For* $f, f' \in \mathcal{P}$, $\mathrm{K}(\succeq_f, \succeq_{f'}) = \mathrm{dist}_{G_{\text{pref}}}(v_f, v_{f'})$.

Proof Sketch. We present a proof by induction over the length ℓ of cardinality-minimal shortest v_f-$v_{f'}$-paths (i.e., a path having minimum number of vertices among all weight-minimal paths between v_f, $v_{f'}$). The proof makes use of the observation that the Kendall-Tau distance between two faces f_X, f_Y corresponds to the symmetric difference $|X \Delta Y|$. The base case $\ell = 2$ is covered by Lemma 4.

Now assume that the statement holds for any cardinality-minimal shortest path of length $\ell - 1$ and observe that each proper subpath of a cardinality-minimal shortest v_f-$v_{f'}$-path consisting of ℓ vertices in G_{pref} is cardinality-minimal; otherwise one can replace the subpath with a cardinality-minimal shortest path, contradicting the assumption on $v_f \dots v_{f'}$. Together with the triangle-inequality for the Kendall-tau distance, we get $\mathrm{K}(\succeq_f, \succeq_{f'}) \leq \mathrm{dist}_{G_{\text{pref}}}(v_f, v_{f'})$.

To show $\mathrm{dist}_{G_{\text{pref}}}(v_f, v_{f'}) \leq \mathrm{K}(\succeq_f, \succeq_{f'})$, we construct a v_f-$v_{f'}$-path of weight $\mathrm{K}(\succeq_f, \succeq_{f'})$ by connecting two arbitrary points $p_f \in f^{\circ}$ and $p_{f'} \in f'^{\circ}$ via a straight line l and extracting a path along the traversal of l from p_f to $p_{f'}$. The path consists of vertices v_g with $l \cap g \neq \emptyset$ such that $g \in \mathcal{P}$ satisfies $\dim(g) < \dim(g')$ for all $g' \in \mathcal{P}$ with $l \cap g = l \cap g'$; also, we connect every two vertices v_i, v_{i+1} which are—w.r.t. the ordering along the line traversal —"adjacent" but not connected via an edge (i.e., $|\dim(f_i) - \dim(f_{i+1})| > 1$ for the corresponding faces f_i, f_{i+1}) via a weight- and vertex-minimal path.

Let $v_f = v_{f_1} \ldots v_{f_s} = v_{f'}$ denote the constructed v_f-$v_{f'}$-path P and let $X_1, \ldots, X_s \subseteq \mathcal{C}^2$ denote the pairs of candidates such that $f_i = f_{X_i}$ according to our notation introduced in Sect. 4. We verify that the constructed path P has the desired weight $K(\succeq_f, \succeq_{f'}) = |X_1 \triangle X_s|$ by showing that a pair $(c, c') \in \mathcal{C}^2$ contributes to the weight of P exactly once if and only if $(c, c') \in X_1 \triangle X_s$. Indeed, it can be shown that there is at most one edge $\{v_{f_i}, v_{f_{i+1}}\} \in P$ satisfying $f_i \cap S_{c,c'} = \emptyset$ but $f_{i+1} \cap S_{c,c'} \neq \emptyset$; also there is at most one edge $\{v_{f_i}, v_{f_{i+1}}\} \in P$ satisfying $f_i \cap S_{c,c'} \neq \emptyset$ but $f_{i+1} \cap S_{c,c'} = \emptyset$; i.e., P "enters" and "exists" a hyperplane $S_{c,c'}$ only once. This follows from the construction and by the fact that a straight line intersects a hyperplane at most once.   □

5.3 The Algorithm

Having established the correspondence between the Kendall-tau distance and the shortest paths in the edge-weighted graph G_{pref} we obtain the following result.

Theorem 2 (Main Theorem). *Determining all p-embeddable Kemeny rankings for a d-Euclidean election $(\mathcal{C}, \mathcal{V}, (\succeq_v)_{v \in \mathcal{V}})$ given by $p : \mathcal{C} \cup \mathcal{V} \rightarrow \mathbb{R}^d$ is possible in time in $\tilde{\mathcal{O}}(|\mathcal{C}|^{2(d \cdot \omega + 1)})$, where $\omega < 2.373$ [4] is the exponent of matrix multiplication.*

Proof. Consider the d-Euclidean preference profile given by the function $p : \mathcal{C} \cup \mathcal{V} \rightarrow \mathbb{R}^d$. We construct the corresponding preference graph G_{pref} using Lemma 3. We then apply the Shoshan-Zwick all-pairs shortest path algorithm for undirected graphs with integer weights (proposed in [37] and corrected in [19]) which returns a matrix $M_{\mathrm{dist}} \in \mathbb{N}^{V(G_{\mathrm{pref}}) \times V(G_{\mathrm{pref}})}$ containing the length of the shortest path between every pair of vertices in G_{pref}. For every vertex $v_f \in V(G_{\mathrm{pref}})$, let $\#(v_f)$ denote the number of voters in f, i.e. $\# : V(G_{\mathrm{pref}}) \rightarrow \mathbb{N}$ with $\#(v_f) = |\{v \in \mathcal{V} \mid p(v) \in f\}|$, or equivalently $\#(v_f) = |\{v \in \mathcal{V} \mid \succeq_v = \succeq_f\}|$. By comparing the corresponding values for each $R \in \mathcal{R}$, we can determine all $R \in \mathcal{R}$ which minimize $\sum_{f' \in \mathcal{P}} \#(v_{f'}) \cdot \mathrm{dist}_{G_{\mathrm{pref}}}(v_{f'}, v_R)$, and denote such an R by $R_{\min}$. We return the (set of) all such $\succ_{R_{\min}}$ as p-embeddable Kemeny rankings.

Correctness follows from the Lemmas 5, 2, and 1.

Runtime. The construction of the preference graph takes time in $\mathcal{O}(|\mathcal{C}|^{2d})$ by Lemma 3. By [19,37], the all-pairs shortest path algorithm for undirected graphs with integer weights runs in time in $\tilde{\mathcal{O}}(M \cdot |V(G_{\mathrm{pref}})|^{\omega})$ where M is the largest edge weight and $\omega < 2.373$ is the exponent of matrix multiplication. Since $M \leq \binom{|\mathcal{C}|}{2}$ we get $\tilde{\mathcal{O}}(M \cdot |V(G_{\mathrm{pref}})|^{\omega}) = \tilde{\mathcal{O}}(|\mathcal{C}|^{2(d\omega + 1)})$ The computation and comparison of the objective function for each $f \in \mathcal{P}$ naively requires time in $\mathcal{O}(|\mathcal{P}|^2) \subseteq \mathcal{O}(|\mathcal{C}|^{4d})$. Thus the overall complexity lies in $\tilde{\mathcal{O}}(|\mathcal{C}|^{2(d\omega + 1)})$.   □

Weak Kemeny Rankings. We remark that whenever we allow p-embeddable Kemeny rankings to be weak rather than strict, we can easily adapt our algorithm by comparing the values of $\sum_{f' \in \mathcal{P}} \#(v_{f'}) \cdot \mathrm{dist}_{G_{\mathrm{pref}}}(v_{f'}, v_f)$ for each $f \in \mathcal{P}$, denoting an f that minimizes this value by $f_{\min}$, and returning $\succeq_{f_{\min}}$ as Kemeny ranking. Correctness then follows immediately from Lemma 1.

Egalitarian Kemeny Rankings. An analogous result for the p-embeddable egalitarian Kemeny method can be obtained by an appropriate adaption of the objective function in the proof of Theorem 2.

Strict Preferences. Conversely whenever we restrict ourselves to instances in which all voters have only strict p-embeddable orders as preferences, we can focus on a proper minor of G_{pref} rather than the whole graph. More specifically we can restrict ourselves to the vertex set given by $\{v \in V(G_{\text{pref}}) \mid \exists R \in \mathcal{R} \quad v = v_R\}$; where edges between the vertices correspond to traversals of single hyperplanes: We contract paths of length 2 in G_{pref} between such vertices to single edges while summing up the weight of contracted edges. More explicitly instead of G_{pref} we can consider the graph H_{pref} given by the following information:

- $V(H_{\text{pref}}) = \{v_R \mid R \in \mathcal{R}\}$;
- $E(H_{\text{pref}}) = \{\{v_R, v_{R'}\} \mid \exists c, c' \in \mathcal{C} \quad \dim(R \cap R' \cap S_{c,c'}) = d - 1\}$; and
- $w : E(H_{\text{pref}}) \to \mathbb{N}, \{v_R, v_{R'}\} \mapsto 2|\{\{c, c'\} \subseteq \mathcal{C} \mid \dim(R \cap R' \cap S_{c,c'}) = d - 1\}|$.

Without weights, this graph is also known as the *region graph* or the *dual graph* of the embedded election induced hyperplane arrangement. Using the representation of the region graph as *medium*, i.e., as a system of *states* and transitions between states via *tokens* [23], we can employ a faster quadratic time all-pairs-shortest-paths algorithm [23] to achieve a better runtime for strict orders.

Theorem 3. *Determining all p-embeddable Kemeny rankings for a d-Euclidean election $(\mathcal{C}, \mathcal{V}, (\succ_v)_{v \in \mathcal{V}})$ in which all voters have strict preferences given by $p : \mathcal{C} \cup \mathcal{V} \to \mathbb{R}^d$ is possible in time in $\mathcal{O}(|\mathcal{C}|^{4d})$.*

6 Approximating the Kemeny Score

Our main algorithm fundamentally rests on the assumption that we are interested in an *embeddable* Kemeny ranking. As we have already seen in Example 1, such an embeddable Kemeny ranking may differ from an optimal Kemeny ranking. It is thus natural to ask

1. how often embeddable Kemeny rankings differ from optimal Kemeny rankings; and
2. how far these rankings can be apart (measured by their Kendall-tau distance).

We investigate these questions via numerical experiments and prove a bound on the worst-case approximation ratio of embeddable Kemeny rankings.

6.1 Approximation

Our goal is to quantify how much an embeddable Kemeny ranking and an optimal Kemeny ranking may differ. This can be phrased as an approximability results for computing Kemeny's voting rule in d-Euclidean elections. We can show that p-embeddable Kemeny rankings 2-approximate optimal Kemeny rankings.

Proposition 2. *Let $\prec$ be an optimal Kemeny ranking, p be a given embedding and $\prec_{res}$ be a p-embeddable Kemeny ranking. Then $\frac{\sum_{v \in \mathcal{V}} K(\prec_{res}, \prec_v)}{\sum_{v \in \mathcal{V}} K(\prec, \prec_v)} \leq 2$.*

However, it is unclear whether our ratio 2 is tight (even for $d = 2$). The largest ratio we are aware of is $8/7$ and arises, e.g., in Example 1.

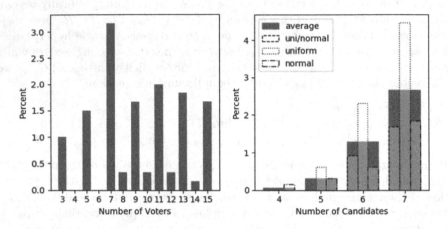

Fig. 4. Percentage of instances with ratio $r > 1$.

6.2   Experiments

We conducted numerical experiments on randomly generated 2-Euclidean elections to test the approximation quality of embeddable Kemeny rankings and to record how often embeddable Kemeny rankings do not achieve an optimal Kemeny score. In brief, our experiments suggest that the optimal Kemeny ranking is p-embeddable in 98.9% of the cases when considering up to 7 candidates.

To compute optimal Kemeny scores, we implemented Kemeny's rule with a trivial brute-force algorithm. The implementation for the p-embeddable Kemeny score used in these experiments[2] does not exploit all runtime improvements from the algorithm for strict orderings described in Sect. 5.3; its runtime currently inhibits experiments on larger instances. We randomly generated instances of 2-Euclidean elections with n voters, $3 \leq n \leq 15$, with strict preferences and m candidates, $4 \leq m \leq 7$, both of which we identify with points in $[0, 1000]^2$. For each pair (m, n), we generated 150 instances: 50 each assuming that (a) candidates and voters are component-wise uniformly distributed; that (b) candidates and voters are component-wise truncated normally distributed with mean 500 and variance 150; and that (c) candidates are uniformly distributed and voters are truncated normally distributed with mean 500 and variance 150.

---

[2] We construct the preference graph H_{pref} by adapting the dual arrangement construction from CGAL (*The CGAL Project*, https://www.cgal.org) and apply Johnson's all-pairs shortest path algorithm to determine the p-embeddable Kemeny rankings.

In total, we ran 7800 tests; among them, only 84 exhibited a p-embeddable Kemeny ranking that differs from the optimal Kemeny ranking. In these 84 instances, the ratio r of embeddable and optimal Kemeny rankings is between 1.0077 and 1.11. A difference in the scores of the optimal and the p-embeddable Kemeny rankings occurred slightly more often in uniformly distributed instances—1.85% of uniformly distributed instances have ratios $r > 1$, which is the case for only $\approx 0.7\%$ for other distributions. Figure 4 gives an overview of the percentage of instances where $r > 1$. The results indicate that an increasing number of voters does not cause a significant rise in the numbers of instances with suboptimal p-embeddable Kemeny rankings. Interestingly, instances with an odd number of voters have suboptimal p-embeddable Kemeny rankings significantly more often (77 out of 84), possibly due to fewer ties. On the other hand, the results indicate a positive correlation between the number of candidates and the number of instances with suboptimal p-embeddable Kemeny ranking (for $m = 4$, there is only one of 1950 instances with $r > 1$ ($\approx 0.05\%$), while for $m = 7$, 52 instances out of 1950 admit ratio $r > 1$ ($\approx 2.66\%$)). This suggests that the low overall percentage is due to the choice of the candidate range. Further tests with a larger number of candidates remains—due to limited computational power and, in terms of runtime, suboptimal implementation of the p-embeddable Kemeny ranking computation—a point on our future agenda.

7   Conclusions and Open Problems

We have shown that p-embeddable Kemeny rankings can be computed in time in $\mathcal{O}(|\mathcal{C}|^{4d})$ for strict orders and $\tilde{\mathcal{O}}(|\mathcal{C}|^{4.746 \cdot d + 2})$ for weak orders. Apart from improving these runtimes, it would be interesting to provide lower bounds on the computational complexity. In particular, a W-hardness result for computing p-embeddable Kemeny rankings could show that the dimension d has to occur in the exponent.

Further, our polynomial time solvability result juxtaposes the NP-hardness for the KEMENY SCORE problem on d-Euclidean elections, i.e., when one assumes p-embeddable preferences (given by p) but allows non-embeddable Kemeny rankings. To slightly relax our embeddability requirement on solutions with the hope of still remaining in P it would also be interesting to consider the problem where one requires a solution to be embeddable together with all voter preferences in the same dimension as the input, but allows the embedding to differ from the input embedding.

Let us end with a conceptual note. While d-Euclidean preferences are well-motivated and used in applications [21,30,31], there have been no successful attempts to leverage their structural properties for tractability results for $d \geq 2$, to the best of our knowledge. A likely reason for this is that combinatorial properties implied by d-Euclidean preferences seem to be difficult to derive. Our constructions of G_{pref} (and H_{pref} for strict preferences) in Sect. 5 may thus be of independent interest as a concise representation of d-Euclidean preferences and their mutual Kendall-tau distances under a fixed embedding. We would like to encourage the study of d-Euclidean preferences also for other computationally hard voting rules (such as Dodgson, Young). On this note, very recently many approval based multiwinner voting rules which are polynomial times solvable on 1-Euclidean elections were shown to be NP-hard on 2-Euclidean elections [25].

Acknowledgments. This work was supported by the Austrian Science Foundation FWF through projects P31890, P31336, W1255-N23, and Y1329.

References

1. Ailon, N., Charikar, M., Newman, A.: Aggregating inconsistent information: ranking and clustering. J. ACM **55**(5), 1–27 (2008)
2. Aledo, J.A., Gámez, J.A., Molina, D.: Tackling the rank aggregation problem with evolutionary algorithms. Appl. Math. Comput. **222**, 632–644 (2013)
3. Ali, A., Meilă, M.: Experiments with Kemeny ranking: what works when? Math. Soc. Sci. **64**(1), 28–40 (2012)
4. Alman, J., Williams, V.V.: A refined laser method and faster matrix multiplication. In: SODA, pp. 522–539. SIAM (2021)
5. Altman, A., Tennenholtz, M.: Axiomatic foundations for ranking systems. J. Artif. Intell. Res. **31**, 473–495 (2008)
6. Azzini, I., Munda, G.: A new approach for identifying the Kemeny median ranking. Eur. J. Oper. Res. **281**(2), 388–401 (2020)
7. Bartholdi, J., Tovey, C.A., Trick, M.A.: Voting schemes for which it can be difficult to tell who won the election. Soc. Choice Welf. **6**(2), 157–165 (1989)
8. Betzler, N., Fellows, M.R., Guo, J., Niedermeier, R., Rosamond, F.A.: Fixed-parameter algorithms for Kemeny rankings. Theor. Comput. Sci. **410**(45), 4554–4570 (2009)
9. Biedl, T., Brandenburg, F.J., Deng, X.: On the complexity of crossings in permutations. Discrete Math. **309**(7), 1813–1823 (2009)
10. Bogomolnaia, A., Laslier, J.F.: Euclidean preferences. J. Math. Econ. **43**(2), 87–98 (2007)
11. Brandt, F., Brill, M., Hemaspaandra, E., Hemaspaandra, L.A.: Bypassing combinatorial protections: polynomial-time algorithms for single-peaked electorates. J. Artif. Intell. Res. **53**, 439–496 (2015)
12. Cohen, M.B., Lee, Y.T., Miller, G., Pachocki, J., Sidford, A.: Geometric median in nearly linear time. In: STOC, pp. 9–21 (2016)
13. Conitzer, V., Davenport, A., Kalagnanam, J.: Improved bounds for computing Kemeny rankings. In: AAAI, pp. 620–626 (2006)
14. Cornaz, D., Galand, L., Spanjaard, O.: Kemeny elections with bounded single-peaked or single-crossing width. In: IJCAI, pp. 76–82 (2013)
15. Davenport, A., Kalagnanam, J.: A computational study of the Kemeny rule for preference aggregation. In: AAAI, pp. 697–702 (2004)
16. Doignon, J.P., Falmagne, J.C.: A polynomial time algorithm for unidimensional unfolding representations. J. Algorithms **16**(2), 218–233 (1994)
17. Dwork, C., Kumar, R., Naor, M., Sivakumar, D.: Rank aggregation methods for the web. In: WWW, pp. 613–622 (2001)
18. Edelsbrunner, H.: Algorithms in Combinatorial Geometry. Springer, Heidelberg (2012)

19. Eirinakis, P., Williamson, M., Subramani, K.: On the Shoshan-Zwick algorithm for the all-pairs shortest path problem. J. Graph Algorithms Appl. **21**(2), 177–181 (2017)
20. Elkind, E., Lackner, M., Peters, D.: Structured preferences. In: Trends in Computational Social Choice, chap. 10, pp. 187–207. AI Access (2017)
21. Enelow, J., Hinich, M. (eds.): Advances in the Spatial Theory of Voting. Cambridge University Press, Cambridge (1990)
22. Ephrati, E., Rosenschein, J.S., et al.: Multi-agent planning as a dynamic search for social consensus. In: IJCAI, pp. 423–429 (1993)
23. Eppstein, D., Falmagne, J.C.: Algorithms for media. Discrete Appl. Math. **156**(8), 1308–1320 (2008)
24. Escoffier, B., Spanjaard, O., Tydrichova, M.: Kemeny ranking is NP-hard for 2-dimensional Euclidean preferences. arXiv:2106.13054, preprint (2021)
25. Godziszewski, M.T., Batko, P., Skowron, P., Faliszewski, P.: An analysis of approval-based committee rules for 2D-Euclidean elections. In: AAAI, pp. 5448–5455 (2021)
26. Goodman, J., O'Rourke, J., Tóth, C.: Handbook of Discrete and Computational Geometry, 3rd edn. (2017)
27. Hemaspaandra, E., Spakowski, H., Vogel, J.: The complexity of Kemeny elections. Theor. Comput. Sci. **349**(3), 382–391 (2005)
28. Kenyon-Mathieu, C., Schudy, W.: How to rank with few errors. In: STOC, pp. 95–103 (2007)
29. Knoblauch, V.: Recognizing one-dimensional Euclidean preference profiles. J. Math. Econ. **46**(1), 1–5 (2010)
30. Laslier, J.F.: Spatial approval voting. Polit. Anal. **14**(2), 160–185 (2006)
31. Londregan, J.: Estimating legislators' preferred points. Polit. Anal. **8**(1), 35–56 (1999)
32. Pennock, D.M., Horvitz, E., Giles, C.L.: Social choice theory and recommender systems: analysis of the axiomatic foundations of collaborative filtering. In: AAAI, pp. 729–734 (2000)
33. Peters, D.: Recognising multidimensional Euclidean preferences. In: AAAI, pp. 642–648 (2017)
34. Peters, D., Lackner, M.: Preferences single-peaked on a circle. J. Artif. Intell. Res. **68**, 463–502 (2020)
35. Popov, V.: Multiple genome rearrangement by swaps and by element duplications. Theor. Comput. Sci. **385**(1), 115–126 (2007)
36. S Badal, P., Das, A.: Efficient algorithms using subiterative convergence for Kemeny ranking problem. Comput. Oper. Res. **98**, 198–210 (2018)
37. Shoshan, A., Zwick, U.: All pairs shortest paths in undirected graphs with integer weights. In: FOCS, pp. 605–614. IEEE (1999)
38. Szufa, S., Faliszewski, P., Skowron, P., Slinko, A., Talmon, N.: Drawing a map of elections in the space of statistical cultures. In: AAMAS, pp. 1341–1349 (2020)
39. Young, H.P., Levenglick, A.: A consistent extension of condorcet's election principle. SIAM J. Appl. Math. **35**(2), 285–300 (1978)
40. Young, P.: Optimal voting rules. J. Econ. Persp. **9**(1), 51–64 (1995)

Iterative Deliberation via Metric Aggregation

Gil Ben Zvi[2], Eyal Leizerovich[1(✉)], and Nimrod Talmon[1]

[1] Ben-Gurion University, Beersheba, Israel
[2] NRGene Ltd., Ness Ziona, Israel

Abstract. We investigate an iterative deliberation process for an agent community wishing to make a joint decision. We develop a general model consisting of a community of n agents, each with their initial ideal point in some metric space (X, d), such that in each iteration of the iterative deliberation process, all agents move slightly closer to the current winner, according to some voting rule $\mathcal{R}$. For several natural metric spaces and suitable voting rules for them, we identify conditions under which such an iterative deliberation process is guaranteed to converge.

Keywords: Deliberation · Social choice · Iterative process · Metric aggregation

1 Introduction

Agent communities wishing to reach joint decisions usually get involved in some voting process: agent preferences wrt. some agreed-upon options are being elicited, and their preferences are being aggregated through the use of some aggregation method. Correspondingly, much of the research in computational social choice [4] evolves around such aggregation methods, usually referred to as voting rules.

If voter preferences are rather diverse, then using a voting rule in a straightforward way might mean that the aggregated result (i.e., the result of the election) is not well-accepted by the agent community (e.g., a large minority may feel that their opinions are not being sufficiently heard). To overcome this issue, it can be useful to precede the voting phase with a deliberation phase, in which agents may interact, mutually hoping to find some common grounds [6]. When taken to the extreme, the best outcome of such a deliberation phase is that it would end in consensus: i.e., in a situation in which all agents eventually hold the same opinion; when all agents are in consensus. Then, informally speaking, the use of a voting rule is not needed, as all agents would be pleased with choosing the consensus opinion (technically, any *unanimous* voting rule – that chooses the consensus opinion whenever it exists – would be accepted).

Naturally, there are many ways by which voting and deliberation may coexist and interact; we discuss some of them in Sect. 2. In this paper, our point of view

© Springer Nature Switzerland AG 2021
D. Fotakis and D. Ríos Insua (Eds.): ADT 2021, LNAI 13023, pp. 162–176, 2021.
https://doi.org/10.1007/978-3-030-87756-9_11

is that the effect of deliberation is a change of the opinions of the agents (as a simplistic example, a right-wing voter may be more centrist after deliberating with a left-wing voter[1]). Correspondingly, in our model we view deliberation as a "black-box" process whose result is the change of agent opinions. In particular, we do not discuss nor model the specifics of deliberation, but rather model the result of using deliberation in diminishing the opinion distances between agents.

In particular, here we consider an iterative process of deliberation: initially, each agent holds to her position, which is modeled as an element of some metric space; then, each iteration consists of an implicit voting step, followed by a discussion step. In the voting step, an aggregated outcome is identified using some voting rule; then, in the discussion step, agents slightly change their opinions, to be more inline with the aggregated outcome computed in the voting step (specifically, agents move slightly closer to the aggregated outcome).

Note that the voting rule ingredient of our model only affects the specifics of how agents change their opinions due to discussion, as it affects the aggregated outcome. Our main interest is to characterize the situations for which such an iterative deliberation process converges, as we view converged processes as successful ones, in particular if the converged configuration is a consensus configuration (i.e., configurations in which all agents share the same opinion).

Specifically, throughout the paper we consider various metric spaces that correspond to certain social choice settings and several voting rules for each of these settings. Then, for each specific realization of our model – that is, for each metric space (X, d) and voting rule $\mathcal{R}$ – we analyze whether our iterative deliberation process is guaranteed to be successful (i.e., whether for any initial configuration it is always the case that the agents will end up in consensus).

For the settings that guarantee convergence, we are also interested in worst-case upper bounds for the time needed for such convergence (i.e., for the number of iterations until convergence, in the worst case). Finally, we are also interested in analyzing the possible results of such iterative deliberation processes, by comparing the initial agent opinions to the consensus opinion reached by such processes, whenever a consensus opinion is reached.

Indeed, our model is very extreme in assuming that, in each iteration, all agents move slightly closer to the aggregated opinion, in a deterministic way. In Sect. 8 we discuss some relaxations to our model. Note that the extremeness of our model means that our negative results – in which we show that an iterative deliberation process need not be successful – are very strong, as such negative results imply that, for such settings, even a very extremely optimistic deliberation process might not succeed. Generally speaking, we believe that our results shed more light on the relation between deliberation and voting by effectively distinguish between metric spaces and voting rules that are more problematic wrt. deliberation as such for which deliberation has greater potential to be successful.

---

[1] Indeed, the result of such deliberation may be the opposite – that the right-winger would be radicalized; we do not focus on such cases, but mention them in Sect. 8.

Due to space constraints, some of the proofs are deferred to the full version (available on the arXiv).

Paper Structure. After discussing related work (Sect. 2) and formally defining our model (Sect. 3), we prove general observations that apply to any metric space (Sect. 4). Then, we consider deliberation in Euclidean spaces (Sect. 5), in hypercubes (Sect. 6), and in ordinal elections (Sect. 7), and conclude with model relaxations and other avenues for future research (Sect. 8). Our results are summarized in Table 1.

Table 1. Summary of our main results. For each model realization – i.e., a metric space (X, d) and a voting rule $\mathcal{R}$ – we report whether convergence is guaranteed, and, if so, what is the upper bound of the number of iterations. VNW (Variable Number of Winners [13]) stands for the set of all subsets of some underlying candidate set, and is modeled via hypercubes; MW (Multi-winner [12]) stands for the set of all k-size subsets of some underlying candidates set, and is modeled via subsets of hypercubes; and SWF (Social Welfare Functions [4]) stands for the set of all rankings over some underlying candidate set.

X	d	$\mathcal{R}$	Convergence	Time	Theorem
$\mathbb{R}^T$	ℓ_1	Mean*	✓	$O\left(\max_{v \in V^0} \frac{d(v,w^0)}{\epsilon}\right)$	8
$\mathbb{R}^T$	ℓ_2	Mean*	✓	$O\left(max_{v \in V^0} \frac{d(v,w^0)}{\epsilon}\right)$	10
$\mathbb{R}^{\geq 3}$	ℓ_∞	Mean*	✗	✗	14
$\mathbb{R}^T$	ℓ_1, ℓ_2	Median*	✓	$max_{v \in V^0} \lceil \frac{d(v,w^0)}{\epsilon} \rceil$	12
$\mathbb{R}^{\geq 3}$	ℓ_∞	Median*	✗	✗	15
VNW	Hamming	Majority	✓	$\max_{v \in V^0} \lceil \frac{d(v,w^0)}{\epsilon} \rceil$	18
VNW	First changed	Monotonic	✓	$\lceil m/\epsilon \rceil$	Omitted
MW	Hamming	Majority	✓	$\max_{v \in V^0} \lceil \frac{d(v,w^0)}{\epsilon} \rceil$	Omitted
MW	First changed	Monotonic	✓	$\lceil m/\epsilon \rceil$	Omitted
SWF	Arbitrary	Kemeny	✓	$max_{v \in V^0} \lceil \frac{d(v,w^0)}{\epsilon} \rceil$	22
SWF	Swap	Monotonic scoring	✓	?	26
SWF	Swap	STV	✓	?	30
SWF	First changed	Monotonic	✓	$\lceil m/\epsilon \rceil$	31

*Mean and median both being element-wise.

2   Related Work

The most relevant literature pointer to our work is the paper of Bulteau et al. [5], in which the authors study aggregation methods for metric spaces. In particular, their model includes a metric space (X, d), where X is the set of elements of the space and d is a metric between pairs of elements of X; the opinion of an agent is an element $x \in X$ – referred to as the agent's *ideal point* – and the distance d determines the ordinal preferences of an agent over all X, where an agent prefers elements that are closer to its ideal point; a voting rule in their framework is a function that takes n points of X and returns an aggregated point in the metric space as the winner of the election. The jargon we use in this paper is largely due to Bulteau et al.; viewed from our angle, Bulteau et al. study a one-time aggregation process in which voters provide their ideal points and an aggregation method is used to find an aggregated point in the space, while we study an iterative process in which the aggregation method is used iteratively, each time causing the agents to move slightly closer to the aggregated point. Note that Bulteau et al. mention that their model may be indeed the basis for studying a process that combines voting and deliberation, as we set to do in the current paper. Indeed, we chose to build on the framework of metric aggregation as it evolves around a notion of distance between opinions (and it is general enough to capture many relevant social choice settings at once); this makes it natural for us to model the effect of deliberation by having each agent change its position to be slightly closer – according to some distance function d – to the aggregated point, in each iteration.

There are other works that consider iterative deliberation processes: e.g., Fain et al. [8] consider a process in which, in each iteration, two agents negotiate and move slightly closer to each other's point in the space; Elkind et al. [7] consider a process of deliberation in a metric space, concentrating on coalitions that may form around compromise points in the metric space; and Garg et al. [14] consider a model in which all agents are moving in the confined radius of a ball around their compromise point. There are also works that consider deliberation and aim at capturing the internal mechanics of deliberation [2,3,18]; we, however, similarly to Elkind et al. [7], abstract away the internal mechanism of deliberation and concentrate on the possibility of reaching consensus by deliberation.

We also mention work on opinion diffusion in social networks [9,15], in which agents are connected via a social network that affects the opinions of neighbors of agents and thus are propagated throughout the network. Technically, our model can be seen as a model of opinion diffusion where the social network is a complete graph (while in standard opinion diffusion the graph is usually not complete), however we prefer to think about our model as a model of deliberation. Furthermore, we mention work on iterative voting [19], in which agents change their votes iteratively after seeing the current votes of other agents. Technically, our model can be also seen as a model of iterative voting where in each iteration all voters change their vote slightly closer to the current aggregated point (while in standard iterative voting, usually voters strategically change their vote), how-ever, again, we prefer to think about our model as a model of deliberation.

3 Formal Model

We describe our formal model, which is parameterized by a metric space (X, d) and a voting rule $\mathcal{R}$. The first three ingredients of our model – namely, the metric space, the agent population, and the voting rule – are adapted from the model of Bulteau et al. [5] – while the discussion ingredient, which is the center of our work, is novel.

Metric Space. Let (X, d) be a metric space with X being a set of elements in the metric space and $d : X \times X \to \mathbb{R}$ being a metric function, so that (1) d is *symmetric*, with $d(x, y) = d(y, x)$ for every pair $x, y \in X$, (2) d is *non-negative*, with $d(x, y) \geq 0$ and $d(x, y) = 0 \leftrightarrow x = y$, and (3) d satisfies the *triangle inequality* i.e., $d(x, z) \leq d(x, y) + d(y, z)$ holds for all $x, y, z \in X$.

Agent Population. Let $V = \{v_0, \ldots, v_{n-1}\}$ be an agent population. Each agent $v \in V$ is associated with its *initial ideal point*, which is an element $x \in X$, understood as the element of X that is most preferred by v. We denote the initial ideal point of agent v_i by v_i^0. (Note that, effectively, the metric space sets the agents' ballot type as well as a distance function between possible ballots.)

Voting Rule. For a metric space (X, d), let $\mathcal{R}$ be a function that takes n elements of X and returns an element $w \in X$. We refer to a set of n elements of X as a *profile* V (of voter ballots) and write $\mathcal{R}(V) = w$ to denote that the $\mathcal{R}$-*winner* of the election (with profile V) is w.

Deliberation. We model deliberation as an iterative process, such that, in each iteration, the positions of the voters might change. Initially, the positions of the voters are given by their initial ideal points. Then, in each iteration, we apply the voting rule $\mathcal{R}$; consequently, all voters move slightly closer to the current $\mathcal{R}$-winner: specifically, denoting by v_i^j the ideal point of voter i at the beginning of the jth iteration (so, in particular, v_i^0 are the initial ideal points), and denoting by $V^j = \{v_0^j, \ldots, v_{n-1}^j\}$ and the $\mathcal{R}$-winner of the jth iteration by w^j (i.e., w^j is the result of applying $\mathcal{R}$ on V^j), we have the following constraints, for some value of ϵ.[2]

Constraint 1. $d(v_i^{j+1}, w^j) = max(0, d(v_i^j, w^j) - \epsilon)$.

Constraint 2. $d(v_i^{j+1}, v_i^j) = \epsilon$ *unless* $d(v_i^{j+1}, w^j) = 0$ *and then* $d(v_i^{j+1}, v_i^j) \leq \epsilon$.

That is, each voter moves an ϵ-closer to the current winner (unless it is already at most an ϵ-close to the current winner, in which case it moves to the winner itself); the second constraint is to make sure that voters do not "jump around" too arbitrarily.

We say that the iterative deliberation process *converges* if all agents cease to move after some finite number of iterations; note that, when an iterative deliberation process converges all agents are in consensus.

We say that convergence is guaranteed for some metric space (X, d) and voting rule $\mathcal{R}$ if all possible deliberation processes *converge*, for every $\epsilon > 0$; note that in each iteration, the agents can have multiple options to move, sometimes even an infinite number of options, and there may be an infinite number of initial profiles, so there may be an infinite amount of different deliberation processes.

---

[2] Indeed, for some sparse spaces these two constraints may not be always satisfiable, as agents moving towards the current winner may need to jump "too far". In the metric spaces we consider in this paper there is always at least a specific ϵ for which these constraints are indeed satisfiable.

Example 1. Let (X, d) be with X being $\mathbb{Z}$ and $d(x, y) = |x - y|$. Let $\mathcal{R}(V) = \sum_{v \in V} \lfloor v_i/n \rfloor$ (so the $\mathcal{R}$-winner is the average, rounded down). Let $V = \{v_0, v_1, v_2\}$ with $v_0^0 = 3$, $v_1^0 = 5$, and $v_2^0 = 8$. Let ϵ be 1. Then, the iterative deliberation process proceeds as follows: (1) at the beginning of the first iteration, v_0 stands on 3, v_1 on 5, and v_2 on 8. The $\mathcal{R}$ winner is $w := \lfloor (3+5+8)/3 \rfloor = 5$. Now, each v_i moves an ϵ-closer to 5; (2) at the beginning of the second iteration, v_0 stands on 4, v_1 remains on 5, and v_2 stands on 7. The $\mathcal{R}$ winner is again $w := 5$; (3) at the beginning of the third iteration, v_0 stands on 5, v_1 remains on 5, and v_2 stands on 6. The $\mathcal{R}$ winner is again $w := 5$; (4) at the beginning of the fourth iteration, v_0 stands on 5, v_1 remains on 5, and v_2 stands on 5. The $\mathcal{R}$ winner is again $w := 5$. In particular, for this example, the iterative deliberation process converges, as, after the fourth iteration, nobody would move. See Fig. 1.

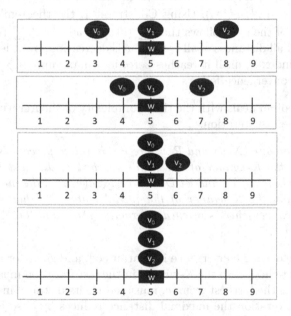

Fig. 1. Illustration for Example 1. The top box shows the initial configuration, with voter v_0 at 3, voter v_1 at 5, and voter v_2 at 8, implying that the aggregated point w is at 5. The second box from the top depicts the situation at the second iteration, each box below shows the situation after another iteration, and the process converges at the fourth iteration, as shown in the bottom box, in which all voters are in consensus at 5.

4   General Observations

We begin with some observations, regarding general sufficient conditions that guarantee convergence (later we will discuss specific metric spaces).

First, we observe that, whenever the process converges, it indeed converges to consensus. This follows as, if the voters are not in consensus, then at least one voter would move in the current iteration.

Observation 1. *In a converged configuration, the profile is consensus.*

The next theorem roughly says that, if the winner does not move too much, then deliberation is guaranteed; the proof follows the intuition that, if all agents move ϵ-closer to the winner, but the winner moves "slower" than this (in particular, by less than ϵ by a $\delta > 0$, for if the demand was only less than ϵ, we could have just approached convergence in the limit where the number of iterations approaches ∞), then eventually the process shall converge.

Theorem 2. *Consider some (X, d) and $\mathcal{R}$. If, for each profile V and for each ϵ, there is an index k such that for each $j \geq k$ there exists $\delta > 0$, and it holds that $d(w^j, w^{j+1}) \leq \epsilon - \delta$, then convergence is guaranteed.*

Proof. The following holds, by the triangle inequality: $\sum_i (d(v_i^{j+1}, w^{j+1})) \leq \sum_i (d(v_i^{j+1}, w^j) + d(w^j, w^{j+1}))$. Using Constraint 1, the theorem's assumption, and contraction of the ϵ, it follows that $\sum_i (d(v_i^{j+1}, w^{j+1})) \leq \sum_i (d(v_i^j, w^j)) - \delta$. Thus, the sum of all distances of all v_i from w is decreasing, by at least a δ in each iteration, from index k, until it reaches zero, in a maximum of $\sum_i (d(v_i^0, w^0))/\delta$ iterations, and convergence follows.   □

The next theorem deals with the time complexity of the iterative process and the winner of the last iteration.

Theorem 3. *For any (X, d) and $\mathcal{R}$ where the iterative process is such that for every profile V, the $\mathcal{R}$-winner of the profile reached in the next iteration V' is equal to the $\mathcal{R}$-winner of V, and let $D = max_{v \in V_0} d(v, w^0)$, the maximal distance of any agent from the $\mathcal{R}$-winner of the initial state, w^0, then the number of iterations until convergence is reached is exactly $\lceil D/\epsilon \rceil$ and the $\mathcal{R}$-winner of the last iteration is w^0.*

Proof. From Theorem 2 convergence is guaranteed, and since for any profile the $\mathcal{R}$-winner doesn't change, so the $\mathcal{R}$-winner in the last iteration must be w_0. Also, the voter which is the farthest from w, must reduce its distance in each iteration by ϵ, so in each iteration the maximal distance reduces by ϵ, so in exactly D/ϵ iterations, convergence shall be reached.   □

The next theorem roughly says that, if the agents move to the center of mass, then convergence is guaranteed.

Theorem 4. *For any (X, d), if $\mathcal{R} = \arg \min_{x \in X} \sum_{v \in V} d(v, x)$, then convergence is guaranteed.*

Proof. We show that in all iterations, the winner stays the same, and therefore, by Theorem 2, convergence follows. Let w denote the winner for some iteration V, so w has the minimum sum of distances from the agents in that iteration - $w = \arg \min_{x \in X} \sum_{v \in V} d(v, x)$. In the next iteration V', suppose that some element y is the winner - $y = \arg \min_{x \in X} \sum_{v \in V'} d(v, x)$ Note that every agent got closer to w by ϵ, unless it was already less than ϵ-far away from w, in which case it moved to w. We denote by V_1 the group of agents that were ϵ or more far away from w, and $V_2 = V - V_1$. So, the new sum of distances from w is $d_w = \sum_{v \in V'} d(v, w) = \sum_{v \in V_1}(d(v, w) - \epsilon) + \sum_{v \in V_2} d(v, w) - \sum_{v \in V_2}(v, w)$ where we replaced V_1' by V_1 by reducing ϵ, and just added and subtracted all elements in V_2. $d_w = \sum_{vinV} d(v, w) - n\epsilon - \sum_{v \in V_2} d(v, w)$, where we joined V_1 and V_2 into V. $d_x = \sum_{v \in V'} d(v, x)$. But by the triangle inequality, for $v' \in V'$ and the corresponding $v \in V$, $d(x, v') + d(v, v') \geq d(x, v)$, and by Constraint 2 $d(v, v') \leq \epsilon$, so $d(x, v') \geq d(x, v) - \epsilon$. Replacing that in the first equation, $d_x \geq \sum_{v \in V}(d(v, x) - \epsilon) = \sum_{v \in V} d(v, x) - n\epsilon$. If we combine the two values, we get $d_x \geq d_w$, because the first sum is just the sum of distances in iteration V, where w was the argmin. Thus, $x = w$ and the claim follows by Theorem 2. $\square$

5   Deliberation in Euclidean Spaces

In this section we consider Euclidean spaces; these are natural spaces that are studied extensively in social choice [1]. Formally, we consider metric spaces (X, d) in which X is a T-dimensional[3] Euclidean space, $X = \mathbb{R}^T$, for some $T = \{1, 2, 3, \ldots\}$; as for the distance function d, we consider ℓ_p norms for $p \in \{1, 2, \infty\}$ (other distances are indeed possible, however here we concentrate on ℓ_p norms). As for the voting rule $\mathcal{R}$, we consider the element-wise mean and median, as formally defined below.

Definition 5. *The* element-wise mean *of n points/voters v_i, $i \in [n]$, in some $\mathbb{R}^T$, is a point in $\mathbb{R}^T$ such that the value in each dimension is the mean of the values of the voters in that dimension; that is, the value at dimension t, $t \in [T]$, is $\sum_{i \in [n]} v_i[t]/n$.*

Definition 6. *The* element-wise median *of n points/voters v_i, $i \in [n]$, in some $\mathbb{R}^T$ is a point in $\mathbb{R}^T$ such that the value in each dimension is the median of the values of the voters in that dimension; that is, the value at dimension t, $t \in [T]$, is $median(v_i[t])$, where we define the median of an even number of real numbers to be the larger between the two middle numbers.*

First, we show that, for ℓ_1, every coordinate of every agent moves closer to the winner and does not pass it.

Lemma 7. *Let (X, d) be such that $X = \mathbb{R}^T$ and d is ℓ_1. Then for any agent v, element i and iteration j, it holds that either $w[i] \leq v^{j+1}[i] \leq v^j[i]$ or $w[i] \geq v^{j+1}[i] \geq v^j[i]$ where w is the winner of the j's iteration.*

Proof. In ℓ_1 the contribution of every coordinate is just added with absolute; thus, if $v^{j+1}[i]$ is not between $v^j[i]$ and $w[i]$, then the contribution of $v^{j+1}[i]$ to $d(v^{j+1}, w)$, will need to be compensated by the other coordinates to accommodate for the ϵ reduction, and then $d(v^j, v^{j+1})$ will be greater than ϵ. $\square$

---

[3] We use "T" and not the standard "d", as "d" is taken by the metric space (X, d).

We use Lemma 7 to show that convergence is guaranteed, for ℓ_1, for element-wise mean.

Theorem 8. *Let (X, d) be such that $X = \mathbb{R}^T$ and d is ℓ_1, and let $\mathcal{R}$ be the element-wise mean. Then, convergence is guaranteed.*

Proof. We know that each coordinate can only approach the mean. So each voter has an ϵ total distance to move to change the mean, because it is ℓ_1. But in each dimension, some voters are larger than the mean, and some voters are smaller than it, so they will contribute contradictory values, and thus the mean will move strictly less than ϵ, and by Theorem 2 convergence is guaranteed.   □

Next we show that, in the case of ℓ_2, each voter, under our constraints, must move directly to the $\mathcal{R}$-winner in a straight line.

Lemma 9. *Let (X, d) be such that $X = \mathbb{R}^T$ and d is ℓ_2, then for every agent v and iteration j, there exists only one point v^{j+1} that adheres to the constraints, and that point is an ϵ closer on a straight line from v^j to the $\mathcal{R}$-winner of the j's iteration.*

We use Lemma 9 to show that, with ℓ_2 and the element wise mean, convergence is guaranteed.

Theorem 10. *Let (X, d) be such that $X = \mathbb{R}^T$ and d is ℓ_2, and let $\mathcal{R}$ be the element-wise mean. Then, convergence is guaranteed.*

Proof. If we look in a coordinate system in which only one axis aligns with the vector between each v and the mean (rotation is invariant to ℓ_2), then we can see that the contribution of v is just ϵ/n in the direction from v to the mean, by Lemma 9, because only one axis has a delta which is not 0. Because it cannot be that all those vectors are in the same direction (in that case, the mean would be closer in the opposite direction until it passed one of them), it follows that the mean moves strictly less than ϵ; by Theorem 2, convergence follows.   □

Next, we look at the time order and the last converged winner in ℓ_1 and ℓ_2 in element wise-mean, and we show that if we look at the smallest T-dimensional ball that contains all the agents, then the time order is of the diameter of that ball divided by ϵ, and the converged winner is inside that ball.

Theorem 11. *Let (X, d) be such that $X = \mathbb{R}^T$ and d is ℓ_1 or ℓ_2, and let $\mathcal{R}$ be the element-wise mean, and let D be the diameter of the smallest T-dimensional ball that contains all n agents. Then, the process would converge to a point inside that ball in $O(D/\epsilon)$ iterations.*

Now we show, using both Lemma 7 and Lemma 9, that for both ℓ_1 and ℓ_2, in the case of element-wise median, convergence is guaranteed.

Theorem 12. *Let (X, d) be such that $X = \mathbb{R}^T$ and d is ℓ_1 or ℓ_2, and let $\mathcal{R}$ be the element-wise median. Then, convergence is guaranteed.*

Proof. In ℓ_1, by Lemma 7, each coordinate stays between the $\mathcal{R}$-winner and the last coordinate, and in ℓ_2, by Lemma 9, also each coordinate stays between the $\mathcal{R}$-winner and the last coordinate, so the median in each coordinate does not change, and because d is the element-wise median, the $\mathcal{R}$-winner does not change, and by Theorem 2, convergence is guaranteed. $\square$

The time order and final winner are known from Theorem 3.

Corollary 13. *Let (X, d) be such that $X = \mathbb{R}^T$ and d is ℓ_1 or ℓ_2, and let $\mathcal{R}$ be the element-wise median, and let $D = max_{v \in V} \circ d(v, w^0)$, then the number of iterations until convergence is reached is exactly $\lceil D/\epsilon \rceil$ and the $\mathcal{R}$-winner of the last iteration is w^0.*

In contrast to the results above, below we show that, for ℓ_∞, for the element-wise mean and element wise median, convergence is not guaranteed, at least for T-dimensional Euclidean spaces with $T \geq 3$. We show this by two counter examples. The first (second) example deals with element-wise mean (respectively, median).

Example 14. Set $\epsilon = 1$, $n = 3$, $v_0^0 = (-4, 2, 2)$, $v_1^0 = (2, -4, 2)$, $v_2^0 = (2, 2, -4)$. Running the iterative process for these initial conditions would result in adding 1 to each voter, in each dimension, in each step of the process. I.e.: $v_0^1 = (-3, 3, 3)$, $v_1^1 = (3, -3, 3)$, $v_2^1 = (3, 3, -3)$; and generally: $v_0^j = (-4 + j, 2 + j, 2 + j)$, $v_1^j = (2 + j, -4 + j, 2 + j)$, $v_2^j = (2 + j, 2 + j, -4 + j)$. Indeed, this is an endless behavior, exploding to infinity. It is possible to adapt both this example and the following one to any dimension $T > 3$, by adding as many zero dimensions as needed.

Example 15. The same pattern repeats for the element wise median $v_0^0 = (0, 0, 0)$, $v_1^0 = (-2, 0, 0)$, $v_2^0 = (0, -2, 0)$, $v_3^0 = (0, 0, -2)$. $v_0^{j+1} = (j, j, j)$, $v_1^{j+1} = (-1 + j, 1 + j, 1 + j)$, $v_2^{j+1} = (1 + j, -1 + j, 1 + j)$, $v_3^{j+1} = (1 + j, 1 + j, -1 + j)$. As we defined the median to be the larger number when there is an even number of agents, the median increases by 1 in every dimension in every iteration.

Remark 16. Generally speaking, non-convergence can be due to two possibilities: (1) Getting stuck in a cycle; or (2) moving to infinity. Note that Example 14 and Example 15 are of the second type, which, in a way, is more dramatic; and, perhaps, less intuitive.

6 Deliberation in Hypercubes

Next we consider T-dimensional hypercubes; these spaces naturally correspond to multiple referenda [17] as well as to multiwinner elections [12] and committee selection with variable number of winners [13]; below, we consider the latter two settings separately (indeed, for convenience, we use the jargon of multiwinner elections). We consider approval ballots here (in the next section we consider ordinal elections).

An important point to make, to all the discrete metric spaces we look at, is that we consider only $\epsilon \in \mathcal{N}$, because otherwise our metric spaces would be too sparse, and our constraints could not be met.

6.1   Committee Elections with Variable Number of Winners

The social choice setting here consists of a set of candidates and a set of agents such that each agent provides a subset of the candidates; then, a subset of the candidates – without restrictions on its size – is to be selected as the winner of the election. This setting is studied under the umbrella of committee elections with variable number of winners [13]; following the literature, we refer to this setting as *VNW*.

Formally, we have a metric space (X, d), where $X = \{0, 1\}^m$ for some integer m and $d(u, v)$ is the Hamming distance. An important class of VNW rules are monotonic rules, as defined next.

Definition 17. *A VNW rule $\mathcal{R}$ is monotonic if the following holds: for each profile V and its $\mathcal{R}$-winner w, it holds that the $\mathcal{R}$-winner w' for V', where V' is similar to V except for one agent that either (1) flips some 1 to 0 for some candidate not in w; or (2) flips some 0 to 1 for some candidate in w; then $w' = w$ (i.e., w stays).*

We show a rather general result next, applying to all monotonic VNW rules. Indeed, many VNW rules are monotonic: in particular, Majority is.

Theorem 18. *Let (X, d) be such that X is VNW and d is the Hamming distance, and let $\mathcal{R}$ be a monotonic VNW rule, then convergence is guaranteed.*

Proof. In each iteration, each agent v must reduce its Hamming distance from w by ϵ. So, that means it must flip ϵ bits that are either 1 in v and 0 in w, or 0 in v and 1 in w. (If $d(v, w) < \epsilon$ then it flips fewer bits, or even 0 bits if it coincides with w.) Now, using – for $m \cdot \epsilon$ times – the fact that $\mathcal{R}$ is *monotonic*, we deduce that the winner stays the same; the result then follows from Theorem 2.   □

The time order and final winner are known from Theorem 3.

Corollary 19. *Let (X, d) be such that X is VNW and d is the Hamming distance, and let $\mathcal{R}$ be a monotonic VNW rule, then $\mathcal{R}$-winner of the last iteration is the $\mathcal{R}$-winner of the first iteration, w^0, and the number of iterations is $\max_{v \in V^0} \lceil d(v, w^0)/\epsilon \rceil$.*

Remark 20. The setting of committee election (i.e., multiwinner elections; MW), is similar to the setting of VNW, thus is omitted (and is available in the full version).

Remark 21. We mention another metric distance, which we developed to imitate the ℓ_∞ property in Euclidean spaces in our quest for non-convergence, however this distance also converges: The *first changed* distance is defined as a function $d : X \times X \to N$ which is equal to $d(v_1, v_2) = \arg\min_i v_1[i] \neq v_2[i]$. And indeed, it turns out that for any (X, d) such that X is MW or VNW and d is the *first changed* distance, and where $\mathcal{R}$ is a monotonic voting rule, then convergence is guaranteed, and the number of iterations is $\lceil m/\epsilon \rceil$. Full details are available in the full version.

7 Ordinal Elections

Here we consider the standard ordinal model of elections [4]: in this setting there is a set of candidates and a set of agents such that each agent provides a linear order (i.e., a ranking, or, equivalently, a permutation) over the set of candidates; then, the result of the aggregation method – that is usually called a *social welfare function* – is an aggregated ranking; following the literature, we refer to this setting as *SWF*. Formally, we have a metric space (X, d) where X is the set of linear orders over some underlying set of candidates and d is the swap distance (of course, other distances are possible, however the swap distance is perhaps the most natural and most popular distance in this context [10,11,16]). In this setting too, we consider only $\epsilon \in \mathcal{N}$, as explained in Sect. 6.

As for the voting rule $\mathcal{R}$, first we observe that, as Kemeny is the realization of $\arg \min_{x \in X} \sum_{v \in V} d(v, x)$ for this context, the next result follows Theorem 4.

Corollary 22. *Let (X, d) be such that X is SWF and d is any distance, and let $\mathcal{R}$ be Kemeny, then convergence is guaranteed.*

And from Theorem 3, the winner of the last iteration is the winner of the first iteration, and the number of iterations is $max_{v \in V_0} d(v, w_0)/\epsilon$.

Corollary 23. *Let (X, d) be such that X is SWF and d is any distance, and let $\mathcal{R}$ be Kemeny, then the $\mathcal{R}$-winner of the final iteration is w^0, and the number of iterations is $max_{v \in V^0} \lceil d(v, w^0)/\epsilon \rceil$.*

As for other voting rules, we provide a rather general result, following the next definition.

Definition 24. *An SWF rule $\mathcal{R}$ is a scoring rule if it corresponds to a function $f : X^n \to \mathbb{R}^m$ (i.e., it takes a profile of n agents, and assigns an individual score (real number) to each candidate - m in the number of candidates), such that it chooses the $\mathcal{R}$-winner by sorting the candidates in decreasing order of their scores (Ties can be handled by an arbitrary, fixed order O over the candidates).*

Definition 25. *A scoring rule is a monotonic scoring rule if for every two profiles V and V', if a candidate c is ranked at least as high in V compared to V' for every agent, then $f(V)[c] \geq f(V')[c]$.*

Theorem 26. *Let (X, d) be such that X is SWF and d is the swap distance, and let $\mathcal{R}$ be a monotonic scoring rule, then convergence is guaranteed.*

Proof. The proof follows a potential function argument. To this end, we define a potential function that assigns a vector to each profile, and we define a lexicographic order on these vectors, and show that each iteration of the deliberation process can only advance in that order in one direction. We also show that the only way that we can stop advancing is if we are in consensus, in which case we have reached a maximum and the process would halt.

More formally, for a profile V, denote by w the $\mathcal{R}$-winner of V. Then, define a triplet for each candidate with index i in w (denoted by w_i), as follows:

$(f(V)[w_i], O[w_i], B(V)[w_i])$, where $B(V)[w_i]$ is the Borda score of candidate w_i. (i.e., for each candidate c, $B(V)[c] := \sum_{v \in V} m - pos_v(c)$, where $pos_v(c)$ is the position of c in the vote of v.) Then, define a vector combining all triplets in the order that their respective candidates appear in w, and consider an order on all profiles V according to the lexicographic order of these vectors.

Example 27. Let $\mathcal{R}$ be Plurality, let the set of candidates be $\{a, b, c\}$, let $O = (a, b, c)$, and let $V = \{v_0, v_1, v_2\}$ with $v_0 = \{a, b, c\}$, $v_1 = \{a, b, c\}$ and $v_2 = \{c, a, b\}$. Then $w = \{a, c, b\}$, and the triplet for a is $(2, 2, 5)$, for c it is $(1, 0, 2)$, and for b it is $(0, 1, 2)$. The combined vector for V is thus $(2, 2, 5, 1, 0, 2, 0, 1, 2)$.

In case of consensus, no swap is made, thus the profile remains the same, so we stay with the same vector and place in the order. Otherwise, we look at the index i in the $\mathcal{R}$-winner (of the iteration before the swap) of the first candidate that was swapped in some agent. First we notice, that it could not have been swapped backwards in any of the agents, because that would violate constraint 1 (that the distance from w must reduce by ϵ; we could not have swapped it back, and make up for it by swapping forward one more candidate, because that would violate constraint 2). Next, its potential must have increased, because at the least, its Borda score must have increased (Borda is strictly monotonic); and because our scoring rule is monotonic, its scoring function did not decrease, and O stayed the same because it is constant. Also, all the candidates in front of candidate indexed i in w did not swap, so their scoring function, Borda score and O did not decrease. Thus, so far we showed that if all the candidates in the new $\mathcal{R}$-winner until place i remained in the same order as the old one, our potential must have increased, and our proof is done. If their order has changed, then we look at the first index $j \leq i$ that has changed (that replaced its candidate). Now, because the switch had occurred, we know that the new candidate must have a higher f score, or the same f score, and a higher O score, by our definition of the *monotonic scoring rule*, and so, our potential increased. So, we proved that the potential must increase in all cases, and so we must advance in our order, until consensus is reached.   □

As Plurality, Borda, and Copeland are all monotonic scoring rules, they all converge.

Corollary 28. *Let (X, d) be such that X is SWF and d is the swap distance, with $\mathcal{R} \in \{Plurality, Borda, Copeland\}$, then convergence is guaranteed.*

We look at another voting rule, STV, with swap distance.

Definition 29. *STV is a SWF rule that chooses the winner as follows: V_0 is set to be V, the input profile of the voting rule; then, in iteration k (of the STV rule procedure, not the iterative process), the Plurality looser of V_k, c, is determined, and $w[m - 1 - k]$ is set to be c. Then V_{k+1} is set to be V_k, with all instances of c removed, leaving $m - 1 - k$ candidates in each agent. This process repeats for m iterations, until w is filled, from the last place to the first. Note that because the definition uses Plurality, then there is also a vector O that defines an order on the candidates for it.*

Theorem 30. *Let (X, d) be such that X is SWF and d is the swap distance, and let $\mathcal{R}$ be STV, then convergence is guaranteed.*

The same proof as with committee elections with *first changed* distance works in Ordinal elections as well.

Corollary 31. *Let (X, d) be such that X is SWF and d is the first changed distance, and let $\mathcal{R}$ be any monotonic voting rule, then convergence is guaranteed, and the number of iterations is exactly $\max_{v \in V^0} \lceil d(v, w^0)/\epsilon \rceil$.*

8 Outlook

We introduced a model of iterative deliberation in metric spaces and instantiated it with several natural social choice settings, by selecting appropriate metric spaces and voting rules. We identified those settings for which convergence of the process is guaranteed, and provided upper bounds regarding the number of iterative steps required for consensus (for those settings in which deliberation is guaranteed to succeed in finding a consensus). Below we mention some directions for future research:

- It is natural to consider further metric spaces, as well as further voting rules; a natural place to look for relevant metric spaces and voting rules is the work of Bulteau et al. [5].
- Another, more relaxed model, that comes to mind, is one where each voter must approach the winner by up to ϵ, instead of exactly by ϵ, and at least one voter must approach the winner by at least δ. This is a more general model, which is a bit closer to reality, where there is only an upper bound on the movement, and a demand that there is movement in each iteration. The demand is at least δ and not just larger than zero, because otherwise we could only reach convergence in the limit when the number of iterations approaches ∞.
- It is natural to study a stochastic model of iterative deliberation, including such that include radicalization (meaning that an agent can move away from the aggregated point, instead of approaching it). A stochastic model, in which such moves happen according to some probability or probability distribution may be closer to reality, thus has the potential of shedding more light on settings for which a deterministic process may converge, but a stochastic process may not. For a stochastic model that incorporates a non-zero probability for radicalizing voters, intuitively, if the probability mass of radicalization is not too large, then convergence shall be maintained.
- Another idea would be to consider coalition structures (such as those of Elkind et al. [7]) in which the agents of each coalition move slightly towards the center of each coalition and study issues of convergence there. (This would be different than the one-coalition setting we consider here; in a way, this would be like several dynamic deliberation groups.)

Acknowledgments. Nimrod Talmon and Eyal Leizerovich were supported by the Israel Science Foundation (ISF; Grant No. 630/19).
Most importantly, we thank the **Hoodska Explosive** for years of fun.

References

1. Arrow, K.: Advances in the Spatial Theory of Voting. Cambridge University Press, Cambridge (1990)
2. Austen-Smith, D., Feddersen, T.: Deliberation and voting rules. In: Austen-Smith, D., Duggan, J. (eds.) Social Choice and Strategic Decisions. SCW, pp. 269–316. Springer, Heidelberg (2005). https://doi.org/10.1007/3-540-27295-X_11
3. Austen-Smith, D., Feddersen, T.J.: Deliberation, preference uncertainty, and voting rules. Am. Polit. Sci. Rev., 209–217 (2006)
4. Brandt, F., Conitzer, V., Endriss, U., Procaccia, A.D., Lang, J.: Handbook of Computational Social Choice. Cambridge University Press, Cambridge (2016)
5. Bulteau, L., Shahaf, G., Shapiro, E., Talmon, N.: Aggregation over metric spaces: proposing and voting in elections, budgeting, and legislation. J. Artif. Intell. Res. **70**, 1413–1439 (2021)
6. Cohen, J., Bohman, J., Rehg, W.: Deliberation and democratic legitimacy, 1997, pp. 67–92 (1989)
7. Elkind, E., Grossi, D., Shapiro, E., Talmon, N.: United for change: deliberative coalition formation to change the status quo. In: Proceedings of AAAI 2021, vol. 35, pp. 5339–5346 (2021)
8. Fain, B., Goel, A., Munagala, K., Sakshuwong, S.: Sequential deliberation for social choice. In: Devanur, N.R., Lu, P. (eds.) WINE 2017. LNCS, vol. 10660, pp. 177–190. Springer, Cham (2017). https://doi.org/10.1007/978-3-319-71924-5_13
9. Faliszewski, P., Gonen, R., Koutecký, M., Talmon, N.: Opinion diffusion and campaigning on society graphs. In: Proceedings of IJCAI 2018, pp. 219–225 (2018)
10. Faliszewski, P., Skowron, P., Slinko, A., Szufa, S., Talmon, N.: How similar are two elections? In: Proceedings of AAAI 2019, vol. 33, pp. 1909–1916 (2019)
11. Faliszewski, P., Skowron, P., Slinko, A., Szufa, S., Talmon, N.: Isomorphic distances among elections. In: Proceedings of CSR 2020, pp. 64–78 (2020)
12. Faliszewski, P., Skowron, P., Slinko, A., Talmon, N.: Multiwinner voting: a new challenge for social choice theory. Trends Comput. Soc. Choice **74**, 27–47 (2017)
13. Faliszewski, P., Slinko, A., Talmon, N.: Multiwinner rules with variable number of winners. In: Proceedings of ECAI 2020 (2020)
14. Garg, N., Kamble, V., Goel, A., Marn, D., Munagala, K.: Iterative local voting for collective decision-making in continuous spaces. J. Artif. Intell. Res. **64**, 315–355 (2019)
15. Grandi, U.: Social choice and social networks. Trends Comput. Soc. Choice, 169–184 (2017)
16. Hogrebe, T.: Complexity of distances in elections: doctoral consortium. In: Proceedings of AAMAS 2019, pp. 2414–2416 (2019)
17. Lang, J., Xia, L.: Voting in combinatorial domains. In: Handbook of Computational Social Choice, pp. 197–222. Cambridge University Press, Cambridge (2016)
18. Lizzeri, A., Yariv, L.: Sequential deliberation (2010). SSRN 1702940
19. Meir, R.: Iterative voting. Trends in Comput. Soc. Choice, 69–86 (2017)

Manipulation in Voting

Obvious Manipulability of Voting Rules

Haris Aziz and Alexander Lam[(✉)]

UNSW Sydney, Kensington, Australia
{haris.aziz,alexander.lam1}@unsw.edu.au

Abstract. The Gibbard-Satterthwaite theorem states that no unanimous and non-dictatorial voting rule is strategyproof. We revisit voting rules and consider a weaker notion of strategyproofness called not obvious manipulability that was proposed by Troyan and Morrill (2020). We identify several classes of voting rules that satisfy this notion. We also show that several voting rules including k-approval fail to satisfy this property. We characterize conditions under which voting rules are obviously manipulable. One of our insights is that certain rules are obviously manipulable when the number of alternatives is relatively large compared to the number of voters. In contrast to the Gibbard-Satterthwaite theorem, many of the rules we examined are not obviously manipulable. This reflects the relatively easier satisfiability of the notion and the zero information assumption of not obvious manipulability, as opposed to the perfect information assumption of strategyproofness. We also present algorithmic results for computing obvious manipulations and report on experiments.

Keywords: Social choice · Voting · Manipulation · Strategyproofness

1 Introduction

Throughout history, voting has been used as a means of making public decisions based on the citizens' preferences. The ancient Greeks would give a show of hands to disclose their most preferred public official, and the winner of the election was chosen as the official with the most first preferences [4]; such a voting system is called the *plurality vote*. Many other voting systems have been developed over time, such as the Borda Count, developed by Jean-Charles de Borda in 1770. The Borda Count gives each candidate a score based on their position in the voters' preference orders. This system was opposed by Marquis de Condorcet, who instead preferred the Condorcet method, which elects the candidate that wins the majority of pairwise head-to-head elections against the other candidates [2]. However, voting systems are not just used in politics; voting theory is frequently used and studied in artificial intelligence to aggregate the preferences of multiple agents into a single decision.

The studies of electoral systems in social choice theory have been wrought with negative results. Arrow's impossibility theorem [1] showed that there exists

© Springer Nature Switzerland AG 2021
D. Fotakis and D. Ríos Insua (Eds.): ADT 2021, LNAI 13023, pp. 179–193, 2021.
https://doi.org/10.1007/978-3-030-87756-9_12

no voting system with three reasonable requirements. In a similar vein, the Gibbard-Sattherthwaite theorem [10, 18] states that when there are at least three alternatives, every unanimous voting rule is either dictatorial, meaning only one voter's preferences are taken into account, or prone to manipulative voting, meaning a voter can give an untruthful ballot to gain a more preferred outcome.

Such strategic behaviour is a commonly studied problem in mechanism design and social choice, as many mechanisms sacrifice efficiency or fairness to ensure strategyproofness. The original notion of strategyproofness fails to explain the variation we observe in voters' tendency to strategically vote in different electoral systems. This has motivated research toward alternative concepts of strategyproofness that may be able to capture such variations. One such notion is *not obvious manipulability*, recently theorized by Troyan and Morrill [23]. Whilst strategyproofness assumes agents have complete information over other agent preferences and the mechanism operation, not obvious manipulability assumes agents are 'cognitively limited' and lack such information. As such, they are only aware of the possible range of outcomes that can result from each mechanism interaction. Put simply, a mechanism satisfies *not obvious manipulability (NOM)* if no agent can improve its best case or worst case outcome under any manipulation. A mechanism is *obviously manipulable (OM)* if either an agent's best case or worst case outcome can be improved by some untruthful interaction.

The assumptions made for *not obvious manipulability* are suitable when applied to voting rules, as ballots are commonly hidden from the voters, restricting their ability to compute a desirable manipulation. In this paper, we explore which voting rules are obviously manipulable, and if so, what the conditions are for obvious manipulability.

Contributions. Our main contribution is to apply the concept of obvious manipulations to the case of voting rules for the first time. We study which voting rules are obviously manipulable, and what conditions are required for obvious manipulability. Whilst many classes of voting rules including Condorcet extensions and strict positional scoring rules with weakly diminishing differences are not obviously manipulable, we show that certain voting rules, including k-approval, are obviously manipulable. We also characterize the conditions under which positional scoring rules are obviously manipulable in the best case. For the class of k-approval voting rules, we characterize the conditions under which the rules are obviously manipulable. Many of our results apply to large classes of voting rules including positional scoring rules or Condorcet extensions. Table 1 summarizes several of our results.

One of our insights is that certain rules are obviously manipulable when the number of alternatives is relatively large compared to the number of voters. We also look at the problem of checking whether a particular instance of a voting problem admits an obvious manipulation. For the class of positional scoring rules, we provide a general polynomial-time reduction to the well-studied *unweighted coalitional manipulation problem*. As a corollary, we show that the problem of checking the existence of an obvious manipulation is polynomial-time solvable

Table 1. List of rules and conditions for voting rules to be NOM or OM.

NOM	OM
Does not admit a voter with veto power	
k-Approval ($n > \frac{m-2}{m-k}$)	**k-Approval** ($n \leq \frac{m-2}{m-k}$)
Plurality	
Almost-unanimous	
Condorcet-extension	
STV	
Plurality with runoff	
Positional scoring rule ($n > \frac{s_1}{s_1 - s_2} + 1$)	Positional scoring rule that admits a voter with veto power (existence)
Positional scoring rule with weakly diminishing differences	
Borda rule	

for the k-approval rule. Finally, we report on experimental results on the fraction of instances that admit obvious manipulations for the k-approval rule.

2 Related Work

Our paper belongs to the rich stream of work in social choice on the manipulability of voting rules. The reader is referred to the book by Taylor [21] that surveys this rich field. A comparison of the susceptibility of voting rules to manipulation has a long history in social choice. For example, one particular approach is to count the relative number of preference profiles under which voting rules are manipulable (see, e.g., [8]). Another approach is analyzing the maximum amount of expected utility an agent can gain by reporting untruthfully [3].

Our work revolves around the concept of obvious manipulations, which was proposed by Troyan and Morrill [23]. This concept was inspired by a paper on 'obviously strategyproof mechanisms' by Li [11]. The latter paper describes the cognitively-limited agent that is only aware of the range of possible outcomes ranging from each report. In the paper, Li then proposes the characterization of 'obvious strategyproofness', a strengthening of strategyproofness. A mechanism is defined as obviously strategyproof if each agent's worst case outcome under a truthful report is strictly better than their best case outcome under any untruthful report. Troyan and Morrill [23] studied obvious manipulations in the context of matching problems. In particular, they showed that whereas the Boston mechanism is obviously manipulable, many stable matching mechanisms (including those that are not strategyproof) are not obviously manipulable.

Other, weaker notions of strategyproofness specific to voting rules have been proposed in the literature. Slinko and White [19,20] considered *safe strategic voting* to represent the coalitional manipulation of scoring rules. Assuming every member of the coalition reports the same ballot, a manipulation is a *safe strategic vote* if it guarantees an outcome which is weakly preferred over truth-telling. Another notion has also been proposed by Conitzer et al. [6], who state that a ballot *dominates* another ballot if it guarantees a weakly more preferred outcome. The authors define a voting rule as being *immune to dominating manipulations* if there are no ballots that dominate any voter's true preferences, and classify the immunity of certain rules under varying levels of information known by the manipulator. In particular relevance to our paper, they find that certain voting rules such as Condorcet-consistent rules and the Borda count are immune to dominating manipulations under zero information. We remark that immunity to dominating manipulations under zero information is a weaker notion than not obvious manipulability, and thus our work investigates a stronger notion defining a voting rule's resistance to manipulation than some existing notions. For further discussion on the strategic aspects of voting with partial information, the reader is referred to Chaps. 6 and 8 of the book by Meir [12], where similar concepts such as local dominance are discussed.

In many elections, voters often lack information of other voters' preferences. This has prompted a probabilistic perspective into the manipulability of voting rules, often assuming a uniform distribution over each preference ordering. In 1985, Nitzan showed that in point scoring rules, a manipulation is more likely to succeed as the number of outcomes increases, and the number of voters decreases [14]. A similar probabilistic perspective was used by Wilson and Reyhani [24]. Computer scientists have also extensively researched the computational complexity of calculating a manipulative ballot; as the number of voters and outcomes becomes large, it can be computationally infeasible to compute a manipulation if the problem is intractable (see, e.g. [5,7]).

3   Preliminaries

We consider the standard social choice voting setting $(N, O, \succ)$ that involves a finite set $N = \{1, 2, \ldots, n\}$ of n voters and a finite set $O = \{o_1, o_2, \ldots, o_m\}$ of m outcomes. We also assume that $n \geq 3$ and $m \geq 3$. Each voter i has a transitive, complete and reflexive preference ordering $\succ_i$ over the set of outcomes O. We denote the preference profile of each voter $i \in N$ as $\succ = (\succ_1, \ldots, \succ_n)$, and use $\mathcal{L}(O)^n$ to denote the set of all such profiles for a given n. For a given voter $i \in N$, we use $\succ_{-i} = (\succ_1, \ldots, \succ_{i-1}, \succ_{i+1}, \ldots, \succ_n)$ to denote the preference profile of the voters in $N \backslash \{i\}$. A voting rule $f \colon \mathcal{L}(O)^n \to O$ is a function that takes as input the preference profile and returns an outcome from O.

An outcome $o \in O$ is called a *possible outcome* under a voting rule f if there exists some preference profile $\succ$ such that $f(\succ) = o$. Since we are considering voting rules that return a single outcome, we will impose tie-breaking over social choice correspondences (voting rules that return more than one outcome) to return a single outcome. Unless specified otherwise, we will assume a fixed tie-break ordering over the outcomes.

Definition 1. *A voting rule f is* manipulable *if there exists some voter $i \in N$, two preference relations $\succ_i, \succ_i'$ of voter i, and a preference profile $\succ_{-i}$ of other* *voters such that* $f(\succ'_i, \succ_{-i}) \succ_i f(\succ_i, \succ_{-i})$. *Such a manipulation is defined as a* profitable manipulation *for voter* i. *A voting rule is* strategyproof (SP) *if it is not manipulable.*

Under voting rule f, a given set of outcomes and a fixed number of voters, we denote by $B_{\succ_i}(\succ'_i, f)$ the best possible outcome (under i's preference $\succ_i$) when she reports $\succ'_i$, over all possible preferences of the other voters. We also denote by $W_{\succ_i}(\succ'_i, f)$ the worst possible outcome (under i's preference $\succ_i$) when she reports $\succ'_i$, over all possible preferences of the other voters. We now present the central concept used in the paper, which has been adapted from the paper by Troyan and Morrill [23] to the field of voting.

Definition 2. *A voting rule* f *is not obviously manipulable (NOM) if for every voter* i *with truthful preference* $\succ_i$ *and every profitable manipulation* $\succ'_i$, *the following two conditions hold:*

$$W_{\succ_i}(\succ_i, f) \succeq_i W_{\succ_i}(\succ'_i, f) \tag{1}$$
$$B_{\succ_i}(\succ_i, f) \succeq_i B_{\succ_i}(\succ'_i, f). \tag{2}$$

If either condition does not hold, then we say the voting rule is *obviously manipulable*. Specifically, if (1) does not hold, then we say the voting rule is *worst case obviously manipulable*. Similarly, if (2) does not hold, then we say it is *best case obviously manipulable*.

4   Sufficient Conditions for Not Being Obviously Manipulable

In this section, we identify certain conditions that imply not obvious manipulability when satisfied by voting rules.

Definition 3. *For a given voting rule* f *and a fixed number of voters* n *and outcomes* m, *a voter* i *has veto power if there exists a possible outcome* $o \in O$ *and report* $\succ_i$ *such that* $f(\succ_i, \succ_{-i}) \neq o$ *for all* $\succ_{-i}$.

Our first result is a sufficient condition for a voting rule being NOM.

Lemma 1. *If a voting rule is obviously manipulable, then it must admit a non-dictatorial voter with veto power.*

However, existence of a voter with veto power does not imply obvious manipulability. We will illustrate this later in the paper.

Definition 4. *A voting rule* f *is almost-unanimous if it returns an outcome* o *when* o *is the most preferred outcome for at least* $n-1$ *voters. Almost-unanimity implies unanimity.*

Theorem 1. *For* $n \geq 3$, *no almost-unanimous voting rule is obviously manipulable.*

Proof. Note that an almost-unanimous voting rule is not dictatorial. By definition, a rule that is almost-unanimous cannot admit a voter with veto power. Hence it follows from Lemma 1 that for $n \geq 3$, no almost-unanimous voting rule is obviously manipulable.   □

Corollary 1. *Any majoritarian (Condorcet extension rule) is NOM.*

Similarly, Theorem 1 applies to several voting rules including STV [22] and Plurality with runoff [13] that are almost-unanimous.

Corollary 2. *STV and Plurality with runoff are NOM.*

We have shown that many voting rules are not obviously manipulable, so we question whether there are any obviously manipulable voting rules. We next investigate positional scoring rules.

5   Positional Scoring Rules

In this section, we consider positional scoring rules, a major class of voting rules which assigns points to candidates based on voter preferences and chooses the candidate with the highest score. A formal definition of a positional scoring rule is given below.

Definition 5. *A positional scoring rule assigns a score to each outcome using the score vector $w = (s_1, s_2, \ldots, s_m)$, where $s_i \geq s_{i+1} \forall i \in \{1, 2, \ldots, m-1\}$ and $\exists i \in \{1, 2, \ldots, m-1\} : s_i > s_{i+1}$. Each voter gives s_i points to their ith most preferred candidate, and the score of a candidate is the total number of points given by all voters. The candidate with the highest number of points is returned by the rule.*

Note that this positional scoring rule definition rules out unreasonable, pathological scoring vectors such as $(1, 2, 3)$. Several well-known rules fall in the class of positional scoring rules. For example if $s_i = m - i$ for all $i \in [m]$, the rule is the Borda voting rule. If $s_1 = 1$ and $s_i = 0$ for all $i > 1$, the rule is plurality. If $s_m = 0$ and $s_i = 1$ for all $i < m$, the rule is anti-plurality.

Next, we identify a sufficient condition for a positional scoring rule to be NOM.

Theorem 2. *A positional scoring rule is NOM if $n > \frac{s_1}{(s_1 - s_2)} + 1$.*

Proof. It is sufficient to show that for $n > \frac{s_1}{(s_1 - s_2)} + 1$, the rule is almost-unanimous. Any outcome a that is the most preferred by $n - 1$ voters has a score of at least $(s_1)(n - 1)$. We show that this score is greater than the score of any other candidate. The maximum score any other outcome b can get is by being in the first position of one voter and second position of all other voters so its score is $(s_2)(n-1) + s_1$. The score of a is greater than the maximum score of b if and only if

$$(s_1)(n-1) > (s_2)(n-1) + s_1$$
$$\Longleftrightarrow (n-1)(s_1 - s_2) > s_1$$
$$\Longleftrightarrow n > \frac{s_1}{(s_1 - s_2)} + 1.$$

$\square$

This result suggests that many positional scoring rules are NOM when there are sufficiently many voters, and that scenarios with few voters may be required for a positional scoring rule to be obviously manipulable.

5.1   k-Approval

The k-approval rule is a subclass of positional scoring rules that lets voters approve of their k most preferred candidates, or voice their disapproval for their $m - k$ least preferred candidates. It is a scoring rule with weight vector $w = (1, \ldots, 1, 0, \ldots, 0)$, where there are k ones, $m - k$ zeroes and $0 < k < m$.

Note that the k-approval rule is the same as the plurality rule when $k = 1$, and it is the same as the anti-plurality rule when $m - k = 1$.

Lemma 2. *The k-approval rule (kApp) is obviously manipulable if $n \leq \frac{m-2}{m-k}$.*

Proof. Suppose there are n voters, the number of outcomes m is at least $n(m - k) + 2$, voter i's true preferences are

$$\succ_i: o_1 \succ_i o_2 \succ_i \cdots \succ_i o_{m-1} \succ_i o_m,$$

and the fixed tie-break ordering is

$$\succ_L: o_k \succ_L o_1 \succ_L o_2 \succ_L \cdots \succ_L o_{k+1} \succ_L o_{k+2} \succ_L \cdots \succ_L o_{m-1} \succ_L o_m.$$

Under a k-approval rule, any voter may disapprove of their $m - k$ least preferred outcomes. Since there are a total of $n(m - k)$ disapprovals and $m \geq n(m-k)+2$, by the pigeonhole principle, there are at least 2 outcomes with zero disapprovals. Therefore the selected outcome must be the tie-break winner of the outcomes with zero disapproval votes, as they are approved by every voter.

Under a truthful ballot $\succ_i$, voter i disapproves of outcomes $\{o_{k+1}, \ldots, o_m\}$, so $W_{\succ_i}(\succ_i, kApp) \notin \{o_{k+1}, \ldots, o_m\}$. We therefore have $W_{\succ_i}(\succ_i, kApp) = o_k$ as at least two outcomes in $\{o_1, \ldots, o_k\}$ must have zero disapproval votes, and o_k has the highest tie-break priority.

If voter i instead disapproves of the outcomes in $\{o_k\} \cup \{o_k+1, \ldots, o_m\} \setminus \{o_{i'}\}$, where $k + 1 \leq i' \leq m$, then the worst case outcome satisfies $W_{\succ_i}(\succ_i', kApp) \succ_i o_{k-1}$, as $o_{i'}$ always loses the tie-break with any outcome from $\{o_1, \ldots, o_{k-1}\}$. We therefore have $W_{\succ_i}(\succ_i', kApp) \succ_i W_{\succ_i}(\succ_i, kApp)$, concluding the proof. $\square$

Lemma 3. *The k-approval rule ($kApp$) is NOM if $n > \frac{m-2}{m-k}$.*

Proof. Suppose that there are n voters, $m \leq \frac{kn-1}{n-1}$ outcomes and without loss of generality that voter i's true preferences are

$$\succ_i: o_1 \succ_i o_2 \succ_i \cdots \succ_i o_m.$$

We note that $m \leq \frac{kn-1}{n-1} \iff n(m-k) \geq m-1$, so there are at least $m-1$ disapproval votes as each of the n voters disapproves of $m-k$ outcomes. We first show that under these conditions, the k-approval rule is not best case obviously manipulable. Under $\succ_i$, voter i's best case outcome of $B_{\succ_i}(\succ_i, kApp) = o_1$ is achievable by the voters voting such that o_1 has zero disapprovals and each of the other outcomes has at least one disapproval. Since i's best case outcome is his first preference, it cannot be strictly improved by any manipulation.

We next show that in this scenario, the k-approval rule is not worst case obviously manipulable. By the pigeonhole principle, there must be at least one outcome with zero disapprovals. Under a truthful ballot, voter i disapproves of outcomes $\{o_{k+1}, \ldots, o_m\}$, so his worst case outcome is $W_{\succ_i}(\succ_i, kApp) = o_k$, achieved by the other voters disapproving of outcomes $\{o_1, \ldots, o_{k-1}\}$. Now under any manipulation, at least one outcome from $\{o_{k+1}, \ldots, o_m\}$ must be approved by voter i. This results in $W_{\succ_i}(\succ_i', kApp) \in \{o_{k+1}, \ldots, o_m\}$, as the other voters can vote such that every outcome except for voter i's least preferred approved outcome has been disapproved at least once. We therefore have $W_{\succ_i}(\succ_i, kApp) \succeq_i W_{\succ_i}(\succ_i', kApp)$, concluding our proof. $\square$

Remark 1. We note that the obvious manipulability of k-approval when $m \geq n(m-k)+2$ and the not obvious manipulability of k-approval when $m = n(m-k)+1$ also holds in the case of weighted voters, as the argument relies on the number of outcomes exceeding the total number of disapprovals.

Based on the two lemmas proved above, we achieve a characterization of the conditions under which the k-approval rule is obviously manipulable.

Theorem 3. *The k-approval rule is obviously manipulable if and only if $n \leq \frac{m-2}{m-k}$.*

Corollary 3. *The plurality rule is NOM.*

Since plurality is generally considered to be one of easiest rules to manipulate, the corollary above underscores the strength of obvious manipulations. We give the following intuition for the result on k-approval. Suppose a small committee is applying the k-approval rule to select a prize winner out of many candidates, and that certain candidates will be approved by every voter. The manipulator may also have a general idea of these candidates conditional on their report. If a fixed tie-break method is used (such as selecting the oldest candidate), the manipulator may disapprove of the oldest candidate who would otherwise win, instead approving a younger candidate who would not be selected regardless.

5.2   Strict Positional Scoring Rules

In the previous section, we noted that the k-approval rule is obviously manipulable. This may lead to the question of whether the lack of strictly decreasing scoring weights contributes to the obvious manipulability of a positional scoring rule. Hence, we focus on strict positional scoring rules in the following section.

Definition 6. *A positional scoring rule with weight vector* $w = (s_1, s_2, \ldots, s_m)$ *is strict if* $s_i > s_{i+1}$ *for all* $i \in \{1, 2, \ldots, m-1\}$.

We first note a strict positional scoring rule can be obviously manipulable.

Lemma 4. *There exists a strict positional scoring rule that can admit a voter with veto power and is obviously manipulable.*

In the following lemma, we also find that a strict positional scoring rule is not necessarily obviously manipulable if it admits a voter with veto power.

Lemma 5. *There exists a class of strict positional scoring rules that can admit a voter with veto power but are NOM.*

Definition 7. *A strict positional scoring rule with* $w = (s_1, s_2, \ldots, s_m)$ *has* diminishing differences *if* $s_i - s_{i+1} > s_{i+1} - s_{i+2}$ *for all* $i \in \{1, 2, \ldots, m-2\}$. *We say it has* weakly diminishing differences *if* $s_i - s_{i+1} \geq s_{i+1} - s_{i+2}$ *for all* $i \in \{1, 2, \ldots, m-2\}$.

An example of such a rule is the Harmonic-Borda/Dowdall system used in Nauru, which has weight vector $w = (1, 1/2, \ldots, 1/m)$ [17]. It is more favourable towards candidates that are the top preference of many voters, and has been described as a scoring rule that "lies between plurality and the Borda count" [9].

Next, we prove that a strict positional scoring rule with weakly diminishing differences is NOM.

Theorem 4. *A strict positional scoring rule with weakly diminishing differences is NOM.*

Corollary 4. *The Borda and Harmonic-Borda/Dowdall rules are NOM.*

Remark 2. Lemma 5 exemplifies a class of strict positional scoring rules which do not satisfy weakly diminishing differences but are NOM.

5.3   Obvious Manipulability in the Best Case

Although our previous results focus on worst case obvious manipulability, it is possible for a positional scoring rule to be best case obviously manipulable.

Lemma 6. *Assuming* $m, n \geq 3$, *a positional scoring rule* f *is best case obviously manipulable if and only if for some* $k > 1$, *the first* k *elements of the scoring vector are the same and* $n \leq \frac{m-2}{m-k}$.

Next, we demonstrate a fundamental connection between best case obvious manipulations and worst case obvious manipulations.

Theorem 5. *Assuming $m, n \geq 3$, for any positional scoring rule, if a voter's preference relation $\succ_i$ admits a best case obvious manipulation, then it also admits a worst case obvious manipulation.*

Proof. Suppose for some positional scoring rule f that a voter's preference relation $\succ_i$ admits a best case obvious manipulation. From Lemma 6, for some $k > 1$, the first k elements of the scoring vector must be the same, and we have $n \leq \frac{m-2}{m-k}$. Consequently, any outcome selected under f must be in the top k outcomes of each voter's report. We say that a voter 'approves' his k most preferred outcomes, and 'disapproves' of his $m - k$ least preferred outcomes. An outcome cannot be chosen by f if it has a disapproval vote from at least one voter.

We now construct the set of feasible outcomes O_f which can be selected under the voter's preference relation $\succ_i$ and some $\succ_{-i}$. Let O_v be the $m - k$ disapproved outcomes by i under $\succ_i$. Since any outcome with at least one disapproval vote cannot be chosen, no outcome in O_v can be selected. Now consider the set $O \backslash O_v$. Suppose without loss of generality that $O \backslash O_v = \{o_1, \ldots, o_k\}$, with tie-break ordering $\succ_L : o_1 \succ_L \cdots \succ_L o_k$. Denote $c := (n-1)(m-k)$ as the number of disapproval votes that the other $n - 1$ voters can distribute. For $j \in \{1, \ldots, c+1\}$, outcome o_j can be selected if the other voters cast disapproval votes for outcomes $\{o_1, \ldots, o_{c+1}\} \backslash \{o_j\}$. Furthermore, outcomes $o_{c+2}, \ldots, o_k$ cannot be selected, regardless of how the other voters report. Therefore the set of feasible outcomes O_f are the $c + 1$ highest tie-breaking ranked outcomes of the set $O \backslash O_v$. Voter i's best case outcome is its most preferred outcome in O_f, whilst its worst case outcome is its least preferred outcome in O_f. We denote $o_b := B_{\succ_i}(\succ_i, f)$ as i's best case outcome, and $o_w := W_{\succ_i}(\succ_i, f)$ as i's worst case outcome.

We now define the set of feasible outcomes O'_f under any preference report by voter i. This is the $n(m-k) + 1$ highest tie-break ranked outcomes of O. Now suppose $\succ_i$ admits a best case obvious manipulation. There must exist an outcome $o'_b \in O'_f \backslash O_f$ that i prefers over o_b. Consider the set $O'_v = \{o_w\} \cup O'_f \backslash (O_f \cup o'_b)$. Since $o_w \notin O'_f \backslash O_f$ and $o'_b \notin O_f$, we have

$$|O'_v| = |\{o_w\}| + |O'_f| - |O_f| - |\{o'_b\}|$$
$$= 1 + n(m-k) + 1 - (n-1)(m-k) - 1 - 1$$
$$= m - k.$$

We now deduce i's worst case outcome $W_{\succ_i}(\succ'_i, f)$ under the manipulation $\succ'_i$ where voter i disapproves of all outcomes from O'_v. Under $\succ'_i$, every outcome in $O'_f \backslash O_f$ except for o'_b has a disapproval vote and therefore cannot be selected. The outcome o'_b satisfies $o'_b \succ_i o_b$ and therefore cannot be the worst case outcome. Finally, o_w has a disapproval vote, so by elimination, we have $W_{\succ_i}(\succ'_i, f) \in O_f \backslash \{o_w\}$. Since o_w is voter i's least preferred outcome in O_f, we have $W_{\succ_i}(\succ'_i, f) \succ_i o_w$, meaning that $\succ'_i$ is a worst case obvious manipulation. $\qquad\square$

6   Computing Obvious Manipulations

In the previous parts of the paper, we focussed on understanding the conditions under which a voting rule is obviously manipulable. Next, we consider the problem of computing an obvious manipulation for a given problem instance. We present algorithmic results for computing obvious manipulations under positional scoring rules.

OBVIOUS MANIPULATION (OM)

Input: Number of voters n, set of outcomes $O = \{o_1, o_2, \ldots, o_m\}$, preference
   relation $\succ_i$ of voter i, tie-break order $\succ_L$ and voting rule f.

Problem: Find a preference relation $\succ'_i$ such that
   $W_{\succ_i}(\succ'_i, f) \succ_i W_{\succ_i}(\succ_i, f)$ or $B_{\succ_i}(\succ'_i, f) \succ_i B_{\succ_i}(\succ_i, f)$.

If we only consider the best case manipulation, we refer to the problem as BEST-CASE OBVIOUS MANIPULATION (BOM). If we only consider the worst case manipulation, we refer to the problem as WORST-CASE OBVIOUS MANIPULATION (WOM).

We present algorithms for the obvious manipulation problems. The algorithms are based on reductions to the Constructive Coalitional Unweighted Manipulation (CCUM) that is well-studied in computational social choice (see e.g., [25, 26]). We now introduce the CONSTRUCTIVE COALITIONAL UNWEIGHTED MANIPULATION (CCUM).

CONSTRUCTIVE COALITIONAL UNWEIGHTED MANIPULATION (CCUM)

Input: Voting rule f, set of outcomes O, distinguished candidate $o \in O$, set
   of voters S that have already cast their votes and set of voters T that
   have not cast their votes.

Problem: Is there a way to cast the votes in T such that o wins the election
   under f?

We show that for any voting rule, there is a polynomial-time algorithm for computing a best case obvious manipulation if CCUM can be solved in polynomial time.

Lemma 7. *For any voting rule, there is a polynomial-time algorithm for BOM if CCUM can be solved in polynomial time.*

Proof. Denote $o_b := B_{\succ_i}(\succ_i, f)$. We can compute o_b as follows. We fix the preference $\succ_i$ of voter i and solve CCUM for each possible outcome while keeping all the other voters as manipulators. This can be checked in $|O|$ calls to an algorithm to solve CCUM. Next, we find i's best possible outcome if she is allowed to report any other preference. This can be checked by solving CCUM for each possible outcome while keeping all the voters as manipulators. Let o^* be the possible outcome that is most preferred with respect to $\succ_i$. The instance is best case obviously manipulable if and only if $o^* \succ_i o_b$.   □

We then show that for any positional scoring rule, there is a polynomial-time algorithm for OM if CCUM can be solved in polynomial time.

Lemma 8. *For any positional scoring rule, there is a polynomial-time algorithm for* WOM *if* CCUM *can be solved in polynomial time.*

Proof. First we compute the worst case outcome $W_{\succ_i}(\succ_i, f)$ of i when she reports the truth. This is easily computed by running an algorithm that solves CCUM with i's report being fixed, and checking which outcomes are possible.

We check whether i can improve her worst case outcome by misreporting. We denote $o_w := W_{\succ_i}(\succ_i, f)$ and $O_{bad} := \{o \in O : o_w \succ_i o\} \cup \{o_w\}$. We also denote $O_{good} := A \setminus O_{bad}$. We want to check whether i can ensure that no outcome from O_{bad} is selected irrespective of how the other voters vote. We define a misreport $\succ_i'$ as follows. In $\succ_i'$, the outcomes of O_{good} are preferred over the outcomes of O_{bad}. In O_{good} the outcomes are ordered so that higher (tie-break) priority outcomes come earlier. In O_{bad} the outcomes are ordered so that higher priority outcomes come later. We solve CCUM with respect to $\succ_i'$ and check whether some outcome in O_{bad} can be selected. If such an alternative cannot be selected, we return yes. Otherwise we return no.   □

Combining the two lemmas above, we get the following.

Theorem 6. *For any positional scoring rule, there is a polynomial-time reduction from solving* OM *to solving* CCUM.

Conitzer and Walsh [5] discuss the computational complexity of CCUM for various different voting rules. In particular, CCUM can be solved in polynomial time for the k-approval problem. For example, Zuckerman et al. [26] present a greedy polynomial-time algorithm for computing CCUM. For the sake of completeness, we explicitly write this algorithm for the k-approval rule with a fixed tie-break ordering. The algorithm assigns approved outcomes to the manipulators as follows. First, it assigns the distinguished outcome as each manipulator's first preference. Each manipulator then approves the $k - 1$ outcomes with the lowest scores. If there are more than $k - 1$ tied outcomes, the ones with the lowest tie-break priority are selected.

Corollary 5. OM *can be solved in polynomial time for* k-*approval.*

Experimental Results

Since the k-approval rule is obviously manipulable and obvious manipulations can be found in polynomial time, we further investigate these manipulations in an experiment. Below, we experimentally determine the effects of k, m and n on the proportion of obviously manipulable voter preferences under the k-approval rule.

Assuming a fixed tie-break ordering, we generate 1 million randomly permuted voter preference orderings and determine what proportion of these orderings admit an OM for a given set of parameters. It suffices to simply consider individual preference orderings as the best- and worst-case outcomes (and therefore obvious manipulability) for an agent's preference relation are over all possible preferences of the other agents. Note that from Theorem 5, the set of WOM-admitting preference orderings is the same as the set of OM-admitting preference orderings.

Effect of n: Figure 1 depicts the results from our experiments determining the effect of the number of voters n on the proportion of obviously manipulable preference orderings. The downwards trend is concurrent with the existing theory that the proportion of individually manipulable voting profiles approaches zero as the number of voters tends to infinity [16]. A significantly lower proportion of preference orderings admit a BOM than those that admit a WOM. These trends are consistent for other values of m and k, though other figures are omitted due to space restrictions.

Effect of m **and** $m - k$: In Fig. 2, we show heat maps of the proportion of OM-admitting preferences for $m \in \{21, \ldots, 30\}$ and $m - k$ values for which the preference profile is obviously manipulable. It is more appropriate to consider the number of disapprovals $m - k$ than the number of approvals k, as the impact of k is relative to its difference from the number of outcomes. For example, it is better to compare $m = 21, k = 20$ with $m = 30, k = 29$ than with $m = 30, k = 20$. For a fixed number of disapprovals, the proportion of OM-admitting preferences increases with the number of outcomes. This is likely because a lower proportion of the outcomes can be 'blocked' by the other voters under the worst case outcome. The proportion increases steadily then rapidly decreases as the number of disapprovals increases, suggesting that an intermediary number

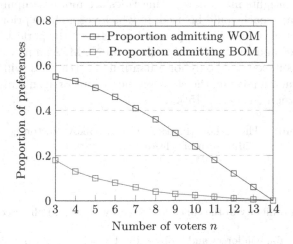

Fig. 1. Effect of n on proportion of preferences that admit WOM and BOM ($k = 14, m = 15$)

Number of outcomes m

No. of disapprovals $m - k$	21	22	23	24	25	26	27	28	29	30
1	0.61	0.61	0.62	0.62	0.63	0.63	0.64	0.64	0.65	0.65
2	0.80	0.81	0.82	0.83	0.84	0.84	0.85	0.85	0.86	0.86
3	0.85	0.87	0.88	0.89	0.90	0.90	0.91	0.92	0.92	0.93
4	0.83	0.86	0.88	0.89	0.91	0.92	0.93	0.94	0.94	0.95
5	0.74	0.80	0.83	0.86	0.89	0.90	0.92	0.93	0.94	0.95
6	0.47	0.60	0.70	0.77	0.82	0.85	0.88	0.90	0.92	0.93
7	0	0	0.28	0.48	0.61	0.71	0.78	0.83	0.87	0.89
8	0	0	0	0	0	0.29	0.49	0.62	0.72	0.79
9	0	0	0	0	0	0	0	0	0.29	0.50

Fig. 2. Effect of m and $m - k$ on proportion of OM-admitting preferences ($n = 3$)

of disapprovals increases individual manipulative power in comparison to the manipulative coalition of the other voters.

7   Conclusion

In this paper, we initiated research on the obvious manipulability of voting rules. One of our key insights is that certain rules are obviously manipulable when the number of outcomes is relatively large as compared to the number of voters. The k-approval rule is an example of such a rule, and we have also shown that under the rule, an obvious manipulation can be computed in polynomial time. Despite all unanimous, non-dictatorial voting rules being manipulable for $n \geq 3$, most commonly used rules are NOM, suggesting that NOM is a significantly weaker notion than strategyproofness. We remark that in the positional scoring rules that we have classified as OM, the obvious manipulations are dependent on a fixed, deterministic tiebreak ordering which is standard in the voting literature. To gain further insights into which voting rules are more manipulable than others, a Bayesian approach could be used, in which voters have prior beliefs on the distribution of other votes. This approach lies between the perfect information of strategyproofness and the lack of information in NOM. As a new concept, NOM has currently been examined only for a handful of settings. It will be interesting to consider it when analyzing the strategic behaviour of agents in other settings such as fair division (see, e.g., [15]).

Acknowledgments. The authors thanks Anton Baychkov, Barton Lee and the anonymous reviewers of ADT 2021 for useful feedback.

References

1. Arrow, K.: A difficulty in the concept of social welfare. J. Polit. Econ. **58**, 328–346 (1950)
2. Black, D.: Borda, condorcet and laplace. In: The Theory of Committees and Elections, chap. 18, pp. 156–162. Springer, Dordrecht (1986). https://doi.org/10.1007/978-94-009-4225-7_18

3. Carroll, G.: A quantitative approach to incentives : application to voting rules (2011)
4. Chisholm, H.: Vote and voting. In: Encyclopaedia Britannica, p. 216 (1911)
5. Conitzer, V., Walsh, T.: Barriers to manipulation in voting. In: Brandt, F., Conitzer, V., Endriss, U., Lang, J., Procaccia, A.D. (eds.) Handbook of Computational Social Choice, chap. 6. Cambridge University Press (2016)
6. Conitzer, V., Walsh, T., Xia, L.: Dominating manipulations in voting with partial information. In: Proceedings of the 25th AAAI Conference (2011)
7. Faliszewski, P., Procaccia, A.D.: AI's war on manipulation: are we winning? AI Mag. **31**, 53–64 (2010)
8. Favardin, P., Lepelley, D., Serais, J.: Borda rule, copeland method and strategic manipulation. Rev. Econ. Des. **7**(2), 213–228 (2002). https://doi.org/10.1007/s100580200073
9. Fraenkel, J., Grofman, B.: The Borda count and its real-world alternatives: comparing scoring rules in Nauru and Slovenia. Aust. J. Polit. Sci. **49**(2), 186–205 (2014)
10. Gibbard, A.: Manipulation of voting schemes: a general result. Econometrica **41**(4), 587–601 (1973)
11. Li, S.: Obviously strategy-proof mechanisms. Am. Econ. Rev. **107**, 3257–3287 (2017)
12. Meir, R.: Strategic voting. Synth. Lect. Artif. Intell. Mach. Learn. **13**, 1–167 (2018)
13. Niou, E.: Strategic voting under plurality and runoff rules. J. Theor. Polit. **13**(2), 209–227 (2001)
14. Nitzan, S.: The vulnerability of point-voting schemes to preference variation and strategic manipulation. Public Choice **47**, 349–370 (1985). https://doi.org/10.1007/BF00127531
15. Ortega, J.: Obvious manipulations in cake-cutting. CoRR abs/1908.02988 (2019)
16. Peleg, B.: A note on manipulability of large voting schemes. Theor. Decis. **11**(4), 401–412 (1979). https://doi.org/10.1007/BF00139450
17. Reilly, B.: Social choice in the south seas: electoral innovation and the Borda count in the Pacific Island countries. Int. Polit. Sci. Rev. **23**(4), 355–372 (2002)
18. Satterthwaite, M.A.: Strategy-proofness and arrow's conditions: existence and correspondence theorems for voting procedures and social welfare functions. J. Econ. Theor. **10**, 187–217 (1975)
19. Slinko, A., White, S.: Non-dictatorial social choice rules are safely manipulable. In: COMSOC 2008, pp. 403–413 (2008)
20. Slinko, A., White, S.: Is it ever safe to vote strategically? Soc. Choice Welfare **43**, 403–427 (2014)
21. Taylor, A.D.: Social Choice and the Mathematics of Manipulation. Cambridge University Press, Cambridge (2005)
22. Tideman, N.: The single transferable vote. J. Econ. Perspect. **9**(1), 27–38 (1995)
23. Troyan, P., Morrill, T.: Obvious manipulations. J. Econ. Theor. **185**, 104970 (2020)
24. Wilson, M.C., Reyhani, R.: The probability of safe manipulation. In: COMSOC 2010 (2010)
25. Xia, L., Zuckerman, M., Procaccia, A.D., Conitzer, V., Rosenschein, J.S.: Complexity of unweighted coalitional manipulation under some common voting rules. In: Proceedings of the 21st IJCAI, pp. 348–353 (2009)
26. Zuckerman, M., Procaccia, A.D., Rosenschein, J.S.: Algorithms for the coalitional manipulation problem. Artif. Intell. **173**(2), 392–412 (2009)

Manipulation in Communication Structures of Graph-Restricted Weighted Voting Games

Joanna Kaczmarek[(✉)] and Jörg Rothe[iD]

Heinrich-Heine-Universität Düsseldorf, Düsseldorf, Germany
{Joanna.Kaczmarek,rothe}@hhu.de

Abstract. Weighted voting games are an important class of compactly representable simple games that can be used to model collective decision-making processes. The influence of players in weighted voting games is measured by power indices such as the Shapley-Shubik and the Penrose-Banzhaf power indices. Previous work has studied how such power indices can be manipulated via actions such as merging or splitting players [1,12], adding or deleting players [13], or tampering with the quota [21]. We study *graph-restricted* weighted voting games [8,9,18], a model in which weighted voting games are embedded into a communication structure (i.e., a graph). We investigate to what extent power indices in such games can be changed by adding or deleting edges in the underlying communication structure and we study the resulting problems in terms of their computational complexity.

1 Introduction

Weighted voting games are one of the most important classes of succinctly representable simple games in cooperative game theory, see, e.g., the books by Chalkiadakis et al. [3] and Taylor and Zwicker [19] and the book chapters by Chalkiadakis and Wooldridge [4] and Elkind and Rothe [6]. Weighted voting games are used in various domains where collective decision-making processes are modeled, such as in legislative bodies or parliaments, but also in less politics-related areas such as shareholder voting in joint stock companies where each shareholder gets votes in proportion to the ownership of a stock. When analyzing weighted voting games, one usually focuses on the strength or influence of a player in a game and uses power indices for that purpose. The best known and most intensively studied power indices are the normalized Penrose-Banzhaf power index [2,11] and its probabilistic variant [5] and the Shapley-Shubik power index [17]; we focus on the probabilistic Penrose-Banzhaf index and the Shapley-Shubik index. We are interested in their values, how they can change in different situations, and in the computational complexity of problems defined on them.

In the definition of weighted voting games it is assumed that any player is able and willing to communicate and to cooperate with all others, which is a really

© Springer Nature Switzerland AG 2021
D. Fotakis and D. Ríos Insua (Eds.): ADT 2021, LNAI 13023, pp. 194–208, 2021.
https://doi.org/10.1007/978-3-030-87756-9_13

strong assumption in real-life situations. Myerson [8] provided a model for cooperative games that does not rely on this assumption, i.e., he introduced *graph-restricted games*: cooperative games embedded into a communication structure (i.e., an undirected graph) that describes which other players any player can form a coalition with. Based on his model, Napel et al. [9] presented *graph-restricted weighted voting games*, a combination of weighted voting games with graph-restricted games generalizing the standard notion of weighted voting games.

For weighted voting games, it has been analyzed how the Penrose-Banzhaf index and the Shapley-Shubik index can change when they are subject to manipulation via actions such as merging or splitting players [1,12] (the latter is a.k.a. *false-name manipulation*) or modifying the quota [21], or when they are subject to structural control by adding or deleting players [13]. For *graph-restricted* weighted voting games, we study how these power indices can change when edges in the graph are added or deleted, i.e., when some players are enabled or disabled to communicate or to cooperate with certain other players, for instance as a part of a political strategy, by legal changes, or simply by opening or closing communication channels between them. In particular, we study the associated problems in terms of their computational complexity.

2   Preliminaries

We start with providing the needed background of cooperative game theory. Let $N = \{1, \ldots, n\}$ denote a set of players. A *coalitional game* is a pair (N, v), where $v : 2^N \to \mathbb{R}_+$ assigns a nonnegative real value to each coalition (i.e., subset) of players; it is said to be *simple* if it is *monotonic* (i.e., $v(A) \leq v(B)$ whenever $A \subseteq B$) and $v(C) \in \{0, 1\}$ for each $C \subseteq N$ (where $v(C) = 1$ means that coalition C *wins*, and $v(C) = 0$ means that C *loses*).

Definition 1. *A* weighted voting game $\mathcal{G} = (w_1, \ldots, w_n; q)$ *is a simple coalitional game that consists of a quota* $q \in \mathbb{N}$ *(i.e., a given threshold) and nonnegative integer weights* w_i, *where* w_i *is the* i-*th player's weight,* $i \in N$. *For each coalition* $S \subseteq N$, *letting* $w_S = \sum_{i \in S} w_i$, S *wins if* $w_S \geq q$, *and loses otherwise:*

$$v(S) = \begin{cases} 1 \text{ if } w_S \geq q, \\ 0 \text{ otherwise.} \end{cases}$$

One of the most important information about players is their significance in the games that is measured usually by so-called *power indices*, which take into consideration how many coalitions a player can make win. We study two of the most popular power indices: the *probabilistic Penrose-Banzhaf power index*, which Dubey and Shapley [5] introduced as an alternative to the original *normalized Penrose-Banzhaf index* [2,11], and the *Shapley-Shubik power index* introduced by Shapley and Shubik [17]. These two indices are defined as follows:

Definition 2. *Let* $n = |N|$ *be the number of players in* $\mathcal{G}$ *and* $i \in N$. *The probabilistic Penrose-Banzhaf power index of player* i *in* $\mathcal{G}$ *is defined by*

$$\beta(\mathcal{G}, i) = \frac{\sum_{S \subseteq N \setminus \{i\}} (v(S \cup \{i\}) - v(S))}{2^{n-1}}.$$

The Shapley-Shubik *power index of player* i *in* $\mathcal{G}$ *is defined by*

$$\varphi(\mathcal{G}, i) = \frac{\sum_{S \subseteq N \setminus \{i\}} |S|!(n - 1 - |S|)!(v(S \cup \{i\}) - v(S))}{n!}.$$

Next, we assume that there is some communication structure among the players and define the corresponding notion of *graph-restricted* weighted voting games, first studied in the early work by Myerson [8] and later on, e.g., by Napel et al. [9] and Skibski et al. [18].

Definition 3. *A* graph-restricted weighted voting game *is a weighted voting game* $\mathcal{G} = (w_1, \ldots, w_n; q)$ *together with a graph* $G = (N, E)$, *where*

$$v(S) = \begin{cases} 1 & \text{if } S \text{ has a connected part } S' \text{ with } w_{S'} \geq q, \\ 0 & \text{otherwise.} \end{cases}$$

Graph-restricted weighted voting games generalize weighted voting games, which are the special cases with a complete graph as their communication structures. In this situation, whether a coalition wins or loses is determined only by its total weight. However, if we limit the possibilities in communication among players, a coalition's weight alone is not enough. Before we define appropriate power indices in graph-restricted weighted voting games, let us present a few useful notions referring to coalitions in the sense of graph restrictions.

Definition 4. *Let* $(\mathcal{G}, G)$ *be a graph-restricted weighted voting game with players* N *and graph* $G = (N, E)$. *For* $S \subseteq N$, *we denote a maximal connected subset of* S *in* G *as* S/G. *The set of all winning connected coalitions is defined as* $\mathcal{WC} = \{S \subseteq N \mid w_S \geq q \text{ and } S \text{ is connected}\}$ *and the set of winning connected coalitions with player* i *is denoted by* $\mathcal{WC}_i$. *The set of all pivotal winning connected coalitions of player* i *is defined as* $\mathcal{PWC}_i = \{S \in \mathcal{WC}_i \mid ((S \setminus \{i\})/G) \cap \mathcal{WC} = \emptyset\}$.

Skibski et al. [18] provided the following general formulas for the analogues of the Penrose-Banzhaf power index and the Shapley-Shubik power index in graph-restricted weighted voting games. Let $\mathcal{N}(i) = \{j \in N \mid \{i, j\} \in E\}$ denote the neighborhood of i in graph G, and let $\mathcal{N}(S) = (\bigcup_{i \in S} \mathcal{N}(i)) \setminus S$ be the set of neighbors of S.

Theorem 1 (Skibski et al. [18]). *Let* $(\mathcal{G}, G)$ *be a graph-restricted weighted voting game with players* N *and the set of winning connected coalitions* $\mathcal{WC}$. *For* $S \subseteq N$, *let* $\gamma^S = \frac{1}{2^{|S|+|\mathcal{N}(S)|-1}}$, $\gamma_1^S = \frac{(|S|-1)!|\mathcal{N}(S)|!}{(|S|+|\mathcal{N}(S)|)!}$, *and* $\gamma_2^S = \frac{|S|!(|\mathcal{N}(S)|-1)!}{(|S|+|\mathcal{N}(S)|)!}$.
The Penrose-Banzhaf *index of player* i *in* $(\mathcal{G}, G)$ *satisfies* $\beta((\mathcal{G}, G), i) = \sum_{S \in \mathcal{PWC}_i} \gamma^S = \sum_{S \in \mathcal{WC}_i} \gamma^S - \sum_{\substack{S \in \mathcal{WC} \\ i \in \mathcal{N}(S)}} \gamma^S$. *The* Shapley-Shubik *index of player* i *in* $(\mathcal{G}, G)$ *satisfies* $\varphi((\mathcal{G}, G), i) = \sum_{S \in \mathcal{PWC}_i} \gamma_1^S = \sum_{S \in \mathcal{WC}_i} \gamma_1^S - \sum_{\substack{S \in \mathcal{WC} \\ i \in \mathcal{N}(S)}} \gamma_2^S$.

We assume the reader to be familiar with the basic concepts of graph theory and computational complexity theory, such as the complexity classes P, NP, and coNP and the notions of completeness and hardness for a complexity class based on polynomial-time many-one reducibility. DP was introduced by Papadimitriou and Yannakakis [10] as the class of sets that can be represented as the difference of two NP sets; it is in the second level of the boolean hierarchy over NP.

In some of our proofs, we will apply the following lemma due to Wagner [20] that previously has proven useful for showing DP-hardness of various "exact" variants of graph problems (such as whether the chromatic number of a given graph is exactly four; see [14–16]).

Lemma 1 (Wagner [20]). *Let A be some NP-complete problem and let B be an arbitrary problem. If there exists a polynomial-time computable function f such that, for all input strings x_1 and x_2 for which $x_2 \in A$ implies $x_1 \in A$, we have that $(x_1 \in A \land x_2 \notin A) \iff f(x_1, x_2) \in B$, then B is DP-hard.*

3   Adding Edges to a Communication Graph

We now consider the impact of adding new edges to the communication structure of a given graph-restricted weighted voting game on changing the Penrose-Banzhaf and the Shapley-Shubik power index of a given player. By this structural change to the game, we allow some players to communicate with each other for whom this was impossible before. Let us start with defining the decision problem in which we ask whether some power index PI can be *increased*:

CONTROL BY ADDING EDGES BETWEEN PLAYERS TO INCREASE PI

Given: A graph-restricted weighted voting game $(\mathcal{G}, G)$ with players $N = \{1, \ldots, n\}$, a communication structure $G = (N, E)$, $|E| < \binom{n}{2}$, a distinguished player $p \in N$, and a nonnegative integer k.

Question: Is it sufficient to add k or fewer edges $E' \subseteq E^c = \{(x, y) \in N \times N \mid (x, y) \notin E\}$ to G to obtain a new game $(\mathcal{G}, G_{\cup E'})$ for which it holds that $\mathrm{PI}((\mathcal{G}, G_{\cup E'}), p) > \mathrm{PI}((\mathcal{G}, G), p)$?

Analogously, we define the decision problems for *decreasing* and *maintaining* a distinguished player's power (by replacing ">" in the question by "<" or "="). Before we present our results, let us see this type of manipulation in the game of Example 1 to better understand the possible changes of the power indices.

Example 1. Let $(\mathcal{G}, G)$ be a graph-restricted weighted voting game with $\mathcal{G} = (1, 2, 3, 4, 5; 8)$ and the following communication structure $G = (N, E)$:

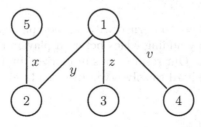

The players have the following values of their Penrose-Banzhaf and Shapley-Shubik power index:

$$\beta((\mathcal{G}, G), 1) = 7/16 \text{ and } \beta((\mathcal{G}, G), i) = 3/16 \text{ for } i \in \{2, \ldots, 5\};$$
$$\varphi((\mathcal{G}, G), 1) = 7/15 \text{ and } \varphi((\mathcal{G}, G), i) = 2/15 \text{ for } i \in \{2, \ldots, 5\}.$$

Thus, for both indices, 1 (even though having the smallest weight) is the most powerful player and the others have the same power, even though their weights are different (also, 2 has a higher degree than 3, 4, and 5 in the graph G).

Let e be an edge between players 3 and 5 and let f be an edge between players 3 and 4. Consider the new game $(\mathcal{G}, G_{\cup\{e\}})$: For both power indices, players 3 and 5 have become the most powerful players by adding the new edge e, whereas the power of the previously strongest player 1 has decreased:

$$\beta((\mathcal{G}, G_{\cup\{e\}}), 1) = 1/4, \quad \beta((\mathcal{G}, G_{\cup\{e\}}), 2) = \beta((\mathcal{G}, G_{\cup\{e\}}), 4) = 1/8, \text{ and}$$
$$\beta((\mathcal{G}, G_{\cup\{e\}}), 3) = \beta((\mathcal{G}, G_{\cup\{e\}}), 5) = 1/2;$$
$$\varphi((\mathcal{G}, G_{\cup\{e\}}), 1) = 1/6, \quad \varphi((\mathcal{G}, G_{\cup\{e\}}), 2) = \varphi((\mathcal{G}, G_{\cup\{e\}}), 4) = 1/12, \text{ and}$$
$$\varphi((\mathcal{G}, G_{\cup\{e\}}), 3) = \varphi((\mathcal{G}, G_{\cup\{e\}}), 5) = 1/3.$$

It is also not hard to see that adding the edge f to G does not change the (Penrose-Banzhaf and Shapley-Shubik) power indices of the players at all.

We will now show to what extent (in terms of the number of new connections) the Penrose-Banzhaf and the Shapley-Shubik power index can change after adding edges to a player in a communication graph. Theorem 2 (whose proof is omitted due to space limitations) presents these change intervals. Moreover, adding edges among other players than the distinguished player can reduce the distinguished player's power indices even to 0.

Theorem 2. *Let $(\mathcal{G}, G)$ be a graph-restricted weighted voting game and let E', $|E'| = m$, be a set of edges that are to be added to G, between a player i and other players, creating a new game $(\mathcal{G}, G_{\cup E'})$. For player i and for $\chi \in \{\beta, \varphi\}$, let $diff_\chi(\mathcal{G}, G, G_{\cup E'}, i) = \chi((\mathcal{G}, G), i) - \chi((\mathcal{G}, G_{\cup E'}), i)$. The old and the new Penrose-Banzhaf index and the old and the new Shapley-Shubik index of player i can differ as follows:*

$$-1 + 2^{-(|\mathcal{N}(i)| + m)} \leq diff_\beta(\mathcal{G}, G, G_{\cup E'}, i) \leq (1 - 2^{-m})\beta((\mathcal{G}, G), i);$$
$$-1 + \frac{1}{|\mathcal{N}(i) + m + 1|} \leq diff_\varphi(\mathcal{G}, G, G_{\cup E'}, i) \leq \left(1 - \frac{(n - m)!}{n!}\right)\varphi((\mathcal{G}, G), i).$$

The next two results are concerned with the complexity of the decision problems regarding control by adding edges between players that we defined at the beginning of this section. Our results are summarized in Table 1 in Sect. 5. Note that these problems are hard to solve even if we add only one edge to the communication graph.

Theorem 3. *Control by adding edges between players to decrease a distinguished player's Penrose-Banzhaf index in a graph-restricted weighted voting game is NP-hard, to maintain this index is coNP-hard, and to increase it is DP-hard.*

Proof. For the first two statements, we use a reduction from the PARTITION problem [7]: Given a set $A = \{1, \ldots, n\}$, a function $a : A \to \mathbb{N} \setminus \{0\}$, $i \mapsto a_i$, such that $\sum_{i=1}^{n} a_i$ is even, does there exist a partition into two subsets of equal weight, that is, does there exist a subset $A' \subseteq A$ such that $\sum_{i \in A'} a_i = \sum_{i \in A \setminus A'} a_i$? Let $(a_1, \ldots, a_n)$ be a PARTITION instance with $n > 1$ and $\alpha = \sum_{i=1}^{n} a_i$, and let $\xi = \#\text{PARTITION}(a_1, \ldots, a_n)$ denote the number of its solutions. Now, construct the control problem instance consisting of a game $\mathcal{G} = (1, a_1, \ldots, a_n, 2\alpha, \frac{\alpha}{2}; \frac{5\alpha}{2} + 1)$ with $n + 3$ players, the distinguished player $p = 1$, and the communication structure $G = (N, E)$, where all the players but $n + 3$ form a complete subgraph and the player $n + 3$ is an isolated vertex. Let E^c be the set of edges not in E, so we can add them to the graph G. Set the addition limit to 1. It holds that

$$(\exists e \in E^c)[\beta((\mathcal{G}, G_{\cup\{e\}}), 1) - \beta((\mathcal{G}, G), 1) < 0] \iff \xi > 0, \tag{1}$$

$$(\exists e \in E^c)[\beta((\mathcal{G}, G_{\cup\{e\}}), 1) - \beta((\mathcal{G}, G), 1) = 0] \iff \xi = 0, \tag{2}$$

$$(\exists e \in E^c)[\beta((\mathcal{G}, G_{\cup\{e\}}), 1) - \beta((\mathcal{G}, G), 1) > 0] \iff \xi = 0. \tag{3}$$

To see this, note that in the structure G, there are only two types of edges that can be added: an edge x between players 1 or $n + 2$ and $n + 3$, and an edge y between player $n + 3$ and a player i, $i \in \{2, \ldots, n + 2\}$.

Let $\xi = 0$. Then $\beta((\mathcal{G}, G), 1) = 0$. If we add the edge x, the players 1 and $n + 3$ will form a winning coalition and the player 1's Penrose-Banzhaf index will increase to $\beta((\mathcal{G}, G_{\cup\{x\}}), 1) = \frac{1}{2^{n+2}}$. If we add any of the y-edges, there will still not be any winning coalition for which the player 1 is pivotal, so the index will not change. Therefore, if there is no solution for PARTITION$((a_1, \ldots, a_n))$, the Penrose-Banzhaf index can either increase or stay unchanged.

Let $\xi > 0$. Then $\beta((\mathcal{G}, G), 1) = \frac{\xi}{2^{n+1}} = \frac{2\xi}{2^{n+2}} = \frac{4\xi}{2^{n+3}}$. It does not matter which edge from E^c we add, player 1's Penrose-Banzhaf index will always decrease: for the edge x because $\beta((\mathcal{G}, G_{\cup\{x\}}), 1) = \frac{\xi}{2^{n+2}} + \frac{1}{2^{n+2}} = \frac{\xi+1}{2^{n+2}}$ and ξ is even, and for y-edges because $\beta((\mathcal{G}, G_{\cup\{y\}}), 1) = \frac{\xi}{2}\frac{1}{2^{n+2}} + \frac{\xi}{2}\frac{1}{2^{n+1}} = \frac{3\xi}{2^{n+3}}$.

By (1), (2) and (3), control by adding edges to decrease a given player's Penrose-Banzhaf index is NP-hard, and to maintain it or to increase it is coNP-hard.

Note that we can similarly show NP-hardness of control by adding edges to *increase* a given player's Penrose-Banzhaf index.[1] However, as claimed in the third statement of the theorem, we can show even more, namely DP-hardness

---

[1] Specifically, construct from the given PARTITION instance $(a_1, \ldots, a_n)$ the game $\mathcal{H} = (1, 2a_1, \ldots, 2a_n, 1; \alpha + 2)$ with $n + 2$ players, the distinguished player $p = 1$, and the communication structure $H = (M, C)$, where as before all players but the $(n + 2)$nd player form a complete subgraph and the $(n + 2)$nd player is an isolated vertex. Set the addition limit to 1. Let C^c be the set of edges not in C, which can be added to H. We can show that $(\exists e \in C^c)[\beta((\mathcal{H}, H_{\cup\{e\}}), 1) - \beta((\mathcal{H}, H), 1) > 0]$ if and only if $\xi > 0$, which gives the desired NP-hardness of our control problem.

of this problem. To this end, we provide a reduction from the NP-complete SUBSETSUM problem (which will play the role of the set A in Lemma 1): Given a sequence $(c_1, \ldots, c_n)$ of positive integers and a positive integer q, do there exist $y_1, \ldots, y_n \in \{0, 1\}$ with $\sum_{i=1}^n y_i c_i = q$?

Let $x_1 = ((c_1, \ldots, c_{n_1}), q)$ and $x_2 = ((d_1, \ldots, d_{n_2}), q)$ be two instances of SUBSETSUM, $\gamma = \sum_{i=1}^{n_1} c_i$, $\delta = \sum_{i=1}^{n_2} d_i$, and let ξ_i be the number of solutions of SUBSETSUM(x_i) for $i \in \{1, 2\}$.

Consider the graph-restricted weighted voting game

$$\mathcal{F} = \left(1, c_1 \cdot 10^s, \ldots, c_{n_1} \cdot 10^s, d_1 \cdot 10, \ldots, d_{n_2} \cdot 10, 10^t, q \cdot 10; 10^t + q \cdot 10^s + q \cdot 10 + 1\right)$$

with $n_1 + n_2 + 3$ players, where $10^s > (\delta + q) \cdot 10 + 2$ and $10^t > \gamma \cdot 10^s + (\delta + q) \cdot 10 + 2$, and the communication graph $F = (L, D)$ looks as follows: All players except the player with weight $q \cdot 10$ form a complete subgraph and this last player is isolated. Thus it is only possible to add an edge between the isolated player and another player. Let the first player (with weight 1) be the distinguished player and let the addition limit be 1.

Assuming that $x_2 \in$ SUBSETSUM implies $x_1 \in$ SUBSETSUM, we show that

$$(\exists e \in D^c)[\beta((\mathcal{F}, F), 1) - \beta((\mathcal{F}, F_{\cup\{e\}}), 1) < 0]$$
$$\Longleftrightarrow (x_1 \in \text{SUBSETSUM} \land x_2 \notin \text{SUBSETSUM}). \tag{4}$$

Indeed, if $x_1 \in$ SUBSETSUM and $x_2 \notin$ SUBSETSUM, then $\beta((\mathcal{F}, F), 1) = 0$, and if we add an edge e between our distinguished first player and the isolated last player, then the index increases to

$$\beta((\mathcal{F}, F_{\cup\{e\}}), 1) = \frac{\xi_1}{2^{n_1 + n_2 + 2}} > 0.$$

On the other hand, if both $x_1 \notin$ SUBSETSUM and $x_2 \notin$ SUBSETSUM, then the index is equal to 0 and adding any edge does not change this.

Finally, if both $x_1 \in$ SUBSETSUM and $x_2 \in$ SUBSETSUM, then

$$\beta((\mathcal{F}, F), 1) = \frac{\xi_1 \xi_2}{2^{n_1 + n_2 + 1}} = \frac{2\xi_1 \xi_2}{2^{n_1 + n_2 + 2}}.$$

Now, if we add an edge between the isolated player and either the distinguished player or the player with weight 10^t, then

$$\beta((\mathcal{F}, F_{\cup\{e\}}), 1) = \frac{\xi_1 \xi_2 + \xi_1}{2^{n_1 + n_2 + 2}} \leq \frac{\xi_1 \xi_2 + \xi_1 \xi_2}{2^{n_1 + n_2 + 2}} = \beta((\mathcal{F}, F), 1),$$

since $\xi_2 \geq 1$, so the index remains unchanged or decreases in those cases. And if we add an edge between the isolated player and a player c_i, then let ξ_{1,c_i} be the number of solutions of SUBSETSUM(x_1) containing the player c_i, and we get

$$\beta((\mathcal{F}, F_{\cup\{e\}}), 1) = \frac{\xi_{1,c_i} \xi_2}{2^{n_1 + n_2 + 2}} + \frac{(\xi_1 - \xi_{1,c_i}) \xi_2}{2^{n_1 + n_2 + 1}} + \frac{\xi_{1,c_i}}{2^{n_1 + n_2 + 2}}$$
$$= \frac{2\xi_1 \xi_2 - \xi_{1,c_i} \xi_2 + \xi_{1,c_i}}{2^{n_1 + n_2 + 2}} \leq \frac{2\xi_1 \xi_2}{2^{n_1 + n_2 + 2}} = \beta((\mathcal{F}, F), 1),$$

so again, the index remains unchanged or decreases. Finally, if we add an edge between the isolated player and a player d_i, then let ξ_{2,d_i} be the number of solutions of SUBSETSUM(x_2) containing the player d_i and in this case we obtain

$$\beta((\mathcal{F}, F_{\cup\{e\}}), 1) = \frac{\xi_1\xi_{2,d_i}}{2^{n_1+n_2+2}} + \frac{\xi_1(\xi_2 - \xi_{2,d_i})}{2^{n_1+n_2+1}}$$

$$= \frac{2\xi_1\xi_2 - \xi_1\xi_{2,d_i}}{2^{n_1+n_2+2}} \leq \frac{2\xi_1\xi_2}{2^{n_1+n_2+2}} = \beta((\mathcal{F}, F), 1),$$

so in this case, the index does not increase either.

Since (4) is satisfied, Lemma 1 implies that the problem of control by adding edges between players to increase a distinguished player's Penrose-Banzhaf index in a graph-restricted weighted voting game is DP-hard. □

We now turn to the Shapley-Shubik index for control by adding edges. The proof of Theorem 4, which makes use of games constructed in the proof of Theorem 3, is omitted due to space limitations.

Theorem 4. *Control by adding edges between players to increase or to decrease a distinguished player's Shapley-Shubik index in a graph-restricted weighted voting game is* NP-*hard and to maintain the index is* coNP-*hard.*

As one could see in the previous proofs (and also by looking at the formulas of the power indices presented by Skibski et al. [18] for graph-restricted weighted voting games), in some situations adding edges can be equivalent to adding new players to the winning connected subgraph in a game. Control by adding players in weighted voting games (without any graph restrictions) were analyzed by Rey and Rothe [13]. Of course, the resulting decision problems (of control by adding edges and by adding players) in these two different settings are not the same problems. Therefore, we need to be careful when comparing their and our results and drawing conclusions.

4   Deleting Edges from a Communication Graph

As in the previous section, we define the problem of control by deleting edges to increase a power index PI; the other two definitions (where the goal is to decrease and to maintain an index) are analogous.

CONTROL BY DELETING EDGES BETWEEN PLAYERS TO INCREASE PI

Given: A graph-restricted weighted voting game $(\mathcal{G}, G)$ with players $N = \{1, \ldots, n\}$, a communication structure $G = (N, E)$, a distinguished player $p \in N$, and a positive integer $k \leq |E|$.

Question: Can at most k edges $E' \subseteq E$ be deleted from G such that for the new game $(\mathcal{G}, G_{\setminus E'})$, it holds that $\mathrm{PI}((\mathcal{G}, G_{\setminus E'}), p) > \mathrm{PI}((\mathcal{G}, G), p)$?

Before presenting our results, we give two short examples.

Example 2. Let $(\mathcal{G}, G)$ be a graph-restricted weighted voting game with $\mathcal{G} = (10, 3, 10; 12)$ and the communication structure $G = (N, E)$ being a complete graph. In this game, the players have equal power with respect to both indices:

$$\beta((\mathcal{G}, G), 1) = \beta((\mathcal{G}, G), 2) = \beta((\mathcal{G}, G), 3) = 1/2;$$
$$\varphi((\mathcal{G}, G), 1) = \varphi((\mathcal{G}, G), 2) = \varphi((\mathcal{G}, G), 3) = 1/3.$$

Now, let us delete from G the edge between the players 1 and 3; let us call it x. In the new game, the power indices of player 2 increase and the indices of the other two players decrease:

$$\beta((\mathcal{G}, G_{\backslash \{x\}}), 2) = 3/4, \quad \text{whereas} \quad \beta((\mathcal{G}, G_{\backslash \{x\}}), 1) = \beta((\mathcal{G}, G_{\backslash \{x\}}), 3) = 1/4;$$
$$\varphi((\mathcal{G}, G_{\backslash \{x\}}), 2) = 2/3, \quad \text{whereas} \quad \varphi((\mathcal{G}, G_{\backslash \{x\}}), 1) = \varphi((\mathcal{G}, G_{\backslash \{x\}}), 3) = 1/6.$$

This illustrates that if we limit communication among players, it will be possible for players with smaller weights to become more powerful than players with larger weights.

Example 3. Consider again the game from Example 1. If we delete either the edge x or the edge y or both, we will get the same Penrose-Banzhaf and Shapley-Shubik power indices of the players, i.e., for $E' \in \{\{x\}, \{y\}, \{x, y\}\}$, we have:

$$\beta((\mathcal{G}, G_{\backslash E'}), 1) = \beta((\mathcal{G}, G_{\backslash E'}), 3) = \beta((\mathcal{G}, G_{\backslash E'}), 4) = 1/4 \quad \text{and}$$
$$\beta((\mathcal{G}, G_{\backslash E'}), 2) = \beta((\mathcal{G}, G_{\backslash E'}), 5) = 0;$$
$$\varphi((\mathcal{G}, G_{\backslash E'}), 1) = \varphi((\mathcal{G}, G_{\backslash E'}), 3) = \varphi((\mathcal{G}, G_{\backslash E'}), 4) = 1/3 \quad \text{and}$$
$$\varphi((\mathcal{G}, G_{\backslash E'}), 2) = \varphi((\mathcal{G}, G_{\backslash E'}), 5) = 0.$$

This illustrates that the power indices can change the same way (increase or decrease) for both the stronger and the weaker players, even after deleting one or more connections.

In the following theorem, whose proof again is omitted, we see how a distinguished player's Penrose-Banzhaf power index and Shapley-Shubik power index can change after deletion of a certain number of edges, i.e., after removing the possibility of communication between a player and his or her neighbors in the graph. Note that after removing edges in another part of a communication structure a player's power indices can increase even if he or she was not pivotal for any coalition (but stays nonpivotal for the two-element coalitions with any neighbor and for the singleton coalition containing only this player).

Theorem 5. *Let $(\mathcal{G}, G)$ be a graph-restricted weighted voting game and let E', $|E'| = m$, be the set of those edges between a player i and his or her neighbors that are to be deleted from G, creating a new game $(\mathcal{G}, G_{\backslash E'})$. The old and the new Penrose-Banzhaf index and the old and the new Shapley-Shubik index of player i can differ as follows:*

$$(1 - 2^m)\beta((\mathcal{G}, G), i) \leq \beta((\mathcal{G}, G), i) - \beta((\mathcal{G}, G_{\backslash E'}), i) \leq \beta((\mathcal{G}, G), i);$$
$$(1 - n!/(n-m)!)\, \varphi((\mathcal{G}, G), i) \leq \varphi((\mathcal{G}, G), i) - \varphi((\mathcal{G}, G_{\backslash E'}), i) \leq \varphi((\mathcal{G}, G), i).$$

The remaining theorems in this section are concerned with the computational complexity of the decision problems defined earlier in this section. Our results are again summarized in Table 1 in Sect. 5. Note that these problems are hard to solve even if we delete only one edge from the communication graph.

Theorem 6. *Control by deleting edges between players to decrease or to increase a distinguished player's Penrose-Banzhaf index in a graph-restricted weighted voting game is DP-hard.*

Proof. We restrict our proof to the goal of decreasing the distinguished player's Penrose-Banzhaf index (the proof for increasing it is similar), and we show DP-hardness by providing a reduction from the NP-complete SUBSETSUM problem (which plays the role of the set A from Wagner's Lemma 1). Let $x_1 = ((c_1, \ldots, c_{n_1}), q)$ and $x_2 = ((d_1, \ldots, d_{n_2}), q)$ be two instances of SUBSETSUM, $\gamma = \sum_{i=1}^{n_1} c_i$, $\delta = \sum_{i=1}^{n_2} d_i$, and let ξ_i be the number of solutions of SUBSETSUM(x_i) for $i \in \{1, 2\}$. Construct the graph-restricted weighted voting game

$$\mathcal{G} = \left(1, c_1 \cdot 10^s, \ldots, c_{n_1} \cdot 10^s, d_1, \ldots, d_{n_2}, 10^t, q, y - q, 2y; 10^t + q \cdot 10^s + 3y + 1\right),$$

where $y > \max(2q, 2\gamma, 2\delta)$ and $s \in \mathbb{N}$ and $t \in \mathbb{N}$ are chosen such that $10^s > 4y$ and $10^t > (\gamma + \delta) \cdot 10^s + 4y$ and large enough for the quota to be greater than half of the total sum of all players' weights. Let the first player with weight 1 be the distinguished player and let the deletion limit be 1. Define the communication graph $G = (V, E)$ as follows:

$$G_1$$

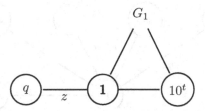

Here, the subgraph G_1 looks as follows: It consists of the players with weights $c_1 \cdot 10^s, \ldots, c_{n_1} \cdot 10^s, d_1, \ldots, d_{n_2}, y - q$, and $2y$, which form a clique. The player with weight $2y$ in this clique is connected to the players with weights 1 and 10^t. Assuming that $x_2 \in$ SUBSETSUM implies $x_1 \in$ SUBSETSUM, we now prove:

$$(\exists e \in E)[\beta((\mathcal{G}, G), 1) - \beta((\mathcal{G}, G_{\backslash\{e\}}), 1) > 0]$$
$$\Longleftrightarrow (x_1 \in \text{SUBSETSUM} \wedge x_2 \notin \text{SUBSETSUM}). \tag{5}$$

First, if $x_1 \notin$ SUBSETSUM and $x_2 \notin$ SUBSETSUM, then

$$(\forall e \in E)[\beta((\mathcal{G}, G), 1) = \beta((\mathcal{G}, G_{\backslash\{e\}}), 1) = 0].$$

And if $x_1 \in$ SUBSETSUM and $x_2 \in$ SUBSETSUM, then

$$\beta((\mathcal{G}, G), 1) = \frac{\xi_1 + \xi_1 \xi_2}{2^{n_1 + n_2 + 4}}.$$

It is easy to see that if we delete any edge other than z, the index stays unchanged. If we delete the edge z, however, then

$$\beta((\mathcal{G}, G_{\setminus\{z\}}), 1) = \frac{\xi_1 \xi_2}{2^{n_1 + n_2 + 3}} = \frac{\xi_1 \xi_2 + \xi_1 \xi_2}{2^{n_1 + n_2 + 4}} \geq \frac{\xi_1 + \xi_1 \xi_2}{2^{n_1 + n_2 + 4}} = \beta((\mathcal{G}, G), 1),$$

since $\xi_1, \xi_2 \geq 1$. So, the power index either increases or stays unchanged after deleting one edge from G in this case.

Conversely, if $x_1 \in$ SUBSETSUM and $x_2 \notin$ SUBSETSUM, then $\beta((\mathcal{G}, G), 1) > 0$ and $\beta((\mathcal{G}, G_{\setminus\{z\}}), 1) = 0$, so the index decreases by deleting the edge z.

Since (5) is satisfied, our control problem is DP-hard by Lemma 1.   □

Next, we show that for the goal of decreasing or increasing the Shapley-Shubik power index, control by deleting edges to limit communication between players is NP-hard.

Theorem 7. *Control by deleting edges between players to decrease or to increase a distinguished player's Shapley-Shubik power index is NP-hard.*

Proof. We give a reduction from the PARTITION problem. Let $(a_1, \ldots, a_n)$ with $a_i \leq \alpha/2$ for $i \in \{1, \ldots, n\}$ be a PARTITION instance with $n > 1$, where $\alpha = \sum_{i=1}^{n} a_i$ and $\xi = \#\text{PARTITION}((a_1, \ldots, a_n))$ denotes the number of its solutions. Consider the game $\mathcal{G} = (1, 2a_1, \ldots, 2a_n, 1; \alpha+2)$ with $n+2$ players, distinguished player $p = 1$ with weight 1, and deletion limit $k = 1$. The communication structure $G = (N, E)$ is defined by:

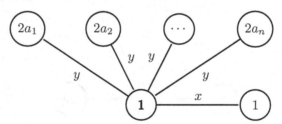

We will now show for this game that

$$(\exists e \in E)[\varphi((\mathcal{G}, G_{\setminus\{e\}}), 1) - \varphi((\mathcal{G}, G), 1) < 0] \iff \xi > 0.$$

Let $\xi = 0$. Then $\varphi((\mathcal{G}, G), 1) = 0$, and if we delete any of the edges, the Shapley-Shubik index of player 1 will still be equal to 0, so it does not decrease.

Let $\xi > 0$. Then $\varphi((\mathcal{G}, G), 1) > 0$. After deleting the edge x, the Shapley-Shubik index of 1 decreases to 0. Therefore, control by deleting edges to decrease a distinguished player's Shapley-Shubik index is NP-hard.

For the goal of *increasing* the distinguished player's Shapley-Shubik index, let us consider the game

$$\mathcal{H} = (1, 4a_1, \ldots, 4a_n, 2, 1; 2\alpha + 3)$$

with $n+3$ players, distinguished player $p=1$ and the communication structure $H=(M,F)$, where all the players but 1 are connected only with the distinguished player. Let x be the edge between 1 and $n+3$. Set the deletion limit to 1.

We now show that for this game, we have

$$(\exists e \in F)[\varphi((\mathcal{H}, H_{\setminus\{e\}}), 1) - \varphi((\mathcal{H}, H), 1) > 0] \iff \xi > 0.$$

Let $\xi = 0$. Then $\varphi((\mathcal{H}, H), 1) = 0$, and if we delete any of the edges, the Shapley-Shubik index of player 1 will still be equal to 0, so it does not increase.

Let $\xi > 0$. Then

$$\varphi((\mathcal{H}, H), 1) = \sum_{S \in PWC_1} \frac{(|S| - 1)!|N(S)|!}{(n+3)!}.$$

Let $r_S = |N(S)|$ from the formula above. After deleting the edge x, the Shapley-Shubik index of player 1 increases to

$$\varphi((\mathcal{H}, H_{\setminus\{x\}}), 1) = \sum_{S \in \mathcal{PWC}_1} \frac{(|S| - 1)!(r_S - 1)!}{(n+2)!} = \sum_{S \in \mathcal{PWC}_1} \frac{(|S| - 1)!r_S!}{(n+3)!} \frac{|S| + r_S}{r_S}$$

$$= \sum_{S \in \mathcal{PWC}_1} \frac{(|S| - 1)!r_S!}{(n+3)!} \left(1 + \frac{|S|}{r_S}\right) > \sum_{S \in \mathcal{PWC}_1} \frac{(|S| - 1)!r_S!}{(n+3)!}.$$

Therefore, control by deleting edges to increase a distinguished player's Shapley-Shubik index is NP-hard as well.   □

Finally, for the Penrose-Banzhaf power index, we show that control by deleting edges between players to *maintain* a distinguished player's power is coNP-hard.

Theorem 8. *Control by deleting edges between players to maintain a distinguished player's Penrose-Banzhaf index in a graph-restricted weighted voting game is* coNP-*hard.*

Proof. We will show both coNP-hardness results by means of a reduction from the PARTITION problem. Let $(a_1, \ldots, a_n)$ be a PARTITION instance with $n > 1$, let $\alpha = \sum_{i=1}^{n} a_i$, and let $\xi = \#\text{PARTITION}((a_1, \ldots, a_n))$ denote the number of its solutions.

For the Penrose-Banzhaf index, construct the control problem instance consisting of a game

$$\mathcal{G} = (1, 4a_1, \ldots, 4a_n, 2, 1, 2\alpha - 2, 2\alpha - 2; 4\alpha + 1)$$

with $n+5$ players, distinguished player $p=1$ with weight 1, deletion limit $k=1$, and the following communication structure $G = (N, E)$:

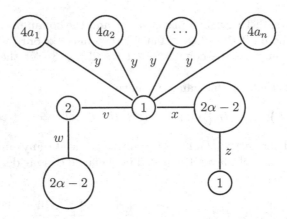

We now show that

$$(\exists e \in E)[\beta((\mathcal{G}, G_{\setminus\{e\}}), 1) - \beta((\mathcal{G}, G), 1) = 0] \iff \xi = 0.$$

Let $\xi = 0$. Then

$$\beta((\mathcal{G}, G), 1) = \frac{1}{2^{n+2}}.$$

If we delete either the edge w or the edge z, the index will not change; therefore, it is possible to maintain the Penrose-Banzhaf power index if there is no solution of PARTITION$((a_1, \ldots, a_n))$.

Let $\xi > 0$. Then

$$\beta((\mathcal{G}, G), 1) = \frac{1}{2^{n+2}} + \frac{\xi}{2^{n+4}} + \frac{\xi}{2^{n+3}} = \frac{4 + 3\xi}{2^{n+4}}.$$

If we delete the edge x, the new Penrose-Banzhaf index will increase:

$$\beta((\mathcal{G}, G_{\setminus\{x\}}), 1) = \frac{1}{2^{n+1}} + \frac{\xi}{2^{n+2}} = \frac{2 + \xi}{2^{n+2}} = \frac{8 + 4\xi}{2^{n+4}}.$$

If we delete the edge z, the index will also increase:

$$\beta((\mathcal{G}, G_{\setminus\{z\}}), 1) = \frac{1}{2^{n+2}} + \frac{2\xi}{2^{n+3}} = \frac{4 + 4\xi}{2^{n+4}}.$$

If we delete any of the y-edges, the index will decrease:

$$\beta((\mathcal{G}, G_{\setminus\{y\}}), 1) = \frac{\xi}{2}\frac{1}{2^{n+3}} + \frac{\xi}{2}\frac{1}{2^{n+2}} = \frac{3\xi}{2^{n+4}}.$$

If we delete the edge v, the new Penrose-Banzhaf index will decrease, too:

$$\beta((\mathcal{G}, G_{\setminus\{v\}}), 1) = \frac{1}{2^{n+1}} = \frac{8}{2^{n+4}} \leq \frac{4 + 2\xi}{2^{n+4}}.$$

And finally, if we delete the edge w, the new Penrose-Banzhaf index will also decrease:

$$\beta((\mathcal{G}, G_{\setminus\{w\}}), 1) = \frac{1}{2^{n+2}} + \frac{\xi}{2^{n+3}} = \frac{2 + \xi}{2^{n+3}} = \frac{4 + 2\xi}{2^{n+4}}.$$

Therefore, control by deleting edges to maintain a distinguished player's Penrose-Banzhaf index is coNP-hard.   □

We leave the complexity of control by deleting edges between players to maintain a distinguished player's Shapley-Shubik index in a graph-restricted weighted voting game open.

5   Conclusions

We have analyzed the (probabilistic) Penrose-Banzhaf and the Shapley-Shubik power index in graph-restricted weighted voting games in terms of manipulation of their communication structures by adding or deleting edges in the graphs. Related to the model due to Skibski et al. [18], we presented upper and lower bounds on how much these power indices can change in Theorems 2 and 5. Further, we have analyzed the resulting control problems related to the goals of increasing, decreasing, or maintaining a power index in terms of their computational complexity. Our complexity results are summarized in Table 1, where the open question mentioned above is marked by a question mark.

Table 1. Overview of complexity results for control problems in graph-restricted weighted voting games with respect to the Shapley-Shubik (φ) and the probabilistic Penrose-Banzhaf index (β)

Goal		Control by adding an edge	Control by deleting an edge
Decrease	β	NP-hard (Theorem 3)	DP-hard (Theorem 6)
	φ	NP-hard (Theorem 4)	NP-hard (Theorem 7)
Increase	β	DP-hard (Theorem 3)	DP-hard (Theorem 6)
	φ	NP-hard (Theorem 4)	NP-hard (Theorem 7)
Maintain	β	coNP-hard (Theorem 3)	coNP-hard (Theorem 8)
	φ	coNP-hard (Theorem 4)	?

Interesting tasks for future research include the question of whether our complexity lower bounds can be raised and whether we can pinpoint the complexity of these problems exactly.

Acknowledgments. This work was supported in part by Deutsche Forschungsgemeinschaft under grant RO 1202/21-1.

References

1. Aziz, H., Bachrach, Y., Elkind, E., Paterson, M.: False-name manipulations in weighted voting games. Journal of Artificial Intelligence Research **40**, 57–93 (2011)

2. Banzhaf, J., III.: Weighted voting doesn't work: a mathematical analysis. Rutgers Law Rev. **19**, 317–343 (1965)
3. Chalkiadakis, G., Elkind, E., Wooldridge, M.: Computational aspects of cooperative game theory. Synth. Lect. Artif. Intell. Mach. Learn. **5**, 1–168 (2011)
4. Chalkiadakis, G., Wooldridge, M.: Weighted voting games. In: Brandt, F., Conitzer, V., Endriss, U., Lang, J., Procaccia, A. (eds.) Handbook of Computational Social Choice, chap. 16, pp. 377–395. Cambridge University Press (2016)
5. Dubey, P., Shapley, L.: Mathematical properties of the Banzhaf power index. Mathematics of Operations Research **4**(2), 99–131 (1979)
6. Elkind, E., Rothe, J.: Cooperative game theory. In: Rothe, J. (ed.) Economics and Computation. An Introduction to Algorithmic Game Theory, Computational Social Choice, and Fair Division, chap. 3, pp. 135-193. Springer, Cham (2015). https://doi.org/10.1007/978-3-662-47904-9_3
7. Garey, M., Johnson, D.: Computers and Intractability: A Guide to the Theory of NP-Completeness. W. H. Freeman and Company, New York (1979)
8. Myerson, R.: Graphs and cooperation in games. Mathematics of Operations Research **2**(3), 225–229 (1977)
9. Napel, S., Nohn, A., Alonso-Meijide, J.: Monotonicity of power in weighted voting games with restricted communication. Mathematical Social Sciences **64**(3), 247–257 (2012)
10. Papadimitriou, C., Yannakakis, M.: The complexity of facets (and some facets of complexity). Journal of Computer and System Sciences **28**(2), 244–259 (1984)
11. Penrose, L.: The elementary statistics of majority voting. Journal of the Royal Statistical Society **109**(1), 53–57 (1946)
12. Rey, A., Rothe, J.: False-name manipulation in weighted voting games is hard for probabilistic polynomial time. J. Artif. Intell. Res. **50**, 573–601 (2014)
13. Rey, A., Rothe, J.: Structural control in weighted voting games. B.E. J. Theor. Econ. **18**(2), 1–15 (2018)
14. Riege, T., Rothe, J.: Completeness in the boolean hierarchy: Exact-Four-Colorability, minimal graph uncolorability, and exact domatic number problems - a survey. Journal of Universal Computer Science **12**(5), 551–578 (2006)
15. Riege, T., Rothe, J.: Complexity of the exact domatic number problem and of the exact conveyor flow shop problem. Theor. Comput. Syst. **39**(5), 635–668 (2006). https://doi.org/10.1007/s00224-004-1209-8
16. Rothe, J.: Exact complexity of Exact-Four-Colorability. Information Processing Letters **87**(1), 7–12 (2003)
17. Shapley, L., Shubik, M.: A method of evaluating the distribution of power in a committee system. The American Political Science Review **48**(3), 787–792 (1954)
18. Skibski, O., Michalak, T., Sakurai, Y., Yokoo, M.: A pseudo-polynomial algorithm for computing power indices in graph-restricted weighted voting games. In: Proceedings of the 24th International Joint Conference on Artificial Intelligence, pp. 631–637. AAAI Press/IJCAI, July 2015
19. Taylor, A., Zwicker, W.: Simple Games: Desirability Relations, Trading, Pseudoweightings. Princeton University Press, Princeton (1999)
20. Wagner, K.: More complicated questions about maxima and minima, and some closures of NP. Theoretical Computer Science **51**(1–2), 53–80 (1987)
21. Zuckerman, M., Faliszewski, P., Bachrach, Y., Elkind, E.: Manipulating the quota in weighted voting games. Artificial Intelligence **180–181**, 1–19 (2012)

Strategic Voting in Negotiating Teams

Leora Schmerler and Noam Hazon[(✉)]

Department of Computer Science, Ariel University, Ariel, Israel
{leoras,noamh}@ariel.ac.il

Abstract. A negotiating team is a group of two or more agents who join together as a single negotiating party because they share a common goal related to the negotiation. Since a negotiating team is composed of several stakeholders, represented as a single negotiating party, there is need for a voting rule for the team to reach decisions. In this paper, we investigate the problem of strategic voting in the context of negotiating teams. Specifically, we present a polynomial-time algorithm that finds a manipulation for a single voter when using a positional scoring rule. We show that the problem is still tractable when there is a coalition of manipulators that uses a x-approval rule. The coalitional manipulation problem becomes computationally hard when using Borda, but we provide a polynomial-time algorithm with the following guarantee: given a manipulable instance with k manipulators, the algorithm finds a successful manipulation with at most one additional manipulator. Our results hold for both constructive and destructive manipulations.

Keywords: Voting · Negotiation · Manipulation.

1 Introduction

Voting is a common way to combine the preferences of several agents in order to reach a consensus. While being prevalent in human societies, it has also played a major role in multi-agent systems for applied tasks such as multi-agent planning [16] or aggregating search results from the web [14]. In its essence, a voting process consists of several voters along with their ranking of the candidates, and a voting rule, which needs to decide on a winning candidate or on a winning ranking of the candidates.

Another common mechanism for reaching an agreement among several agents is a negotiation [19]. In a negotiation there is a dialogue between several agents in order to reach an agreement that is beneficial for all of them. Extensive work has been invested in developing negotiation protocols for many settings, but bilateral negotiations, where there are only two negotiating parties, is the most common type of negotiations [6]. Many works have focused on the case where each negotiating party represents a single agent. However, there are many cases in which a negotiating party represents more than one individual.

© Springer Nature Switzerland AG 2021
D. Fotakis and D. Ríos Insua (Eds.): ADT 2021, LNAI 13023, pp. 209–223, 2021.
https://doi.org/10.1007/978-3-030-87756-9_14

For example (motivated by Sánchez-Anguix et al. [25]), consider an agricultural cooperative that negotiates with the government. Even though the members of the cooperative have a common goal, they may have different preferences regarding the prohibition of importing products, government supervision of prices, insect control, tax concessions, etc. As another example, consider the government of the United Kingdom that negotiates with the European Union (EU) regarding withdrawal from the EU (i.e., the Brexit). The members of the EU have similar interests and objectives, and thus they are considered a single party in the negotiation process. Nevertheless, the EU is composed of different countries, and they may have different preferences regarding sovereignty, migrants and welfare benefits, economic governance, competitiveness, etc. These situations are denoted by social scientists as negotiating teams, in which *a group of two or more interdependent persons join together as a single negotiating party because their similar interests and objectives relate to the negotiation* [8].

Since a negotiating team comprises several stakeholders represented as a single negotiating party, there is need for a coordination mechanism, and a voting rule is a natural candidate. Ideally, the voters report their true preferences so that the voting rule will be able to choose the most appropriate outcome. However, as shown by Gibbard [20] and Satterthwaite [26], every reasonable voting rule with at least 3 candidates is prone to strategic voting. That is, voters might benefit from reporting rankings different from their true ones. Clearly, this problem of manipulation also exists in a negotiating team. For example, suppose that there is a EU council committee that negotiates with the UK on agricultural and fishery policies. The committee may decide that the UK will be excluded from the agricultural policy due to Brexit, or the UK will still be included. Similarly, the committee may decide that the fishery policy no longer applies to the UK or include the UK. Therefore, there are 4 possible outcomes, denoted by o_1, o_2, o_3 and o_4. Now, suppose that Germany prefers o_1 over o_2, o_2 over o_3, and o_3 over o_4. We may assume that the preferences of the UK government are publicly known, and it is also possible that Germany, which currently holds the presidency of the EU council, is familiar with the preferences of the other EU council members. Since the negotiation protocol usually is also known, Germany might be able to reason that o_3 is the negotiation result, but if Germany will vote strategically and misreport its preferences then o_2 will be the negotiating result. To the best of our knowledge, the analysis of manipulation in the context of negotiating teams has not been investigated to date.

In this paper, we investigate manipulation in the context of negotiating teams. We assume that there is a negotiation process between two parties. One of the parties is a negotiating team, and the team uses a voting rule to reach a decision regarding its negotiation strategy. Specifically, the negotiating team uses a positional scoring rule as a *social welfare function (SWF)*, which outputs a complete preference order. This preference order represents the negotiating party, and is the input in the negotiation process. We thus assume that there is a negotiation protocol that can work with ordinal preferences. We use the *Voting by Alternating Offers and Vetoes (VAOV)* protocol [1], since it is intuitive, easy to understand, and the negotiation result is Pareto optimal. Moreover, Erlich et al. [17] have shown that we can identify the negotiation result of the VAOV protocol if both parties follow a sub-game perfect equilibrium with an intuitive procedure.

We analyze two types of manipulation, constructive and destructive. We begin by studying constructive manipulation by a single voter, where there is a single manipulator that would like to manipulate the election so that a preferred candidate will be the negotiation result. We show that placing the preferred candidate in the highest position in the manipulative vote is not always the optimal strategy, unlike in the traditional constructive manipulation of scoring rules, and we provide a polynomial-time algorithm to find a manipulation (or decide that such a manipulation does not exist). We then analyze the constructive coalitional manipulation problem, where several voters collude and coordinate their votes so that an agreed candidate will be the negotiation result. We show that this problem is still tractable for any x-approval rule, but it becomes computationally hard for Borda. However, we provide a polynomial-time algorithm for the coalitional manipulation of Borda with the following guarantee: given a manipulable instance with k manipulators, the algorithm finds a successful manipulation with at most one additional manipulator. Finally, we show that our hardness result and algorithms can be adapted for destructive manipulation problems, where the goal of the manipulation is to prevent a candidate from being the result of the negotiation.

The contribution of this work is twofold. First, it provides an analysis of a voting manipulation in the context of negotiating teams, a problem that has not been investigated to date. Our analysis also emphasizes the importance of analyzing voting rules within an actual context, because it leads to new insights and a deeper understanding of the voting rules. Second, our work concerns the manipulation of SWF, which has been scarcely investigated.

2   Related Work

The computational analysis of voting manipulation was initially performed by Bartholdi, Tovey, and Trick [3], and Bartholdi and Orlin [2], who investigated constructive manipulation by a single voter. Following these pioneer works, many researchers have investigated the computational complexity of manipulation, and studied different types of manipulation with different voting rules in varied settings. We refer the reader to the survey provided by [18], and more recent survey by [11]. All of the works that are surveyed in these papers analyze the manipulation of voting rules as *social choice functions*, that is, the voting rules are used to output one winning candidate (or a set of tied winning candidates). In our work we investigate manipulation of a resolute SWF, i.e., it outputs a complete preference order of the candidates.

There are very few papers that investigate the manipulation of SWFs. This is possibly since the opportunities for manipulation are not well-defined without additional assumptions. That is, since the output of a SWF is an order, and voters do not report their preferences over all possible orderings, some assumptions have to be made on how the voters compare possible orders. Indeed, the first work that directly deals with the manipulation of SWF was by [5], who assumed that a voter prefers one order over another if the former is closer to her own preferences than the latter according to the Kemeny distance, and mainly presented impossibility results. Bossert and Sprumont [4] assumed that a voter prefers one order over another if the former is strictly between the latter and the voter's own preferences. Built on this definition their work studies three classes of SWF that are not prone to manipulation (i.e., strategy-proof). Dogan and Lainé [13] characterized the conditions to be imposed on SWFs so that if we extend the preferences of the voters to preferences over orders in specific ways the SWFs will not be prone to manipulation. Our work also investigates the manipulation of SWF, but we analyze the SWF in the specific context of a negotiation. Therefore, unlike all of the above works, the preferences of the manipulators are well-defined and no additional assumptions are needed.

Our work is also connected to committee elections or multi-winner elections, where manipulation of scoring rules has been considered [7,22,24]. However, in committee election we are given the size of the committee as an input. In our setting the output of the voting rule (i.e., the ranking) essentially determines the point in which RC terminates (see Sect. 3 for the definition of RC). Using the model of committee election in our setting we can say that the ranking determines the size of the committee. That is, each possible manipulation determines not only the position of each candidate but also the size of the committee.

The work that is closest to ours is the paper by Sánchez-Anguix et al. [25], which involves the use of voting rules for the decision process of a negotiating team, i.e., the same basic scenario that we consider. The paper presents several strategies they developed, which use some specific, tailored-made, voting rules, and experimentally analyzes them in different environments. Our work analyzes voting in the context of a negotiation from a theoretical perspective. We formally define the general problem, show polynomial-time algorithms for some cases, and provide hardness results and approximations for others.

Finally, we note that in our setting there is a SWF, which outputs an order over the candidates, and this order is used as an input for the negotiation process. In Sect. 3 we note that there is a connection between the sub-game perfect equilibrium of the negotiation and the Bucklin voting rule. Therefore, our setting is also related to a multi-stage voting. Several variants of multi-stage voting have been considered [9,12,15,23]. All of these works did not consider the case of SWF in the first round, as we do. More importantly, in all of these works the set of voters remains the same throughout the application of the voting rules. In our case the set of the voters in the first stage is different from the set in the second stage. In the first stage the voters are the agents in the negotiating team, and they use a scoring rule as a SWF. In the second stage there are only two voters, which are the negotiating parties, and they use an equivalent of Bucklin on their full preference orders.

3   Preliminaries

We assume that there is a set of outcomes, O, $|O| = m$ and a set of voters $V = \{1, ..., n\}$. Each voter i is represented by her preference p_i, which is a total order over O. We write $o \succ_{p_i} o'$ to denote that outcome o is preferred over outcome o' according to p_i. The position of outcome o in preference p_i, denoted by $pos(o, p_i)$, is the number of outcomes that o is preferred over them in p_i. That is, the most preferred outcome is in position $m - 1$ and the least preferred outcome is in position 0^1. We also refer to the outcomes of O as candidates, and to the total orders over O as votes.

A preference profile is a vector $\vec{p} = (p_1, p_2, ..., p_n)$. In our setting we are interested in a *resolute social welfare function*, which is a mapping of the set of all preference profiles to a single strict preference order. A scoring vector for m candidates is $\vec{s} = (s_{m-1}, \ldots, s_0)$, where every s_i is a real number, $s_{m-1} \geq \ldots \geq s_0 \geq 0$, and $s_{m-1} > s_0$. A scoring vector essentially defines a voting rule for m candidates: each voter awards s_i points to the candidate in position i. Then, when using the rule as a SWF, the candidate with the highest aggregated score is placed in the top-most position, the candidate with the second highest score is placed in the second highest position, etc. Since ties are possible, we assume that a lexicographical tie-breaking rule is used. We study *positional scoring rules*, where each rule in this family applies an appropriate scoring vector for each number of candidates. That is, a scoring rule is represented by an efficiently compu table function f such that for each $m \in \mathbb{N}$, $f(m) = (s_{m-1}^m, \ldots, s_0^m)$ is a scoring vector for m candidates. Some of our results hold only for x-approval rules, in which $f(m) = (1, \ldots, 1, 0, \ldots, 0)$, where the number of 1's is x. Note that the well-known *Plurality* rule (where each voter awards one point to her favorite candidate) is 1-approval and the *Veto* rule (where each voter awards one point to all the candidates, except for the least preferred one) is $(m-1)$-approval and they are thus both x-approval rules. We also analyze the Borda rule, where each voter awards the candidate a score that equals the candidate's position, i.e., $f(m) = (m - 1, m - 2, \ldots, 1, 0)$. In general, we denote the resulting social welfare function $\mathcal{F}$.

In the negotiation process we assume there are two parties: t is the negotiating team, which comprises a set of voters, and there is another party. The parties negotiate over the set of outcomes O, and their preferences are also total orders over O. However, since t is a negotiating team that comprises several stakeholders, the preference order of t, p_t, is determined by the social welfare function over the preference profile of the members of t, that is, $p_t = \mathcal{F}(\vec{p})$. We denote by p_o the preference order of the other party.

We assume that negotiating parties use the *Voting by Alternating Offers and Vetoes (VAOV)* protocol [1], which is a negotiation protocol that works

---

1   Our definition of a candidate's position in a voter's ranking is the opposite of the commonly used, and we chose it to enhance the readability of the proofs: $pos(o, p_i) \geq pos(o', p_j)$ is naturally translated to "o is ranked in p_i higher than o' is ranked in p_j".

with ordinal preferences. The protocol works as follows. Let p_1 be the party that initiates the negotiation and let p_2 be the other party. At round 1, party p_1 offers an outcome $o \in O$ to p_2. If p_2 accepts, the negotiation terminates successfully with o as the result of the negotiation. Otherwise, party p_2 offers an outcome $o' \in O \setminus \{o\}$. If p_1 accepts, the negotiation terminates successfully with o' as the result of the negotiation. Otherwise, p_1 offers an outcome $o'' \in O \setminus \{o, o'\}$ to p_2, and so on. If no offer was accepted until round m then the last available outcome is accepted in the last round as the result of the negotiation. We further assume that the negotiating parties are rational and each party has full information on the other party's preferences. Therefore, the parties will follow a sub-game perfect equilibrium (SPE) during the negotiation. Anbarci [1] showed that if both parties follow an SPE the negotiation result will be unique. We can thus also call this outcome the SPE result. The SPE result depends on p_t, p_o, and on the identity of the party that initiates the negotiation, and we thus denote by $\mathcal{N}_t(p_t, p_o)$ the SPE result if the negotiation team t initiates the negotiation, and by $\mathcal{N}_o(p_t, p_o)$ the SPE result if the other party initiates the negotiation.

In some negotiation settings there is a central authority that can force the parties to offer specific outcomes in a specific order. In this case it is common to use a bargaining rule, which is a function that assigns each negotiation instance a subset of the outcomes that is considered the result of the negotiation. One such bargaining rule is the *Rational Compromise* (*RC*) bargaining rule [21]. Let $A^j_{(p_t)} = \{$the j most preferred outcomes in $p_t\}$. $A^j_{(p_o)}$ is defined similarly for p_o. RC is computed as follows:

1. Let $j = 1$
2. If $|A^j_{(p_t)} \cap A^j_{(p_o)}| > 0$ then return $A^j_{(p_t)} \cap A^j_{(p_o)}$.
3. Else, $j \leftarrow j + 1$ and go to line 2.

Note that the RC bargaining rule is equivalent to Bucklin voting with two voters and no tie-breaking mechanism. An important finding of [17] shows that the negotiation result of the VAOV protocol if both parties follow an SPE (i.e., the SPE result) is always part of the set returned by the RC rule. We use this connection between RC and the VAOV negotiation protocol whenever we need to identify the SPE result. Specifically, if RC returns one outcome, this is also the SPE result. If RC returns two outcomes then the SPE result depends on the number of outcomes and on the party that initiates the negotiation.

4   Constructive Manipulation by a Single Voter

We begin by studying the problem of constructive manipulation by a single voter. In this setting a manipulator v' would like to manipulate the election so that a preferred candidate p will be the SPE result. We assume that the decision of which party initiates the negotiation is not always known in advance. Therefore, we require that both $\mathcal{N}_t$ and $\mathcal{N}_o$ returns the preferred candidate. The Constructive Manipulation in the context of Negotiations (C-MaNego) is defined as follows:

Definition 1. (**C-MaNego**) *We are given social welfare function $\mathcal{F}$, a preference profile $\vec{p}$ of honest voters on the negotiating team t, the preference of the other party p_o, a specific manipulator v', and a preferred candidate $p \in O$. We are asked whether a preference order $p_{v'}$ exists for the manipulator v' such that $\mathcal{N}_t(\mathcal{F}(\vec{p} \cup p_{v'}), p_o) = \mathcal{N}_o(\mathcal{F}(\vec{p} \cup p_{v'}), p_o) = p$.*

We first observe that manipulation problems in the context of negotiations are inherently different from the traditional voting manipulation problems. First, in voting manipulation there is one set of voters in which their preferences are the inputs of the voting rule. The manipulator only needs to take these preferences into account when she decides on her manipulative vote. In our case there are two stages: in the first stage there is a set of voters and in the second stage there are two negotiating parties, and the manipulator needs to consider the preferences of all of these agents when she decides on her manipulative vote. In addition, unlike constructive manipulation in many voting rules, placing the preferred candidate p in the highest position in the manipulative vote is not always the optimal strategy, since constructive manipulation in our case requires sometimes also destructive actions. Indeed, the following example describes a scenario where there is no manipulation where p is placed in the highest position. However, manipulation is possible if p is placed in the second highest position, since this placement allows for a destructive action against another candidate.

Example 1. Assume that p_o is the following preference order: $p_o = b \succ p \succ a \succ c$. There is one manipulator v', and $\vec{p}$ comprises 4 voters with the following preferences: $p \succ c \succ a \succ b, p \succ b \succ a \succ c, b \succ p \succ a \succ c, b \succ a \succ c \succ p$. Assume that we use the Borda rule, and thus the voters of $\vec{p}$ give the following scores: b gets 8 points, p gets 8 points, a gets 5 points and c gets 3 points. Since we assume that the tie-breaking rule is a lexicographical order, $p_t = b \succ p \succ a \succ c$. In order to find a successful manipulation v' needs to make sure that b will not be in the two highest positions in $\mathcal{F}(\vec{p} \cup p_{v'})$. Now, if the manipulator places p in the highest position then p gets 11 points. Then, placing the other candidates in every possible order results in b in the second highest positions in $\mathcal{F}(\vec{p} \cup p_{v'})$. Alternatively, if v' votes as follows: $a \succ p \succ c \succ b$, then p gets 10 points, a and b get 8 points, and c gets 4 points; thus $\mathcal{F}(\vec{p} \cup p_{v'}) = p \succ a \succ b \succ c$. Now the SPE result is p.

We now present a polynomial-time algorithm for C-MaNego with any scoring rule. Let p^a be the order that the algorithm finds (i.e., p^a is a possible $p_{v'}$), and let $p_t^a = \mathcal{F}(\vec{p} \cup p^a)$. Note that during the algorithm we use $\mathcal{F}(\vec{p} \cup p^a)$, where p^a is not a complete preference order, i.e., p^a comprises m' candidates, $m' < m$, that are placed in specific positions. In these situations we assume that all of the candidates that are not in p^a get a score of 0 from p^a. Given i, $1 \leq i \leq \lceil m/2 \rceil$, let H^i be the set that contains p, and the $i - 1$ most preferred outcomes in p_t that do not belong to $A_{(p_o)}^i$.

Our algorithm works as follows. It uses the connection between RC and the negotiation protocol to identify the SPE result. Clearly, if the position of p in p_o is less than $\lceil m/2 \rceil$ then for any possible p^a RC does not return p. Therefore, there is no manipulation and the algorithm returns false (lines 1–2). Otherwise, we use the variable i to indicate the iteration number in which RC terminates. Thus, the algorithm iterates over the values of i from 1 to $\lceil m/2 \rceil$ (line 3). For a given i, the algorithm tries to ensure that no outcome from $A^i_{(p_o)}$ will be placed in the i highest positions in p^a_t. Consequently, the algorithm places the outcomes from H^i in the highest positions and they receive the highest scores. Moreover, the outcomes are placed in a reverse order (with regards to their order in p_t) to ensure that even the least preferred outcome in H^i will receive a score that is as high as possible (in order to be included in the highest positions in p^a_t). Then, the algorithm places the remaining outcomes, denoted C, so that they will not prevent p from being the negotiation result (lines 5–10). Specifically, the algorithm places the outcomes of C in the lowest positions in p^a and the outcomes are placed in a reverse order, with regards to their order in p_t. Then, if p^a is a successful manipulation the algorithm returns it (line 12). Otherwise, the algorithm proceeds to the next iteration.

ALGORITHM 1: Constructive manipulation by a single voter

---

1   **if** $pos(p, p_o) < \lceil m/2 \rceil$ **then**
2   **return** false
3   **for** $i = 1$ to $\lceil m/2 \rceil$ **do**
4     $p^a \leftarrow H^i$ in a reverse order of the positions in p_t
5     $C \leftarrow O \setminus H^i$
6   **for** $j = 1$ to $|O \setminus H^i|$ **do**
7       $c \leftarrow$ the most preferred outcome from C under p_t
8   place c in p^a such that $pos(c, p^a) = j - 1$
9       $j \leftarrow j + 1$
10   remove c from C
11   **if** $\mathcal{N}_t(\mathcal{F}(\vec{p} \cup p^a), p_o) = \mathcal{N}_o(\mathcal{F}(\vec{p} \cup p^a), p_o) = p$ **then**
12   **return** p^a
13 **return** false

Theorem 1. *Algorithm 1 correctly decides the C-MaNego problem with any positional scoring rule in polynomial time.*

Proof. Clearly, the algorithm runs in polynomial time since there are two loops, where each loop iterates at most m times. In addition, if the algorithm successfully constructs a manipulation order, p will be the negotiation result. We need to show that if an order that makes p the negotiation result exists, then our algorithm will find such an order. Assume that we have a manipulative vote, p^m, that makes p the negotiation result, and let $p^m_t = \mathcal{F}(\vec{p} \cup p^m)$. Thus, $\mathcal{N}_t(p^m_t, p_o) = \mathcal{N}_o(p^m_t, p_o) = p$. In addition, given a set H^i let $L^i = \{\ell | \exists h \in H^i \text{ s.t. } h \prec_{p_t} \ell\}$ and $R^i = \{o | o \in O, o \notin H^i \text{ and } o \notin L^i\}$.

We show that Algorithm 1 returns p^a in line 12, when i equals the iteration in which RC terminates given p^m_t and p_o. There are two possible cases to consider:

- $A^i_{(p^a)} = A^i_{(p^m)}$: according to Algorithm 1, $A^i_{(p^a)} = H^i$, and since $A^i_{(p^a)} = A^i_{(p^m)}$, $A^i_{(p^m)} = H^i$. By definition, $\forall r \in R^i$ and $\forall h \in H^i$, $r \prec_{p_t} h$ and $r \prec_{p^m} h$. Since we use a scoring rule, $\forall r \in R^i$ and $\forall h \in H^i$, $r \prec_{p_t^m} h$. Since p^m is a successful manipulation and RC terminates at iteration i, then $\forall \ell \in L^i$ where $\ell \in A^i_{(p_o)}$, $\ell \notin A^i_{(p_t^m)}$. For any other $\ell \in L^i$ we know that $p \prec_{p_t} \ell$ and for any $h \in H^i \setminus \{p\}$, $\ell \prec_{p_t} h$. Since p^m is a successful manipulation and RC terminates at iteration i, $p \in A^i_{(p_t^m)}$. Overall, $A^i_{(p_t^m)} = H^i$.

We first assume that all the candidates that are not in H^i get a score of 0 from p^a, and we show that $A^i_{(p_t^a)} = H^i$. For any $h \in H^i$, if $pos(h, p^a) \geq pos(h, p^m)$ then $pos(h, p_t^a) \geq pos(h, p_t^m)$. Otherwise, let $h \in H^i$ be a candidate such that $pos(h, p^a) < pos(h, p^m)$ and let $s = pos(h, p^a)$. There are $m - s - 1$ candidates from H^i above h in p^a. According to the pigeonhole principle, at least one of them, denoted h', that is placed in p^m at position s or lower. That is, $pos(h', p^m) \leq pos(h, p^a)$. By the algorithm construction, all of the candidates that are ranked higher than h in p^a are ranked lower than h in p_t. That is, $h' \prec_{p_t} h$. However, $h' \in A^i_{(p^m)}$ and thus $h \in A^i_{(p_t^a)}$. Overall, $A^i_{(p_t^a)} = H^i$.

We now show that Algorithm 1 (lines 6–10) can assign scores to all the candidates in $O \setminus H^i$ such that p^a is a successful manipulation. For any $o \in O \setminus H^i$, if $pos(o, p^a) \leq pos(o, p^m)$ then $pos(o, p_t^a) \leq pos(o, p_t^m)$. Since $o \notin A^i_{(p_t^m)}$ then $o \notin A^i_{(p_t^a)}$. Otherwise, let $o \in O \setminus H^i$ be a candidate such that $pos(o, p^a) > pos(o, p^m)$ and let $s = pos(o, p^a)$. There are s candidates from $O \setminus H^i$ below o in p^a. According to the pigeonhole principle, at least one of them, denoted o', is placed in p^m at position s or higher. That is, $pos(o', p^m) \geq pos(o, p^a)$. By the algorithm construction, all of the candidates $c \in O \setminus H^i$ that are ranked lower than o in p^a are ranked higher than o in p_t. That is, $o \prec_{p_t} o'$. However, $o' \notin A^i_{(p_t^m)}$ and thus $o \notin A^i_{(p_t^a)}$. Overall, after placing the candidates from $O \setminus H^i$ in p^a, $\forall o \in O \setminus H^i$, $o \notin A^i_{(p_t^a)}$. That is, $A^i_{(p_t^a)} = H^i$, and thus $\mathcal{N}_t(\mathcal{F}(\vec{p} \cup p^a), p_o) = \mathcal{N}_o(\mathcal{F}(\vec{p} \cup p^a), p_o) = p$.

- $A^i_{(p^a)} \neq A^i_{(p^m)}$: let $p^{m'}$ be the manipulation p^m with the following changes: each $r \in A^i_{(p_t^m)} \setminus H^i$ is replaced with a candidate $h_r \in H^i \setminus A^i_{(p_t^m)}$. That is, $pos(r, p^{m'}) = pos(h_r, p^m)$ and $pos(h_r, p^{m'}) = pos(r, p^m)$. Since p^m is a successful manipulation, if $r \in A^i_{(p_t^m)} \setminus H^i$ then $r \notin A^i_{(p_o)}$. Thus, by the definition of H^i, $\forall r \in A^i_{(p_t^m)} \setminus H^i$ and $\forall h \in H^i \setminus A^i_{(p_t^m)}$, $pos(r, p_t) < pos(h, p_t)$. Therefore, since each $r \in A^i_{(p_t^m)} \setminus H^i$ is ranked in the highest i positions in p_t^m, then h_r is ranked in the highest i positions in $p_t^{m'}$. Similarly, since each h_r is not ranked in the highest i positions in p_t^m, then r is not ranked in the highest i positions in $p_t^{m'}$. That is, $h_r \in A^i_{(p_t^{m'})}$ and $r \notin A^i_{(p_t^{m'})}$, and thus, $H^i = A^i_{(p_t^{m'})}$. Let $p^{m''}$ be the manipulation $p^{m'}$ with the following changes: each $r \in A^i_{(p^{m'})} \setminus H^i$ is replaced with a candidate $h_r \in H^i \setminus A^i_{(p^{m'})}$. That is, $A^i_{(p^{m''})} = H^i$. Note that $c \notin A^i_{(p_t^{m'})}$ for every $c \in O \setminus H^i$, and therefore $c \notin A^i_{(p_t^{m''})}$. Thus, $A^i_{(p_t^{m''})} = H^i$. That is, $p^{m''}$ is a successful manipulation, and

$A^i_{(p^{m\prime\prime})} = A^i_{(p^a)}$. This brings us back to the first case we already considered and showed that p^a is a successful manipulation.

□

5   Constructive Coalitional Manipulation

We now consider the problem of constructive manipulation by a coalition of voters. That is, several manipulators, denoted by M, might decide to collude and coordinate their votes in such a way that an agreed candidate p will be the SPE result. The constructive coalitional manipulation problem is defined as follows:

Definition 2 (CC-MaNego). *Given a social welfare function $\mathcal{F}$, a preference profile $\vec{p}$ of honest voters on the negotiating team t, the preference of the other party p_o, a number of manipulators k, and a preferred candidate $p \in O$, we check whether a preference profile $\vec{p}_M$ for the manipulators exists such that $\mathcal{N}_t(\mathcal{F}(\vec{p} \cup \vec{p}_M), p_o) = \mathcal{N}_o(\mathcal{F}(\vec{p} \cup \vec{p}_M), p_o) = p$.*

We show that CC-MaNego can be decided in polynomial time for any x-approval rule using Algorithm 2, which works as follows. Similarly to Algo-

---

ALGORITHM 2: Coalitional manipulation

---

1 **if** $pos(p, p_o) < \lceil m/2 \rceil$ **then**
2 | **return** false
3 **for** $i = 1$ *to* $\lceil m/2 \rceil$ **do**
4 | $\vec{p}_M \leftarrow []$
5 | **for** $\ell = 1$ *to* $|M|$ **do**
6 | | $p^a \leftarrow$ empty preference order
7 | | $C \leftarrow H^i$
8 | | **for** $j = 1$ *to* $|H^i|$ **do**
9 | | | $c \leftarrow$ the least preferred outcome from C under $\mathcal{F}(\vec{p} \cup \vec{p}_M)$
10 | | | place c in p^a such that $pos(c, p^a) = m - j$
11 | | | $j \leftarrow j + 1$
12 | | | remove c from C
13 | | $C \leftarrow O \setminus H^i$
14 | | **for** $j = 1$ *to* $|O \setminus H^i|$ **do**
15 | | | $c \leftarrow$ the most preferred outcome from C under $\mathcal{F}(\vec{p} \cup \vec{p}_M)$
16 | | | place c in p^a such that $pos(c, p^a) = j - 1$
17 | | | $j \leftarrow j + 1$
18 | | | remove c from C
19 | | add p^a to $\vec{p}_M$
20 | **if** $\mathcal{N}_t(\mathcal{F}(\vec{p} \cup \vec{p}_M), p_o) = \mathcal{N}_o(\mathcal{F}(\vec{p} \cup \vec{p}_M), p_o) = p$ **then**
21 | | **return** $\vec{p}_M$
22 **return** false

---

rithm 1, the algorithm iterates over the possible values of i, where i indicates the iteration number in which RC terminates. For any given i, the algorithm iterates over the number of manipulators and determines their votes (Lines 5–19). We refer to each of these iterations as a *stage* of the algorithm. In each stage, a vote of one manipulator is determined, denoted by p^a. We begin with an empty set of votes, $\vec{p}_M$. Then, the algorithm places the outcomes from H^i in the highest positions in p^a. The outcomes are placed in a reverse order, with regards to their order in $\mathcal{F}(\vec{p} \cup \vec{p}_M)$. Similarly, the algorithm places all the other outcomes in the lowest positions in p^a and the outcomes are placed in a reverse order, with regards to their order in $\mathcal{F}(\vec{p} \cup \vec{p}_M)$. Note that the set H^i does not change throughout the algorithm's stages. However, the order of the outcomes in H^i and $O \setminus H^i$ according to $\mathcal{F}(\vec{p} \cup \vec{p}_M)$ may change when we update $\vec{p}_M$, which implies that the order in which we place the outcomes from H^i and $O \setminus H^i$ in p^a may differ from one vote to another. Due to space constraints, the full proofs of this theorem and all subsequent theorems are provided in the full version of this paper [27].

Theorem 2. *Algorithm 2 correctly decides the CC-MaNego problem with x-approval rule in polynomial time.*

Proof. (*sketch*) In order to prove the theorem we use the following definitions. Given i, $1 \leq i \leq \lceil m/2 \rceil$, let $U_0^i = \arg\min_{h \in H^i} pos(h, p_t)$. For each $s = 1, 2, ...,$ let $U_s^i \subseteq H^i$ be $U_s^i = U_{s-1}^i \cup \{u : u$ was ranked above some $u' \in U_{s-1}^i$ in some stage l, $1 \leq l < k$, but u was ranked below some $u' \in U_{s-1}^i$ in stage $l+1\}$. Now, let $U^i = \bigcup_{0 \leq s} U_s^i$. The set D^i is defined similarly. Specifically, let $D_0^i = \arg\max_{d \in O \setminus H^i} pos(d, p_t)$. For each $s = 1, 2, ...,$ let $D_s^i \subseteq O \setminus H^i$ be $D_s^i = D_{s-1}^i \cup \{d : d$ was ranked below some $d' \in D_{s-1}^i$ in some stage l, $1 \leq l < k$, but d was ranked above some $d' \in D_{s-1}^i$ in stage $l+1\}$. Now, let $D^i = \bigcup_{0 \leq s} D_s^i$. We first show that the scores in $\mathcal{F}(\vec{p} \cup \vec{p}_M)$ of the candidates in U^i (D^i) are extremely dense. That is, the difference between the scores of every two candidates from U^i (D^i) is at most 1. Therefore, the difference between the minimal (maximal) score of a candidate from U^i (D^i) and the average score of the candidates in U^i (D^i) is less than 1. Using this property we show that if for every given i, $1 \leq i \leq \lceil m/2 \rceil$, $\exists d \in D^i$ and $\exists u \in U^i$, $u \prec_{p^a} d$, then there is no manipulation. Now, $\forall h \in H^i \setminus U^i$ and $\forall u \in U^i$, $u \prec_{p^a} h$, and $\forall o \in O \setminus (H^i \cup D^i)$ and $\forall d \in D^i$, $o \prec_{p^a} d$. Therefore, if there exists i, $1 \leq i \leq \lceil m/2 \rceil$, such that $\forall d \in D^i$ and $\forall u \in U^i$, $d \prec_{p_t^a} u$, then $\forall h \in H^i$ and $\forall o \in O \setminus H^i$, $o \prec_{p_t^a} h$. By definition of H^i, $A_{(p_o)}^i \cap H^i = \{p\}$. Therefore, $A_{(p_o)}^i \cap A_{(p_t^a)}^i = \{p\}$. That is, Algorithm 2 finds a successful manipulation. $\square$

Unlike with the family of x-approval rules, CC-MaNego is computationally hard with Borda. The reduction is from the Permutation Sum problem [29].

Theorem 3. *CC-MaNego is NP-Complete with Borda.*

Even though CC-MaNego with Borda is NP-complete, it might be still possible to develop an efficient heuristic algorithm that finds a successful coalitional manipulation. We now show that Algorithm 2 is such a heuristic, and show its theoretical guarantee. Specifically, the algorithm is guaranteed to find a coalitional manipulation in many instances, and we characterize the instances in which it may fail. Formally,

Theorem 4. *Given an instance of CC-MaNego with Borda,*

1. *If there is no preference profile making p the negotiation result, then Algorithm 2 will return false.*
2. *If a preference profile making p the negotiation result exists, then for the same instance with one additional manipulator, Algorithm 2 will return a preference profile that makes p the negotiation result.*

That is, Algorithm 2 will succeed on any given instance such that the same instance but with one less manipulator is manipulable. Thus, it can be viewed as a 1-additive approximation algorithm (this approximate sense was introduced by [30] when analyzing Borda as a social choice function (SCF)).

Proof. (sketch) Interestingly, this proof is in the same vein as the proof of Theorem 2, and we again use the sets U^i and D^i. However, the proof here is more involved. Let $s_\ell(c)$ be the score of candidate c in $\mathcal{F}(\vec{p} \cup \vec{p}_M)$ after stage ℓ. We first show that, given i, $1 \leq i \leq \lceil m/2 \rceil$, the sets of scores $\{s_{k-1}(u) : u \in U^i\}$ and $\{s_{k-1}(d) : d \in D^i\}$ are 1-dense, which is the following:

Definition 3 (due to [30]). *A finite non-empty set of integers B is called 1-dense if when sorting the set in a non-increasing order $b_1 \geq b_2 \geq \cdots \geq b_i$ (such that $\{b_1, \ldots, b_i\} = B$), $\forall j, 1 \leq j \leq i - 1$, $b_{j+1} \geq b_j - 1$ holds.*

Let $q(U^i)$ and $q(D^i)$ be the average score of candidates in U^i and D^i, respectively, after $k - 1$ stages. Using the 1-dense property we show that $q(U^i) \leq \min_{u \in U^i}\{s_k(u)\} - m + |U^i|$, and similarly, $\max_{d \in D^i}\{s_k(d)\} \leq q(D^i) + |D^i| - 1$. That is, we bound the distance between the minimal score in U^i (the maximal score in D^i) after stage k and the average score in U^i (D^i) after stage $k-1$. Now, suppose that there is a successful manipulation for Borda with $k - 1$ manipulators. Then, we show that there exists i, $1 \leq i \leq \lceil m/2 \rceil$, such that $q(D^i) \leq q(U^i)$. Now, by definition of U^i and D^i, $|U^i| + |D^i| \leq m$. Combining all the inequalities we get that $\max_{d \in D^i}\{s_k(d)\} < \min_{u \in U^i}\{s_k(u)\}$. Then, it is possible to show that by adding one additional manipulator the algorithm will find a successful manipulation. $\square$

6 Destructive Manipulation

In this section we study the *destructive* manipulation problem, where the goal of the manipulation is to prevent an outcome from being the SPE result. We begin with the destructive variant of manipulation by a single voter.

Definition 4 (D-MaNego). *We are given a social welfare function $\mathcal{F}$, a preference profile $\vec{p}$ of honest voters on the negotiating team t, the preference of the other party p_o, a specific manipulator v', and a disliked candidate $e \in O$. We are asked whether a preference order $p_{v'}$ exists for the manipulator v' such that $e \neq \mathcal{N}_t(\mathcal{F}(\vec{p} \cup p_{v'}), p_o)$ and $e \neq \mathcal{N}_o(\mathcal{F}(\vec{p} \cup p_{v'}), p_o)$.*

Recall that C-MaNego is in P for any scoring rule, but this does not immediately imply that D-MaNego is also in P. Indeed, it is possible to run Algorithm 1 for each candidate $c \neq e$. However, since Algorithm 1 returns a manipulation only when $\mathcal{N}_t(\mathcal{F}(\vec{p} \cup p_{v'}), p_o) = \mathcal{N}_o(\mathcal{F}(\vec{p} \cup p_{v'}), p_o) = c$, it does not find a solution where $\mathcal{N}_t(\mathcal{F}(\vec{p} \cup p_{v'}), p_o) = c$ and $\mathcal{N}_o(\mathcal{F}(\vec{p} \cup p_{v'}), p_o) = c'$, $c \neq c'$, and both $c, c' \neq e$, which is a possible solution for D-MaNego. Nevertheless, we can use a slightly modified version of Algorithm 1 for D-MaNego.

Theorem 5. *D-MaNego with any positional scoring rule can be decided in polynomial time.*

We now continue with the destructive coalitional manipulation problem, where several manipulators might decide to collude and coordinate their votes in such a way that an agreed candidate e will not be the SPE result. The problem is defined as follows:

Definition 5 (DC-MaNego). *Given a social welfare function $\mathcal{F}$, a preference profile $\vec{p}$ of honest voters on the negotiating team t, the preference of the other party p_o, a number of manipulators k, and a disliked candidate $e \in O$, we check whether a preference profile $\vec{p}_M$ exists for the manipulators such that $e \neq \mathcal{N}_t(\mathcal{F}(\vec{p} \cup \vec{p}_M), p_o)$ and $e \neq \mathcal{N}_o(\mathcal{F}(\vec{p} \cup \vec{p}_M), p_o)$.*

Similar to C-MaNego, we show that a slightly modified version of Algorithm 2 decides DC-MaNego with any x-approval rule.

Theorem 6. *DC-MaNego with any x-approval rule can be decided in polynomial time.*

Indeed, DC-MaNego with Borda is computationally hard. Note that this result is surprising, since the destructive coalitional manipulation problem when using Borda as an SCF is in P [10].

Theorem 7. *DC-MaNego with Borda is NP-Complete.*

Finally, similar to CC-MaNego, we show that the modified Algorithm 2 is an efficient heuristic algorithm that finds a successful destructive manipulation, and we guarantee the same approximation. That is, the algorithm succeeds in finding a destructive manipulation for any given instance such that success for the same instance with one less manipulator is possible.

Theorem 8. *There is a 1-additive approximation algorithm for DC-MaNego with Borda.*

7 Conclusion and Future Work

In this paper we analyze the problem of strategic voting in the context of negotiating teams. Specifically, a scoring rule is used as a SWF, which outputs an
order over the candidates that is used as an input in the negotiation process
with the VAOV protocol. We show that the single manipulation problem is in
P with this two stage procedure, and the coalitional manipulation is also in
P for any x-approval rule. The problem of coalitional manipulation becomes
hard when using Borda, but we provide an algorithm that can be viewed as
a 1-additive approximation for this case. Interestingly, our complexity results
hold both for constructive and destructive manipulations, unlike the problems
of manipulation when using Borda as an SCF. Note also that our algorithms are
quite general. Algorithm 1 provides a solution with any scoring rule. Algorithm 2
solves the coalitional manipulation problem with any x-approval rule and it is
also an efficient approximation with Borda.

For future work we would like to extend our analysis to other voting rules.
In addition, designing FPT algorithms for CC-MaNego and DC-MaNego with
Borda is a promising open research direction, since there is an FPT algorithm
for the constructive coalitional manipulation of Borda as a SCF with respect to
the number of candidates [28].

Acknowledgments. This research was supported in part by the Ministry of Science,
Technology & Space, Israel.

References

1. Anbarci, N.: Noncooperative foundations of the area monotonic solution. Q. J. Econ. **108**(1), 245–258 (1993)
2. Bartholdi, J.J., Orlin, J.B.: Single transferable vote resists strategic voting. Soc. Choice Welfare **8**(4), 341–354 (1991). https://doi.org/10.1007/BF00183045
3. Bartholdi, J.J., Tovey, C.A., Trick, M.A.: The computational difficulty of manipulating an election. Soc. Choice Welfare **6**(3), 227–241 (1989). https://doi.org/10.1007/BF00295861
4. Bossert, W., Sprumont, Y.: Strategy-proof preference aggregation: possibilities and characterizations. Games Econ. Behav. **85**, 109–126 (2014)
5. Bossert, W., Storcken, T.: Strategy-proofness of social welfare functions: the use of the kemeny distance between preference orderings. Soc. Choice and Welfare **9**(4), 345–360 (1992). https://doi.org/10.1007/BF00182575
6. Brams, S.J.: Negotiation Games: Applying Game Theory to Bargaining and Arbitration, vol. 2. Psychology Press, Hove (2003)
7. Bredereck, R., Kaczmarczyk, A., Niedermeier, R.: On coalitional manipulation for multiwinner elections: shortlisting. In: Proceedings of IJCAI-2017, pp. 887–893 (2017)
8. Brodt, S., Thompson, L.: Negotiating teams: a levels of analysis approach. Group Dyn. Theor. Res. Pract. **5**(3), 208–219 (2001)
9. Conitzer, V., Sandholm, T.: Universal voting protocol tweaks to make manipulation hard. In: Proceedings of IJCAI, pp. 781–788 (2003)

10. Conitzer, V., Sandholm, T., Lang, J.: When are elections with few candidates hard to manipulate? J. ACM (JACM) **54**(3), 14 (2007)
11. Conitzer, V., Walsh, T.: Barriers to manipulation in voting. In: Brandt, F., Conitzer, V., Endriss, U., Lang, J., Procaccia, A.D. (eds.) Handbook of Computational Social Choice, pp. 127–145. Cambridge University Press (2016)
12. Davies, J., Narodytska, N., Walsh, T.: Eliminating the weakest link: making manipulation intractable? In: Proceedings of AAAI, pp. 1333–1339 (2012)
13. Dogan, O., Lainé, J.: Strategic manipulation of social welfare functions via strict preference extensions. In: The 13th Meeting of the Society for Social Choice and Welfare, p. 199 (2016)
14. Dwork, C., Kumar, R., Naor, M., Sivakumar, D.: Rank aggregation methods for the web. In: Proceedings of WWW, pp. 613–622 (2001)
15. Elkind, E., Lipmaa, H.: Hybrid voting protocols and hardness of manipulation. In: Deng, X., Du, D.Z. (eds.) Algorithms and Computation, vol. 3827, pp. 206–215. Springer, Heidelberg (2005). https://doi.org/10.1007/11602613_22
16. Ephrati, E., Rosenschein, J.S., et al.: Multi-agent planning as a dynamic search for social consensus. In: Proceedings of IJCAI, pp. 423–429 (1993)
17. Erlich, S., Hazon, N., Kraus, S.: Negotiation strategies for agents with ordinal preferences. In: Proceedings of IJCAI, pp. 210–218 (2018)
18. Faliszewski, P., Procaccia, A.D.: AI's war on manipulation: are we winning? AI Mag. **31**(4), 53–64 (2010)
19. Fatima, S., Kraus, S., Wooldridge, M.: Principles of Automated Negotiation. Cambridge University Press, Cambridge (2014).
20. Gibbard, A.: Manipulation of voting schemes: a general result. Econometrica **41**(4), 587–601 (1973)
21. Kıbrıs, Ö., Sertel, M.R.: Bargaining over a finite set of alternatives. Soc. Choice Welfare **28**(3), 421–437 (2007). https://doi.org/10.1007/s00355-006-0178-z
22. Meir, R., Procaccia, A.D., Rosenschein, J.S., Zohar, A.: Complexity of strategic behavior in multi-winner elections. J. Artif. Intell. Res. **33**, 149–178 (2008)
23. Narodytska, N., Walsh, T.: Manipulating two stage voting rules. In: Proceedings of AAMAS, pp. 423–430 (2013)
24. Obraztsova, S., Zick, Y., Elkind, E.: On manipulation in multi-winner elections based on scoring rules. In: Proceedings of AAMAS, pp. 359–366 (2013)
25. Sánchez-Anguix, V., Botti, V., Julián, V., García-Fornes, A.: Analyzing intra-team strategies for agent-based negotiation teams. In: Proceedings of AAMAS, pp. 929–936 (2011)
26. Satterthwaite, M.A.: Strategy-proofness and arrow's conditions: existence and correspondence theorems for voting procedures and social welfare functions. J. Econ. Theor. **10**(2), 187–217 (1975)
27. Schmerler, L., Hazon, N.: Strategic voting in the context of negotiating teams. arXiv preprint arXiv:2107.14097 (2021)
28. Yang, Y., Guo, J.: Exact algorithms for weighted and unweighted borda manipulation problems. Theor. Comput. Sci. **622**, 79–89 (2016)
29. Yu, W., Hoogeveen, H., Lenstra, J.K.: Minimizing makespan in a two-machine flow shop with delays and unit-time operations is NP-hard. J. Sched. **7**(5), 333–348 (2004). https://doi.org/10.1023/B:JOSH.0000036858.59787.c2
30. Zuckerman, M., Procaccia, A.D., Rosenschein, J.S.: Algorithms for the coalitional manipulation problem. Artif. Intell. **173**(2), 392–412 (2009)

The Nonmanipulative Vote-Deficits
of Voting Rules

Yongjie Yang[✉][iD]

Chair of Economic Theory, Saarland University, Saarbrücken, Germany
yyongjiecse@gmail.com

Abstract. We introduce a new parameter which we call the nonmanipulative vote-deficit (NMVD) for single-winner voting rules. In particular, the NMVD of a voting rule at an election is the minimum number of votes needed to be added to transform this election into a nonmanipulable one yet without changing the winner. A voting rule has a bounded NMVD if the NMVDs of this rule at all elections are bounded from above by a constant. We show that the prevalent voting rules Borda, Plurality with Runoff, and Maximin have bounded NMVDs. In addition, we show that the NMVD of r-Approval, r-Veto, and Bucklin at every election can be bounded by a function of the number of candidates. For Copeland$^\alpha$, though that in general the NMVDs at elections cannot be bounded by a function of the number of candidates, we show that many special elections are still expected to have small NMVDs. Many of our results are tight.

Keywords: Voting rules · Manipulation · Nonmanipulative vote-deficits · Borda · Maximin · Copeland · Plurality with runoff · Tournaments

1 Introduction

We consider single-winner voting where every voter holds a linear preference over a given set of candidates, and a voting rule is applied to select one candidate as the winner. An election is a tuple consisting of a set of candidates and a multiset of preferences (votes) over the candidates. We put forward the notion of *nonmanipulative vote-deficit* (NMVD). In particular, we define the NMVD of a voting rule at an election as the minimum number of votes needed to be added so that

(1) the resulting election yields the same winner as the original one; and
(2) it is impossible to change the preference of any one voter in the election to improve the result in favor of the voter.

A voting rule has a bounded NMVD if the NMVDs of this rule at all elections are bounded from above by a constant.

Our notion of NMVD is related to the classic problem CONSTRUCTIVE/DESTRUCTIVE COALITION MANIPULATION (CCM/DCM) which has been

© Springer Nature Switzerland AG 2021
D. Fotakis and D. Ríos Insua (Eds.): ADT 2021, LNAI 13023, pp. 224–240, 2021.
https://doi.org/10.1007/978-3-030-87756-9_15

extensively and intensively studied in the literature [1,4,6,16,24–26]. Recall that in the CCM (resp. DCM) problem, we are given an election and an integer k (the number of manipulators), and the question is whether we can add k new votes (cast by the manipulators) so that a distinguished candidate wins (resp. does not win) the election. Instead of changing the winner in a particular way, our notion is concerned with enhancing the winning status of the current winner by adding the minimum number of votes, assuming that everyone is self-interested and may manipulate the election as long as doing so makes herself better off.

The NMVD of a voting rule at an election is related to the margin of victory of the rule, which is defined as the minimum number of votes needed to be changed in order to change the winner (who is the new winner does not matter) [8,14,23]. It is easy to see that the margin of victory of a voting rule at an election is at least 2 if and only if the NMVD of this rule at the election is 0, and the margin of victory of a voting rule at an election is 1 if and only if the NMVD of this rule at the election is at least 1. It should be pointed out that the problem of computing margin of victory has been also studied under the name DESTRUCTIVE BRIBERY (see [11] for further details on DESTRUCTIVE BRIBERY).

NMVD is also related to the renowned Gibbard-Satterthwaite (G-S) theorem [12,20], which implies that every nondictorial (deterministic) and onto voting rule is manipulable at some election, i.e., for every voting rule there exists at least one election such that at least one voter can benefit from misreporting her true preference under this rule. Since the publication of this fundamental result, much effort has been made to circumvent the G-S theorem, including the adoption of randomized rules, the restriction of preference domains, the establishment of the complexity barrier against manipulations, etc. [3–5,7,13,17,19]. Our notion more or less provides a different approach to prohibiting manipulations via the operation of adding dummy votes.

Our Contributions. We show that several commonly used voting rules have their NMVDs bounded by two. Moreover, for many voting rules, a constant number of votes satisfying the two conditions given above can be constructed without knowing the exact votes. What we need to know is merely the candidate set and who is the current winner, or head-to-head comparisons among candidates, or the majority graph of the election, depending on which rules are considered and which bounds are used.

Motivated by the fact that in many real-world applications the number of candidates is often small, for voting rules whose NMVDs are not bounded, we also investigate their NMVDs in this special case and obtain many interesting results.

For most of the upper bound results, we also show their tightness by providing concrete elections at which the NMVD of a voting rule matches these bounds.

Due to space limitations, proofs of several theorems (labeled by $\star$) are omitted.

2 Preliminaries

For an integer $i > 0$, let $[i]$ denote the set of all positive integers not greater than i.

An *election* is a tuple (C, V) where C is a set of candidates and V is a multiset of votes. Each vote $\succ \in V$ is defined as a linear order over C, indicating the preference of the vote where a candidate a is preferred to another candidate b if a is ordered before b, i.e., $a \succ b$. Sometimes we omit the notation of a vote if we are only interested in the preference. For instance, when we say a vote with the preference $a\,b\,c$ we mean a vote who prefers a to b to c. The *position* of a candidate in a vote is the number of candidates ordered before this candidate plus one. A *voting rule* φ maps each nonempty election (C, V) to a candidate $\varphi(C, V) \in C$ which is called the *winner*. For an election (C, V) and two candidates a and b in C, let $n_{(C,V)}(a, b)$ be the number of votes preferring a to b. We drop (C, V) from the notation when it is clear which election is considered. We say that a *beats* b if $n(a, b) > n(b, a)$, and a *ties* b if $n(a, b) = n(b, a)$. We consider the following voting rules.

Table 1. Some important positional scoring rules. For r-Approval and r-Veto we have that $0 < r < m$. 1-Approval is also referred to as Plurality in the literature.

Rules	Vectors
Borda	$\langle m - 1, m - 2, \ldots, 0 \rangle$
r-Approval	$\langle 1, \ldots, 1, 0, \ldots, 0 \rangle$ (exactly r many 1s)
r-Veto	$\langle 1, \ldots, 1, 0, \ldots, 0 \rangle$ (exactly r many 0s)

Positional scoring rules. A positional scoring rule is characterized by a function mapping a nonnegative integer m to a vector $\langle \beta_1, \beta_2, \ldots, \beta_m \rangle$ of m rational numbers such that $\beta_i \geq \beta_{i+1}$ for each $i \in [m - 1]$. Here, m is considered as the number of candidates. Each vote gives β_i points to the candidate in the i-th position. The winner is a candidate with the maximum total score. Table 1 summarizes some important positional scoring rules.

Plurality with runoff (PluRun). This rule has two stages. First, all candidates except the first and the second candidates with the maximum Plurality score(s) are eliminated. (We may need some tie-handling rule to select exactly two candidates in this stage) In the second stage, between the two candidates surviving the first stage, the one who is preferred by a majority of votes is selected as the winner. (We may also need to break ties in this stage if the two candidates are tied)

Maximin. The Maximin score of a candidate c is defined as $\min_{c' \in C \setminus \{c\}} n(c, c')$. The winner is a candidate with the maximum score.

Copeland $^\alpha$ $(0 \leq \alpha \leq 1)$. For a candidate $c \in C$, let $n^{\mathrm{B}}(c) = |\{c' \in C \setminus \{c\} : n(c, c') > n(c', c)\}|$ and $n^{\mathrm{T}}(c) = |\{c' \in C \setminus \{c\} : n(c, c') = n(c', c)\}|$. The

Copeland$^\alpha$ score of c is $n^B(c) + \alpha \cdot n^T(c)$. The winner is a candidate with the maximum score.

Bucklin. The Bucklin score of a candidate $c \in C$ is defined as the minimum integer i such that a majority of votes rank c in the top-i positions. A candidate with the minimum score is the winner.

In the above descriptions, when several candidates have the same maximum score (for all except PluRun and Bucklin) or have the minimum score (for Bucklin), a particular tie-breaking rule is used to determine the winner. In this paper, unless stated otherwise, ties are broken by a predefined order $\lhd$ over the candidates. This is one of the most important tie-breaking rules studied in the literature (see, e.g., [9]). However, many of our results also hold no matter which tie-breaking rules are used.

Manipulation. Let (C, V) be an election and $w = \varphi(C, V)$ be the winner of (C, V) with respect to a voting rule φ. We say that a vote $\succ \in V$ is *manipulable* (or is a manipulative vote) at (C, V) with respect to φ if there exists a candidate $c \in C \setminus \{w\}$ and a linear order $\succ'$ over C such that $c \succ w$ and c becomes the winner under φ if $\succ$ is replaced with $\succ'$ in V. More precisely, we call $\succ$ a *c-manipulator* at (C, V) with respect to φ.

For an election $E = (C, V)$ and a manipulative vote $\succ \in V$, let $C_\varphi(\succ, E)$ be the set of candidates $c \in C$ such that $\succ$ is a c-manipulator at E with respect to φ. For a nonmanipulative vote $\succ$, we define $C_\varphi(\succ, E) = \emptyset$. Let $C_\varphi(E) = \bigcup_{\succ \in V} C_\varphi(\succ, E)$.[1]

Nonmanipulative Vote-Deficit. For an election (C, V) and a voting rule φ, the NMVD of φ at (C, V), denoted $\mathsf{NMVD}_{(C,V)}(\varphi)$, is defined as the minimum integer ℓ such that there exists a multiset U of ℓ votes over C such that $\varphi(C, V) = \varphi(C, V \cup U)$ and no vote in V is manipulable at $(C, V \cup U)$. In particular, we say that the NMVD of φ is *bounded* if there exists a constant t such that for all elections (C, V) it holds that $\mathsf{NMVD}_{(C,V)}(\varphi) \le t$.

3 NMVDs in General

In this section, we study the NMVDs of many well-studied voting rules in the general case. Our main results are summarized in Table 2.

First, it is easy to check that the NMVD of PluRun is at most six no matter which tie-breaking rule is used in the first stage. Suppose that w and b are the two candidates surviving the first stage. Then, after adding three votes ranking b in the top, and adding three votes ranking w in the top, the winner remains unchanged and, moreover no manipulative vote exists.

Observation 1 . *The NMVD of PluRun is bounded by six and, moreover, this holds no matter which tie-breaking rule is used in the first stage.*

---

[1] Based on a polynomial-time algorithm for a manipulation problem (CCM with exactly one manipulator) presented in [4,18], $C_\varphi(E)$ can be calculated in polynomial time for all φ considered in this paper.

Table 2. The upper bounds of NMVDs of many voting rules. Here, m and n are the numbers of candidates and votes, respectively. Results marked by ✓ hold regardless of the tie-breaking rules, and marked by ♡ mean that the bounds are tight.

Rules	NMVDs	References
Borda	$2\ (\checkmark, \heartsuit)$	Theorem 2
Maximin	$6\ (\checkmark)$	Theorem 7
PluRun	$6\ (\checkmark, \heartsuit)$	Observation 1
r-Approval	$\left\lfloor \frac{m-1}{m-r} \right\rfloor$	Theorem 3
r-Veto	$\left\lceil \frac{m-1}{r} \right\rceil$	Theorem 4
Bucklin	$\min\{n+1, 2(m-1)\}$	Corollary 1
Copeland$^\alpha$	$n+1\ (\checkmark)$	Theorem 8

The result of the above observation is tight, in the sense that there exist tie-breaking rules and elections at which the NMVD of PluRun is exactly six, as illustrated by the following example.

Example 1. Consider an election with three candidates a, b, w, and with 30 votes defined as follows (number of votes: preferences):

$$10:\ w\ b\ a\ \mid\ 10:\ b\ a\ w\ \mid\ 5:\ a\ w\ b\ \mid\ 5:\ a\ b\ w$$

In the first stage, ties are broken so that (1) when all candidates have the same highest Plurality score, b and w survive the first stage; (2) when w has the unique highest Plurality score, and a and b have the same Plurality score, a is the one who survives the first stage with w; (3) when a or b has the highest Plurality score and the other two have the same Plurality score, a and b survive the first stage. In the second stage, ties are broken so that when w ties b, w is the winner. One can check that the NMVD of PluRun at this election is six.

If we use a fixed linear order to break ties in the first stage, the NMVD of PluRun decreases to four.

Theorem 1. *The NMVD of PluRun is at most four if in the first stage a pre-defined linear order is used to break ties.*

The result of the above lemma is also tight, which can be illustrated by an election obtained from the one in Example 1 by adding one more vote with the preference $w\ b\ a$ and one more vote with the preference $b\ a\ w$, and the tie-breaking order is $a \lhd b \lhd w$ for the first stage. Now we study the NMVD of Borda.

Theorem 2. *The NMVD of Borda is bounded by two for all tie-breaking rules.*

Proof. Let (C, V) be an election with $w \in C$ being its Borda winner. Additionally, let $(c_1, c_2, \ldots, c_{m-1})$ be any arbitrary but fixed order of $C \setminus \{w\}$. Let U be a set consisting of two votes with respectively the preferences $w \ c_1 \ c_2 \ \cdots \ c_{m-1}$ and $w \ c_{m-1} \ c_{m-2} \ \cdots \ c_1$. The addition of these two votes increases the score gap between w and every other candidate by exactly m. We claim that in the election $(C, V \cup U)$ no vote in V is manipulable. Assume for contradiction that there is a c-manipulator in V for some $c \in C \setminus \{w\}$. Recall that c is ranked before w in this vote. Hence, changing this vote can only decrease the score gap between w and c by at most $m - 2$, implying that c has Borda score smaller than that of w no matter how this vote is changed, a contradiction. $\square$

The upper bound of NMVD for Borda in Theorem 2 is tight, in the sense that no matter which tie-breaking rules are used, there are elections at which the NMVD of Borda is exactly two, as shown in the following example.

Example 2. Consider an election (C, V) where $C = \{c_1, c_2, c_3, c_4\}$. We have the following votes in V. In particular, for each $c_i \in C$, where $i \in [4]$, there are 12 votes in V as follows ((c_x, c_y, c_z) is the order over $C \setminus \{c_i\}$ such that $x < y < z$):

3: $c_x \ c_y \ c_i \ c_z$ | 3: $c_y \ c_z \ c_i \ c_x$ | 3: $c_z \ c_x \ c_i \ c_y$ | 1: $c_i \ c_x \ c_y \ c_z$ | 1: $c_i \ c_x \ c_z \ c_x$ | 1: $c_i \ c_z \ c_x \ c_y$

In total, V consists of 48 votes. Obviously, all candidates have the same Borda score 72. Without loss of generality, assume that c_i for some $i \in [4]$ is selected as the Borda winner according to a tie-breaking rule. It is easy to see that there is at least one manipulator in this election. For instance, the vote with the preference $c_x \ c_y \ c_i \ c_z$ could be changed into $c_y \ c_x \ c_i \ c_z$ to make the candidate c_y the winner. Hence, to preclude manipulation we need to add at least one additional vote. However, we show that adding one vote does not do the job. First, as we request that the original winner should remain as the winner after adding additional votes in our model, the added vote must rank c_i in the top. Without loss of generality, assume that the candidate in the second position of the added vote is c_x. Then, a vote with the preference $c_z \ c_x \ c_i \ c_y$ is a manipulator, because by changing it into $c_x \ c_z \ c_y \ c_i$, the candidate c_x becomes the Borda winner.

Notice that to find the two votes in the proof of Theorem 2, we need only to know the candidate set C and the current winner. Now we study r-Approval.

Theorem 3. *For φ being r-Approval, NMVD of φ at every election $E = (C, V)$ is bounded by $\left\lceil \frac{|\mathsf{C}_\varphi(E)|}{m-r} \right\rceil \leq \left\lceil \frac{m-1}{m-r} \right\rceil$, where $m = |C|$.*

Proof. Let $\lhd$ be the predefined tie-breaking order for E. Let w be the current r-Approval winner of E with respect to φ. Let $b = \left\lceil \frac{|\mathsf{C}_\varphi(E)|}{m-r} \right\rceil$. We construct b votes as follows. In particular, we partition $\mathsf{C}_\varphi(E)$ into b subsets denoted by $C(w, 1), C(w, 2), \ldots, C(w, b)$ such that every subset contains at most $m - r$ candidates. Then, we create b votes where the i-th vote ranks w in the top, and ranks all candidates in $C(w, i)$ and any arbitrary $m - r - |C(w, i)|$ candidates in $C \setminus (C(w, i) \cup \{w\})$ in the last $m - r$ positions. The relative orders among candidates that are not specified do not matter. Let E' be the election obtained from E by adding the above votes. Apparently, w remains as the winner in E'. In the following, we show that none of V is a manipulative vote at E'.

For the sake of contradiction, assume that there is a c-manipulator $\succ \in V$ where $c \in C \setminus \{w\}$ in E'. Observe that, when ties are broken by a predefined linear order over the candidates, if a vote is not a c-manipulator in E, it cannot be a c-manipulator in E' too. In other words, it holds that $\mathsf{C}_\varphi(E') \subseteq \mathsf{C}_\varphi(E)$. As a consequence, c must be from $\mathsf{C}_\varphi(E)$. Then, note that all added votes approve w but at least one of them does not approve c (the vote corresponding to $C(w, i)$ such that $c \in C(w, i)$). Therefore, adding the above b votes increases the score gap between w and c by at least one. Observe also that because $\succ$ is a c-manipulator, $c \succ w$ holds and, moreover, either both w and c are ranked in the top-r positions, or both are ranked in the last $m - r$ positions in $\succ$. Hence, changing $\succ$ decreases the score gap between w and c by at most one. The proof proceeds by distinguishing between two cases.

Case 1. Candidates c and w have the same r-Approval score in E.

In this case, as w is the winner in E, it holds that $w \lhd c$. By the above discussion, w receives at least one more point than c in E'. However, as changing $\succ$ only decreases the score gap between w and c by at most one, no matter how $\succ$ is changed, c has at most the same score as w in E'. Given $w \lhd c$, we know that $\succ$ cannot be a c-manipulator in E', a contradiction.

Case 2. The winner w has one more point than that of c in E.

In this case, w has at least two more points than c in the new election E'. Changing $\succ$ only decreases the score gap between w and c by at most one, as discussed above. Therefore, no matter how $\succ$ is changed, c has at least one fewer point than w in E', contradicting that $\succ$ is a c-manipulator in E' too. $\square$

In the proof of Theorem 3, we can replace $\mathsf{C}_\varphi(E)$ by $C \setminus \{w\}$ to get the bound $\left\lceil \frac{m-1}{m-r} \right\rceil$. An advantage of using $C \setminus \{w\}$ is that in this case, to construct the desired added votes, we need only to know the current winner other than calculating $\mathsf{C}_\varphi(E)$, which needs the definition of all votes. Theorem 3 also shows that the NMVD of r-Approval is bounded for all constants r. Additionally, as r-Veto is exactly $(m-r)$-Approval, Theorem 3 also offers us the following result.

Theorem 4. *For φ being r-Veto, the NMVD of φ at an election $E = (C, V)$ is at most $\left\lceil \frac{|\mathsf{C}_\varphi(E)|}{r} \right\rceil \leq \left\lceil \frac{m-1}{r} \right\rceil$, where $m = |C|$.*

Now we consider Maximin and Copeland$^\alpha$. Unlike other rules studied in the paper, Maximin and Copeland$^\alpha$ are Condorcet-consistent[2]. We show that the NMVDs of these rules differ largely—Maximin has a bounded NMVD while Copeland$^\alpha$ does not.

---

[2] A *Condorcet winner* is a candidate which beats all the other candidates. A voting rule is *Condorcet-consistent* if it selects the Condorcet winner as the winner whenever the Condorcet winner exists. Condorcet-consistency is a significant axiomatic property of voting rules (see [22, Figure 9.3] or [21, Table 2] for voting rules and their axiomatic properties).

Theorem 5. ($\star$). *The NMVD of Maximin is bounded by three.*

We omit the proof of Theorem 5 because it relies on several lemmas on certain special directed graphs whose proofs take some space. However, we would like to point out that to construct the three votes needed to be added, we need to know not only the original winner but also the head-to-head comparisons among all candidates. If we are only aware of the original winner, can we still prevent manipulation by adding a constant number of votes? The following theorem answers the question.

Theorem 6. ($\star$). *Let (C, V) be an election. If V is hidden but we know the Maximin winner of (C, V), we can add four votes to prevent manipulation without changing the winner.*

Furthermore, if we add two more votes, we could even prohibit manipulation without minding the tie-breaking rules.

Theorem 7. ($\star$). *The NMVD of Maximin is bounded by six for all tie-breaking rules.*

Additionally, we would like to show by the following example that the upper bound given in Theorem 5 is tight.

Example 3. Let us consider an election with four candidates a, b, c, and w, and with the following votes (the head-to-head comparisons among the candidates are summarized in the table on the right side, where each entry indexed by a row candidate x and a column candidate y is $n(x, y)$):

	a	b	c	w
a	-	6	9	9
b	9	-	6	9
c	6	9	-	9
w	6	6	6	-

3: $b\ a\ w\ c$ | 1: $w\ a\ b\ c$ | 1: $a\ b\ c\ w$
3: $c\ b\ w\ a$ | 1: $w\ b\ a\ c$ | 1: $c\ b\ a\ w$
3: $a\ c\ w\ b$ | 1: $w\ c\ a\ b$ | 1: $c\ b\ a\ w$

In the tie-breaking order, w is ranked in the first place. Hence, w is the Maximin winner. Clearly, there are manipulators in the election. For instance, if one vote with the preference $b\ a\ w\ c$ shifts a one position up, a becomes the Maximin winner. We show that the NMVD of Maximin at this election cannot be smaller than 3. For the sake of contradiction, assume that we can add t votes where $t \in [2]$ so that w remains the winner, and there are no manipulators after adding the vote(s). Observe first that we may assume that w is ranked in the top in the added vote(s), and hence the Maximin score of w in the new election is $6 + t$. Let $x \in \{a, b, c\}$ be a candidate that is in the second place of at least one of the added vote(s). Therefore, in the new election, x has Maximin score at least 7. If $x = a$ (resp. $x = b$, $x = c$), one vote with the preference $b\ a\ w\ c$ (resp. $c\ b\ w\ a$, $a\ c\ w\ b$) can improve the result in its favor by changing it into $a\ b\ c\ w$ (resp. $b\ c\ a\ w$, $c\ a\ b\ w$). Precisely, after the change, the Maximin score of w decreases from $6 + t$ to $5 + t \leq 7$, but the Maximin score of x increases to at least 8, making x the new winner, a contradiction.

Now we study Copeland$^\alpha$.

Theorem 8. ($\star$). *Regardless of tie-breaking rules, the NMVD of Copeland$^\alpha$ at an election (C, V) is bounded by $n + 3 - 2 \cdot \min_{c \in C \setminus \{w\}} n(w, c) \leq n + 1$, where n is the number of votes and w is the Copeland$^\alpha$ winner of (C, V).*

In the next section, we shall see that there are elections with only three candidates at which the NMVD of Copeland$^\alpha$ is not bounded by any constant.

Now we consider the NMVD of Bucklin. A trivial upper bound is $n + 1$, because after adding $n + 1$ votes ranking the current winner w in the top, the Bucklin score of w is one and this holds even if some original vote is changed. In the following, we give a bound with respect to the number of candidates. First, we have the following lemma regarding the Bucklin score of Bucklin winners.

Lemma 1. *For every election (C, V) with $m > 0$ candidates, the Bucklin score of the Bucklin winner is at most $\lceil \frac{m+1}{2} \rceil$.*

Proof. For each $c \in C$ and $i \in [m]$, let n_c^i be the number of votes in V ranking c in the i-th position, and let $n_c^{\leq i} = \sum_{j=1}^i n_c^j$ be the number of votes ranking c in the top-i positions. Clearly, for each $i \in [m]$, it holds that

$$\sum_{c \in C} n_c^i = n. \tag{1}$$

Let $t = \lceil \frac{m+1}{2} \rceil$. For the sake of contradiction, assume that the Bucklin score of the winner is larger than t. Then, we have $n_c^{\leq t} \leq n/2$ for every candidate $c \in C$. It follows that $\sum_{c \in C} n_c^{\leq t} \leq \frac{n \cdot m}{2}$. However, due to Eq. (1), we also have

$$\sum_{c \in C} n_c^{\leq t} = \sum_{c \in C} \sum_{i=1}^t n_c^i = \sum_{i=1}^t \sum_{c \in C} n_c^i = n \cdot t > \frac{n \cdot m}{2},$$

a contradiction. $\qquad\square$

Based on Lemma 1, we can prove the following theorem.

Theorem 9. *Let (C, V) be an election with $m > 0$ candidates and $n > 0$ votes such that the Bucklin winner has score at most $\lceil m/2 \rceil$. Then, the NMVD of Bucklin at (C, V) is bounded by $2(m - 1)$ if n is even and by $m - 1$ if n is odd.*

Proof. Let $C = \{c_1, c_2, \ldots, c_m\}$. We prove the theorem by distinguishing the following cases. Without loss of generality, let us assume that $w = c_m$ is the Bucklin winner of (C, V). Moreover, let $x \leq \lceil \frac{m}{2} \rceil$ denote the Bucklin score of w. Finally, let A (resp. B) denote the set of candidates ranking before (resp. after) w in the tie-breaking order.

Case 1. $n = 2k + 1$ and $m = 2t$ for some k and t.

In this case, for each integer $i \in [2t - 1]$ we add one vote with the preference $w\ c_i\ c_{i+1}\ \cdots\ c_{2t-1}\ c_1\ \cdots\ c_{i-1}$. Hence, we add in total $2t - 1$ votes such that each candidate $c_i \in C \setminus \{w\}$ is ranked in the top-x positions at most $t - 1$ times. Let E' denote the new election after adding these votes. Note that for every $c \in A$, at most k votes in V rank c in the top-x positions. Therefore, at most $k + t - 1$ votes rank c in the top-x positions in the new election E'. In this case, even if some vote is changed there can be at most $k + t$ votes ranking c in the top-x positions, implying that there exist no c-manipulators in E' for any $c \in A$. Analogously, we can show that there is no c-manipulator for any $c \in B$.

Case 2. $n = 2k$ and $m = 2t$ for some k and t.

In this case, for each $i \in [2t-1]$ we add two votes with the same preference as in the first case, and analogously we can show that there are no c-manipulators for any $c \in C \setminus \{w\}$ after adding these $2(2t - 1)$ votes.

Case 3. $n = 2k + 1$ and $m = 2t + 1$ for some k and t.

If the Bucklin score of w is at most t, the proof is analogous to the above cases. So, let us assume that $x = t + 1$. If (C, V) is not manipulable, we don't need to add any vote. Otherwise, V contains at least one vote which ranks w in the last $m - t - 1$ positions. Hence, the $2t$ candidates in $C \setminus \{w\}$ occupy at least $n \cdot t + 1$ top-$(t + 1)$ positions of votes in V, implying that there exists at least one candidate $c \in C \setminus \{w\}$ whose Bucklin score is $t + 1$ too. Moreover, c must be ranked after w in the tie-breaking order. Then, we add $2t - 1$ votes so that (1) w is ranked in the top, and c is ranked in the $(t + 1)$-th positions in all these votes; and (2) every candidate in $C \setminus \{c, w\}$ is ranked in the top-t positions at most $t - 1$ times. Such votes clearly exist (they can be constructed in a way similar to the one in Case 1). Analogous to Case 1, it can be shown that there are no manipulators after adding these votes.

Case 4. $n = 2k$ and $m = 2t + 1$ for some k and t.

In this case, after adding $4t - 2$ votes where w is ranked in the top, some candidate $c \neq w$ who has Bucklin score $t + 1$ is ranked in the $(t + 1)$-th positions, and every other candidate is ranked in the top-t positions at most $2t - 2$ times, there are no manipulators. $\qquad\square$

Now we consider the case where the Bucklin winner w has score exactly $\lceil \frac{m+1}{2} \rceil$. Notice that when m is odd, we have $\lceil \frac{m}{2} \rceil = \lceil \frac{m+1}{2} \rceil$. Hence, we need only to consider the case where m is even. We have the following result.

Theorem 10. $(\star)$. *Let (C, V) be an election of m candidates and n votes such that $m = 2t > 0$ is even and the Bucklin winner w has score $t + 1$. Then, it holds that*

1. *$|V|$ is even; and*
2. *the NMVD of Bucklin at (C, V) is at most $2(m - 2)$.*

Due to Lemma 1, Theorems 9 and 10, and the trivial bound $n + 1$, we have the following corollary.

Corollary 1. *The NMVD of Bucklin at every election with m candidates and n votes is at most $\min\{n+1, 2(m-1)\}$.*

4 NMVD with a Small Candidate Set

In many real-world applications, the number of candidates is a small constant (see, e.g., [15]). Therefore, it makes much sense to study the NMVDs of voting rules at elections with only a few candidates. This is the focus of this section. For Borda, Maximin, and PluRun, their NMVDs are bounded, regardless of the number of candidates and tie-breaking rules (see Table 2). Due to Theorems 3 and 4, and Corollary 1, if m is a constant, r-Approval, r-Veto, and Bucklin also have constant bounded NMVDs. So, the most interesting rules in this section are Copeland$^\alpha$ where $\alpha \in [0,1]$.

One may wonder whether the NMVD of Copeland$^\alpha$ can be bounded by a constant too, or at least some function of m. Our answers are somewhat interesting. For Copeland^1, the answer is "Yes". In particular, we obtain a bound $\gamma \cdot \frac{m}{\log m}$, where γ is a constant. However, for Copeland$^\alpha$ where $\alpha \in [0,1)$, our answer is in the negative: even when there are only three or four candidates, the NMVD of Copeland$^\alpha$ cannot be bounded by a constant. However, if we consider only elections without ties in head-to-head comparisons (e.g., when the number of votes is odd), we again have a positive answer. Our main results in this section are summarized in Tables 3 and 4.

Table 3. Upper bounds of the NMVDs of Copeland$^\alpha$. Here, n denotes the number of votes. "odd" and "even" are with respect to n.

		Number of candidates m				
		3	4	$m \leq 8$	$m \leq 11$	general
$\alpha = 0$	Odd	2	3	3	5	$O(\frac{m}{\log m})$
	Even	$n-1$	$n+1$	$n+1$	$n+1$	$n+1$
$\alpha = 1$	Odd	2	6	6	10	$O(\frac{m}{\log m})$
	Even	2	3	3	5	$O(\frac{m}{\log m})$
$\alpha \in (0,1)$	Odd	2	5	6	10	$O(\frac{m}{\log m})$
	Even	2	$n+1$	$n+1$	$n+1$	$n+1$

Table 4. Upper bounds of the NMVDs of Copeland$^\alpha$ at every election with n votes whose majority graph without the current winner is ℓ-inducible. Results in a row labeled with $\heartsuit$ hold only when the majority graph without the current winner is a tournament.

	Odd n	Even n
$\alpha = 0$ ($\heartsuit$)	$\max\{\ell, 2\}$	ℓ
$\alpha \in (0,1)$ ($\heartsuit$)	2ℓ	2ℓ
$\alpha = 1$	2ℓ	$\max\{\ell, 2\}$

The *majority graph* of an election $E = (C, V)$ (or simply V), denoted G_E, is a digraph with C being the vertex set, and there is an arc from $c \in C$ to $c' \in C$ if and only if c beats c' in E. A *tournament* is a digraph such that between every pair of candidates there exists exactly one arc. A digraph G with vertex set C is ℓ-*inducible* if there exists a multiset V of ℓ linear orders over C such that the majority graph of (C, V) is G. In particular, the minimum integer ℓ such that G is ℓ-inducible is called the *dimension* of G. For a digraph G and a subset S of vertices of G, $G[S]$ is the subgraph of G induced by S. The following lemma is due to [10].

Lemma 2. *There exists a constant γ such that every digraph with m vertices is $\gamma \cdot \frac{m}{\log m}$-inducible.*

4.1   Copeland[1]

In this section, we study Copeland[1]. Our main result is as follows.

Theorem 11. *Let $E = (C, V)$ be an election and $w \in C$ the Copeland[1] winner of E. If the majority graph of E without w (i.e., $G_E[C \setminus \{w\}]$) is ℓ-inducible, then the NMVD of Copeland[1] at E is at most $\max\{\ell, 2\}$ if n is even, and is at most 2ℓ if n is odd.*

Proof. Let T be the majority graph of E without w. If $\ell = 0$, all candidates are tied in head-to-head comparisons and, moreover, w is the first candidate in the tie-breaking order. In this case, it is easy to check that after adding two votes ranking w in the top, there does not exist any manipulator.

For $\ell \geq 1$, let $\succ_1, \ldots, \succ_\ell$ be ℓ linear orders over $C \setminus \{w\}$ whose majority graph is T. If n is even, say $n = 2k$ for some integer $k > 0$, and $\ell \geq 2$, we add ℓ votes obtained from $\succ_1, \ldots, \succ_\ell$ by inserting w in the top. Let E' be the new election. Clearly, the score of w in E' is at least that in E and, moreover, in the new election E' the score of w does not decrease even if one vote in V is changed. We claim that the score of every candidate $a \in C \setminus \{w\}$ in E' is at most that in E, even if one vote in V is changed. To this end, it suffices to show that every candidate $b \in C \setminus \{a\}$ who beats a in E still beats a in the new election E' even after some vote in V is changed. If b beats a in E, at least $k+1$ votes in V prefer b to a. In the above added votes, there are at least $\lceil \frac{\ell+1}{2} \rceil$ votes preferring b to a. It follows that in the new election E', there are at least $k + 1 + \lceil \frac{\ell+1}{2} \rceil \geq \lceil \frac{2k+\ell}{2} \rceil + 1$ votes who prefer b to a. Then, given that $\ell \geq 2$, even if some vote is changed, there are still a majority of votes preferring b to a.

For all the other remaining cases (i.e., the case where n is even and $\ell = 1$, and the case where n is odd), we add one more copy of the ℓ votes added in the above case. Similarly, we can show that after adding these votes there are no manipulators.   □

The above theorem together with Lemma 2 directly give us the following corollary.

Corollary 2. *The NMVDs of Copeland^1 at elections with m candidates are bounded by $\gamma \cdot \frac{m}{\log m}$, where γ is a constant.*

For tournaments of small sizes, Bachmeier et al. [2] studied the following lemma.

Lemma 3. *All tournaments of up to 7 and 10 vertices are respectively 3-inducible and 5-inducible.*

The largest integer i such that all tournaments of up to i vertices are 5-inducible is still unknown in the literature. Due to Theorem 11, Lemma 3, and the fact that all tournaments of up to two vertices are 1-inducible, we have the following corollary.

Corollary 3. *Let (C, V) be an election with $m \leq 11$ candidates and n votes. The NMVD of Copeland^1 at (C, V) is bounded by 10 if n is odd, and bounded by 5 if n is even. In addition, if $m \leq 8$, it is bounded by 6 if n is odd and by 3 if n is even. Moreover, if $m = 3$, it is bounded by 2.*

4.2 Copeland$^{\alpha < 1}$ with at Most Four Candidates

Now we study NMVD of Copeland$^\alpha$ where $\alpha \in [0, 1)$. Somewhat surprisingly, even when there are only three candidates, the NMVD of Copeland^0 cannot be bounded by a constant, standing in a sharp contrast to Copeland^1.

Theorem 12. $(\star)$. *Let (C, V) be an election with three candidates and let $n = |V| > 0$. Then, the NMVD of Copeland$^\alpha$ where $\alpha \in [0, 1)$ at (C, V) is bounded by*

$$\begin{cases} n - 1 & \text{if } n \text{ is even and } \alpha = 0, \\ 2 & \text{otherwise.} \end{cases}$$

For Copeland$^\alpha$ where $\alpha \in (0, 1)$, we can also show that if we just have one more candidate, the NMVD cannot be bounded by a constant anymore if the number of votes is even, as shown in the following example.

Example 4. Consider an election with four candidates w, a, b, c, and with $2k$ votes as follows. Here, k can be any integer greater than three. In the digraph on the right side, the weight of an arc from x to y is $n(x, y)$.

$$
\begin{array}{l}
k - 2: c \succ w \succ a \succ b \\
k - 3: b \succ a \succ c \succ w \\
2: c \succ w \succ b \succ a \\
2: a \succ b \succ c \succ w \\
1: w \succ b \succ a \succ c
\end{array}
$$

The tie-breaking order is $a \lhd b \lhd c \lhd w$. One can check that we cannot preclude manipulation without changing the winner by adding less than $2k - 2$ votes.

4.3   Copeland$^{\alpha<1}$ with ℓ-Inducible Majority Graphs

A major reason why the NMVD of Copeland$^{\alpha}$, $\alpha \in [0,1)$, is not bounded in several theorems studied so far is that there are ties in head-to-head comparisons among nonwinning candidates. This motivates us to study elections without such ties. Consider an election (C,V) whose majority graph restricted to $C \setminus \{w\}$ is an ℓ-inducible tournament T, where $\ell \geq 2$, and w is the current winner. If we add ℓ votes ranking w in the top and whose induced majority graph restricted to $C \setminus \{w\}$ is T, the comparisons between candidates are enhanced, in the sense that changing one vote in V is unable to reverse any arc among candidates in $C \setminus \{w\}$ but may make some arcs be removed. However, as in Copeland^0, two tied candidates do not get any point from their comparison, changing one vote does not increase the scores of candidates in $C \setminus \{w\}$. This observation leads to the following theorem.

Theorem 13. *Let $E = (C,V)$ be an election and w the Copeland^0 winner of E. If the majority graph of E without w is an ℓ-inducible tournament, then the NMVD of Copeland^0 at E is at most ℓ if $\ell \geq 2$ or $|V|$ is even, and is at most 2 otherwise.*

Proof. Let T denote the majority graph of E restricted to $C \setminus \{w\}$, which is ℓ-inducible. Let $U = \{\succ_1, \ldots, \succ_\ell\}$ be a multiset of ℓ votes over $C \setminus \{w\}$ such that majority graph of $(C \setminus \{w\}, U)$ is exactly T.

We consider first the case where $\ell \geq 2$ or $|V|$ is even. In this case, we add to the election ℓ votes obtained from $\succ_1, \ldots, \succ_\ell$ by inserting w into the top position. After adding these votes, the majority graph of the election restricted to $C \setminus \{w\}$ is still T. As every added vote ranks w above other candidates, one can check that if a candidate a is beaten by w in advance, then a is still beaten by w after adding the above votes, even if one vote in V is changed. Hence, the score of w in the new election remains at least the same as that in the original election. Moreover, if there is an arc from a candidate $a \in C \setminus \{w\}$ to another candidate $b \in C \setminus \{w\}$ in T, then, after adding the above votes a beats b by at least

$$\left\lceil \frac{n+1}{2} \right\rceil + \left\lceil \frac{\ell+1}{2} \right\rceil \geq \left\lceil \frac{n+\ell}{2} \right\rceil + 1.$$

This implies that when one vote in V is changed, a either still beats b or ties with b. So, the change of any vote does not increase the score of a. As this holds for every candidate $a \in C \setminus \{w\}$, we can conclude that changing one vote in V does not prevent w from winning.

For the case where $\ell = 1$ and $|V|$ is odd, the above argument does not hold, because in this case if a candidate a is beaten by w in advance, then after adding one vote, it may be that w ties a if some vote in V is changed. However, if we add two votes obtained from $\succ_1$ by inserting w in the first place, w remains as the winner and there is no manipulator in V. □

Theorem 13 and Lemma 3 yield the following result.

Corollary 4. *The NMVD of Copeland^0 at every election with m candidates and an odd number of votes is at most 3 if $m \leq 8$ and is at most 5 if $m \leq 11$. Moreover, the bound is tight for every m such that $4 \leq m \leq 8$.*

Similar to Theorem 13, for Copeland$^\alpha$, $\alpha \in (0,1)$, we obtain the following result.

Theorem 14. ($\star$). *Let $E = (C, V)$ be an election, and let w be the Copeland$^\alpha$ winner of E where $\alpha \in (0,1)$. If the majority graph of E restricted to $C \setminus \{w\}$ is an ℓ-inducible tournament, then the NMVD of Copeland$^\alpha$ at E is at most 2ℓ.*

Theorem 14 and Lemma 3 yield the following result.

Corollary 5. *The NMVD of Copeland$^\alpha$ where $\alpha \in (0,1)$ at every election (C, V) with an odd number of votes is at most 6 if $|C| \leq 8$ and is at most 10 if $|C| \leq 11$.*

For elections with four candidates, we can improve the bound by one.

Theorem 15. ($\star$). *The NMVD of Copeland$^\alpha$ where $\alpha \in (0,1)$ at every election (C, V) with four candidates is at most 5 if $|V|$ is odd.*

We would like to remark that Theorems 11, 13, and 14 suggest using the notion of dimension of digraphs to find the votes whose addition precludes election manipulation. Unfortunately, calculating the dimension of a digraph seems to be a time-consuming task. To the best of our knowledge, the complexity of this problem in general still remains open so far. But for the case where there are only a few candidates, calculating the dimension can be done in an acceptable time. We refer to [2] for a more detailed discussion on this issue.

5   Future Research

It is interesting to study the NMVDs of many other voting rules such as the ranked pairs, Schulze's, Baldwin's, etc. In addition, one can investigate the NMVDs of voting rules at elections with a bit larger but still a constant number of candidates. Finally, it is intriguing to investigate our notion in the more general setting where we allow multiple manipulators to form coalitions and coordinate their actions.

Acknowledgements. The author would like to thank the anonymous reviewers of ADT-2021 for their constructive comments.

References

1. Aziz, H., Gaspers, S., Mattei, N., Narodytska, N., Walsh, T.: Ties matter: complexity of manipulation when tie-breaking with a random vote. In: AAAI, pp. 74–80 (2013)

2. Bachmeier, G., et al.: k-majority digraphs and the hardness of voting with a constant number of voters. J. Comput. Syst. Sci. **105**, 130–157 (2019)
3. Bartholdi III, J.J., Orlin, J.B.: Single transferable vote resists strategic voting. Soc. Choice Welfare **8**(4), 341–354 (1991). https://doi.org/10.1007/BF00183045
4. Bartholdi III, J.J., Tovey, C.A., Trick, M.A.: The computational difficulty of manipulating an election. Soc. Choice Welfare **6**(3), 227–241 (1989). https://doi.org/10.1007/BF00295861
5. Bartholdi III, J.J., Tovey, C.A., Trick, M.A.: How hard is it to control an election? Math. Comput. Model. **16**(8–9), 27–40 (1992)
6. Betzler, N., Niedermeier, R., Woeginger, G.J.: Unweighted coalitional manipulation under the Borda rule is NP-hard. In: IJCAI, pp. 55–60 (2011)
7. Black, D.: On the rationale of group decision-making. J. Polit. Econ. **56**(1), 23–34 (1948)
8. Cary, D.: Estimating the margin of victory for instant-runoff voting. In: EVT/WOTE (2011)
9. Elkind, E., Grandi, U., Rossi, F., Slinko, A.: Gibbard-Satterthwaite games. In: IJCAI, pp. 533–539 (2015)
10. Erdős, P., Moser, L.: On the representation of directed graphs as unions of orderings. Math. Inst. Hung. Acad. Sci. **9**, 125–132 (1964)
11. Faliszewski, P., Rothe, J.: Control and bribery in voting. In: Brandt, F., Conitzer, V., Endriss, U., Lang, J., Procaccia, A. (eds.) Handbook of Computational Social Choice, chap. 7, pp. 146–168. Cambridge University Press (2016)
12. Gibbard, A.: Manipulation of voting schemes: a general result. Econometrica **41**(4), 587–601 (1973)
13. Gibbard, A.: Manipulation of schemes that mix voting with chance. Econometrica **45**(3), 665–681 (1977)
14. Magrino, T.R., Rivest, R.L., Shen, E.: Computing the margin of victory in IRV elections. In: EVT/WOTE (2011)
15. Mattei, N., Walsh, T.: PREFLIB: a library for preferences HTTP://WWW.PREFLIB.ORG. In: Perny, P., Pirlot, M., Tsoukiàs, A. (eds.) ADT 2013. LNCS (LNAI), vol. 8176, pp. 259–270. Springer, Heidelberg (2013). https://doi.org/10.1007/978-3-642-41575-3_20
16. Menton, C., Singh, P.: Manipulation can be hard in tractable voting systems even for constant-sized coalitions. Comput. Sci. Rev. **6**(2–3), 71–87 (2012)
17. Moulin, H.: On strategy-proofness and single peakedness. Public Choice **35**(4), 437–455 (1980)
18. Obraztsova, S., Elkind, E., Hazon, N.: Ties matter: complexity of voting manipulation revisited. In: AAMAS, pp. 71–78 (2011)
19. Procaccia, A.D.: Can approximation circumvent Gibbard-Satterthwaite? In: AAAI, pp. 836–841 (2010)
20. Satterthwaite, M.: Strategy-proofness and Arrow's conditions: existence and correspondence theorems for voting procedures and social welfare functions. J. Econ. Theory **10**(2), 187–217 (1975)
21. Schulze, M.: A new monotonic, clone-independent, reversal symmetric, and Condorcet-consistent single-winner election method. Soc. Choice Welfare **36**(2), 267–303 (2011). https://doi.org/10.1007/s00355-010-0475-4
22. Smith, W.D.: Descriptions of single-winner voting systems (2006). http://m-schulze.9mail.de/votedesc.pdf
23. Xia, L.: Computing the margin of victory for various voting rules. In: EC, pp. 982–999 (2012)

24. Xia, L., Zuckerman, M., Procaccia, A.D., Conitzer, V., Rosenschein, J.S.: Complexity of unweighted coalitional manipulation under some common voting rules. In: IJCAI, pp. 348–353 (2009)
25. Yang, Y.: Manipulation with bounded single-peaked width: a parameterized study. In: AAMAS, pp. 77–85 (2015)
26. Yang, Y., Guo, J.: Exact algorithms for weighted and unweighted Borda manipulation problems. Theor. Comput. Sci. **622**, 79–89 (2016)

Fair Division and Resource Allocation

Allocating Indivisible Items with Minimum Dissatisfaction on Preference Graphs

Nina Chiarelli[1,2], Clément Dallard[1,2], Andreas Darmann[3(✉)], Stefan Lendl[3], Martin Milanič[1,2], Peter Muršič[1], Nevena Pivač[1,2], and Ulrich Pferschy[3]

[1] FAMNIT, University of Primorska, Koper, Slovenia
{nina.chiarelli,clement.dallard,peter.mursic}@famnit.upr.si
[2] IAM, University of Primorska, Koper, Slovenia
martin.milanic@upr.si, nevena.pivac@iam.upr.si
[3] Department of Operations and Information Systems,
University of Graz, Graz, Austria
{andreas.darmann,stefan.lendl,ulrich.pferschy}@uni-graz.at

Abstract. We consider the task of allocating indivisible items to agents, when each agent's preferences are captured by means of a directed acyclic graph. The vertices of such a graph represent items and an arc (a, b) means that the respective agent prefers item a over item b. The dissatisfaction of an agent is measured by the number of non-assigned items which are desired by the agent and for which no more preferred item is given to the agent. The aim is to allocate the items to the agents in a way that minimizes (i) the total dissatisfaction over all agents, or (ii) the maximum dissatisfaction among the agents. For both problems we study the status of computational complexity and obtain NP-hardness results as well as polynomial algorithms with respect to different underlying graph structures, such as trees, stars, paths, and matchings.

Keywords: Fair division · Partial order · Preference graph

1 Introduction

Consider the situation in which a set of indivisible presents should be divided among a set of kids. The kids may be overwhelmed with the task of comparing all available presents among each other, but they are able to state certain preferences such as disapproval of certain presents or strict preference of a certain present over another present. A kid will have difficulties to keep an overview of the complicated preference structure resulting from these pairwise comparisons, but it is well versed in complaining when it sees a present given to another kid and it receives no present it likes better than that present.

In such a scenario, the parents want to allocate the presents to the kids in a way that minimizes the (total or maximum) dissatisfaction. The dissatisfaction of a kid is here measured by the number of desired presents not received and for

© Springer Nature Switzerland AG 2021
D. Fotakis and D. Ríos Insua (Eds.): ADT 2021, LNAI 13023, pp. 243–257, 2021.
https://doi.org/10.1007/978-3-030-87756-9_16

which it does not get any other more preferred present. Note that, in this setting, adding less preferred presents will not improve the happiness of a kid. This can occur in a more general situation of preferences implied by skills or abilities, where the effect of an object with certain skills is not improved by adding an object with lesser skills. We introduce a model for such a setting in which the preferences of each kid i are captured by a preference graph, i.e., a directed acyclic graph G_i, where arc (a, b) means that a is preferred over b; presents that are not contained in graph G_i are disapproved by kid i.

In general, we are hence concerned with the problem of fairly dividing a set of indivisible items among a set of agents who have preferences over the items (for surveys see, e.g., Bouveret et al. [5] and Thomson [19]). Various approaches towards allocating indivisible items have been taken. In the model considered by Herreiner and Puppe [15], the agents rank all possible subsets of items, and the objective is to find an allocation that maximizes the minimum rank among the assigned sets in the agents' rankings. Another, common approach avoiding the tedious task of eliciting all such preferences is to consider cardinal or ordinal preferences over the single items instead of subsets of items (see, e.g., Aziz et al. [1], Brams et al. [6], or Baumeister et al. [4]). Different desirable criteria representing fairness ideas such as envy-freeness, equitability, or proportionality have been in the focus of research. In addition, the social welfare—in particular, utilitarian, egalitarian, or Nash social welfare—induced by an allocation has been analysed, often from a computational perspective (Baumeister et al. [4], Bansal and Sviridenko [3], Chiarelli et al. [7], Darmann and Schauer [9], Garg and McGlaughlin [11], Roos and Rothe [17]). Recently, however, the case of dichotomous preferences over the items has received particular attention (see, e.g., Babaioff et al. [2], Duddy [10], and Halpern et al. [14]).

In the model presented in this paper, each agent expresses preferences over the items by means of a directed acyclic graph, where the vertex set of the graph is a subset of the set of items. Vertices (= items) not contained in the graph of an agent are regarded as disapproved by the agent (and we do not allocate such an item to that agent); arc (a, b) means the agent prefers a over b. Assuming transitivity of the preferences, arcs (a, b) and (b, c) imply that the agent also prefers a over c, regardless of whether arc (a, c) is contained in the graph or not. Observe that the graph of an agent hence induces a partial order over a subset of items.

Given such an input, the goal is to allocate the items to the agents in a way such that the dissatisfaction of the agents is minimized. We take into account both the total dissatisfaction, i.e., summing up the dissatisfaction of each agent, and the maximum dissatisfaction of an agent. We consider an agent to be dissatisfied with not receiving an item if she does not receive another more preferred item. The dissatisfaction of an agent is then determined by the number of such items.

We provide a computational complexity study of finding allocations minimizing the maximum and total dissatisfaction with respect to different kinds of preference graphs and also consider the case of identical preferences. In Sect. 2

we present the formal framework of the paper including problem definitions. We then show that the corresponding decision problems MIN-MAX DISSATIS-FACTION and MIN-SUM DISSATISFACTION are NP-complete when the preference graph of each agent is an out-tree (Sect. 3), whereas an allocation minimizing the total dissatisfaction can be found in polynomial time when each preference graph is a directed matching (Sect. 4). However, when each preference graph is a path, both an allocation minimizing the maximum and an allocation minimizing total dissatisfaction can be found in polynomial time; in particular, for the task of minimizing total dissatisfaction we provide a fixed-parameter tractability result with respect to the number of vertices with in-degree or out-degree greater than 1 (see Sect. 5). Section 6 deals with the case of identical preferences: in particular, we show that for two agents, both an allocation minimizing the maximum and an allocation minimizing total dissatisfaction can be found in polynomial time; for three agents, however, the corresponding decision problems turn out to be NP-complete.

2 Formal Framework

An *undirected graph* is a pair $G = (V, E)$ with vertex set V and edge set $E \subseteq \{\{u, v\} \mid u, v \in V, u \neq v\}$. A *directed graph* is a pair $G = (V, A)$ with vertex set V and set of arcs $A \subseteq V \times V$.

Consider a directed graph $G = (V, A)$. For $a = (u, v) \in A$, vertex u is called tail of a and vertex v is called head of a. The *in-degree* of a vertex u is the number of arcs in A for which u is head and the *out-degree* of u is the number of arcs in A for which u is tail. The *degree* of a vertex u is the number of arcs in A for which u is either head or tail. A sequence $p = (v_0, v_1, v_2, \ldots, v_\ell)$ with $\ell \geq 0$ and $(v_i, v_{i+1}) \in A$ for each $i \in \{0, \ldots, \ell - 1\}$ is called a *walk* of length ℓ from v_0 to v_ℓ; it is a *path* (of length ℓ from v_0 to v_ℓ) if all its vertices are pairwise distinct. A walk from v_0 to v_ℓ is *closed* if $v_0 = v_\ell$. A *cycle* is a closed walk of positive length in which all vertices are pairwise distinct, except that $v_0 = v_\ell$. A *directed acyclic graph* is a directed graph with no cycle. An *out-tree* is a directed acyclic graph $G = (V, A)$ with a dedicated vertex r (called root) such that for each vertex $v \in V \setminus \{r\}$ there is exactly one path from r to v. An *out-star* is an out-tree in which each such path is of length 1. A *matching* in an undirected graph G is a set of pairwise disjoint edges. A *directed matching* is a directed acyclic graph such that each vertex has degree exactly one (i.e., the edges of the underlying undirected graph G form a matching in G).

A binary relation $\succ \subseteq V \times V$ is a *strict partial order* over V if it is asymmetric (for all $u, v \in V$, if $u \succ v$ then $v \succ u$ does not hold) and transitive (for all $u, v, w \in V$, $u \succ v$ and $v \succ w$ imply $u \succ w$). Observe that a directed acyclic graph $G = (V, A)$ induces a strict partial order $\succ$ on V by setting $u \succ v$ for each pair $(u, v) \in V \times V$ such that $u \neq v$ and there is a path from u to v. In particular, $u \succ v$ for every arc $(u, v) \in A$.

In what follows, we will consider a vertex set V and a set of agents K along with directed acyclic graphs $G_i = (V_i, A_i)$ for all $i \in K$, where $V_i \subseteq V$ represents the items desired by agent i. Let $pred_i(v) \subseteq V_i$ denote the set of predecessors of v in graph G_i, i.e., the set of all vertices $u \neq v$ such that there is a path from u to v in G_i. Observe that $pred_i(v)$ corresponds to the set of items which agent i *prefers* over v under relation $\succ$. For $u, v \in V_i$ we say that item u is *dominated* by item v if $u = v$ or $v \in pred_i(u)$. In addition, let $succ_i(v) \subseteq V_i$ denote the set of successors of v in graph G_i, i.e., the set of all vertices $u \neq v$ such that there is a path from v to u in G_i. Hence, $succ_i(v)$ denotes the set of all items to which agent i prefers item v.

An *allocation* π is a function $K \to 2^V$ that assigns to the agents pairwise disjoint sets of items, i.e., for $i, j \in K$, $i \neq j$, we have $\pi(i) \cap \pi(j) = \emptyset$. To measure the attractiveness of an allocation we will count the number of items, which an agent **does not receive** and for which she receives no other more preferred item.

Formally, for an allocation π the *dissatisfaction* $\delta_\pi(i)$ of agent i is defined as number of items in G_i not dominated by any item in $\pi(i)$. A *dissatisfaction profile* is a $|K|$-tuple $(d_i \mid i \in K)$ with $d_i \in \mathbb{N}_0$ for all $i \in K$ such that there is an allocation π with $\delta_\pi(i) = d_i$ for each $i \in K$.

Note that the underlying undirected graphs of the directed graphs G_i are not necessarily connected and may also contain isolated vertices. According to the definition of $\delta_\pi(i)$, every isolated vertex in G_i contributes one unit to $\delta_\pi(i)$, if it is not allocated to agent i. Vertices in $V \setminus V_i$ are irrelevant for agent i. They do not have any influence on the dissatisfaction function $\delta_\pi(i)$.

In this paper, we focus on the following two problems aiming to minimize the maximum and total dissatisfaction among the agents.

MIN-MAX DISSATISFACTION:
Given: A set K of agents, a set V of items, a directed acyclic graph $G_i = (V_i, A_i)$ for each $i \in K$ with $V_i \subseteq V$, and an integer d.
Question: Is there an allocation π of items to agents such that the dissatisfaction $\delta_\pi(i)$ is at most d for each agent $i \in K$?

MIN-SUM DISSATISFACTION:
Given: A set K of agents, a set V of items, a directed acyclic graph $G_i = (V_i, A_i)$ for each $i \in K$ with $V_i \subseteq V$, and an integer d.
Question: Is there an allocation π of items to agents such that the total dissatisfaction $\sum_{i \in K} \delta_\pi(i)$ is at most d?

The graphs G_i in the above problem definitions are called *preference graphs*. Throughout the paper, we denote the number of agents by $k = |K|$ and the number of items by $n = |V|$.

While in this work the focus is laid on the minimization of (maximum or total) dissatisfaction, some remarks on the associated dual problem of maximizing satisfaction are in order. In this context, satisfaction $s_\pi(i)$ of agent i with respect to allocation π is measured by means of the number of items in V_i that are dominated by some item in $\pi(i)$. Observe that the items in $V \setminus V_i$ are irrelevant for the satisfaction of agent i.

Observe that minimizing the total dissatisfaction is equivalent to maximizing the total satisfaction of all agents. In order to verify this, it is sufficient to note that $\delta_\pi(i) = |V_i| - s_\pi(i)$, which implies that $\sum_{i \in K} \delta_\pi(i) = \sum_{i \in K} |V_i| - \sum_{i \in K} s_\pi(i)$ is minimized if and only if $\sum_{i \in K} s_\pi(i)$ is maximized.

On the other hand, we point out that minimizing the maximum dissatisfaction in general does not correspond to maximizing the minimum satisfaction among the agents, as the following example shows.

Example 1. Let $V = \{a, b, c\}$ and $K = \{1, 2, 3\}$, with each graph G_i consisting of isolated vertices only. Graph G_1 consists of all three vertices a, b, c, graphs G_2 and G_3 are made up of b and c respectively. An allocation with dissatisfaction of at most 1 per agent is given only by allocations that give at least two items to agent 1. On the other hand, the allocation π with $\pi(1) = \{a\}$, $\pi(2) = \{b\}$, $\pi(3) = \{c\}$ is the only allocation that yields a satisfaction of 1 for each agent, resulting in a dissatisfaction of 2 for agent 1.

In what follows, we provide NP-completeness results on the one hand and positive results, i.e., polynomial-time solvable cases, on the other. Concerning these positive results we point out that we are not only able to answer the corresponding decision question but also to solve the associated optimization problem, i.e., we can find an allocation that minimizes total resp. maximum dissatisfaction in polynomial time. With respect to the NP-completeness results, note that from the definition of MIN-MAX DISSATISFACTION and MIN-SUM DISSATISFACTION it follows that any NP-hardness result implies NP-hardness in the strong sense. Several of the hardness results presented in this paper reduce from 3X3C, an NP-complete variant of EXACT COVER BY 3-SETS (see Gonzalez [12]).

Due to the page limit, some of the proofs are omitted or shortly sketched.

3 Hardness Results for Out-Trees

We begin our computational complexity study with out-trees as preference graphs. It turns out that in this case both MIN-MAX DISSATISFACTION and MIN-SUM DISSATISFACTION are NP-complete; for the former, NP-completeness even holds when restricted to out-stars.

Theorem 1. MIN-MAX DISSATISFACTION *is NP-complete, even if each graph* G_i *is an out-star.*

Proof. Given an instance $\mathcal{J}$ of 3X3C with a set X of elements, $|X| = 3q$, and a collection $C = \{C_1, \ldots, C_p\}$ of 3-element subsets of X, we construct an instance $\mathcal{I}$ of MIN-MAX DISSATISFACTION as follows. Recall that we have $p = 3q$. Set $\ell = \frac{2}{3}p + 1$, and let the set of items $V = X \cup \{1, \ldots, p\} \cup \{h_1, \ldots, h_{\ell+1}\}$. The set of agents K is made up of the agents $D_1, \ldots, D_{\ell+1}$, agent A, and agents

$C_1, \ldots, C_p$ (the latter agents are identified with the sets of the same label in instance $\mathcal{J}$). The graphs G_i are out-stars displayed in Fig. 1; the graph of agent C_j has root vertex j, and contains the edges (j, h_r), for $r \in \{1, \ldots, \ell - 1\}$ and (j, a) for $a \in C_j$ (for the figure, we assume set $C_1 = \{x, y, z\}$ and $C_2 = \{x, u, v\}$). We ask if there is an allocation with dissatisfaction at most ℓ per agent.

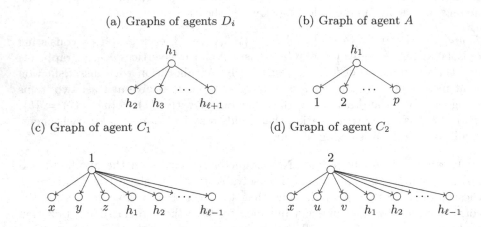

(a) Graphs of agents D_i (b) Graph of agent A

(c) Graph of agent C_1 (d) Graph of agent C_2

Fig. 1. Graphs of agents in the proof of Theorem 1.

We show that $\mathcal{J}$ is a yes-instance of 3X3C if and only if $\mathcal{I}$ is a yes-instance of MIN-MAX DISSATISFACTION. Assume first that there is an allocation with dissatisfaction at most ℓ per agent. Agents $D_1, \ldots, D_{\ell+1}$ all have the same graph (displayed in Fig. 1). To respect the dissatisfaction bound ℓ, each of the agents $D_1, \ldots, D_{\ell+1}$ has to get exactly one of the items $h_1, \ldots, h_{\ell+1}$. Hence, also item h_1 is already allocated. Therefore, agent A has to get at least $\frac{1}{3}p$ items of $\{1, \ldots, p\}$ in order to respect the bound ℓ. Thus, for the agents $C_1, \ldots, C_p$ there are only at most $\frac{2}{3}p$ items of $\{1, \ldots, p\}$ left. As a consequence, in order to respect bound ℓ, at least $\frac{1}{3}p$ of these agents need to get all three items which make up the respective set in instance $\mathcal{J}$ (e.g., agent C_1 gets all three items x, y, z). As each present can be given to one agent and the set X contains exactly p elements, this means that the collection of sets C_j such that the agent of the same label gets all three items of C_j forms an exact cover of X.

For the converse direction, let B be a collection of sets of C that forms an exact cover of X. Observe that B contains exactly $\frac{p}{3}$ sets. Give h_i to agent D_i for each i. For collection B, (i) give, for each $C_j \in B$ its three elements (items) to agent C_j (there are $\frac{p}{3}$ of such sets/agents), and (ii) give all items j such that $C_j \in B$ to agent A (these are $\frac{p}{3}$ such items), (iii) give item j such that $C_j \notin B$ to agent C_j. Then, for each agent the dissatisfaction bound ℓ is respected. $\quad\square$

Using further reductions from 3X3C we can show that both MIN-SUM DIS-SATISFACTION and MIN-MAX DISSATISFACTION are computationally hard, even when all graphs G_i are out-trees each containing all vertices of V.

Theorem 2. MIN-SUM DISSATISFACTION *and* MIN-MAX DISSATISFACTION *are NP-complete, even if each graph* G_i *is an out-tree containing all vertices of* V.

4   Each Preference Graph is a Directed Matching

After the strikingly negative results of Sect. 3, where it was shown that even elementary graphs, such as out-stars and out-trees, imply NP-completeness of our two problems, we now consider the basic graph structure of directed matchings. Indeed, the pairwise comparison of two items with no connection to any other items seems to be one of the most basic possibilities of considering any preferences at all. Also in decision science, the pairwise comparison of options constitutes the elementary building block for multi-criteria decision making methods, such as outranking methods [13].

We exhibit an interesting difference between the two objectives. While MIN-SUM DISSATISFACTION is shown to be polynomially solvable if all G_i are directed matchings, the same situation turns out to be still NP-complete for MIN-MAX DISSATISFACTION. Nonetheless, we obtain a positive result for the Min-Max objective for the special case of $k = 2$ agents.

Theorem 3. *When each graph* $G_i = (V_i, A_i)$ *is a directed matching,* MIN-SUM DISSATISFACTION *can be solved in polynomial time.*

Proof. To solve MIN-SUM DISSATISFACTION we will compute a maximum weight matching on an auxiliary undirected bipartite graph H. Its vertex set $V(H) = X \cup Y \cup Z$ consists of a vertex for every vertex in V_i, i.e., $X = \{x_j^i \mid i \in K, j \in V_i\}$, a vertex for every arc in A_i, i.e., $Y = \{y_a^i \mid i \in K, a \in A_i\}$, and a vertex for every item, i.e., $Z = \{z_\ell \mid \ell \in V\}$. The edge set $E(H) = S \cup T$ contains edges connecting every item vertex in Z with all its copies in X, i.e., $S = \{\{x_j^i, z_j\} \mid i \in K, j \in V_i\}$, and edges connecting the two endpoints of a matching arc in A_i with the corresponding vertex in Y, i.e., for each agent $i \in K$, and each arc $(a, a') \in A_i$, set T contains edges $\{y_{(a,a')}^i, x_a^i\}$ and $\{y_{(a,a')}^i, x_{a'}^i\}$. We claim that every matching M in H implies a feasible allocation of items to the agents by assigning item j to agent i if $e = \{x_j^i, z_j\} \in M$. Since there can be at most one edge in M joining a vertex z_j in Z to a vertex x_j^i in X, every item is allocated at most once. To avoid that both endpoints of an arc in A_i are allocated to i, the edges in T are assigned a very high weight. Then, every maximum weight matching will contain one of the two edges in T incident with a vertex y_a^i in Y, which forbids that the other endpoint in X corresponds to an item allocated to i.

The following weights are assigned to each $e \in E(H)$:

$$
w(e) = \begin{cases} 1, & \text{if } e = \{x_j^i, z_j\} \in S \text{ and } j \text{ is the head of an edge in } A_i, \\ 2, & \text{if } e = \{x_j^i, z_j\} \in S \text{ and } j \text{ is the tail of an edge in } A_i, \\ 2|V|, & \text{if } e \in T. \end{cases}
$$

The weights on the edges in S correspond to the number of vertices that each vertex from V_i dominates in G_i. Hence, a maximum weight matching will maximize the total satisfaction and thus minimize the total dissatisfaction. It holds that the maximum total satisfaction is equal to $w(M) - 2|T||V|$ and hence the minimum total dissatisfaction is equal to $\sum_{i \in K} |V_i| + 2|T||V| - w(M)$.   □

Theorem 4. MIN-MAX DISSATISFACTION *is NP-complete, even if each graph G_i is a directed matching.*

Proof Sketch. We again reduce from 3X3C. Given an instance $\mathcal{J}$ of 3X3C with a set X of elements and a collection $C = \{C_1, \ldots, C_p\}$ of 3-element subsets of X, let $\ell = \frac{4p}{3}$. We may assume without loss of generality that $p \geq 6$ and thus $\ell \geq 8$. We construct an instance $\mathcal{I}$ of the MIN-MAX DISSATISFACTION by introducing the items $V = X \cup \{h_j \mid 1 \leq j \leq \ell + 1\} \cup \{j, b_j^0, b_j^1, e_j \mid 1 \leq j \leq p\} \cup \{a_j \mid 1 \leq j \leq \ell + 1\}$, and the set K of agents made up of agents D_j for $1 \leq j \leq \ell + 1$, agent F, and the agents B_j, C_j for $1 \leq j \leq p$; their graphs are displayed in Fig. 2 where w.l.o.g. we assume set $C_j = \{x, y, z\}$. We ask whether there is an allocation with dissatisfaction of at most ℓ per agent. Observe that ℓ is even since p is a multiple of 3. Using this construction it can be shown that C contains an exact cover of X if and only if $\mathcal{I}$ admits an allocation π with $\max_{i \in K} \delta_\pi(i) \leq \ell$.   □

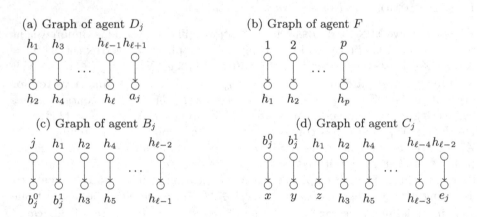

Fig. 2. Graphs of agents in the proof of Theorem 4.

Given the negative result of the above theorem we give a complementing positive result for MIN-MAX DISSATISFACTION below. Namely, if we restrict the number of agents to two, i.e., $k = 2$, we have a positive counterpart to Theorem 4.

Theorem 5. *When $k = 2$ and both preference graphs are directed matchings,* MIN-MAX DISSATISFACTION *can be solved in polynomial time.*

Proof Sketch. First note that since G_1 and G_2 are directed matchings, the underlying undirected graph of $G = G_1 \cup G_2$ (including possible multi-edges) is a collection of (agent 1)-(agent 2) alternating cycles (including cycles with two vertices) and paths (including single-edge paths).

We can obtain the set of all possible dissatisfaction profiles of the two agents for one path and then also for one cycle using a dynamic programming approach. After that, these sets of dissatisfaction profiles for all the paths and cycles can be combined to obtain all dissatisfaction profiles with respect to all items.   □

5   Further Polynomially Solvable Special Cases

Recall that in Sect. 3 it was shown that both MIN-MAX DISSATISFACTION and MIN-SUM DISSATISFACTION are computationally hard, even on special variants of out-trees. In the following we show that MIN-SUM DISSATISFACTION becomes polynomially solvable if the arborization of the graphs is restricted, namely by having only a constant number of *junction vertices* (vertices with in- or out-degree greater than 1). The corresponding Theorem 6 also implies a polynomial algorithm for the case where all G_i are directed matchings, but we already described a simpler approach in Theorem 3.

From Theorem 4 we know that MIN-MAX DISSATISFACTION remains NP-hard even for the special case where all G_i are directed matchings, and thus do not contain any junction vertices at all. However, for a different setting without junction vertices where all G_i are paths, MIN-MAX DISSATISFACTION is solvable in polynomial time (see Theorem 7).

Let us now turn to MIN-SUM DISSATISFACTION and the above-mentioned restriction of the preference graphs. Formally, we denote by $J_i \subseteq V(G_i)$ for each $i \in K$ the set of junction vertices in G_i, i.e., vertices in G_i with in- or out-degree greater than 1, and by $\gamma = \sum_{i \in K} |J_i|$, the total number of junction vertices (counted with multiplicities). Also, we call a vertex with in-degree 0 and out-degree 1 a *simple source* and a vertex with in-degree 1 and out-degree 0 a *simple sink*. Note that γ constant implies that all G_i, except constantly many, consist only of collections of paths. For background on fixed-parameter tractability, we refer to [8].

Theorem 6. MIN-SUM DISSATISFACTION *is fixed-parameter tractable with respect to* γ.

Proof Sketch. We introduce an algorithm to solve the maximization problem for the total satisfaction, which implies the solution of MIN-SUM DISSATISFACTION. Note that there exists an optimal allocation π which fulfills the *minimality condition*, meaning it is minimal with respect to the property that for each agent i no item allocated to agent i is dominated by any other item allocated to i. Hence, we restrict our search and feasibility test to allocations fulfilling this condition. The main observation used in the algorithm is that for an allocation π fulfilling the minimality condition, for each vertex $v \in J_i$ exactly one of the following four cases occurs.

(1) π allocates v to agent i.
(2) π allocates some item in $pred_i(v)$ to i.
(3) π allocates some item in $succ_i(v)$ to i.
(4) π does not allocate any item in $pred_i(v) \cup succ_i(v) \cup \{v\}$ to agent i.

In our algorithm, we enumerate all 4^γ possible assignments of cases (1)–(4) for all vertices in each J_i. Note that if $v \in V$ is a junction vertex in multiple agent graphs G_i we enumerate all possible assignments for each of the agents independently. This is possible in time 4^γ, since in the definition of γ such junction vertices are counted with multiplicity with respect to the graphs G_i. Since feasible allocations π fulfilling the minimality condition may not exist for all of the 4^γ assignments, we derive a method to test efficiently whether this is the case. The most involved setting concerns internal vertices on paths connecting junction vertices with guesses (2) and (3). Here, it is not directly clear on which of of these paths it is mandatory to assign items to agent i and on which it is just optional. We resolve this by additional guesses that can be bounded by a function depending only on γ.

Using these methods we test for each assignment whether it is feasible and if so we determine an optimal allocation of the remaining items subject to the conditions of cases (1)–(4), by reducing this subproblem to an instance of the minimum-cost circulation problem, which is solvable in strongly polynomial time (see, e.g., [18, Chapter 12]). Among all those allocations we take the one maximizing the total satisfaction.   □

For MIN-MAX DISSATISFACTION we now consider the following case without junction vertices, namely the case where each G_i is a single path. Observe that Theorem 6 implies that MIN-SUM DISSATISFACTION is solvable in polynomial time in that setting.

Theorem 7. *When each graph G_i is a path,* MIN-MAX DISSATISFACTION *can be solved in polynomial time.*

Proof. We reduce the problem to at most n matching problems by applying a threshold approach. The idea is to consider, in turn for increasing values of t, the first t vertices of every path G_i and try to assign one of these items to each agent. If this is possible, the dissatisfaction level is at most $t - 1$; otherwise, we continue with $t + 1$. Observe that if an agent has fewer than t vertices on her path, her dissatisfaction level is less than t anyways, and we can ignore that agent in future iterations.

Hence, starting with $t = 1$, for each possible value of $t = 1, 2, \ldots, n$, for which at least one path contains at least t vertices, we construct a bipartite undirected graph $H_t = (A_t \cup B_t, E_t)$, where agent $i \in A_t$ if G_i contains at least t vertices and item $j \in B_t$ if j is among the first t vertices on at least one of the paths G_i for which $i \in A_t$ holds; there is an edge between $i \in A_t$ and $j \in B_t$ if the respective vertex j is among the first t vertices of G_i. Compute a maximum cardinality matching in H_t. If the size of the matching is $|A_t|$, there is an allocation with dissatisfaction of at most $t - 1$: assign item j to agent i whenever the edge $\{i, j\}$

is contained in the maximum cardinality matching; the remaining items (if there are any) are assigned arbitrarily.

If the size of the matching is less than $|A_t|$, we check whether there is at least one path containing $t + 1$ vertices. If this is not the case, the maximum dissatisfaction corresponds to t and we can assign the items arbitrarily; otherwise, we increment t (i.e., set $t := t + 1$) and continue with the next graph H_t. On the other hand, given an allocation with dissatisfaction of at most t per agent, the matching induced by pairing each agent to the most-preferred among the received items results in a maximum cardinality matching in H_t.   □

6   Same Preferences for All Agents

Finally, we consider the situation in which all agents share the same preferences. In that scenario we refer to the single directed acyclic graph G representing the agents' preferences as the *common graph*. Our first result in this section states that for two agents with identical preferences both MIN-MAX DISSATISFACTION and MIN-SUM DISSATISFACTION can be solved in polynomial time.

Theorem 8. *When* $k = 2$ *and* $G_1 = G_2$, MIN-MAX DISSATISFACTION *and* MIN-SUM DISSATISFACTION *can be solved in polynomial time.*

Proof. Let us denote by $G = (V, A)$ the common directed acyclic graph representing the preferences of the two agents. Let S be the set of *sources* of G, that is, vertices of G with in-degree zero (these are the vertices corresponding to items that are not dominated by any other item). Let G' be the graph $G - S$ and let S' be the set of sources of G'. Furthermore, let S_1 be any subset of S with cardinality $\left\lfloor \frac{|S|}{2} \right\rfloor$ and $S_2 = S \setminus S_1$. We denote by S'_1 the vertices in S' that are not dominated by any vertex in S_1, and, similarly, by S'_2 the vertices in S' that are not dominated by any vertex in S_2. More formally, $S'_1 = S' \setminus \{succ_1(v) : v \in S_1\}$ and $S'_2 = S' \setminus \{succ_1(v) : v \in S_2\}$. We claim that the sets S_1, S_2, S'_1, and S'_2 are pairwise disjoint. The disjointness of any pair follows immediately from the definitions, except for S'_1 and S'_2. Suppose for a contradiction that there exists a vertex $u \in S'_1 \cap S'_2$. Then u belongs to S' and, hence, is a source in $G - S$. Since u is not a source in G, it must have a predecessor v in G. Furthermore, since u is a source in $G - S$, vertex v must belong to S; in particular, we must have $v \in S_i$ for some $i \in \{1, 2\}$. However, this implies that $u \notin S'_i$, a contradiction.

Since the sets $S_1 \cup S'_1$ and $S_2 \cup S'_2$ are disjoint, setting $\pi(1) = S_1 \cup S'_1$ and $\pi(2) = S_2 \cup S'_2$ defines a valid allocation of items to the two agents. The undominated items for agent 1 are exactly the items in S_2, and, similarly, the undominated items for agent 2 are exactly the items in S_1. Clearly, π can be computed in polynomial time and the corresponding dissatisfaction of the two agents is $\delta_\pi(1) = |S_2| = \left\lceil \frac{|S|}{2} \right\rceil$ and $\delta_\pi(2) = |S_1| = \left\lfloor \frac{|S|}{2} \right\rfloor$. To complete the proof, we show that π in fact optimally solves both MIN-MAX DISSATISFACTION and MIN-SUM DISSATISFACTION problems for the given input instance $G_1 = G_2 = G$.

First, consider the MIN-SUM DISSATISFACTION problem. Assignment π has a total dissatisfaction of $|S|$. Let π^* be an allocation that minimizes the total dissatisfaction. Notice that for every $v \in S$, either $v \notin \pi^*(1)$ or $v \notin \pi^*(2)$. We conclude that each item $v \in S$ contributes to the dissatisfaction at least once in the sum $\delta_{\pi^*}(1) + \delta_{\pi^*}(2)$, and hence that $\delta_{\pi^*}(1) + \delta_{\pi^*}(2) \geq |S|$. Thus, π is an optimal solution for the MIN-SUM DISSATISFACTION problem.

Now consider the MIN-MAX DISSATISFACTION problem. Under assignment π, dissatisfaction of each agent is at most $\left\lceil \frac{|S|}{2} \right\rceil$. Let π^* be an allocation that minimizes the maximum dissatisfaction. We may assume without loss of generality that $\delta_{\pi^*}(1) \geq \delta_{\pi^*}(2)$. Since the minimum total dissatisfaction equals $|S|$, we have $\delta_{\pi^*}(1) + \delta_{\pi^*}(2) \geq |S|$, and therefore $\max\{\delta_{\pi^*}(1), \delta_{\pi^*}(2)\} = \delta_{\pi^*}(1) \geq \left\lceil \frac{|S|}{2} \right\rceil = \max\{\delta_\pi(1), \delta_\pi(2)\}$. We conclude that π is an optimal solution for the MIN-MAX DISSATISFACTION problem. $\square$

The result of Theorem 8 is best possible regarding the number of agents, unless P = NP. Indeed, we show next that MIN-MAX DISSATISFACTION and MIN-SUM DISSATISFACTION are NP-complete for three agents with identical preferences, even if the common directed acyclic graph has no directed path of length two. In the proof of the below theorem, we will make use of the following concepts for undirected graphs. Let G be an undirected graph. We say that G is *cubic* if every vertex of G is incident with exactly three edges. A matching M in G is said to be *perfect* if every vertex of G is an endpoint of an edge in M. Note that every matching M in G satisfies $|M| \leq |V(G)|/2$, with equality if and only if M is a perfect matching. The *line graph* of G is the graph $L(G)$ with vertex set $E(G)$, with $e, f \in E(G)$ adjacent if and only if e and f are distinct edges in G that share an endpoint.

Theorem 9. MIN-MAX DISSATISFACTION *and* MIN-SUM DISSATISFACTION *are NP-complete, even when $k = 3$ and $G_1 = G_2 = G_3$.*

Proof. We reduce from the following variant of 3-EDGE-COLORING problem: given an undirected cubic graph H, is there a partition $\{M_1, M_2, M_3\}$ of $E(H)$ into three matchings? As shown by Holyer [16], this problem is NP-complete. Take any undirected cubic graph H. Note that if H is 3-edge-colorable and $\{M_1, M_2, M_3\}$ is a partition of $E(H)$ into three matchings, then $3|V(H)| = 2|E(H)| = 2(|M_1| + |M_2| + |M_3|) \leq 3|V(H)|$ since $|M_i| \leq |V(H)|/2$ for all $i \in \{1, 2, 3\}$; hence, equalities must hold and each M_i is a perfect matching in H. Let G be the line graph of H. We now describe an instance $\mathcal{I}_1$ of the MIN-MAX DISSATISFACTION problem and an instance $\mathcal{I}_2$ of the MIN-MAX DISSATISFACTION problem. In both instances, the set of agents is $K = \{1, 2, 3\}$, the directed acyclic graph of each agent $i \in K$ is the same, namely the graph G_i given by $V(G_i) = V(G) \cup E(G)$ and $E(G_i) = \{(v, e) \mid v \in V(G), e \in E(G), v \in e\}$, and the set of items is $V(G_1)(= V(G_2) = V(G_3))$. See Fig. 3 for an example.

The only difference in the two instances is in the corresponding upper bounds b_1 and b_2 on the maximum and total dissatisfaction, respectively. We set $b_1 = |V(H)|$ in instance $\mathcal{I}_1$ and $b_2 = 3|V(H)|$ in instance $\mathcal{I}_2$. Note that $b_2 = 3b_1$ and

$b_2 = 2|V(G)|$. We complete the proof by showing that the following statements are equivalent.

1. H is 3-edge-colorable.
2. There exists an allocation π of items to agents such that dissatisfaction $\delta_\pi(i)$ of each agent $i \in K$ is at most b_1.
3. There exists an allocation π of items to agents with total dissatisfaction at most b_2.

Suppose first that H is 3-edge-colorable, and let $\{M_1, M_2, M_3\}$ be a partition of $E(H)$ into three perfect matchings. Recall that $3|V(H)| = 2|E(H)|$ and hence each matching M_i has cardinality $\frac{|V(H)|}{2} = \frac{|E(H)|}{3} = \frac{|V(G)|}{3}$. Thus, $\{M_1, M_2, M_3\}$ is a partition of the vertices of G into three equally sized independent sets. We construct an allocation π of items to agents in K as follows. For each $1 \leq i < j \leq 3$, let E_{ij} denote the set of edges of G with one endpoint in M_i and the other one in M_j. Then $\{E_{12}, E_{13}, E_{23}\}$ is a partition of $E(G)$. Since also $\{M_1, M_2, M_3\}$ is a partition of $V(G)$, setting $\pi(1) = M_1 \cup E_{23}, \pi(2) = M_2 \cup E_{13}, \pi(3) = M_3 \cup E_{12}$ yields an allocation of items to agents in K. See Fig. 3 for an example.

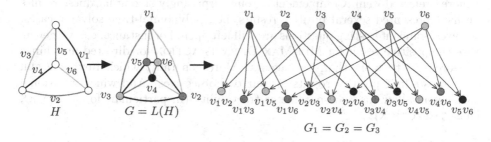

$$G_1 = G_2 = G_3$$

Fig. 3. Transforming a cubic graph H (with edges labeled as $v_1, \ldots, v_6$) into $G_1 = G_2 = G_3$ and mapping a 3-edge coloring of H into an allocation of items to agents. Let $M_i = \{v_i, v_{i+3}\}$ for all $i \in K$. Then, edges of G that belong to E_{12}, E_{13}, and E_{23} are colored light grey, grey, and black, respectively.

The dissatisfaction of agent 1 is equal to $\delta_\pi(1) = |M_2| + |M_3| = |V(H)| = b_1$. By symmetry, we also have $\delta_\pi(2) = \delta_\pi(3) = b_1$. Thus, the dissatisfaction of each agent is at most b_1.

Since $b_2 = 3b_1$, any allocation π such that the dissatisfaction of each agent is at most b_1 has total dissatisfaction at most b_2.

Finally, suppose that there exists an allocation π of items to agents with total dissatisfaction at most $b_2 = 2|V(G)|$. For $i \in K$, let $V_i = \pi(i) \cap V(G)$ and $E_i = \{e \in E(G) \mid e \subseteq V_i\}$. For each $1 \leq i < j \leq 3$, let E_{ij} denote the set of edges of G with one endpoint in V_i and the other one in V_j. Note that each $v \in V(G)$ contributes to the total dissatisfaction at least twice, since it is not going to be in at least two of the sets $\pi(1), \pi(2), \pi(3)$ and hence not dominated by them. We infer that the total dissatisfaction is at least $2|V(G)|$. Since we assumed that the total dissatisfaction is at most $2|V|$, equality must hold. This implies that each $v \in V(G)$ contributes to the total dissatisfaction exactly twice, and hence $\{V_1, V_2, V_3\}$ is a partition of $V(G)$. Furthermore, $\{E_1, E_2, E_3, E_{12}, E_{13}, E_{23}\}$ is a partition of E. Furthermore, for $\{i, j, k\} = \{1, 2, 3\}$, each $e \in E_i$ contributes at least one unit to the total dissatisfaction, since either $e \notin \pi(j)$ or $e \notin \pi(k)$. Hence we can estimate the total dissatisfaction as $\sum_{i \in K} \delta_\pi(i) \geq (|V_2| + |V_3|) + (|V_1| + |V_3|) + (|V_1| + |V_2|) + (|E_1| + |E_2| + |E_3|) = 2|V(G)| + (|E_1| + |E_2| + |E_3|)$. Since we already showed that the total dissatisfaction is equal to $2|V(G)|$ we infer that $E_1 = E_2 = E_3 = \emptyset$. Hence V_1, V_2, V_3 are independent sets and they form a 3-coloring of G, or, equivalently, a 3-edge-coloring of H.   □

7   Conclusion

We have considered a model in which agents' preferences over indivisible items are captured by means of directed acyclic graphs (preference graphs). In that respect, we have analysed the task of allocating the items in a way that minimizes total and maximum dissatisfaction, where the latter is measured by the number of desired items an agent does not receive and for which she does not get a more preferred item. Complementing our surprisingly strong hardness results we have presented several positive results, i.e., polynomial-time solvable cases; however, some interesting questions are still left open. For instance, can we generalize the positive result for MIN-MAX DISSATISFACTION for directed matchings (Theorem 5) to more than two agents? More generally, which further graph structures admit positive results for our two objectives? And which additional parameters allow for fixed-parameter tractability (in particular, for MIN-MAX DISSATISFACTION)?

Acknowledgements. The authors wish to thank Matjaž Krnc for valuable discussions. The work of this paper was done in the framework of two bilateral projects between University of Graz and University of Primorska, financed by the OeAD (SI 22/2018 and SI 31/2020) and the Slovenian Research Agency (BI-AT/18-19-005 and BI-AT/20-21-015). The authors acknowledge partial support of the Slovenian Research Agency (I0-0035, research programs P1-0404, P1-0285, P1-0383, research projects N1-0102, N1-0160, J1-9110, J1-1692, J1-9187 and a Young Researchers Grant) and by the Field of Excellence "COLIBRI" at the University of Graz.

References

1. Aziz, H., Gaspers, S., Mackenzie, S., Walsh, T.: Fair assignment of indivisible objects under ordinal preferences. Artif. Intell. **227**, 71–92 (2015)
2. Babaioff, M., Ezra, T., Feige, U.: Fair and truthful mechanisms for dichotomous valuations. In: Proceedings of the 35th AAAI Conference on Artificial Intelligence (AAAI'20), pp. 5119–5126. AAAI Press (2021)
3. Bansal, N., Sviridenko, M.: The santa claus problem. In: Proceedings of the 38th Annual ACM Symposium on Theory of Computing (STOC'06), pp. 31–40 (2006)

4. Baumeister, D., et al.: Positional scoring-based allocation of indivisible goods. Auton. Agents Multi-Agent Syst. **31**(3), 628–655 (2016). https://doi.org/10.1007/s10458-016-9340-x

5. Bouveret, S., Chevaleyre, Y., Maudet, N.: Fair division of indivisible goods. In: Brandt, F., et al. (eds.) Handbook of Computational Social Choice, chapter 12. Cambridge University Press, Cambridge (2016)

6. Brams, S., Kilgour, M., Klamler, C.: Two-person fair division of indivisible items: an efficient, envy-free algorithm. Notices AMS **61**(2), 130–141 (2014)

7. Chiarelli, N., Krnc, M., Milanič, M., Pferschy, U., Pivač, N., Schauer, J.: Fair packing of independent sets. In: Gasieniec, L., Klasing, R., Radzik, T. (eds.) IWOCA 2020. LNCS, vol. 12126, pp. 154–165. Springer, Cham (2020). https://doi.org/10.1007/978-3-030-48966-3_12

8. Cygan, M., et al.: Parameterized Algorithms. Springer, Heidelberg (2015). https://doi.org/10.1007/978-3-319-21275-3

9. Darmann, A., Schauer, J.: Maximizing Nash product social welfare in allocating indivisible goods. Eur. J. Oper. Res. **247**(2), 548–559 (2015)

10. Duddy, C.: Fair sharing under dichotomous preferences. Math. Soc. Sci. **73**, 1–5 (2015)

11. Garg, J., McGlaughlin, P.: Improving Nash social welfare approximations. In: Proceedings of the 28th International Joint Conference on Artificial Intelligence (IJCAI'19), pp. 294–300 (2019)

12. Gonzalez, T.F.: Clustering to minimize the maximum intercluster distance. Theor. Comput. Sci. **38**, 293–306 (1985)

13. Greco, S., Ehrgott, M., Figueira, J.R.: Multiple Criteria Decision Analysis, 2nd edn. Springer, Heidelberg (2016). https://doi.org/10.1007/978-1-4939-3094-4

14. Halpern, D., Procaccia, A.D., Psomas, A., Shah, N.: Fair division with binary valuations: one rule to rule them all. In: Chen, X., Gravin, N., Hoefer, M., Mehta, R. (eds.) WINE 2020. LNCS, vol. 12495, pp. 370–383. Springer, Cham (2020). https://doi.org/10.1007/978-3-030-64946-3_26

15. Herreiner, D., Puppe, C.: A simple procedure for finding equitable allocations of indivisible goods. Soc. Choice Welfare **19**(2), 415–430 (2002). https://doi.org/10.1007/s003550100119

16. Holyer, I.: The NP-completeness of edge-coloring. SIAM J. Comput. **10**(4), 718–720 (1981)

17. Roos, M., Rothe, J.: Complexity of social welfare optimization in multiagent resource allocation. In: Proceedings of the 9th International Conference on Autonomous Agents and Multiagent Systems (AAMAS'10), pp. 641–648 (2010)

18. Schrijver, A.: Combinatorial optimization. In: Polyhedra and Efficiency, vol. 24. Springer-Verlag, Berlin (2003)

19. Thomson, W.: Introduction to the theory of fair allocation. In: Brandt, F., et al. (eds.) Handbook of Computational Social Choice, chapter 11. Cambridge University Press, Cambridge (2016)

On Fairness via Picking Sequences
in Allocation of Indivisible Goods

Laurent Gourvès[1], Julien Lesca[1], and Anaëlle Wilczynski[2(✉)]

[1] Université Paris-Dauphine, Université PSL, CNRS, LAMSADE,
75016 Paris, France
{laurent.gourves,julien.lesca}@dauphine.fr
[2] MICS, CentraleSupélec, Université Paris-Saclay, Gif-sur-Yvette, France
anaelle.wilczynski@centralesupelec.fr

Abstract. Among the fairness criteria for allocating indivisible resources to a group of agents, some are based on minimum utility levels. These levels can come from a specific allocation method, such as maximin fair-share criterion which is based on the cut-and-choose protocol. We propose to analyze criteria whose minimum utility levels are inspired by picking sequences, a well-established protocol for allocating indivisible resources. We study these criteria and investigate their connections with known fairness criteria, enriching the understanding of fair allocation of indivisible goods.

Keywords: Fair division · Resource allocation · Computational social choice

1 Introduction

Fair division of indivisible goods is a fundamental and challenging question in collective decision making that has been widely investigated [6,10,22]. Many criteria have been proposed in the literature in order to evaluate the fairness of an allocation when agents express preferences over bundles of goods via additive utilities. A very natural criterion is *envy-freeness (EF)* [18,30], a comparison-based criterion which asks that no agent prefers the bundle assigned to another agent over her own assigned bundle of goods. This criterion notably requires that agents are aware of the other agents' allocation. Alternatively, many criteria simply impose, for an allocation to be considered fair, that each agent gets a utility for her assigned bundle that is greater than or equal to a predefined minimum utility level, called a *fair guarantee* [5]. As defined by Bogomolnaia et al. [5], a fair guarantee for an agent is a utility level defined only according to the utility function of the agent and the number of agents n. One can cite the *proportionality (Prop)* fair guarantee [28] where each agent must get at least a

This research benefited from the support of the FMJH Program PGMO under grant DAMPER and from the support of EDF, Thales, and Orange.

D. Fotakis and D. Ríos Insua (Eds.): ADT 2021, LNAI 13023, pp. 258–272, 2021.
https://doi.org/10.1007/978-3-030-87756-9_17

utility equal to her value for the whole set of goods divided by n. In addition, a fair guarantee can be defined according to a given allocation procedure like, e.g., the *maximin share (MMS)* [13] or the *min-max-fair-share (mFS)* [9] which are computed thanks to the cut-and-choose protocol. In this article, we define several fairness criteria whose fair guarantee can be computed thanks to picking sequences.

In the well-established allocation protocol of *picking sequences (PS)* [7,12,20], all goods are initially available and, given a sequence of agents (a.k.a. *policy*), each agent picks at her turn an object among the remaining ones. Understanding which allocations emerge from such a mechanism has been done, for example, by Brams et al. [11], and Aziz et al. [4]. Moreover, picking sequences have been widely studied in a strategic perspective where agents may choose not to pick their best object [2,8,19,29]. Non strategic agents are said to be *sincere*.

One of the main assets of picking sequences is their simplicity: everyone can quickly understand how they work and they are easy to implement. Thus, they are good candidates for sharing resources. If the final allocation is not built with a picking sequence, then an agent may advocate for it and claim that her utility must be as good as the one resulting from a picking sequence that she has in mind. However, the number of possible policies is huge and, on top of this, every agent can have her own policy in mind. This offers a number of combinations which is undoubtedly too large. For a positive integer p bounded by the number of agents, we propose a simple criterion named PS_p in which the fair guarantee of every agent is her utility for a subset of objects built as follows. Rank the objects from best to worst under the agent's preference, and keep the items whose ranks are multiples of p. An agent would be endowed such a set in a sincere picking sequence if her positions in the policy were multiples of p, and if the other agents had the same preference. Indeed, without knowing the others' preferences, an agent may suppose that, in the worst case, everyone has the same object ranking as hers. In PS_p, the parameter p makes it possible to move gradually between a very optimistic scenario where all the agents choose first ($p = 1$), and a more pessimistic one where they all choose last ($p = n$). PS criteria only rely on a very simple sequential allocation protocol, which is commonly known (think about composition of sports teams at school). Moreover, agents only need to know their assigned bundle, the number of agents and their own preferences over goods. Therefore, these criteria are easy to understand and can be naturally expressed as requirements by an agent.

The fact that the agent appears recursively in the policy is inspired by *round robin*, a well-known method for allocating resources [3,26]. Round robin falls into the class of *recursively balanced* (RB) policies [4], where each sequence of agents can be divided into rounds during which all the agents pick an object exactly once (all rounds are identical in round robin). At any step of the sequence, the agents have chosen almost the same number of objects. Without any prior knowledge on the agents' utility functions, letting the agents pick the same number of times, leading then to an *even-shares division*, constitutes a natural first argument for equity [10]. Moreover, it is known from Aziz et al. [3] that picking sequences with RB policies generate allocations that are *envy-free up to one good (EF1)* [13,23], a well-accepted fairness criterion which relaxes envy-freeness.

The *round robin share criterion* introduced by Conitzer et al. [17] in the context of *public* decision making corresponds to PS_n. The round robin share provides a one half approximation to *Prop1*, a relaxed version of proportionality. Conitzer et al. [17] focus on mechanisms that satisfy this criterion among others, whereas we focus on the properties of the PS criteria.

Contribution and Organization. Section 2 contains a formal definition of the model, a review of classical fairness concepts, some common relaxations up to some goods, and a map explaining how all these notions relate. We notably complement the state of the art on the relations between relaxations of envy-freeness and proportionality. The PS criteria are introduced in Sect. 3. Analogously to many relaxed criteria based on the satisfaction of the fairness requirement up to the addition of some goods in the agent's bundle (like, e.g., EF1 for envy-freeness or Prop1 for proportionality), we also study relaxations of the PS criteria up to some goods. For a given allocation of goods, the satisfaction of a PS criterion can be checked in polynomial time. We identify in Sect. 4 the PS criteria for which a satisfying allocation always exists, and when it is not the case, we settle the complexity of deciding the existence of a satisfying allocation in a given instance. Contrary to many classical criteria, we identify two non-trivial PS criteria, namely PS_n and PS_11 (the relaxation up to one good of PS_1), for which a satisfying allocation always exists. Afterwards, we provide a complete picture of the implications that relate the PS criteria and the classical fairness concepts (Sect. 5), as well as their relaxations up to some goods. All these results are summarized in Fig. 1. Finally, we complement our study with experiments which give an intuition on how well fairness criteria can be compatible with efficiency. Due to space limitation, some proofs are omitted.

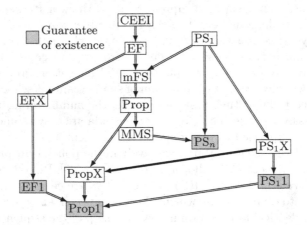

Fig. 1. Summary of the relations among fairness criteria and their existence guarantee (*crit*X stands for the relaxation of the criterion *crit* up to any good). An arrow from criterion A to criterion B means that A implies B (A is stronger than B). If there is no path from A to B then A is not stronger than B.

2 Fair Division of Indivisible Goods

2.1 The Setting

We are given a set $N = \{1, \ldots, n\}$ of $n \geq 2$ agents and a set $M = \{x_1, \ldots, x_m\}$ of m indivisible resources (or objects) which are goods. The agents have cardinal preferences over the bundles of objects, expressed via utility function $u_i : 2^M \to \mathbb{R}^+$ for each agent i. We assume that the utilities are additive, i.e., for each bundle of objects O and each agent i, $u_i(O) = \sum_{x_j \in O} u_i(\{x_j\})$. For the sake of simplicity, we denote $u_i(\{x\})$ by $u_i(x)$. We represent the preferences by an $(n \times m)$-matrix where the value in row i and column j corresponds to $u_i(x_j)$. Preferences are *strict (on the objects)* whenever $u_i(x) \neq u_i(y)$ for every agent i and pair of objects x and y. We denote by o_i^k the kth most preferred object of agent i, for $1 \leq k \leq m$ (an arbitrary order over the objects is used in case of ties). We suppose, w.l.o.g., that the number of objects m is a multiple of the number of agents n (dummy objects with utility 0 can be added if it is not initially the case), and q denotes the quotient m/n.

An *allocation* σ is a mapping $\sigma : N \to 2^M$ such that $\sigma(i) \cap \sigma(j) = \emptyset$ for all agents i and j, and $\bigcup_{i \in N} \sigma(i) = M$, where $\sigma(i)$ is the bundle assigned to agent i. $\mathcal{A}$ denotes the set of all allocations. The n-vector $u(\sigma) = (u_1(\sigma(1)), \ldots, u_n(\sigma(n)))$ describes the utilities that the agents obtain from allocation σ.

In this article, $[t] := \{1, \ldots, t\}$ for all positive integers t.

2.2 Classical Fairness Criteria

For the sake of self-containedness, we recall some classical fairness criteria.

- *Maximin share guarantee (MMS)* [13]: Allocation σ is MMS iff $u_i(\sigma(i)) \geq mms_i$ for every agent i, where $mms_i = \max_{\sigma' \in \mathcal{A}} \min_{j \in N} u_i(\sigma'(j))$.
- *Proportionality (Prop)* [28]: Allocation σ is Prop iff $u_i(\sigma(i)) \geq \frac{1}{n} \sum_{x \in M} u_i(x)$, for every agent i.
- *Min-max-fair-share guarantee (mFS)* [9]: Allocation σ is mFS iff $u_i(\sigma(i)) \geq mfs_i := \min_{\sigma' \in \mathcal{A}} \max_{j \in N} u_i(\sigma'(j))$, for every agent i.
- *Envy-freeness (EF)* [18, 30]: Allocation σ is EF iff $u_i(\sigma(i)) \geq u_i(\sigma(j))$ for all agents i and j.
- *Competitive Equilibrium with Equal Incomes (CEEI)* (see, e.g., Moulin [24]): Allocation σ is CEEI iff there exists a price vector $pr \in [0, 1]^m$ such that $\sigma(i) \in \arg\max_{O \subseteq M} \{u_i(O) : \sum_{o \in O} pr_o \leq 1\}$ for every agent i.

An implication $A \Rightarrow B$ between criteria means that if an allocation satisfies A for a given instance, then the same allocation also satisfies B for the same instance. When such an implication holds, we say that criterion A is stronger, or more demanding, than criterion B. All previous fairness criteria can be connected w.r.t. such implications to form the following "scale of fairness" [9].

$$CEEI \Rightarrow EF \Rightarrow mFS \Rightarrow Prop \Rightarrow MMS \tag{1}$$

An allocation satisfying MMS is guaranteed to exist for two agents [9]. However, starting from 3 agents, there may not exist an MMS allocation [21,27].

Common relaxations of envy-freeness and proportionality are based on satisfying the criterion up to a fixed number c of goods, for c a positive integer. Allocation σ is *proportional up to c goods (Propc)* [16] iff for every agent i, there exists $X^i \subseteq M \setminus \sigma(i)$ such that $|X^i| \leq c$ and $u_i(\sigma(i)) + u_i(X^i) \geq u_i(M)/n$. Allocation σ is *envy-free up to c goods (EFc)* iff for all agents i and j, there exists $X^i \subseteq \sigma(j)$ such that $|X^i| \leq c$ and $u_i(\sigma(i)) + u_i(X^i) \geq u_i(\sigma(j))$. By definition, Prop $\Rightarrow$ Propc (resp., EF $\Rightarrow$ EFc) holds for all c, and Propc $\Rightarrow$ Propc' (resp., EFc $\Rightarrow$ EFc') whenever $c \leq c'$.

In addition, we show that EFc $\Rightarrow$ Propc holds for all c.

Proposition 1. *EFc $\Rightarrow$ Propc.*

It follows that EFc $\Rightarrow$ Propc' whenever $c \leq c'$. The existence of an allocation satisfying EF1 [13,23] or Prop1 [17] i.e., $c = 1$, is guaranteed for every instance. Relaxations up to one good have been strengthened to *any* good. Allocation σ is *proportional up to any good (PropX)* [25] iff for every agent i, $u_i(\sigma(i)) + u_i(x) \geq u_i(M)/n$ holds *for all* $x \in M \setminus \sigma(i)$. By definition we have Prop $\Rightarrow$ PropX $\Rightarrow$ Prop1. Allocation σ is *envy-free up to any good (EFX)* [15,23] iff $u_i(\sigma(i)) + u_i(x) \geq u_i(\sigma(j))$ for every pair of agents i and j and *any* object $x \in \sigma(j)$. It holds that EF $\Rightarrow$ EFX $\Rightarrow$ EF1.

Though EFc implies Propc, we show that EFX does not imply PropX when $n > 2$. Up to our best knowledge, this fact has not been explicitly stated before.

Proposition 2. *If $n = 2$ then EFX $\Rightarrow$ PropX, but EFX $\not\Rightarrow$ PropX when $n > 2$.*

Proof. Suppose there are two agents. Take an instance and an EFX allocation σ. Take the viewpoint of an agent, say agent 1. We have $u_1(\sigma(1)) + u_1(x) \geq u_1(\sigma(2))$ for all $x \in \sigma(2)$. Add $u_1(\sigma(1)) + u_1(x) \geq u_1(\sigma(2))$ to $u_1(\sigma(1)) + u_1(x) \geq u_1(\sigma(1))$ in order to get that $2(u_1(\sigma(1)) + u_1(x)) \geq u_1(\sigma(1)) + u_1(\sigma(2))$. By definition, we also have $u_1(\sigma(1)) + u_1(\sigma(2)) = u(M)$. Therefore σ satisfies PropX.

Suppose there are 3 agents and 5 objects $\{x_1, \ldots, x_5\}$ valued $(1, 1, 1, 0.25, 0.05)$ by agent 1. The utility of agents 2 and 3 is 1 for all objects. Agents 1, 2 and 3 get $\{x_1\}$, $\{x_2, x_3\}$, and $\{x_4, x_5\}$, respectively. This allocation is EFX, but not PropX because $1.05 = u_1(x_1) + u_1(x_5) \not\geq u_1(M)/3 = 1.1$. We can extend this instance to any number of agents $n > 3$. $\qquad\square$

3   Picking Sequence (PS) Fairness Criteria

We present new fairness criteria inspired by some *picking sequences*. A *policy* $\pi : \{1, \ldots, m\} \rightarrow N$ is a sequence of agents of size m, denoted by $\pi = \langle \pi(1), \ldots, \pi(m) \rangle$. A picking sequence is a sequential protocol asking agent $\pi(t)$ to pick an object within the set of remaining objects at stage t. A policy π is *recursively balanced (RB)* [4] if π can be decomposed into $q = \frac{m}{n}$ rounds, and each agent chooses an object exactly once at each round. Round robin is a special RB policy where all rounds are identical [3].

The PS fairness criteria use fair guarantees [5]. For every $p \in [n]$, PS_p imposes that the utility of an agent i for her share is at least $ps_p(i)$ where

$$ps_p(i) := \sum_{k=1}^{q} u_i(o_i^{(k-1)n+p}).$$

For example, agent i would get utility $ps_p(i)$ in a sincere[1] picking sequence if her turns in π were all the multiples of p, and if the other agents had identical preferences. Without knowing the preferences of the others, agent i considers the worst case where all the other agents have the same induced ordinal preferences as hers. In such a case, at each turn k, agent i can only get her $((k-1)n+p)^{\text{th}}$ most preferred available object, i.e., $o_i^{(k-1)n+p}$.

Let ps_p be the n-vector $(ps_p(1), \ldots, ps_p(n))$. An allocation satisfies a PS criterion if it fulfills the PS fair guarantees for every agent and some common position p. It is important to note that allocations satisfying a PS criterion do not need to be generated by a picking sequence.

Definition 1 (PS_p allocation). *An allocation $\sigma \in \mathcal{A}$ is PS_p if for every agent $i \in N$, $u_i(\sigma(i)) \geq ps_p(i)$.*

By definition, $PS_p \Rightarrow PS_{p'}$ holds for every $1 \leq p \leq p' \leq n$. In this article, we pay particular attention to positions $p = 1$ and $p = n$, which correspond to an optimistic and pessimistic view, respectively.

The PS fair guarantees are computable in polynomial time, by definition. Therefore, checking whether a given allocation satisfies a PS criterion is computationally easy. Whereas this polynomial-time verification also holds for proportionality and envy-freeness, this is not the case for CEEI, nor for MMS and mFS [9], although the two latter notions are also based on fair guarantees. Note that, contrary to envy-based criteria, the verification of satisfaction of a PS criterion does not even need to have access to other agents' allocation.

Like EF and Prop, PS_p can be relaxed up to some goods. Allocation σ satisfies PS_p up to c goods ($PS_p c$) iff for every agent i, there exists $X^i \subseteq M \setminus \sigma(i)$ such that $|X^i| \leq c$ and $u_i(\sigma(i)) + u_i(X^i) \geq ps_p(i)$. By definition we have $PS_p c \Rightarrow PS_{p'} c \Rightarrow PS_{p'} c'$ whenever $p \leq p'$ and $c \leq c'$. However, a PS_p allocation may not satisfy $PS_{p-1} 1$, as stated below.

Proposition 3. *PS_2 does not imply $PS_1 1$.*

Proof. Consider an instance where $n = 2$ and $m = 6$. The utilities are:

$$\begin{pmatrix} ㉔ & 16 & 12 & 6 & 5 & 2 \\ 10 & ⑤ & ④ & ③ & ② & ① \end{pmatrix}$$

Allocation σ (circles) is PS_2 since $ps_2 = (24, 9)$ and $u(\sigma) = (24, 15)$. However, it is not $PS_1 1$ because $u_1(\sigma(1)) + \max_{x \notin \sigma(1)} u_1(x) = 24 + 16 < 41 = ps_1(1)$. $\square$

---

[1] Agents always pick their favorite object.

Allocation σ satisfies PS_p up to *any* good ($PS_p X$) iff for every agent i, $u_i(\sigma(i)) + u_i(x) \geq ps_p(i)$ holds *for all* $x \in M \setminus \sigma(i)$. It holds that $PS_p \Rightarrow PS_p X \Rightarrow PS_p 1$. However, no relaxation up to any good implies a PS criterion with no relaxation, as stated below.

Proposition 4. *$PS_1 X$ does not imply PS_n.*

4   Allocations Satisfying PS Criteria

Observe first that a PS_p allocation may not exist if $p < n$: Consider an instance where $n = m$ with agents having the same induced preference order and no object with zero utility. Every agent should receive one object but no agent wants the common least preferred object. However, when $p = n$, the existence is guaranteed for every number of goods m because every allocation resulting from a picking sequence with an RB policy is PS_n.

Proposition 5. *Every allocation resulting from a sincere picking sequence with an RB policy is PS_n.*

Proof. Consider an allocation σ resulting from a sincere picking sequence with an RB policy π, and take an arbitrary agent i. For each round k of π, let $p_i(k)$ denote the position occupied by agent i in π during round k, while x_k is the object picked by agent i in round k. By definition, we have $u_i(\sigma(i)) = \sum_{1 \leq k \leq q} u_i(x_k)$. Since agent i is sincere and $p_i(k) - 1$ objects have been taken before agent i picks at round k, it follows that $u_i(x_k) \geq u_i(o_i^{p_i(k)})$ for every round k. Thus, $u_i(\sigma(i)) \geq \sum_{1 \leq k \leq q} u_i(o_i^{p_i(k)})$. By definition of an RB sequence, $(k-1)n + 1 \leq p_i(k) \leq kn$ holds. Therefore, we get that $u_i(\sigma(i)) \geq \sum_{1 \leq k \leq q} u_i(o_i^{kn}) = ps_n(i)$. $\square$

The converse of Proposition 5 is not true. That is, not every PS_n allocation can result from a picking sequence with an RB policy, as shown in the next example. This notably shows that allocations satisfying the PS criteria do not necessarily emerge from a picking sequence (in particular, agents do not necessarily get the same number of objects).

Example 1. Consider an instance where $n = 2$ and $m = 4$. The utilities are:

$$\begin{pmatrix} \boxed{20} & 3 & 2 & 1 \\ 5 & \boxed{4} & \boxed{3} & \boxed{2} \end{pmatrix}$$

The encircled allocation σ is PS_n since $ps_n = (4, 6)$ and $u(\sigma) = (20, 9)$. However, this allocation cannot result from a picking sequence with an RB policy since the two agents do not have the same number of objects.

Nevertheless, checking the existence of a PS_p allocation is hard for every constant $p < n$, even when $m = 2n$.

Theorem 1. *Determining whether a PS_p allocation exists is NP-complete, even when $m = 2n$ and $p < n$ is a constant.*

Moreover, checking the existence of a PS_1 allocation is hard even when $n = 2$, showing that even checking the existence of a PS_{n-1} allocation is hard.

Theorem 2. *Determining whether a PS_1 allocation exists is NP-complete, even when $n = 2$.*

However, an allocation satisfying the relaxation up to one good of PS_1 always exists.

Proposition 6. *Every allocation resulting from a sincere picking sequence with an RB policy is $PS_1 1$.*

Proof. Consider the allocation σ built with the sincere picking sequence that uses an RB policy. Take an agent i. Her objects are $\{o_i^{f(1)}, o_i^{f(2)}, \ldots, o_i^{f(q)}\}$ for some increasing function $f : [q] \rightarrow [m]$ where $f(j)$ is the rank in the preference order of i of the object picked by i during round j. Let r be the smallest index such that $o_i^r \notin \sigma(i)$. Agent i has in her share every object o_i^j with $j < r$. Thus, $o_i^{f(j)} = o_i^j$ for all $j < r$. We deduce that $\sum_{j<r} u_i(o_i^{f(j)}) = \sum_{j<r} u_i(o_i^j) \geq \sum_{j<r} u_i(o_i^{1+(j-1)n})$ (2). The policy being RB, we also have $u_i(o_i^{f(j)}) \geq u_i(o_i^{jn}) \geq u_i(o_i^{1+jn})$ for all $j \in [q-1]$, from which we deduce that $\sum_{j=r}^{q-1} u_i(o_i^{f(j)}) \geq \sum_{j=r}^{q-1} u_i(o_i^{1+jn}) = \sum_{j=r+1}^{q} u_i(o_i^{1+(j-1)n})$ (3). Combine (2) and (3) with $u_i(o_i^r) \geq u_i(o_i^{1+(r-1)n})$ to get that $u_i(\sigma(i)) + u_i(o_i^r) \geq \sum_{k=1}^{q} u_i(o_i^{1+(k-1)n}) = ps_1(i)$. In other words, σ is $PS_1 1$ for agent i. □

Propositions 5 and 6, together with Proposition 1 from Aziz et al. [3], imply that a sincere picking sequence with an RB policy produces an allocation that simultaneously satisfies EF1, Prop1, $PS_1 1$ and PS_n.

By Proposition 6, a $PS_p 1$ allocation exists for every $p \in [n]$. It is not the case for $PS_p X$, even when $p = n-1$ for any number n of agents: Consider an instance where $m = 2n$ with the following preferences for every agent i: $u_i(x_j) = 1$ for every $j \in [n-1]$, $u_i(x_j) = 1/n$ for every $n \leq j \leq 2n-1$ and $u_i(x_{2n}) = 0$. We have $ps_{n-1}(i) = 1 + 1/n$ for every agent i. To satisfy $PS_{n-1}X$, each agent i must be in one of the following situations: $u_i(\sigma(i)) \geq 1 + 1/n$, or $\sigma(i) = \{x_k, x_{2n}\}$ for some $k \in [n-1]$, or $\sigma(i) = \{x_j : n \leq j \leq 2n\}$. Making n disjoint bundles under such conditions is impossible.

Since a $PS_1 1$ allocation always exists, there is no need to consider relaxations of PS_p up to c goods for $p > 1$ and $c > 1$. Combined with the fact that the existence of $PS_p X$ allocations is not guaranteed even for $p = n-1$, we can focus on stronger relaxations and only consider, as relaxed criteria, $PS_1 1$ and $PS_1 X$.

5   Relations Between Fairness Criteria

We compare in this section the PS criteria with the classical fairness criteria of the literature given in Sect. 2.2. We will show that the ordered scale of fairness (1), completed with known relaxations of envy-freeness and proportionality, can be connected with the PS criteria as shown in Fig. 1.

Surprisingly, the strongest requirement CEEI in the fairness scale (1) does not even imply PS_{n-1} or PS_11, which are among the least demanding PS criteria.

Proposition 7. *CEEI $\not\Rightarrow$ PS_{n-1} for any number of agents n and CEEI $\not\Rightarrow$ $PS_1 1$.*

Proof. Consider an instance where $m = 2n$, and two integers α and β such that $\beta n > \alpha > \beta(n-1)$ and $\beta > 1$. The utilities are such that $u_i(x_i) = \alpha$ and $u_i(x_j) = 0$ for every index $j \neq i$ and every agent $i \in [n-2]$, and $u_{n-1}(x_{n-1}) = u_{n-1}(x_{2n}) = \alpha$ and $u_{n-1}(x) = 0$ for every object $x \in M \setminus \{x_{n-1}, x_{2n}\}$. The utility function of agent n is such that $u_n(x_j) = \alpha$ for every index $j \in [n-1]$, $u_n(x_j) = \beta$ for every index $j \in \{n, \ldots, 2n-1\}$, and $u_n(x_{2n}) = 0$. Let us denote by σ the allocation assigning object x_i to every agent $i \in [n-2]$, the bundle of objects $\{x_{n-1}, x_{2n}\}$ to agent $n-1$ and the bundle $\{x_n, x_{n+1}, \ldots, x_{2n-1}\}$ to agent n. Observe that allocation σ is CEEI w.r.t. price vector pr given by $pr_i = 1$ for every $i \in [n-2]$, $pr_{n-1} = \frac{n}{n+1}$, $pr_i = \frac{1}{n}$ for every $i \in \{n, n+1, \ldots, 2n-1\}$ and $pr_{2n} = \frac{1}{n+1}$. However, allocation σ is not PS_{n-1} because $u_n(\sigma(n)) = \beta n < \alpha + \beta = u_n(o_n^{n-1}) + u_n(o_n^{2n-1}) = ps_{n-1}(n)$, and thus is not PS_p for any position $p \in [n-1]$. For $PS_1 1$, it suffices to remark that the encircled allocation in the proof of Proposition 3 is CEEI w.r.t. price vector $pr = (1, 0.75, 0.6, 0.2, 0.2, 0.25)$ but not $PS_1 1$. □

It follows that none of the criteria of the scale of fairness implies PS_p when $p < n$, meaning that PS_p is not always "weaker" than any criterion of the scale of fairness. However, all criteria of the scale of fairness imply the PS_n criterion.

Proposition 8. *MMS $\Rightarrow$ PS_n.*

Proof. We prove that $mms_i \geq ps_n(i)$ for every agent i. Take an allocation where the ℓth bundle ($1 \leq \ell \leq n$) gathers all the $(\ell + kn)^{\text{th}}$ most preferred objects of agent i for $0 \leq k < q$. The nth bundle, whose value is $ps_n(i)$, is the least preferred. Thus, $mms_i \geq ps_n(i)$ because mms_i is agent i's maximum value for the worst bundle for every possible allocation. □

Conversely, for $p > 1$, the PS criteria are not stronger than any classical criterion either. Indeed, PS_2 does not imply MMS, the least demanding criterion in the fairness scale (1), or Prop1, thus no PS criterion with $p > 1$ does.

Proposition 9. *1. $PS_2 \not\Rightarrow$ MMS for any n,*
2. $PS_2 \not\Rightarrow$ Prop1, even under strict preferences on the objects,
3. $PS_2 \not\Rightarrow$ Propc for large enough m and any c.

Sketch of Proof. We only present case 2. here. Consider an instance where $n = 3$ and $m = 6$. The utilities are:

$$\begin{pmatrix} \text{㉔} & 16 & 15 & 14 & 8 & 7 \\ 1 & ② & ③ & ④ & 5 & 6 \\ 2 & 4 & 6 & 8 & ⑩ & ⑫ \end{pmatrix}$$

The encircled allocation σ is PS_2 since $ps_2 = (24, 7, 14)$. However, σ is not Prop1 because $u_1(\sigma(1)) + \max_{x \notin \sigma(1)} u_1(x) = 24 + 16 < 42 = u_1(M)/2$.   □

However, PS_1 implies the mFS criterion.

Proposition 10. $PS_1 \Rightarrow mFS$.

Proof. For every agent i, $ps_1(i) = \max_{\sigma \in \mathcal{A}'} \max_{j \in N} u_i(\sigma(j))$ where $\mathcal{A}' \subseteq \mathcal{A}$ is the set of allocations giving to each agent exactly one object within $\{o_i^{kn+1}, o_i^{kn+2}, \ldots, o_i^{kn+n}\}$ for each $0 \le k < m/n$. Thus, $ps_1(i) \ge \min_{\sigma \in \mathcal{A}'} \max_{j \in N} u_i(\sigma(j)) \ge \min_{\sigma \in \mathcal{A}} \max_{j \in N} u_i(\sigma(j)) = mfs_i$.   □

Nevertheless, PS_1 is not stronger than EF since it does not even imply EF1.

Proposition 11. $PS_1 \not\Rightarrow EF1$, even under strict preferences on the objects.

Proof. Consider an instance where $n = 4$ and $m = 12$. The utilities are:

$$\begin{pmatrix} ⑳ & 19 & 18 & 17 & ⑧ & 7 & 6 & 5 & ④ & 3 & 2 & 1 \\ 15 & ⑫ & ⑪ & ⑩ & 6 & 3 & 9 & 5 & 13 & 8 & 7 & 2 \\ 1 & 6 & 9 & 8 & 12 & ⑱ & ⑳ & ⑲ & 5 & 10 & 14 & 3 \\ 2 & 5 & 6 & 15 & 11 & 12 & 9 & 8 & 4 & ⑳ & ⑱ & ⑲ \end{pmatrix}$$

Allocation σ (circles), with $u(\sigma) = (32, 33, 57, 57)$, is PS_1 since $ps_1 = (32, 31, 38, 38)$. But agent 1 envies agent 2, even if any object is removed from $\sigma(2)$.   □

The fact that $PS_1 \not\Rightarrow EF1$ may look surprising since fair guarantees of PS criteria can be interpreted via RB sequences, which generate EF1 allocations. However, satisfying a PS criterion does not impose to be the outcome of an RB sequence but focuses on the fulfillment of associated minimum utility levels, that are personal to each agent and do not need the inter-comparison between agents.

In their relaxed versions, PS_1 and Prop remain connected.

Proposition 12. $PS_1c \Rightarrow Propc$.

Proof. Take an instance, an agent i, and a PS_1c allocation σ. Recall that, w.l.o.g., $m = qn$. There exists $X^i \subseteq M \setminus \sigma(i)$ such that $|X^i| \le c$ and $u_i(\sigma(i)) + u_i(X^i) \ge u_i(\{o_i^1, o_i^{1+n}, \ldots, o_i^{1+(q-1)n}\})$. Since $u_i(\{o_i^1, o_i^{1+n}, \ldots, o_i^{1+(q-1)n}\}) \ge u_i(M)/n$, we get that $u_i(\sigma(i)) + u_i(X^i) \ge u_i(M)/n$. Thus, σ satisfies Propc.   □

From Proposition 9, PS_n cannot imply Propc when the number of objects m is large enough. However, PS_n implies Propc when m is at most $(1 + c)n$.

Proposition 13. If $m \le (1 + c)n$, then $PS_n \Rightarrow Propc$.

Conversely, no "up to" relaxation of envy-freeness implies a PS criterion.

Proposition 14. EFX does not imply any PS criterion.

Proof. By Proposition 7, we have that EFX $\not\Rightarrow$ PS$_1$1. To prove that EFX $\not\Rightarrow$ PS$_n$, consider an instance where $n = 2$ and $m = 4$. The utilities are:

$$\begin{pmatrix} 12 & \circledS 8 & \boxed{12} & \boxed{8} \\ \boxed{\circledS 14} & 1 & \circledS 13 & 0 \end{pmatrix}$$

Allocation σ (circles) is EFX: agent 2 is not envious and agent 1 is not envious if one good is removed from $\sigma(2)$. However, σ is not PS$_n$ since $ps_n = (20, 13)$. $\square$

Proposition 2 states that EFX $\not\Rightarrow$ PropX though EFc $\Rightarrow$ Prop$c$. We know that PS$_1$1 $\Rightarrow$ Prop1 and it turns out that PS$_1$X $\Rightarrow$ PropX.

Proposition 15. *PS$_1$X $\Rightarrow$ PropX.*

Proof. Take an instance, an agent i, and a PS$_1$X allocation σ. Recall that, w.l.o.g., $m = qn$. We have $u_i(\sigma(i)) + u_i(x) \geq u_i(\{o_i^1, o_i^{1+n}, \ldots, o_i^{1+(q-1)n}\})$ for all $x \in M \setminus \sigma(i)$. Since $u_i(\{o_i^1, o_i^{1+n}, \ldots, o_i^{1+(q-1)n}\}) \geq u_i(M)/n$, $u_i(\sigma(i)) + u_i(x) \geq u_i(M)/n$ holds for all $x \in M \setminus \sigma(i)$. Hence, σ satisfies PropX. $\square$

EFX $\Rightarrow$ PS$_1$X cannot hold because EFX $\not\Rightarrow$ PropX and PS$_1$X $\Rightarrow$ PropX. Moreover, PS$_1$X $\Rightarrow$ EFX cannot hold because it would contradict Proposition 11.

Note that PS$_n$ gives a $\frac{1}{n}$-approximation for MMS, like EF1 [1]. The approximation ratio of a PS$_p$ allocation can be generalized to $\frac{n-p+1}{n}$ for all $p \in [n]$.

Proposition 16. $\forall (p, i) \in [n] \times N, ps_p(i) \geq \frac{n-p+1}{n} mms_i.$

6   Efficiency of Fair Allocations

It is long known that there is a tension between the two goals of computing efficient and fair allocations [14,26]. In this section we propose an empirical analysis of how the PS criteria go together with efficiency, and a comparison with the classical fairness criteria.

An allocation is *Pareto-efficient* if there is no other allocation σ' that Pareto-dominates it, i.e., such that $u_i(\sigma'(i)) \geq u_i(\sigma(i))$ for each agent i, and the inequality is strict for at least one agent. The *social welfare (SW)* is another efficiency measure. The *utilitarian SW* of allocation σ is equal to $\sum_{i \in N} u_i(\sigma(i))$, the *egalitarian SW* to $\min_{i \in N} u_i(\sigma(i))$, and the *Nash SW* to $\prod_{i \in N} u_i(\sigma(i))$.

It is easy to see that a Pareto-efficient allocation may not satisfy a PS criterion, and vice versa. Similarly, maximizing the social welfare and achieving PS fairness may be disconnected, as illustrated in the next example.

Example 2. Consider the instance given in the proof of Proposition 14. The allocation σ (circles), which is the unique allocation maximizing the utilitarian, egalitarian and Nash SW, is not even PS$_n$ since $u_1(\sigma(1)) = 16 < 20 = ps_n(1)$. It is surprising since an allocation maximizing the Nash SW is known to be EF1 and to provide a good approximation to MMS [15]. Alternatively, the unique PS$_1$ allocation (frames) is different and thus does not maximize any social welfare.

We study how often "fair" allocations are efficient. We run 1,000 instances with $n = 3$ agents where m ranges from 6 to 9 (adding dummy objects can make m a multiple of n). The valuations of the agents over the objects are integers between 0 and 100 generated following a uniform distribution and then normalized. All the n^m possible allocations are considered. We compare in Table 1 the percentage of allocations satisfying classical and PS fairness criteria, and the proportion of these allocations that are Pareto-efficient. Moreover, we compare in Fig. 2 the average of the utilitarian and egalitarian SW among the allocations satisfying a given fairness criterion (the behavior for Nash SW is similar).

In practice, the set of PS_1X allocations is almost the same as the set of PS_1 allocations, where PS_1 is very demanding, even though less than CEEI. A significant proportion of allocations are PS_n ($\approx 20\%$). This is less than the proportion of PS_11 allocations ($\approx 40\%$), and even significantly less than the proportion of Prop1 allocations ($\approx 70\%$). Among these 3 criteria which can be satisfied for any instance, our experiments show that PS_n is the most selective one.

Table 1. Percentage of allocations satisfying fairness criteria and percentage of these allocations that are Pareto-efficient for $n = 3$

m	% fair allocations				% Pareto-eff. alloc./fair alloc.			
	6	7	8	9	6	7	8	9
MMS	9.73	6.69	5.35	4.53	20.54	13.50	8.37	4.80
Prop	3.64	3.67	3.67	3.51	30.85	18.29	10.38	5.56
mFS	2.20	2.60	2.94	3.06	38.00	21.77	11.73	6.05
EF	0.83	0.86	0.82	0.77	51.15	32.44	20.22	11.67
CEEI	0.39	0.25	0.13	0.07	86.31	80.34	80.21	82.28
Prop1	79.28	72.99	68.50	64.49	7.94	4.43	2.33	1.19
PropX	4.87	4.5	4.37	4.13	23.02	14.72	8.58	4.71
EF1	18.65	17.04	15.42	14.21	13.65	7.90	4.30	2.22
EFX	2.85	2.33	1.88	1.66	29.03	18.85	12.25	7.11
PS_3	30.69	18.43	17.46	18.94	11.77	8.03	4.31	2.12
PS_2	4.01	4.27	5.42	3.73	28.38	16.76	8.25	5.61
PS_1	0.15	0.43	0.22	0.26	83.67	50.26	44.63	26.02
PS_11	46.76	49.53	36.48	34.60	9.16	5.13	2.97	1.51
PS_1X	0.16	0.50	0.26	0.30	82.36	41.72	33.52	22.14

We observe that the proportion of efficient allocations among fair allocations seems to be dependent on the number of fair allocations. In particular, the proportion of PS_1 allocations that are Pareto-efficient is superior to the same quantity for EF allocations, and the utilitarian SW for PS_1 is also better in average than the SW for EF. EF1 and PS_n, which are both always satisfiable, seem to be equivalent regarding efficiency.

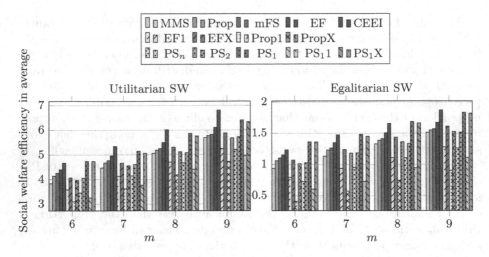

Fig. 2. Average of the utilitarian and egalitarian social welfares within the set of fair allocations for $n = 3$ and $m \in \{6, 7, 8, 9\}$

7 Conclusion

We have introduced some criteria whose fair guarantees are inspired by picking sequences where every agent supposes that her turns are the multiples of p. By the simplicity of the picking sequence protocol and the definition of the PS fair guarantee, PS criteria can be easily expressed by an agent as a fairness requirement. Moreover, even without knowing how the rest of goods is allocated, it is easy for an agent to check whether she can be satisfied with her own assigned bundle according to a PS criterion. The two extreme criteria (PS_1 and PS_n) are well connected with the ordered scale of fairness. More precisely, PS_1 implies mFS, MMS implies PS_n, whereas $PS_1 1$ and $PS_1 X$ imply Prop1 and PropX, respectively. In light of the fact that EFX $\not\Rightarrow$ PropX, the connection of relaxed PS_1 with relaxed proportionality is interesting and highlights that PS criteria and proportionality are conceptually close. We have proved that allocations satisfying PS_n and $PS_1 1$ can always be found whereas, as shown by our extensive comparative study of PS criteria with classical fairness criteria, these two criteria are far from being trivial. This positive result regarding the possibility of satisfaction is appealing since only a few known fairness criteria (EF1 and Prop1 among those studied here) are always satisfiable. The whole picture of existence and interactions between fairness criteria is depicted in Figure 1. By this picture combined with our experiments which explore the compatibility of the PS criteria with efficiency, our work contributes to the understanding of how the fairness criteria for allocating indivisible goods interact.

The fact that EFX $\not\Rightarrow$ PropX when $n > 2$ (Proposition 2) calls for the following less restrictive notion of envy-freeness: Allocation σ is *broadly* EFc (bEFc) iff for every pair of agents i and j, there exists $X^i \subseteq M \setminus \sigma(i)$ such that $|X^i| \leq c$ and $u_i(\sigma(i)) + u_i(X^i) \geq u_i(\sigma(j))$. Allocation σ is *broadly* EFX (bEFX) iff for every pair of agents i and j, $u_i(\sigma(i)) + u_i(x) \geq u_i(\sigma(j))$ holds for any $x \in M \setminus \sigma(i)$. Then, EFc $\Rightarrow$ bEFc $\Rightarrow$ Propc, bEFX $\Rightarrow$ EFX, and bEFX $\Rightarrow$ PropX hold. Obviously, EFX and bEFX coincide when $n = 2$.

Several research directions can be derived from this work. Alternative definitions of fair guarantees can be explored like what some agent i would get in a picking sequence with a *balanced alternation policy* $\langle 1, \ldots, n \mid n, \ldots, 1 \mid \ldots \rangle$. Another possibility would be to consider a probability distribution over the possible positions taken by the agent in round robin. By using a uniform distribution, we fall back to the definition of proportionality. One can also suppose that p in PS_p is not the same for all agents.

We have focused on a particular type of relaxation of a fairness criterion F, namely when F can be satisfied up to some good(s). Other relaxations can be studied. For example, F can be satisfied for any given subset of t *privileged* agents. Another type of relaxation consists of satisfying F up to a multiplicative factor. We know that PS_n can always be satisfied (Proposition 5) but when $p < n$, no positive α guarantees the existence of an allocation σ such that $u_i(\sigma(i)) \geq \alpha ps_p(i)$ for all $i \in N$.

References

1. Amanatidis, G., Birmpas, G., Markakis, E.: Comparing approximate relaxations of envy-freeness. In: Proceedings of IJCAI'18, pp. 42–48 (2018)
2. Aziz, H., Bouveret, S., Lang, J., Mackenzie, S.: Complexity of manipulating sequential allocation. In: Proceedings of AAAI'17, pp. 328–334 (2017)
3. Aziz, H., Huang, X., Mattei, N., Segal-Halevi, E.: The constrained round robin algorithm for fair and efficient allocation. arXiv preprint arXiv:1908.00161 (2019)
4. Aziz, H., Walsh, T., Xia, L.: Possible and necessary allocations via sequential mechanisms. In: Proceedings of IJCAI'15, pp. 468–474 (2015)
5. Bogomolnaia, A., Moulin, H., Stong, R.: Guarantees in fair division: general or monotone preferences. arXiv preprint arXiv:1911.10009 (2019)
6. Bouveret, S., Chevaleyre, Y., Maudet, N.: Fair allocation of indivisible goods. In: Handbook of Computational Social Choice, chap. 12, pp. 284–310. Cambridge University Press, Cambridge (2016)
7. Bouveret, S., Lang, J.: A general elicitation-free protocol for allocating indivisible goods. In: Proceedings of IJCAI'11, pp. 73–78 (2011)
8. Bouveret, S., Lang, J.: Manipulating picking sequences. In: Proceedings of ECAI'14, pp. 141–146 (2014)
9. Bouveret, S., Lemaître, M.: Characterizing conflicts in fair division of indivisible goods using a scale of criteria. Auton. Agents Multi-Agent Syst. **30**(2), 259–290 (2015). https://doi.org/10.1007/s10458-015-9287-3
10. Brams, S.J., Edelman, P.H., Fishburn, P.C.: Fair division of indivisible items. Theory Decis. **55**(2), 147–180 (2003)
11. Brams, S.J., King, D.L.: Efficient fair division: help the worst off or avoid envy? Ration. Soc. **17**(4), 387–421 (2005)
12. Brams, S.J., Taylor, A.D.: The Win-Win Solution: Guaranteeing Fair Shares to Everybody. WW Norton & Company, New York (2000)

13. Budish, E.: The combinatorial assignment problem: approximate competitive equilibrium from equal incomes. J. Polit. Econ. **119**(6), 1061–1103 (2011)
14. Caragiannis, I., Kaklamanis, C., Kanellopoulos, P., Kyropoulou, M.: The efficiency of fair division. Theory Comput. Syst. **50**(4), 589–610 (2012)
15. Caragiannis, I., Kurokawa, D., Moulin, H., Procaccia, A.D., Shah, N., Wang, J.: The unreasonable fairness of maximum Nash welfare. ACM Trans. Econ. Comput. **7**(3), 12:1-12:32 (2019)
16. Chakraborty, M., Igarashi, A., Suksompong, W., Zick, Y.: Weighted envy-freeness in indivisible item allocation. In: Proceedings of AAMAS'20, pp. 231–239 (2020)
17. Conitzer, V., Freeman, R., Shah, N.: Fair public decision making. In: Proceedings of EC'17, pp. 629–646 (2017)
18. Foley, D.K.: Resource allocation and the public sector. Yale Econ. Essays **7**(1), 45–98 (1967)
19. Kalinowski, T., Narodytska, N., Walsh, T., Xia, L.: Strategic behavior when allocating indivisible goods sequentially. In: Proceedings of AAAI'13, pp. 452–458 (2013)
20. Kohler, D.A., Chandrasekaran, R.: A class of sequential games. Oper. Res. **19**(2), 270–277 (1971)
21. Kurokawa, D., Procaccia, A.D., Wang, J.: When can the maximin share guarantee be guaranteed? In: Proceedings of AAAI'16, pp. 523–529 (2016)
22. Lang, J., Rothe, J.: Fair division of indivisible goods. In: Rothe, J. (ed.) Economics and Computation. STBE, pp. 493–550. Springer, Heidelberg (2016). https://doi.org/10.1007/978-3-662-47904-9_8
23. Lipton, R.J., Markakis, E., Mossel, E., Saberi, A.: On approximately fair allocations of indivisible goods. In: Proceedings of EC'04, pp. 125–131 (2004)
24. Moulin, H.: Cooperative Microeconomics: A Game-Theoretic Introduction. Princeton University Press, Princeton (1995)
25. Moulin, H.: Fair division in the internet age. Ann. Rev. Econ. **11**(1), 407–441 (2019)
26. Procaccia, A.D.: Cake cutting algorithms. In: Handbook of Computational Social Choice, pp. 311–330. Cambridge University Press, Cambridge (2016)
27. Procaccia, A.D., Wang, J.: Fair enough: guaranteeing approximate maximin shares. In: Proceedings of EC'14, pp. 675–692 (2014)
28. Steinhaus, H.: The problem of fair division. Econometrica **16**(1), 101–104 (1948)
29. Tominaga, Y., Todo, T., Yokoo, M.: Manipulations in two-agent sequential allocation with random sequences. In: Proceedings of AAMAS'16, pp. 141–149 (2016)
30. Varian, H.R.: Equity, envy, and efficiency. J. Econ. Theory **9**(1), 63–91 (1974)

On Reachable Assignments in Cycles

Luis Müller and Matthias Bentert[✉]

TU Berlin, Algorithmics and Computational Complexity, Berlin, Germany
luis.mueller@campus.tu-berlin.de, matthias.bentert@tu-berlin.de

Abstract. The efficient and fair distribution of indivisible resources among agents is a common problem in the field of *Multi-Agent-Systems*. We consider a graph-based version of this problem called REACHABLE ASSIGNMENT, introduced by Gourvès, Lesca, and Wilczynski [IJCAI, 2017]. The input for this problem consists of a set of agents, a set of objects, the agent's preferences over the objects, a graph with the agents as vertices and edges encoding which agents can trade resources with each other, and an initial and a target distribution of the objects, where each agent owns exactly one object in each distribution. The question is then whether the target distribution is reachable via a sequence of rational trades. A trade is rational when the two participating agents are neighbors in the graph and both obtain an object they prefer over the object they previously held. We show that REACHABLE ASSIGNMENT is solvable in $\mathcal{O}(n^3)$ time when the input graph is a cycle with n vertices.

Keywords: Multi-Agent Systems · Resource allocation · Polynomial-time algorithm · Reduction to 2-SAT

1 Introduction

The efficient distribution of resources among agents is a frequent problem in *Multi-Agent-Systems* [6] with e. g. medical applications [1]. These resources are often modeled as objects and sometimes they can be divided among agents and sometimes they are indivisible. One famous problem in this field is called HOUSING MARKET: Each of the n participating agents initially owns one house (an indivisible object) and the agents can trade their houses in trading cycles with other agents [15,17]. Versions of this problem were considered with different optimization criteria like Pareto-Optimality [2,14,18] or envy-freeness [5,7,9]. Gourvès et al. [11] studied two similar problems where agents are only able to perform (pairwise) trades with agents they trust. This is modeled by a social network of the participating agents where an edge between two agents means that they trust each other. The first version is called REACHABLE ASSIGNMENT: Can one reach a given target assignment by a sequence of rational swaps? A swap is rational if both participating agents obtain an object they prefer over their current object and the agents share an edge in the social network. The second version is called REACHABLE OBJECT and the question is whether there

© Springer Nature Switzerland AG 2021
D. Fotakis and D. Ríos Insua (Eds.): ADT 2021, LNAI 13023, pp. 273–288, 2021.
https://doi.org/10.1007/978-3-030-87756-9_18

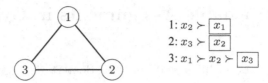

Fig. 1. Example for REACHABLE ASSIGNMENT on a triangle with preference lists on the right-hand side. We use the notation "1: $x_2 \succ \boxed{x_1}$" to denote that agent 1 prefers object x_2 the most and object x_1 the second most. Moreover, agent 1 initially holds object x_1 and since we only consider rational swaps and hence agent x_1 will never hold an object it prefers less than object x_1, we do not list these objects for agent 1. In the target assignment, each agent shall hold its most preferred object. First, agents 2 and 3 can swap their currently held objects. Afterwards, agent 1 can trade object x_1 to agent 3 and receive object x_2 in return.

is a sequence of rational swaps such that a given agent obtains a given target object. Figure 1 displays an example of REACHABLE ASSIGNMENT.

Gourvès et al. [11] showed that REACHABLE ASSIGNMENT and REACHABLE OBJECT are both NP-hard on general graphs. They further proved that REACHABLE ASSIGNMENT is decidable in polynomial time if G is a tree. Huang and Xiao [13] showed that if the underlying graph is a path, then REACHABLE OBJECT can be solved in polynomial time. Moreover, they studied a version of REACHABLE OBJECT that allows weak preference lists, i.e. an agent can be indifferent between different objects and show that this problem is NP-hard even if the input graph is a path. Contributing to the REACHABLE OBJECT problem, Saffidine and Wilczynski [16] proposed an alternative version of REACHABLE OBJECT, called GUARANTEED LEVEL OF SATISFACTION, where an agent is guaranteed to obtain an object at least as good as a given target object. They showed that GUARANTEED LEVEL OF SATISFACTION is co-NP-hard. Finally, REACHABLE OBJECT is polynomial-time solvable if the input graph is a cycle and NP-hard if the input graph is a clique, that is, all agents can trade with each other [3].

In our work, we focus on REACHABLE ASSIGNMENT for the special cases where the input graph is a cycle and provide an $\mathcal{O}(n^3)$-time algorithm. The algorithm is shown in three steps. In the first step, we will show that once an object is swapped into either clockwise or counter-clockwise direction, it is impossible to swap it back into the opposite direction. Moreover, assigning a direction to each object is equivalent to providing a sequence of rational swaps. We will also show how to verify in polynomial time whether such an assignment corresponds to a solution. In this case, we will say that the assignment of directions *yields the target assignment*. In a second step, we show a characterization of assignments of directions that yield the target assignment. We call these direction assignments *valid*. In the third and final step, we will iterate over all edges in the input graph and construct for each iteration a 2-SAT formula that is satisfiable if and only if there exists a valid assignment of directions that corresponds to a solution in which the first swap is done over the iterated edge. Besides a novel characterization of instances that have a solution, our main technical contribution is a non-trivial reduction to 2-SAT. The approach of reducing a problem to 2-SAT for showing polynomial running times was used before [3,4,12,13], but our reduction is not based on any of the earlier ones and we believe that the potential of reducing to 2-SAT is still relatively unexplored. Due to space constraints, some formal proofs are omitted. Affected results are marked with a $(\star)$.

2   Preliminaries and Preprocessing

We use standard graph-theoretical notation similar to Diestel [10]. Let $G := (V, E)$ be a graph and let $W \subseteq V$ be a set of vertices. We use $G[W]$ to denote the induced subgraph of W in G, that is, the graph $G' := (W, E')$ where $\{v, w\} \in E'$ if and only if $\{v, w\} \in E$ and $v, w \in W$. If graphs H and K are two subgraphs of a graph G, then we denote with $H \cup K$ the induced graph $G[V(H) \cup V(K)]$ and with $H \cap K$ we denote the induced graph $G[V(H) \cap V(K)]$.

For two integers a and b we denote by $[a, b]$ the set of integers $\{a, a+1, ..., b\}$. If G is a cycle, then we always assume that the agents are numbered from 0 to $n-1$, where $n = |V|$, such that agent i shares an edge with agents $i+1 \bmod n$ and with agent $i-1 \bmod n$. We denote this cycle by C_n. We also say that agent $i+1 \bmod n$ is the clockwise neighbor of agent i and that agent $i-1 \bmod n$ is the counter-clockwise neighbor of agent i. We denote by $\mu_{a,b}$ the sequence of clockwise neighbors starting from a and ending in b, that is,

$$\mu_{a,b} = \begin{cases} (a, a+1, ..., n-1, 0, 1, ..., b) & \text{if } b < a \\ (a, a+1, ..., b) & \text{otherwise} \end{cases} \tag{1}$$

We now formally define REACHABLE ASSIGNMENT. Let N be a set of n agents and let X be a set of n indivisible objects. Each agent $i \in N$ has a *preference list* $\succ_i$ over a non-empty subset X_i of the set of objects X. Each preference list is a strict ordering on X_i. The set of all preference lists is called a *preference profile* $\succ$. A bijection $\sigma : N \to X$ is called an *allocation or assignment*. Akin to the problem HOUSING MARKET [2], each agent is initially assigned exactly one object. We denote this initial assignment by σ_0 and for the sake of simplicity, we will interchangeably use $\sigma(i) = o$ and $(i, o) \in \sigma$ for an agent i, an object o, and an assignment σ. Let $G = (N, E)$ be a graph where the set of vertices is the set N of agents. We will use the term agents interchangeably with vertices of G. A trade between agents i and j is only possible if their corresponding vertices share an edge in G and if both i and j receive an object that they prefer over the objects they hold in the current assignment σ. We can express this formally by $\sigma(i) \succ_j \sigma(j)$ and $\sigma(j) \succ_i \sigma(i)$ and we call such a trade a *rational swap*. A *sequence of rational swaps* is a sequence of assignments $(\sigma_s, ..., \sigma_t)$ where σ_i is the result of performing one rational swap in assignment σ_{i-1} for all $i \in \{s+1, s+2, ..., t\}$. We call an assignment σ *reachable* from σ_0 if there exists a sequence of rational swaps $(\sigma_0, ..., \sigma)$. Gourvès et al. [11] introduced the problem REACHABLE ASSIGNMENT as follows.

REACHABLE ASSIGNMENT

Input: A set N of agents, a set X of objects, a preference profile $\succ$, a
graph G, an initial assignment σ_0, and a target assignment σ.

Question: Is σ reachable from σ_0?

Recall that assignments are bijections. This allows us for any assignment σ,
any object y and the agent i with $\sigma(i) = y$ to denote i by $\sigma^{-1}(y)$.

We end this section with a simple data reduction rule for REACHABLE
ASSIGNMENT. Whenever an agent prefers an object q over the object p it is
assigned in the target assignment σ, then we can simply remove q from its preference lists. This is due to the fact that once the agent possesses q, then it cannot
receive p anymore. We will assume that every instance has already been prepro-
cessed by the following data reduction rule, which is equivalent to assuming
that σ assigns each agent its most preferred object.

Reduction Rule 1. *Let $\mathcal{I}$ be an instance of* REACHABLE ASSIGNMENT *with
preference profile $\succ$ and target assignment σ. If for any agent i with object $p :=
\sigma(i)$ there exists an object q such that $q \succ_i p$, then remove q from $\succ_i$.*

3   A Polynomial-Time Algorithm for REACHABLE ASSIGNMENT on Cycles

In this section we develop a polynomial-time algorithm for REACHABLE ASSIGN-
MENT on cycles. To the best of our knowledge, this is the first polynomial-time
algorithm for REACHABLE ASSIGNMENT beyond the initial algorithm for trees by
Gourvès et al. [11]. Our algorithm generalizes several ideas used in the algorithm
for trees. Note however, that a solution might need to swap objects over all edges
in the cycle and hence we cannot simply remove one of the edges and directly
use the algorithm by Gourvès et al. for trees. Moreover, there are up to $n/2$
swaps over a single edge in a sequence of swaps and we therefore cannot iterate over all $O(n^{n/2})$ possible swaps over one edge in polynomial-time. Thus,
our approach for solving REACHABLE ASSIGNMENT on cycles is new and uses
a novel characterization of solutions. We divide the approach into three parts
that are covered in the following three subsections. In Subsect. 3.1, we formally
define what we mean by swapping an object in a certain direction and provide
a polynomial-time algorithm to verify whether a given assignment of directions
to all objects corresponds to a solution. In this case, we will say that the assignment of directions *yields* σ. In Subsect. 3.2, we define a property we call *validity*
and show that this characterizes the assignments of directions yielding σ. Finally
in Subsect. 3.3, we reduce the problem of deciding whether there exists a valid
assignment of directions to 2-SAT.

3.1   Swapping Directions in a Cycle

In this subsection we will formally introduce assignments of directions.
We will refer to them as *selections* and always denote them by γ.

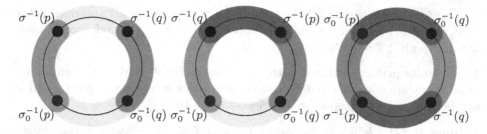

Fig. 2. Given a selection γ and two objects p and q with $\gamma(q) \neq \gamma(p) = 1$, the two marked paths in are the respective paths of p (red) and q (blue) for γ. Since they do not intersect in the left figure, the set $\xi_\gamma(p, q)$ of shared paths of p and q for γ is empty. In the center figure $\xi_\gamma(p, q)$ contains a single intersection between $\sigma^{-1}(q)$ and $\sigma^{-1}(p)$ (the purple path) and in the right figure it contains the two intersections in the top and in the bottom. The two objects are opposite in the left and the right figure, but not in the center. (Color figure online)

Let $\mathcal{I} := (N, X, \succ, G, \sigma_0, \sigma)$ be an instance of REACHABLE ASSIGNMENT. Let p be an object, let $j = \sigma^{-1}(p)$, and let $i = \sigma_0^{-1}(p)$. Since the underlying graph is a cycle, there are exactly two paths between agents i and j for object p. By definition of rational swaps, once p has been swapped, say from agent i to agent $i + 1 \bmod n$ then p is not able to return to agent i since agent i just received an object that it prefers over p and will therefore not accept p again. Hence, if p is swapped again, then it is given to agent $i + 2 \bmod n$ and the argument can be repeated for agent $i + 1 \bmod n$. As there are only two paths between agents i and j, there are also only two directions, namely clockwise and counter-clockwise. We will henceforth encode these directions into a binary number saying that the direction of p is 1 if p is swapped in clockwise direction and 0 otherwise. A *selection* γ of $\mathcal{I}$ is a function that assigns each object $p \in X$ a direction $\gamma(p) \in \{0, 1\}$.

We say that an object p is *closer* to ℓ in direction d than another object q, if starting from ℓ and going in direction d, one finds object p before one finds object q. Therein, ℓ can be an edge, an agent, or a third object.

We will see that for a pair of objects (with assigned directions) there is a unique edge over which they can be swapped. To show this, we need the following definition of *a path of an object*. These paths consist of all agents that will hold the respective object in any successful sequence of swaps that respects the selection.

Definition 1. *Let γ be a selection for instance $\mathcal{I}$ and let p and q be two objects such that $\gamma(p) \neq \gamma(q)$. Let $i := \sigma_0^{-1}(p)$, let $j = \sigma^{-1}(p)$, let $I := \mu_{i,j}$, and let $J := \mu_{j,i}$. The path of$p$ for γ is*

$$P_\gamma(p) := \begin{cases} C_n[I] & \text{if } \gamma(p) = 1 \\ C_n[J] & \text{otherwise.} \end{cases}$$

The set $\xi_\gamma(p, q)$ of shared paths of p and q for γ is the set of all connected paths in $P_\gamma(p) \cap P_\gamma(q)$.[1] *Finally, p and q are* opposite *if there exists a selection γ' such that $|\xi_{\gamma'}(p, q)| \geq 2$ and $\gamma'(p) \neq \gamma'(q)$.*

Examples for paths and shared paths are given in Fig. 2. With these definitions at hand, we are now able to determine unique edges where two objects p and q can be swapped with respect to all selections γ with $\gamma(q) \neq \gamma(p) = 1$.

Lemma 1. *Let γ be a selection for an instance $\mathcal{I}$ and let p and q be two objects with $\gamma(p) \neq \gamma(q)$. If for some path $P \in \xi_\gamma(p, q)$ there is not exactly one edge $e = \{u, v\}$ in P such that the preference lists of u and v allow p and q to be swapped between agents u and v, then γ does not yield target assignment σ.*

Proof. Note that p and q have to be swapped somewhere on each path in $\xi_\gamma(p, q)$ as otherwise at least one of the objects cannot reach its target agent. For each path $P \in \xi_\gamma(p, q)$ there has to be at least one edge on P where p and q can be swapped or γ does not yield the target assignment σ. Assume towards a contradiction that there are at least two edges e and f on some path $P \in \xi_\gamma(p, q)$ where p and q can be swapped according to the corresponding preference lists and γ yields σ. Suppose p and q are swapped over e in a sequence of rational swaps that yields σ (the case for f is analogous). Then one of p or q must have already passed the other edge f, say p (again, the case for q is analogous). Since p and q could be swapped over f, it holds for the two agents a and b incident to f that $(p \succ_a q)$ and $(q \succ_b p)$. Since p already passed f, agent a already held p. Hence, agent a will not accept object q in the future and hence q can not reach its destination as a is on the path of q. This contradicts the assumption that γ yields σ.   □

We next show that two objects p and q are swapped exactly once on each path in $\xi_\gamma(p, q)$ in any sequence of rational swaps and that $|\xi_\gamma(p, q)| \leq 2$. Figure 2 depicts examples with $|\xi_\gamma(p, q)| \in \{0, 1, 2\}$.

Lemma 2. *Let γ be a selection and let p and q be two objects with $\gamma(q) \neq \gamma(p)$. There are at most two edges on $P_\gamma(p) \cup P_\gamma(q)$ such that p and q can only be swapped over these edges or for every selection γ' where $\gamma'(p) = \gamma(p)$ and $\gamma'(q) = \gamma(q)$, γ' does not yield a target assignment σ. Each of these edges is on a different path in $\xi_\gamma(p, q)$.*

Proof. Let γ be any selection such that $\gamma(q) \neq \gamma(p)$. Note that p and q cannot be swapped over any edge that is not on a path in $\xi_\gamma(p, q)$ and that by Lemma 1 they have to be swapped over a specific edge for each path in $\xi_\gamma(p, q)$. Hence the number of edges where p and q can be swapped over in any sequence of rational swaps that yield σ and that respects γ is equal to $|\xi_\gamma(p, q)|$.

Now assume that $|\xi_\gamma(p, q)| \geq 3$. Then, p and q are swapped at least thrice. For p and q to perform a first swap, both of them must have passed at least one agent each. Observe that if p and q are swapped for a second time, then

---

[1] If $\{\sigma_0^{-1}(p), \sigma^{-1}(p)\} \in E \cap P_\gamma(q)$, then $\xi_\gamma(p, q)$ contains two disjoint paths, one with $\sigma_0^{-1}(p)$ as endpoint and one with $\sigma^{-1}(p)$.

each agent has held p or q between the first and the second swap of p and q. Hence, after the second swap, they must have passed at least $n + 2$ agents combined. Repeating this argument once again, we get that they must have passed passed $2n + 2$ agents after the third swap. This means that at least one of the agents has passed more than n agents, a contradiction to the fact that an agent does not accept an object once the agent traded that object away. Thus, two objects can only be swapped twice and we can use Lemma 1 to find at most two unique edges.   □

Observe that $\xi_\gamma(p, q)$ only depends on $\gamma(p)$ and $\gamma(q)$ and hence for each selection γ' with $\gamma'(p) = \gamma(p) \neq \gamma'(q) = \gamma(q)$ holds $\xi_\gamma(p, q) = \xi_{\gamma'}(p, q)$ and the edges specified in Lemma 1 are the same for γ and γ'. We will denote the set of edges specified in Lemma 1 by $E_\gamma(p, q)$. If $\gamma(p) = \gamma(q)$, then we define $E_\gamma(p, q) := \emptyset$. Note that Lemma 2 also implies that if γ yields σ, then $|E_\gamma(p, q)| = |\xi_\gamma(p, q)| \leq 2$.

We next show that the order in which objects are swapped is irrelevant once a selection is fixed. We use the following definition to describe an algorithm that checks in polynomial time whether a selection yields σ.

Definition 2. Let $\mathcal{I} := (N, X, \succ, C_n, \sigma_0, \sigma)$ be an instance of REACHABLE ASSIGNMENT and let γ be selection of $\mathcal{I}$. Let p and q be two objects with $\gamma(q) \neq \gamma(p) = 1$. Let i be the agent currently holding p. If q is held by agent $i + 1 \mod n$, then p and q are facing each other and if also $\sigma(i) \neq p$ and $\sigma(i + 1) \neq q$, then p and q are in swap position.

Using the notion of swap positions, we are finally able to describe a polynomial-time algorithm that decides whether a given selection yields σ. We refer to our algorithm as `Greedy Swap` and it is a generalization of the polynomial-time algorithm for REACHABLE ASSIGNMENT on trees by Gourvès et al. [11]. `Greedy Swap` arbitrarily swaps any pair of objects that is in swap position until no such pair is left. If σ is reached in the end, then it returns true and otherwise it returns false.

Proposition 1. Let $\mathcal{I} := (N, X, \succ, C_n, \sigma_0, \sigma)$ be an instance of REACHABLE ASSIGNMENT and let γ be a selection of $\mathcal{I}$. `Greedy Swap` returns true if and only if γ yields σ.

Proof. Observe that `Greedy Swap` only performs rational swaps and only returns true if σ is reached. Hence, if it returns true, then γ yields σ.

Now suppose that a selection γ yields assignment σ, but `Greedy Swap` returns false. Then either $\sigma_0 \neq \sigma$ and every object is assigned the same direction, or there are two objects p and q, which are in swap position at some edge e, but the corresponding preference lists do not allow a swap. In the former case γ clearly does not yield σ. In the latter case, consider the initial positions of p and q and the corresponding shared paths $\xi_\gamma(p, q)$. Since p and q are in swap position at the point where `Greedy Swap` returns false, it holds that $\gamma(p) \neq \gamma(q)$ and there is a shared path $P \in |\xi_\gamma(p, q)|$ such that e is on P.

If p and q face each other at edge e for the first time, then let m be the number of objects that are closer to q than p and that are assigned direction $\gamma(p)$. If p

and q face each other at edge e for the second time, then let m be the number of objects that are assigned direction $\gamma(p)$. We will now show that q has to be swapped with exactly $m - 1$ objects[2] before q and p can be swapped (for the first or between the first and the second swap, respectively). We will only show the case for p and q being swapped over e for the first time as the other case is analogous when considering the instance after the first swap of p and q has just happened. Starting from the agent that initially holds q we can then calculate the edge e' where q and p face each other in every sequence of rational swaps that results in reaching the target assignment σ. Suppose q can be swapped with more than $m - 1$ objects before facing p at edge e'. Then q must either be swapped with at least one object q' that is not closer to q than p or it must be swapped with the same object q' twice before being swapped with p for the first time. This, however, means that q will face q' after p, a contradiction. Now suppose that q can be swapped with less than $m - 1$ objects before facing p. Then there is an object r that is closer to q than p and that is assigned direction $\gamma(p)$ and which is not swapped with q before q and p are swapped. However, then q will not be able to face p as at least r is between them, a contradiction.

Now that we have shown that q will swap exactly m objects before facing p, we can also determine the edge f where, given γ, they must necessarily face each other, in every sequence of rational swaps where objects are swapped according to the directions assigned to them by γ. Now since we assumed that σ is reachable with selection γ, there must be a sequence of rational swaps where q and p must be swapped at edge f. But since p and q met, by assumption, at edge e, either $e = f$, which is a contradiction because at e the preference lists of the incident agents do not allow a swap between p and q, or p and q cannot have met at edge e if all swaps were performed according to the directions assigned by selection γ, a contradiction to the definition of **Greedy Swaps**.   □

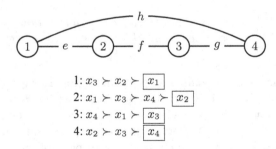

Fig. 3. Example for REACHABLE ASSIGNMENT on a C_4 with edges e, f, g, h. The candidate lists of x_1 are $C(x_1, e) := \{x_2, x_3\}$, $C(x_1, h) := \emptyset$, $C(x_1, g) := \emptyset$, and $C(x_1, f) := \{x_4\}$. Note that since $\sigma^{-1}(x_1) = 2$, the candidate list $C(x_1, e)$ is defined such that x_1 is swapped in counter-clockwise direction while the candidate list of x_1 for all other edges is defined for clockwise direction.

---

[2] Note that p is one of the m objects.

3.2   Validity of Selections

We will next define a set of properties that a *selection* needs to fulfill in order for Greedy Swap to return true and then show that these properties are both necessary and sufficient. The idea is to have one property that guarantees that for each object and for each edge on its path there exists exactly one object that should be swapped with that object over this edge and a second property to guarantee that for each two objects p and q there are exactly $|\xi_\gamma(p,q)|$ edges where p and q should be swapped. Before we can formally state these properties, we define some notation.

For two objects p and q, let $\mathcal{Z}(p,q)$ be the set of edges e such that p and q can be swapped over $e := \{i,j\}$ in the direction from their initial to their final agent. Let $H := X \times E(C_n)$ be the set of all *object-edge-pairs* and $d : H \to \{0,1\}$ be a function that assign to each object-edge-pair $(p,e) \in H$ the direction c such that e lies on the path of p from its initial agent to its destination in direction c. Using the notion of object-edge-pairs, we define *candidate lists*. These will be used to eliminate possible choices for selections.

Definition 3. *Let $(p,e) \in H$ be an object-edge-pair and let γ be a selection. The candidate list $C(p,e)$ of p at edge e and the size $f_\gamma(p,e)$ of $C(p,e)$ with respect to γ are*

$$C(p,e) := \{q \in X \mid e \in \mathcal{Z}(p,q)\} \ and$$

$$f_\gamma(p,e) := \begin{cases} |\{q \in C(p,e) \mid \gamma(p) \neq \gamma(q)\}| & if \ d(p,e) = \gamma(p) \\ 1 & otherwise. \end{cases}$$

Figure 3 presents an example of candidate lists. Note that the size of $C(p,e)$ with respect to γ is set to one if $d(p,e) \neq \gamma(p)$. This is due to the fact that we will later search for a selection γ^* such that $f_{\gamma^*}(p,e) = 1$ for all objects p and all edges e with $d(p,e) = \gamma^*(p)$. This definition then avoids a case distinction. The following observation follows from Lemma 2 and the observation that $E_\gamma(p,q) = E_{\gamma'}(p,q)$ for all γ' with $\gamma(p) = \gamma'(p)$ and $\gamma(q) = \gamma'(q)$ with a simple counting argument.

Observation 1. *For each $p \in X$, it holds that $\sum_{e \in E} |C(p,e)| \leq 4|X|$.*

We can finally define the first property called *exact* from the set of properties that characterizes the selections for which Greedy Swap returns true and which, by Proposition 1, correspond to a solution for REACHABLE ASSIGNMENT.

Definition 4. *Let $\mathcal{I} := (N, X, \succ, C_n, \sigma_0, \sigma)$ be an instance of* REACHABLE ASSIGNMENT. *A selection γ is* unambiguous *if for all $(p,e) \in H$ it holds that $f_\gamma(p,e) \leq 1$; γ is* complete *if for all $(p,e) \in H$ it holds that $f_\gamma(p,e) \geq 1$, and* exact *for all $(p,e) \in H$ if $f_\gamma(p,e) = 1$.*

Where exactness guarantees that for each object and for each edge on its path for γ there exists a possible swapping partner for this edge, the second property will guarantee that no pair of objects blocks one another in a selection. We start with the case in which an object blocks another object that is swapped in the same direction.

Definition 5. *Let γ be a selection and let p and q be two objects with $\gamma(p) = \gamma(q)$. Then q shields p in direction $\gamma(p)$ if $P_\gamma(p) \cap P_\gamma(q) \neq \emptyset$, the object q is closer to $\sigma^{-1}(p)$ than to p in direction $\gamma(p)$, and there exists an agent i on $P_\gamma(p) \cap P_\gamma(q)$ with $q \succ_i p$.*

In Fig. 3 the object x_1 shields x_4 in clockwise direction. Note that if q shields p in direction c, then p and q cannot be both swapped in direction c as the agent i obtains q first and then will not accept p afterwards. However, by definition, i is on the path of p and hence has to obtain p or the target assignment σ cannot be reached.

We continue with the case where an object blocks another object that is swapped in the opposite direction. Based on Lemma 1, we define compatibility of two objects.

Definition 6. *Let γ be a selection and let p and q be two objects with $\gamma(p) \neq \gamma(q)$. Then, p and q are compatible in selection γ if for each $P \in \xi_\gamma(p, q)$ there exists exactly one edge on P that is also contained in $E_\gamma(p, q)$.*

Based on Definitions 5 and 6 we formalize the second of the properties that characterize valid selections as follows.

Definition 7. *A selection γ is harmonic if for every object p there is no object q moving in direction $\gamma(p)$ that shields p in that direction, and every object r with $\gamma(r) = 1 - \gamma(p)$ is compatible with p. A selection is valid, if it is both exact and harmonic.*

We end this subsection with the statement that valid selections are exactly the selections that correspond to a solution of REACHABLE ASSIGNMENT.

Proposition 2 ($\star$). *Let $\mathcal{I} := (N, X, \succ, C_n, \sigma_0, \sigma)$ be an instance of REACH-ABLE ASSIGNMENT and let γ be a selection for $\mathcal{I}$. Greedy Swap returns true on input γ if and only if γ is valid.*

3.3   Reduction to 2-SAT

In this subsection we will construct a 2-SAT formula ϕ such that a selection is valid if and only if it corresponds to a satisfying truth assignment of ϕ. We will use a variable for each object and say that setting this variable to true corresponds to swapping this object in clockwise direction. We will use objects interchangeably with their respective variables. The formula ϕ will be the conjunction of two subformulas ψ_h and ψ_e. The first subformula ψ_h corresponds to harmonic selections and is constructed as follows.

Construction 2. *Let $\mathcal{I} := (N, X, \succ, C_n, \sigma_0, \sigma)$ be an instance of REACHABLE ASSIGNMENT. For every pair p, q of objects for which there exists a direction c such that q shields p in direction c, we add the clause $p \rightarrow \neg q$[3] if $c = 1$,*

---

[3] In this paper, we represent 2-SAT clauses by implications, equalities, and inequalities. Note that $p \rightarrow q \equiv (\neg p \lor q)$, $p = q \equiv (p \lor q) \land (\neg p \lor \neg q)$, and $p \neq q \equiv (p \lor \neg q) \land (\neg p \lor q)$.

and $\neg p \rightarrow q$ otherwise. Further, for every pair p, q of objects for which there exists a direction c such that p and q are not compatible for any selection γ where $c = \gamma(p) \neq \gamma(q)$, we add the clause $p \rightarrow q$ if $c = 1$, and $\neg p \rightarrow \neg q$ otherwise. We use ψ_h to denote the constructed formula.

Each satisfying truth assignment of ψ_h corresponds to a harmonic selection.

Lemma 3. Let $\mathcal{I} := (N, X, \succ, C_n, \sigma_0, \sigma)$ be an instance of REACHABLE ASSIGNMENT. A selection γ is harmonic if and only if it corresponds to a satisfying truth assignment of ψ_h.

Proof. We will prove both directions of the statement by contradiction. For the first direction suppose that γ is a harmonic selection but it does not correspond to a satisfying truth assignment of ψ_h. Then there must exist a clause c in ψ_h such that c is not satisfied by the truth assignment corresponding to γ. We distinguish between two cases. In the first case, c is a clause for q shielding p in direction c_p for every selection γ' where $\gamma'(p) = c_p = \gamma'(q)$ but it holds that $\gamma(p) = c_p = \gamma(q)$. But then, by definition, γ is not harmonic, a contradiction. In the second case, c is a clause for two non-compatible objects p and q for every selection γ where $\gamma(p) = c_p = 1 - \gamma(q)$ but it holds that $\gamma(p) = c_p = 1 - \gamma(q)$, a contradiction to γ being harmonic.

We will now show the other direction of the statement. Suppose that γ corresponds to a satisfying truth assignment of ψ_h but it γ is not harmonic. We again distinguish between two cases. In the first case there exist two objects p, q and a direction $c_p \in \{0, 1\}$ such that q shields p in direction c_p for γ where $\gamma(p) = c_p = \gamma(q)$. But since γ corresponds to a satisfying truth assignment of ψ_h, due to the clauses for shielding objects, if $\gamma(p) = c_p$, then $\gamma(q) = 1 - c_p$ and thus, p does not shield q, a contradiction.

In the second case there exist two objects p and q and a direction $c_p \in \{0, 1\}$ such that p and q are not compatible in direction c_p for each selection γ' where $\gamma'(p) = c_p$ and $\gamma'(q) = 1 - c_p$. Since γ corresponds to a satisfying truth assignment of ψ_h, due to the clause for compatibility, if $\gamma(p) = c_p$, then $\gamma(q) = c_p$ and thus, p and q are compatible, a contradiction. □

A formula for exact selections requires much more work. To this end, we first introduce a variant of REACHABLE ASSIGNMENT.

FIRST SWAP REACHABLE ASSIGNMENT
Input: An instance $\mathcal{I} := (N, X, \succ, C_n, \sigma_0, \sigma)$ of REACHABLE ASSIGN-
         MENT and an edge $e \in E(C_n)$
Question: Is σ reachable if the first swap is performed over edge e?

Observe that $\mathcal{I}$ is a yes-instance of REACHABLE ASSIGNMENT if and only if there is an edge e in C_n such that $(\mathcal{I}, e)$ is a yes-instance of FIRST SWAP REACHABLE ASSIGNMENT. We will later iterate over all edges and check for each whether it yields a solution. Suppose that some instance $(\mathcal{I}, e_i)$ of FIRST SWAP REACHABLE ASSIGNMENT is a yes-instance. Let $e_i = \{i, i + 1 \mod n\}$ and $j = i + 1 \mod n$. Then, the object $\sigma_0(i)$ is swapped into clockwise direction and the object $\sigma_0(j)$ is swapped into counter-clockwise direction. Thus, in every selection γ that yields the target assignment σ as a solution to $(\mathcal{I}, e_i)$, it must hold that $\gamma(\sigma_0(i)) = 1$ and $\gamma(\sigma_0(j)) = 0$. We refer to the pair of objects $x := \sigma_0(i)$ and $y := \sigma_0(j)$ as the *guess* of that instance and denote it $\Phi := (x, y)$, where x refers to the object that is swapped into clockwise direction and y refers to the object that is swapped into counter-clockwise direction. For each selection γ where $\gamma(x) = 1 = 1 - \gamma(y)$, we say that γ *respects* the guess $\Phi := (x, y)$. We also denote the unique path of x in clockwise direction from its initial agent to its destination with P_x and the unique path of y in counter-clockwise direction from its initial agent to its destination with P_y. If we can prove that for a guess Φ and the corresponding instance of FIRST SWAP REACHABLE ASSIGNMENT there exists no valid selection that respects Φ, then we say that the guess Φ is *wrong*. We will now define a property of objects which cannot be swapped in one direction given a certain guess Φ.

Definition 8. *Let* $\mathcal{I} := ((N, X, \succ, C_n, \sigma_0, \sigma), e)$ *be an instance of* FIRST SWAP REACHABLE ASSIGNMENT *with guess* $\Phi := (x, y)$ *and let* p *be an object such that there exists a direction* d *such that there is no selection* γ *with* $\gamma(p) = d$ *that respects* Φ *and that is valid. Then* p *is* decided *in direction* $1 - d$.

We find that an object $p \notin \{x, y\}$ is decided by means of six rules listed below. An object p is decided if at least one of the rules applies and the correctness of these rules is omitted due to space constraints. Note that we also always obtain the direction in which an object is decided. Therein, q is an object in $\{x, y\}$ and e is some edge.

1. The objects p and q are opposite.
2. The object p occurs in none of q's candidate lists.
3. The object p occurs in a candidate list $C(q, e)$ of q together with another object r that is decided in direction $1 - d(p, e)$.
4. The agent $\sigma(p)$ is neither on P_x nor on P_y.
5. The object p starts between two decided and opposite objects.
6. The object p occurs in a candidate list $C(q, e)$ of q where every $q' \in C(q, e)$, such that $q' \neq p$, is decided.

We mention that there are a couple of additional technical conditions for the penultimate rule and that there may be decided objects that are not found by either of the six rules. For the sake of simplicity, we say that an object is decided only if it fulfills at least one of the rules above. All other objects are said to be *undecided*. After performing these rules exhaustively, there are at most two undecided objects in each candidate list of the two guessed objects.

Proposition 3 ($\star$). *Let* $\Phi := (x, y)$ *be a guess. Let further* $(p, e) \in H$, *such that* $p \in \Phi$. *The number of undecided objects in* $C(p, e)$ *is either zero or two or* Φ *is wrong.*

The last ingredient for the construction of a 2-SAT formula for exact selection is given in the next lemma. From this we will generalize our notion of decided objects, which will also cover objects that are decided, relative to the direction of other objects. The lemma is quite similar to Lemma 1, but where Lemma 1 states that there is a unique edge where two objects *can be swapped*, the following lemma states that there is a unique edge where the two objects *face each other* (Defintion 2).

Lemma 4 ($\star$). *Let $\mathcal{I} := ((N, X, \succ, C_n, \sigma_0, \sigma), e')$ be an instance of* FIRST SWAP REACHABLE ASSIGNMENT. *Let $\Phi := (x, y)$ be the guess of $\mathcal{I}$, let $(p, e) \in H$ be an object-edge-pair and let $q \in C(p, e)$ where q is an undecided object. Let P be the shared path of p and q such that $e \in E(P)$. After an $O(n^3)$-time preprocessing, we can compute in constant time an edge f on P such that for every valid selection γ that respects Φ and for which $\gamma(q) \neq \gamma(p) = d(p, e)$ it holds that p and q face each other at edge f in the execution of* Greedy Swap *on input γ.*

We use Lemma 4 to define successful tuples. Intuitively, a tuple is successful if the edge where two objects face each other according to Lemma 4 is the edge over which they can be swapped according to Lemma 1.

Definition 9. *Let $(p, e) \in H$ be an object-edge-pair and let $q \in C(p, e)$ such that q is undecided. Let P be their shared path with $e \in E(P)$. Let f be the edge according to Lemma 4 where p and q face each other in every valid selection γ that respects Φ and for which it holds that $\gamma(q) \neq \gamma(p) = d(p, e)$. If q is closer to p in direction $d(p, e)$ than guessed object $\Phi_{d(p,e)}$, then let $c := 1$ and $c := 0$ otherwise. Then, $S(p, e, q)$ denotes the tuple (f, c) and it is successful if $e = f$ and unsuccessful otherwise.*

We will now introduce conditionally decided objects and construct the 2-SAT formula afterwards. Let $C(p, e)$ be a candidate list and let $q \in C(p, e)$ be an object such that there exists a direction d such that there is no selection γ with $\gamma(p) = d(p, e)$ and $\gamma(q) = d$ that respects Φ and that is valid. Then q is *p-decided* in direction $1 - d$.

Observe that the class of decided objects is a sub-class of p-decided objects. Analogously to decided objects, we present rules to find conditionally decided objects and say that objects are conditionally undecided if none of the rules apply for them. Let $C(p, e)$ be a candidate list, let $q \in C(p, e)$ be an object, and let d be the direction such that e is on the path of p in direction d. Then q is p-decided in direction $1 - d$ if one of the following rules apply:

1. The object q shields p in direction $d(p, e)$.
2. The tuple $S(p, e, q)$ is unsuccessful.
3. There exists an object q' with $q \neq q' \in C(p, e)$, such that q' is decided in direction $1 - d(p, e)$.

The proof of correctness is again omitted.

We now construct a 2-SAT formula for exact selections using decided, conditionally decided and undecided objects.

Construction 3. *Let* $\mathcal{I} := ((N, X, \succ, C_n, \sigma_0, \sigma), e')$ *be an instance of* FIRST SWAP REACHABLE ASSIGNMENT, *let* $\Phi := (x, y)$ *be the guess of* $\mathcal{I}$, *and let* $C(p, e) \in H$ *be a candidate list.*

If the number of p-decided objects in direction $1 - d(p, e)$ is not one and there exists no p-undecided object in $C(p, e)$, then we add the clause[4]

$$(p \neq d(p, e)). \tag{2}$$

For every decided object q in direction d we add the clause

$$q = d. \tag{3}$$

Further, for every p-decided object q in direction d that is not also a decided object, we add the clause

$$(p = d(p, e)) \rightarrow (q = d). \tag{4}$$

Lastly, if $p \in \{x, y\}$ and there exists an undecided object $q \in C(p, e)$, then we distinguish between two cases. If there exists exactly one other object $q' \in C(p, e)$ that is undecided, then we add the clause

$$q \neq q'. \tag{5}$$

Otherwise we add the the clause

$$\bot, \tag{6}$$

i.e. the clause that always evaluates to false, making the formula unsatisfiable. The constructed formula is $\phi = \psi_h \wedge \psi_e$.

The following proposition states that a selection is valid if and only if it corresponds to a satisfying truth assignment for ϕ.

Proposition 4 ($\star$). *Let* $\mathcal{I} := ((N, X, \succ, C_n, \sigma_0, \sigma), e')$ *be an instance of* FIRST SWAP REACHABLE ASSIGNMENT. *Let* $\Phi := (x, y)$ *be the guess of* $\mathcal{I}$ *and let* γ *be a selection that respects* Φ. *Then* γ *is valid if and only if it corresponds to a satisfying truth assignment of ϕ.*

The main theorem follows from this and Observation 1.

Theorem 4. REACHABLE ASSIGNMENT *on cycles is decidable in $\mathcal{O}(n^3)$ time.*

Proof. Let $\mathcal{I} := (N, X, \succ, C_n, \sigma_0, \sigma)$ be an instance of REACHABLE ASSIGNMENT. Given the instance, we compute the 2-SAT formula ψ_h according to Construction 2. We first precompute in overall $\mathcal{O}(n^3)$ time for each pair of objects whether they are opposite and for each possible assignment of directions for these two objects whether one object shields the other object and whether they are

---

[4] We use the notation $q \neq d(p, e)$ for some object q to avoid case distinctions. Since $d(p, e)$ is precomputed, this clause is equivalent to $\neg q$ if $d(p, e) = 1$ and q otherwise.

compatible. Constructing ψ_h then takes $\mathcal{O}(n^2)$ time as it requires to check for each pair p, q of objects whether they are compatible and whether one shields the other in any direction. Afterwards we divide $\mathcal{I}$ into n instances $(\mathcal{I}, e)$ of FIRST SWAP REACHABLE ASSIGNMENT. This further determines the guess $\Phi := (x, y)$. We then compute the set of all decided objects in $O(n^2)$ time. Moreover, we iterate over all objects p and compute for each possible direction d of p the set of all p-decided objects in $O(n)$ time. We then create a 2-SAT formula according to Construction 3 to check for an exact selection for the given instance of FIRST SWAP REACHABLE ASSIGNMENT.

Since correctness follows from Proposition 4, we focus on the running time here. For constructing ϕ, we add clauses for each object-edge-pair and due to Observation 1, there are $\mathcal{O}(n^2)$ such clauses. The time for computing each clause is constant and solving ϕ takes linear time in the number of clauses, that is, $O(n^2)$ time. Thus the procedure takes overall $\mathcal{O}(n^2)$ time per instance of FIRST SWAP REACHABLE ASSIGNMENT and $\mathcal{O}(n^3)$ time in total. □

4 Conclusion

In this work, we have investigated a version of HOUSING MARKET called REACHABLE ASSIGNMENT. This problem was first proposed by Gourvès et al. [11] and we presented an $\mathcal{O}(n^3)$-time algorithm for cycles.

The key to solving REACHABLE ASSIGNMENT on trees and on cycles was to exploit the number of unique paths an object can be swapped along. Finding graph classes in which this number is bounded and solving REACHABLE ASSIGNMENT for these graph classes is a natural next step for further research. Moreover, since cycles are paths with one additional edge, it seems promising to investigate graphs with constant feedback edge number or graphs of treewidth two next. Afterwards, one may study the parameterized complexity of REACHABLE ASSIGNMENT with respect to the parameters feedback edge number or treewidth. Other possibilities for parameters are related to the agent's preferences as studied by Bentert et al. [3] for REACHABLE OBJECT. One might also consider generalized settings such as allowing ties in the preference lists as studied by Huang and Xiao [13]. They showed that REACHABLE OBJECT is polynomial-time solvable on paths, but it is NP-hard on paths when ties in the preference lists are allowed.

Finally, we mention that Cechlárová and Schlotter [8] studied the parameterized complexity of a version of HOUSING MARKET that allows for approximation. Finding meaningful versions of REACHABLE ASSIGNMENT that allow for approximation is another alley for future research.

References

1. Abraham, D.J., Blum, A., Sandholm, T.: Clearing algorithms for barter exchange markets: enabling nationwide kidney exchanges. In: Proceedings of the 8th ACM Conference on Electronic Commerce (EC '07), pp. 295–304. ACM (2007)

2. Abraham, D.J., Cechlárová, K., Manlove, D.F., Mehlhorn, K.: Pareto optimality in house allocation problems. In: Deng, X., Du, D.-Z. (eds.) ISAAC 2005. LNCS, vol. 3827, pp. 1163–1175. Springer, Heidelberg (2005). https://doi.org/10.1007/11602613_115

3. Bentert, M., Chen, J., Froese, V., Woeginger, G.J.: Good things come to those who swap objects on paths. CoRR abs/1905.04219 (2019)

4. Bentert, M., Malík, J., Weller, M.: Tree containment with soft polytomies. In: Proceedings of the 16th Scandinavian Symposium and Workshops on Algorithm Theory (SWAT '18), LIPIcs, vol. 101, pp. 9:1–9:14. Schloss Dagstuhl - Leibniz-Zentrum für Informatik (2018)

5. Beynier, A., et al.: Local envy-freeness in house allocation problems. Auton. Agents Multi-Agent Syst. **33**(5), 591–627 (2019). https://doi.org/10.1007/s10458-019-09417-x

6. Brandt, F., Wilczynski, A.: On the convergence of swap dynamics to pareto-optimal matchings. In: Caragiannis, I., Mirrokni, V., Nikolova, E. (eds.) WINE 2019. LNCS, vol. 11920, pp. 100–113. Springer, Cham (2019). https://doi.org/10.1007/978-3-030-35389-6_8

7. Bredereck, R., Kaczmarczyk, A., Niedermeier, R.: Envy-free allocations respecting social networks. In: Proceedings of the 17th International Conference on Autonomous Agents and Multiagent Systems (AAMAS '18), pp. 283–291. International Foundation for Autonomous Agents and Multiagent Systems (ACM) (2018)

8. Cechlárová, K., Schlotter, I.: Computing the deficiency of housing markets with duplicate houses. In: Raman, V., Saurabh, S. (eds.) IPEC 2010. LNCS, vol. 6478, pp. 72–83. Springer, Heidelberg (2010). https://doi.org/10.1007/978-3-642-17493-3_9

9. Chevaleyre, Y., Endriss, U., Maudet, N.: Allocating goods on a graph to eliminate envy. In: Proceedings of the 22nd Conference on Artificial Intelligence (AAAI '07), pp. 700–705. AAAI Press (2007)

10. Diestel, R.: Graph Theory. Graduate Texts in Mathematics, vol. 173, 4th edn. Springer, Heidelberg (2012)

11. Gourvès, L., Lesca, J., Wilczynski, A.: Object allocation via swaps along a social network. In: Proceedings of the 26th International Joint Conference on Artificial Intelligence (IJCAI '17), pp. 213–219 (2017)

12. Gusfield, D., Wu, Y.: The three-state perfect phylogeny problem reduces to 2-SAT. Commun. Inf. Syst. **9**(4), 295–302 (2009)

13. Huang, S., Xiao, M.: Object reachability via swaps under strict and weak preferences. Auton. Agents Multi-Agent Syst. **34**(2), 1–33 (2020). https://doi.org/10.1007/s10458-020-09477-4

14. Igarashi, A., Peters, D.: Pareto-optimal allocation of indivisible goods with connectivity constraints. In: Proceedings of the 33rd AAAI Conference on Artificial Intelligence (AAAI '19), pp. 2045–2052. AAAI Press (2019)

15. Roth, A.E.: Incentive compatibility in a market with indivisible goods. Econ. Lett. **9**(2), 127–132 (1982)

16. Saffidine, A., Wilczynski, A.: Constrained swap dynamics over a social network in distributed resource reallocation. In: Deng, X. (ed.) SAGT 2018. LNCS, vol. 11059, pp. 213–225. Springer, Cham (2018). https://doi.org/10.1007/978-3-319-99660-8_19

17. Shapley, L., Scarf, H.: On cores and indivisibility. J. Math. Econ. **1**(1), 23–37 (1974)

18. Sönmez, T., Ünver, M.U.: House allocation with existing tenants: a characterization. Games Econ. Behav. **69**(2), 425–445 (2010)

Minimizing and Balancing Envy Among Agents Using Ordered Weighted Average

Parham Shams[1]([✉]), Aurélie Beynier[1], Sylvain Bouveret[2], and Nicolas Maudet[1]

[1] Sorbonne Université, CNRS, LIP6, 75005 Paris, France
{parham.shams,aurelie.beynier,nicolas.maudet}@lip6.fr
[2] Univ. Grenoble Alpes, CNRS, LIG, Grenoble, France
sylvain.bouveret@imag.fr

Abstract. In fair resource allocation, envy freeness (EF) is one of the most interesting fairness criteria as it ensures no agent prefers the bundle of another agent. However, when considering indivisible goods, an EF allocation may not exist. In this paper, we investigate a new relaxation of EF consisting in minimizing the Ordered Weighted Average (OWA) of the envy vector. The idea is to choose the allocation that is fair in the sense of the distribution of the envy among agents. The OWA aggregator is a well-known tool to express fairness in multiagent optimization. In this paper, we focus on fair OWA operators where the weights of the OWA are decreasing. When an EF allocation exists, minimizing OWA envy will return this allocation. However, when no EF allocation exists, one may wonder how fair min OWA envy allocations are.

After defining the model, we show how to formulate the computation of such a min OWA envy allocation as a Mixed Integer Program. Then, we investigate the link between the min OWA allocation and other well-known fairness measures such as max min share and EF up to one good or to any good. Finally, we run some experiments comparing the performances of our approach with MNW (Max Nash Welfare) on several criteria such as the percentage of EF up to one good and any good.

Keywords: Social choice · Multiagent resource allocation · Fair allocation · Fair division of indivisible goods

1 Introduction

In this paper, we investigate fair division of indivisible goods. In this context, several approaches have been proposed to model fairness. Amongst these models, one prominent solution concept is to look for *envy-free* allocations [12]. In such allocations, no agent would swap her bundle with the bundle of any other agent.

Envy-freeness is an attractive criterion: the fact that each agent is better off with her own share than with any other share is a guarantee of social stability. Besides, it does not rely on any interpersonal comparability. Unfortunately, envy-freeness is also a demanding notion as soon as we require all goods to be

© Springer Nature Switzerland AG 2021
D. Fotakis and D. Ríos Insua (Eds.): ADT 2021, LNAI 13023, pp. 289–303, 2021.
https://doi.org/10.1007/978-3-030-87756-9_19

allocated, and it is well-known that in many situations, no such allocation exists (consider for instance the situation where the number of items to allocate is strictly less than the number of agents at stake). Hence several relaxations of envy-freeness have been studied in recent years. Two orthogonal approaches have been considered. A first possibility is to "forget" some items when comparing the agents' shares. This leads to the definition of envy-freeness up to one good [17] and envy-freeness up to any good [7]. Recently, Amanatidis *et al.* [3] explored how different relaxations of envy-freeness relate to each other. Another possible approach is to relax the Boolean notion of envy and to introduce a quantity of envy that we seek to minimize. This is the path followed by Lipton *et al.* [17] or Endriss *et al.* [9] for instance. Several approximation algorithms dedicated to minimize these measures were subsequently designed – see e.g. Nguyen *et al.* [18]. Of course such approaches always rely on a specific choice to measure the degree of envy, in particular regarding the aggregation of agents' envies, which can be disputed: is it more appropriate to minimize the maximum envy experienced by some agent in the society, or to minimize the sum of agents' envies?

In this paper, we elaborate on this idea of minimizing the degree of envy but seek to offer a broader perspective. More precisely, we explore the possibility of finding allocations where envy is "fairly balanced" amongst agents. For that purpose, we start from the notion of individual degree of envy and use a *fair* Ordered Weighted Average operator (by "fair", we mean an OWA where weights are non-increasing.) to aggregate these individual envies into a collective one, that we try to minimize. This family of operators contains both the egalitarian and utilitarian operators mentioned previously. But doing so also sometimes allows us to draw results valid for the whole family of fair operators. Along our way, we shall for instance see that no algorithm fairly minimizing envy can be guaranteed to return an envy-free allocation up to any good, even though such allocation does exist. More generally, we provide several insights regarding the behaviour of such fair minimizing operators, comparing their outcomes with alternative approaches, either analytically or experimentally. Technically, this is made possible through to the use of linearization techniques which alleviate the burden of computing these outcomes.

The remainder of this paper is as follows. After giving some preliminary definitions in Sect. 2, we formally introduce our fairness minOWA envy criterion (Sect. 3) and we show that OWA minimization problems can be formulated as linear programs. We then investigate the link between minimizing the OWA of the envy vector and other fairness notions (Sect. 4). We thus study fairness guarantees of the minOWA solutions. Finally, we present some experimental results investigating the fairness of min OWA solutions (Sect. 5).

2   Model and Definitions

We will consider a classic multiagent resource allocation setting, where a finite set of *objects* $\mathcal{O} = \{o_1, \ldots, o_m\}$ has to be allocated to a finite set of *agents* $\mathcal{N} = \{a_1, \ldots, a_n\}$. In this setting, an *allocation* is a vector $\boldsymbol{\pi} = \langle \pi_1, \ldots, \pi_n \rangle$ of *bundles* of objects, such that $\forall a_i, a_j \in \mathcal{N}$ with $i \neq j : \pi_i \cap \pi_j = \emptyset$ (preemption: a given object cannot be allocated to more than one agent) and $\bigcup_{a_i \in \mathcal{N}} \pi_i = \mathcal{O}$ (no free-disposal: all the objects are allocated). $\pi_i \subseteq \mathcal{O}$ is called agent a_i's *share*. The set of all the possible allocations will be denoted $\mathcal{P}(I)$.

A crucial aspect of fair division problem is how the agents express their preferences over bundles. Here, we assume that these preferences are *numerically additive*: each agent a_i has a *utility function* $u_i : 2^{\mathcal{O}} \rightarrow \mathbb{R}$ measuring her satisfaction $u_i(\pi_i)$ when she obtains share π_i, which is defined as $u_i(\pi_i) \overset{def}{=} \sum_{o_k \in \pi_i} w(a_i, o_k)$, where $w(a_i, o_k)$ is the weight given by agent a_i to object o_k. This assumption, as restrictive as it may seem, is made by a lot of authors [4,17, for instance] and is considered a good compromise between expressivity and conciseness.

Definition 1. *An instance of the* additive multiagent resource allocation problem *(add-MARA instance for short)* $I = \langle \mathcal{N}, \mathcal{O}, w \rangle$ *is a tuple with* $\mathcal{N}$ *and* $\mathcal{O}$ *as defined above and* $w : \mathcal{N} \times \mathcal{O} \rightarrow \mathbb{R}$ *is a mapping with* $w(a_i, o_k)$ *being the weight given by* a_i *to object* o_k. *We will denote by* $\mathcal{P}(I)$ *the set of allocations for* I.

In the following, we denote by $\mathcal{I}$ the set of all add-MARA instances. Furthermore, different domain restrictions will be of interest: we denote by $\mathcal{I}^p$ the set of add-MARA instances involving only two agents (pairwise instances), and by $\mathcal{I}^b$ the set of add-MARA instances where agents have binary utilities.

Unless stated otherwise, we will only consider MARA instances with *commensurable* preferences, such that: $\exists K \in \mathbb{N}$ s.t $\forall i \in [\![1, n]\!], \sum_{j=1}^{m} w(a_i, o_j) = K$.

2.1 Envy-Free Allocations

A prominent fairness notion in multiagent resource allocation is *envy-freeness*. Envy-freeness (EF) can be defined as follows:

Definition 2. *Let* $I = \langle \mathcal{N}, \mathcal{O}, w \rangle$ *be an add-MARA instance and* $\boldsymbol{\pi}$ *be an allocation of* I. $\boldsymbol{\pi}$ *is* envy-free *if and only if* $\forall a_i, a_j \in \mathcal{N}, u_i(\pi_i) \geq u_i(\pi_j)$.

In other words, every agent a_i weakly prefers her own share to the share of any other agent a_j. In the context of fair division of indivisible goods, this notion is very demanding and there exists a lot of add-MARA instances for which no envy-free allocation exists. To relax envy-freeness, a possibility is to introduce a notion of degree of envy based on pairwise envy [17].

Definition 3. *Let* $I = \langle \mathcal{N}, \mathcal{O}, w \rangle$ *be an add-MARA instance and* $\boldsymbol{\pi}$ *be an allocation of* I. *The* pairwise envy *between* a_i *and* a_j *is defined as:*

$$pe(a_i, a_j, \boldsymbol{\pi}) \overset{def}{=} \max\{0, u_i(\pi_j) - u_i(\pi_i)\}.$$

In other words, the pairwise envy between a_i and a_j is 0 if a_i does not envy a_j, and otherwise is equal to the difference between a_i's utility for agent a_j's bundle and her actual utility in $\boldsymbol{\pi}$. It can be interpreted as how much a_i envies a_j's bundle.

From that notion of pairwise envy, we can derive a notion of global envy of an agent, that we define as the maximal pairwise envy that this agent experiences:

Definition 4. *Let $I = \langle \mathcal{N}, \mathcal{O}, w \rangle$ be an add-MARA instance and π be an allocation of I. a_i's envy: $e(a_i, \pi) \stackrel{def}{=} \max_{a_j \in \mathcal{N}} pe(a_i, a_j, \pi)$. The vector $e(\pi) = \langle e(a_1, \pi), ..., e(a_n, \pi) \rangle$ will be called* envy vector *of allocation π.*

Here, the max operator is rather a standard choice in the context where one seeks for allocations with bounded envy [17]. Note that an allocation π is envy-free if and only if $e(\pi) = \langle 0, ..., 0 \rangle$.

2.2 Weaker Notions of Envy-Freeness

Besides minimizing a degree of envy, different relaxations of the envy-freness notions have also been proposed to cope with situations where there is no envy-free solution. *Envy-freness up to one good* (EF1) [6,17] is one of the most studied relaxations. An allocation is said to be envy-free up to one good if, for each envious agent a_i, the envy of a_i towards an agent a_j can be eliminated by removing an item from the bundle of a_j.

Definition 5. *Let $I = \langle \mathcal{N}, \mathcal{O}, w \rangle$ be an add-MARA instance and π be an allocation of I. π is envy-free up to one good if and only if $\forall a_i, a_j \in \mathcal{N}$, either $u_i(\pi_i) \geq u_i(\pi_j)$ or $\exists o_k \in \pi_j$ such that $u_i(\pi_i) \geq u_i(\pi_j \backslash \{o_k\})$.*

It has been proved that an EF1 allocation always exists and, in the additive case, can be obtained using a round-robin protocol [7].

Caragiannis *et al.* [7] proposed another relaxation of the notion of envy-freeness which is stronger than EF1. An allocation is said to be *envy-free up to any good* (EFX) if for all envious agents a_i, the envy of a_i towards a_j can be eliminated by removing *any* item from a_j's bundle.

Definition 6. *Let $I = \langle \mathcal{N}, \mathcal{O}, w \rangle$ be an add-MARA instance and π be an allocation of I. π is envy-free up to any (strictly positively valued) good if and only if $\forall a_i, a_j \in \mathcal{N}$, either $u_i(\pi_i) \geq u_i(\pi_j)$ or $\forall o_k \in \pi_j$ for which $w(a_i, o_k) > 0$, $u_i(\pi_i) \geq u_i(\pi_j \backslash \{o_k\})$.*

An even more demanding notion called EFX_0 [15,21] differs on the fact that an agent can forget any object even the ones valued to 0:

Definition 7. *Let $I = \langle \mathcal{N}, \mathcal{O}, w \rangle$ be an add-MARA instance and π be an allocation of I. π is envy-free up to any good if and only if $\forall a_i, a_j \in \mathcal{N}$, either $u_i(\pi_i) \geq u_i(\pi_j)$ or $\forall o_k \in \pi_j$, $u_i(\pi_i) \geq u_i(\pi_j \backslash \{o_k\})$.*

Clearly, we have $EF \implies EFX_0 \implies EFX \implies EF1$. While an EF1 allocation can be computed in polynomial time, the guarantee of existence of an EFX allocation remains an open issue in the general settings [7]. The existence guarantee of an EFX solutions has been proved for few agents (at most 3 agents) and specific utility functions. For instance it has been proved that an EFX_0 allocation always exists for instances with identical valuations and for instances involving two agents with general and possibly distinct valuations [21], as well as for three agents with additive valuations [8]. When the objects have only two possible valuations, Amanatidis *et al.* [2] proved that any allocation maximizing the Nash Social Welfare is EFX_0. This result provides a polynomial algorithm for computing EFX_0 allocations in the two-agent setting.

Other notions of fairness have been introduced in the literature. Bouveret *et al.* [5] for instance exhibited some connections between widely used notions among which the max-min share (MMS, also known as I cut you chose). An allocation is MMS if every agent gets at least her max-min share. As shown by Bouveret et al. [5], MMS is less demanding than EF and every EF allocation also satisfies MMS.

Example 1. Let us consider the add-MARA instance with 3 agents and 4 objects:

	o_1	o_2	o_3	o_4
a_1	2*	6	1	1
a_2	2	5*	2	1
a_3	1	5	2*	2*

Note that there is no EF allocation in this instance. The squared allocation π and the starred allocation π' are both EF1 and EFX. Both allocations satisfy MMS. Allocation π leads to the envy vector $e(\pi) = \langle 0, 3, 2 \rangle$ while allocation π' leads to the envy vector $e(\pi') = \langle 4, 0, 1 \rangle$. Both allocations have the same global envy when considering the sum of the individual envies. However, the envy in π' is mainly supported by a_1. To promote fairness, it is natural to prefer the allocations where the envy is balanced among the agents. In this example, π should be considered as more fair than π'.

Recently, the Nash social welfare (which maximizes the product of utilities) was celebrated as a particularly good trade-off between efficiency and fairness [7] because it guarantees to return an EF1 and Pareto-optimal allocation, among others. Finally, some authors have proposed to explore inequality indices in multiagent fair division settings [1,11,23]. However, this differs from our proposal since in these approaches inequality is (more classically) evaluated at the level of utilities, while we apply it to envies, as we detail in the next section.

3   MinOWA Envy

Our approach elaborates on minimizing the degree of envy of the agents while balancing the envy among the agents as suggested by Lipton *et al.* [17]. The general idea would be to look for allocations that minimize this vector of envy in some sense: the lower this vector is, the less envious the agents are. This corresponds to a multiobjective optimization problem where each component of the envy vector is a different objective to minimize.

3.1   Fair OWA

There are different ways to tackle this minimization problem, each approach conveying a different definition of minimization. Our approach, guided by the egalitarian notion of fairness [22], is to ensure that, while being as low as possible, the envy is also distributed as equally as possible amongst agents. To this end, we use a prominent aggregation operator that can convey fairness requirements: order weighted averages.

Ordered Weighted Averages (OWA) have been introduced by Yager [25] with the idea to build a family of aggregators that can weight the importance of objectives (or agents) according to their relative utilities, instead of their identities. In this way, we can explicitly choose to favour the poorest (or richest) agents, or to concentrate the importance of the criterion on the middle-class agents. Formally, the OWA operator is defined as follows:

Definition 8. *Let* $\alpha = \langle \alpha_1, \ldots, \alpha_n \rangle$ *be a vector of weights. In the context of minimization, the* ordered weighted average *parameterized by* α *is the function* $owa^\alpha : \boldsymbol{x} \mapsto \sum_{i=1}^n \alpha_i \times x_i^\downarrow$, *where* $\boldsymbol{x}^\downarrow$ *denotes a permutation of* $\boldsymbol{x}$ *such that* $x_1^\downarrow \geq x_2^\downarrow \geq \ldots \geq x_n^\downarrow$.

Amongst all OWA, only those giving more weight to the unhappiest agents can be considered fair in the egalitarian sense. This property can be formalized as follows. Let $\boldsymbol{x}$ be a vector such that $x_j \geq x_i$ (a_i is better off than a_j) and let ε be such that $0 \leq \varepsilon \leq 2(x_j - x_i)$. Then, for any non-increasing vector α: $owa^\alpha(\boldsymbol{x}) \leq owa^\alpha(\langle x_1, \ldots, x_i + \varepsilon, \ldots, x_j - \varepsilon, \ldots, x_n \rangle)$.

In other words, such an OWA favours any transfer of wealth from a happier agent to an unhappier agent. Such a transfer is called a *Pigou-Dalton* transfer, and the OWA with non-increasing weight vectors α are called *fair OWA*. Moreover, we have considered wlog in this paper that the weight vector sums to 1 so we will make no difference between weights $\langle 1, 1, 1 \rangle$ and $\langle \frac{1}{3}, \frac{1}{3}, \frac{1}{3} \rangle$. Note that fair OWA is also referred to as *Generalized Gini Index* [24] in the literature. In matching problems [16] and multiagent allocation problems [14], fair OWA has been applied to the utility vector so as to maximize a global utility function while reducing inequalities. However, we can note that maximizing the OWA of the utility vector does not necessarily return an EF allocation even when such an allocation exists:

Example 2. Consider this add-MARA instance with 3 agents and 4 objects:

	o_1	o_2	o_3	o_4
a_1	1	2*	3	4*
a_2	2	2	5*	1
a_3	4*	0	4	2

The squared allocation is the allocation that maximizes the value of the OWA of the utility vector with weight $\langle 1, 0, 0 \rangle$. We can easily notice that this allocation is not envy free as a_1 envies a_2. Moreover, the star allocation is obviously an EF one. Note that in the context of maximization, a fair OWA is also defined with non-increasing weights but by sorting the components by decreasing value.

Since our motivation is to return an EF allocation when there is one and otherwise minimize the envy while equally distributing it between the agents, we propose to minimize the fair OWAs of the envy vector.

Definition 9. *Let* $I = \langle \mathcal{N}, \mathcal{O}, w \rangle$ *be an add-MARA instance and* α *be a non-increasing vector. An allocation* $\hat{\pi}$ *is an* α-*minOWA Envy allocation if:*

$$\hat{\pi} \in \arg \min_{\pi \in \mathcal{P}(I)} (owa^\alpha(e(\pi))).$$

It is important to note that a major advantage of this solution is that it always exists as it is the result of an optimization process. Moreover, this optimization problem can be modeled as an Integer Linear Program, which will give a way to compute optimal allocations. Keep also in mind that there can be several allocations with the same OWA envy value.

Let us now see some helpful properties of fair OWA. Note that we will consider here that we are in a minimization context.

Definition 10. *By denoting* $v_k^\downarrow$ *the* k^{th} *biggest component of a given vector* v, *the Lorenz vector* L *of* v *is defined as* $L(v) = \langle v_1^\downarrow, v_1^\downarrow + v_2^\downarrow, ..., \sum_{i=1}^n v_i^\downarrow \rangle$.

Definition 11. *Let* x *and* y *be two vectors of the same size and* x_i *(respectively* y_i) *be the* i^{th} *component of* x *(respectively* y). *We say that* x *Pareto dominates* y *iff for every component* $x_i \leq y_i$ *and there is one component* x_j *for which* $x_j < y_j$ *and* x *strongly Pareto dominates* y *iff for every component* $x_i < y_i$.

Definition 12. *We say* x *(strongly) Lorenz dominates* y *iff* $L(x)$ *(strongly) Pareto dominates* $L(y)$.

Theorem 1. *Perny and Spanjaard 2003*
If x *Lorenz dominates* y *then for any non-increasing weight* α: $owa^\alpha(x) \leq owa^\alpha(y)$. *Similarly if* x *strongly Lorenz dominates* y *then for any non-increasing weight* α: $owa^\alpha(x) < owa^\alpha(y)$.

This helpful property is shown in [20]. As (strong) Pareto dominance implies (strong) Lorenz dominance, the same theorem holds with (strong) Pareto dominance.

By using a linearization introduced by Ogryczak [19] we can model our problem of minimizing the OWA of the envy vector as a linear program. Moreover we consider decreasing OWA weights (fair OWA) so $\alpha_1 \geq \alpha_2 .... \geq \alpha_n$ and we denote by $\alpha' = \langle \alpha_1 - \alpha_2, \alpha_2 - \alpha_3, ..., \alpha_n \rangle$. We introduce a set of $n \times m$ Boolean variables z_i^j: z_i^j is 1 iff o_j is allocated to a_i while r_k and b_i^k are the dual variables (of the LP computing the Lorenz components) and e_i the envy of a_i.

$$\min owa(e(\pi)) = \min \sum_{k=1}^n \alpha_k'(kr_k + \sum_{i=1}^n b_i^k)$$

$$\begin{cases} r_k + b_i^k \geq e_i & \forall i, k \in [\![1, n]\!] \\ e_i \geq \sum_{j=1}^m w(a_i, o_j)(z_h^j - z_i^j) & \forall i, h \in [\![1, n]\!] \\ \sum_{i=1}^n z_i^j = 1 & \forall j \in [\![1, m]\!] \\ z_i^j \in \{0, 1\} \quad \forall j \in [\![1, m]\!] & \forall i \in [\![1, n]\!] \\ b_i^k \geq 0, \quad e_i \geq 0 & \forall i, k \in [\![1, n]\!] \end{cases}$$

4   Link with Other Fairness Measures

We focus here on the possible links between the min OWA allocation and other fairness measures. We recall that if an envy-free allocation exists, it will be returned by the min OWA optimization. For any instance I, we denote by $PROP(I)$ the set of allocations satisfying $PROP \in \{EF1, EFX, EFX_0, MMS\}$. We also denote by $\boldsymbol{\alpha}$-min OWA(I) the set of all min OWA optimal allocation for the specific weight vector $\boldsymbol{\alpha}$, and by $\forall$-min OWA(I) the set of $\boldsymbol{\alpha}$-min OWA, for *all* (fair) weight vectors $\boldsymbol{\alpha}$.

4.1   Warm-Up: $n = 2$

In the special case where the allocation problem involves only two agents, we highlight strong connections between min OWA allocations and other fairness measures (MMS, EF1 and EFX).

Proposition 1. $\forall I \in \mathcal{I}^p : \forall$-*min OWA*$(I) \subseteq MMS(I) \subseteq EFX(I)$

Proof. For add-MARA instances where an envy-free allocation exists, our proof is straightforward as min OWA returns the EF allocation. It is thus also MMS, EF1 and EFX.

We now focus on add-MARA instances for which there is no EF allocation. In the presence of only 2 agents any min OWA allocation π is such that only one of the two agents is envious. Indeed, if no agent is envious then it means the add-MARA instance has an envy-free allocation (which is a contradiction). Similarly, if both agents are envious it means there is an envy-free allocation (which is again a contradiction) as the agents would just have to exchange their bundles to obtain that allocation. Consequently, the sorted envy vector will be of the form $(e, 0)$. Suppose for the sake of contradiction that such an allocation is not MMS. The agent that is envy-free (let us say w.l.o.g it is a_2) obviously has her max-min share. So, under the assumption that the allocation is not MMS, a_1 does not have her max-min share. It means that there is an allocation π' such that $\min(u_1(\pi_1'), u_1(\pi_2')) > u_1(\pi_1)$ and a_2 is still not envious (if a_2 is envious in π', just swap her share with a_1's). Obviously, a_1's pairwise envy for a_2 has decreased in π' compared to that of π, and a_2's envy is still 0. This contradicts the fact that π is the optimal min-OWA envy allocation. Finally, it is known [7] in the two-agents setting that MMS implies EFX, which completes the proof.

However, even though an MMS allocation is EFX, this does not hold for EFX_0 even for 2 agents as we can see in Example 3.

Example 3. Consider this add-MARA instance with 2 agents and 3 objects:

	o_1	o_2	o_3
a_1	1	0	2
a_2	0	1	2

It is easy to see that the squared allocation is MMS as the max-min share of each agent is 1. Moreover, we can see that this allocation is EFX (a_1 can forget o_3) whereas it is not EFX_0 (because a_1 has to forget o_2 which does not make here becoming envy-free).

However, we show that we can very easily build an EFX_0 allocation from an arbitrary min OWA envy one.

Proposition 2. *For any instance $I \in \mathcal{I}^p$ and for any weight vector $\vec{\alpha}$: $\alpha\text{-}minOWA(I) \cap EFX_0(I) \neq \emptyset$. Furthermore, it can be obtained from an arbitrary $\alpha\text{-}min\text{-}OWA$ envy optimal allocation in linear time.*

Proof. Let us call π an arbitrary min OWA allocation. If π is envy-free then it is obviously EFX_0 and the proof concludes. Note that envy-freeness is checked in $O(1)$ as we just have to check the values of both variables e_1 and e_2. Otherwise, it means that one and exactly one agent is envious, by using a same argument as in the proof of Proposition 1. W.l.o.g. we consider a_1 is the envious agent. We start from π and transfer to a_1 all the objects that she values with utility zero. The resulting allocation is called π'. We show that π' is EFX_0. a_1 still envies a_2 in π' but is EFX by Proposition 1. By transferring all zero-valued objects to her share, she becomes EFX_0 in π'. Now consider a_2. If a_2 envies a_1 in π' then by swapping their bundles, we can obtain an envy-free allocation. This contradicts the fact that π is min-OWA envy optimal. Hence, a_2 still does not envy a_1 in π', and thus is also EFX_0 obviously. Since in π' a_2 is still envy-free and the pairwise envy from a_1 to a_2 has not changed, π' is still min-OWA envy optimal. The complexity is linear in the number of objects since we have to implement the transfer of zero-valued objects to a_1's bundle.

On Example 3, this means that a_1 should receive o_2. This adjustment is inefficient: by construction, it returns an allocation which is Pareto-dominated by the original min OWA envy optimal allocation. Intuitively, it can be seen as the price to pay to get EFX_0: by assigning those items that the agent does not value to her, the mechanism offers the strongest possible fairness guarantees.

4.2 General Case: $n \geq 3$

We now turn to more general settings involving at least 3 agents. Since an EF1 allocation is guaranteed to exist, we more specifically focus on the relation between min OWA and EF1. Unfortunately, we notice that in the general case these two sets can be disjoint, i.e. there are instances for which no allocation is both EF1 and min-OWA, for any weight vector:

Proposition 3. $\exists I \in \mathcal{I} : EF1(I) \cap \forall\text{-}min\ OWA(I) = \emptyset$

Proof. Let us consider the add-MARA instance with 4 agents and 5 objects:

	o_1	o_2	o_3
a_1	1	0	2
a_2	0	1	2

In order to prove the proposition we will show that the squared allocation is the only min OWA envy allocation (for any given weight vector) and that it is (obviously) not EF1. First note that as a_1 and a_2 have similar preferences the allocation derived from the squared allocation where we swap the bundles of these agents will be the same in terms of Lorenz envy vector. The squared allocation has a vector of envy $e = \langle 0, 14, 14, 0 \rangle$ and $L(e) = \langle 14, 28, 28, 28 \rangle$. First consider the allocations in which a_4 does not possess o_5. We have $e_1 = \langle e_1, e_2, e_3, 30 \rangle$ and $L(e_1) = \langle 30, L_2, L_3, L_4 \rangle$ with L_2, L_3, L_4 being greater than or equal to 30. e_1 is thus strongly Lorenz dominated by e. Let us now consider the other possible allocations (in which a_4 possesses o_5): if a_3 has o_1 instead of a_1 then $e_2 = \langle 20, 14, 1, 0 \rangle$ and $L(e_2) = \langle 20, 34, 35, 35 \rangle$. e_2 is thus strongly Lorenz dominated by e. Finally, we focus on allocations in which a_3 has one to three items from the set of objects $\{o_1, o_2, o_3\}$. If a_3 has one of these items we have $e_3 = \langle 0, 16, 13, 0 \rangle$ and $L(e_3) = \langle 16, 29, 29, 29 \rangle$. If a_3 has two of these items we have $e_4 = \langle 0, 18, 12, 0 \rangle$ and $L(e_4) = \langle 18, 30, 30, 30 \rangle$. Finally if a_3 has all these items we have $e_5 = \langle 0, 20, 11, 0 \rangle$ and $L(e_5) = \langle 20, 31, 31, 31 \rangle$. All e_3 e_4 and e_5 are strongly Lorenz dominated by e. As we know that minimizing fair OWA of a vector is consistent with the Lorenz dominance (see Theorem 1), it means that if a solution strongly Lorenz dominates another, then its fair OWA value will be strictly lower (in a minimization problem such as ours) for any non-creasing weight. We can then conclude that the squared allocation is indeed the only min OWA envy one and it is not EF1.

However, a significant number of experiments actually suggest that for almost any instance, some EF1 allocation is also min-OWA, either for the weight vector $\langle 1, 0, \ldots 0 \rangle$, or for the weight vector $\langle 1, 1, \ldots 1 \rangle$. Moreover, we have a positive result in the restricted domain where agents have binary utilities.

Proposition 4. $\forall I \in \mathcal{I}^b : EFX_0(I) \cap \forall\text{-}min\ OWA(I) \neq \emptyset$

Proof. First note that if the instance is EF then the min OWA envy allocation will be EF and thus EF1 and the proof concludes. Hence we will consider instances that are not EF. As we consider binary utilities, we know thanks to [8] that an EFX_0 allocation always exists. We can easily notice that any such allocation is such that the envy of each agent is at most 1. Hence, as with the weight vector $\langle 1, 0, \ldots 0 \rangle$ the OWA envy value of an EFX_0 allocation is 1 (as we supposed no EF allocation exists), it is the minimum OWA envy value possible. It can thus be returned by minimizing the OWA envy value.

5   Experimental Results

We drew some experiments to compare the performances of the allocations obtained by min OWA envy with the Maximization of Nash Welfare. More precisely we implemented the linearization described in [7] that returns an allocation approximating MNW but closely enough to keep interesting properties such as EF1 and Pareto Optimality. As we have seen through this paper the range of possibilities offered by the fact that OWA is parameterized is interesting. We will see how three different weights $\alpha_1 = \langle 1, 0 \ldots 0 \rangle$, $\alpha_2 = \langle \frac{1}{2}, \frac{1}{4}, \ldots \frac{1}{2^n} \rangle$ and $\alpha_3 = \langle 1, 1, \ldots 1 \rangle$ compare to each other. α_1 and α_3 correspond to respectively minimize the max envy and the sum of the envies. α_2 is somewhere in the middle of those two extrema with a strictly decreasing weight vector.

All the tests presented in this section have been run on an Intel(R) Core(TM) i7-2600K CPU with 16 GB of RAM and using the Gurobi solver to solve Mixed Integer Programs[1]. We have tested our methods on two types of instances: Spliddit instances [13] and synthetic instances under uniformly distributed commensurable preferences (that is, for each agent a_i and object o_j, utilities are drawn i.i.d. following the uniform distribution on some interval $[x, y]$ and such that the utilities of each agent sums to $5\,\mathrm{m}$).

We evaluate the performances of the OWA envy minimization outcome for both types of instances through the following criterion: EF, EFX_0, EFX, EF1 and Pareto dominance. Tables 1 and 2 present the percentage of min OWA envy outcomes that satisfy each criterion. We also study how the vector of weights of the OWA influences the characteristics of the outcomes. The computation time (in seconds) of each approach is also mentioned. We recall the strong connections between the 4 first fairness notions as EF $\implies$ EFX_0 $\implies$ EFX $\implies$ EF1. As it can be checked in Tables 1 and 2, the percentage of EF allocations should always be lower than or equal to the number of EFX_0 ones which should be lower than or equal to the number of EFX allocations and so on.

5.1   Spliddit Instances

Our first set of experiments has been performed on real-world data from the fair division website Spliddit [13]. There is a total of 3535 instances from 2 agents to 15 agents and up to 93 items. Note that 1849 of these instances involve 3 agents and 6 objects. By running the MIPs minimizing the OWA envy with the three different weights' vectors described above with a timeout of 1 min (after this duration the best current solution, if it exists, is returned) we were able to solve all the instances to optimal. The results of these experiments are presented in Table 1. The first three columns respectively correspond to the results of minimization of the OWA envy with respectively α_1, α_2 and α_3, while the fourth column presents the results of the optimization of MNW.

Minimizing the OWA envy provably returns an EF allocation if there exists one. Hence, among the Spliddit instances 65.4% are envy-free. Note that only

---

[1] The code is available at https://gitlab.com/MrPyrom/balancing-envy.

Table 1. Performances for minimizing the OWA envy (with weights α_1, α_2, α_3) or maximizing the Nash Welfare on Spliddit instances

	α_1	α_2	α_3	MNW
%EF	65.4	65.4	65.4	57.2
%EFX$_0$	90.0	93.0	92.7	90.9
%EFX	98.5	99.4	99.0	94.9
%EF1	99.4	99.8	99.3	100
%Pareto	77.1	78.7	79.2	100
%EF+PO	45.7	45.6	46.0	57.2
Time(s)	$3.5*10^{-3}$	$5.7*10^{-7}$	$6.9*10^{-7}$	$1.1*10^{-6}$

57.2% of the allocations returned by MNW are EF which means that for around 8.2% of the Spliddit instances, an EF allocation exists but MNW failed to return it. Moreover, without any surprise as Pareto optimality (PO) of the MNW allocations is guaranteed, minimizing OWA envy returns fewer PO allocations than MNW. However, around slightly less than 80% of the min OWA envy allocations are PO. It is also guaranteed that MNW returns an EF1 solution. However, we can observe, for every weight, that more than 99% of the allocations returned by min OWA envy are EF1. This balances the negative result in Proposition 3. Moreover, it can be very interestingly observed that the percentage of EFX$_0$ is greater for α_2 and α_3 than for MNW. The same holds for the percentage of EFX but for the 3 weights' vectors and by a more noticeable margin of around 5%. However, MNW performs slightly better than min OWA when we consider EF alongside with PO. Finally, we can see that all the optimization programs run very quickly in average with a slightly longer time for α_1.

5.2   Synthetic Instances

For each couple $(|\mathcal{N}|,|\mathcal{O}|)$ from $(3, 4)$ to $(10, 12)$, we generated 100 synthetic add-MARA instances with uniformly distributed preferences. We then ran the four optimization methods described above on the generated instances. We considered such couples of values in order to produce settings where few EF allocations exist as suggested in [10]. Although it is interesting to consider EF instances to compare with MNW, minimizing OWA envy is even more relevant when no EF allocation exists. Due to lack of space, Table 2 presents the results for only 4 couples (n, m) but similar trends can be observed for the other couples of values. As witnessed for the Spliddit instances, MNW often fails to return an EF allocation even when there exists one. As shown in Table 2, the number of EF allocations missed by MNW can be quite important as shown by the gap between the percentage of EF allocations returned by min OWA envy and the percentage for MNW. This is exemplified in Table 2 for 2 agents and 5 agents where the gap is respectively of 16% and 31%. Even more significantly, it turns out that min OWA outperforms MNW when we consider EF together with PO. Once again and in an even stronger way than for the Spliddit instances, these results heavily balance the result of Proposition 3: in practice the allocations returned by the min OWA envy were always EF1. Concerning EFX_0 and EFX we also obtained very positive results. Indeed, min OWA envy returns around 10% more EFX_0 and EFX instances than MNW. Note that we confirm Proposition 1 as we have 100% of EFX allocations when $n = 2$. Note that we did not adjust the allocation returned by the min OWA optimization to break ties as discussed in the proof of Proposition 2. Thus, we get 97% of EFX_0 but this percentage could be even higher. However, these positive results about EF, EFX_0 and EFX come with a price on efficiency as we can see that PO is not guaranteed and the percentage gets lower as the number of agents increases but is still above 60% for α_2 and α_3. This highlights the inherent compromise and tension between efficiency and fairness. Besides, as it was the case for the Spliddit instances we can see that the computation is overall quite fast. We can notice that the MNW computation never surpasses 0.02 s whereas for 10 agents, min OWA envy optimization is slightly faster than a second for α_1 and α_3 and around 2 s for α_2. Finally, we can see that the three different weights considered here lead to quite similar performances. We can globally notice more encouraging results for α_3 except for EFX. However, keep in mind that the advantage of using a parameterized function is its rich expressiveness so we could see our method as a combination of the results of the 3 weights.

Table 2. Performances for minimizing the OWA envy (with weights α_1, α_2, α_3) or maximizing the Nash Welfare on synthetics instances (as a function of the number of agents and objects (n, m) ($\epsilon \leq 10^{-3}$)).

	(2,3)				(5,7)				(8,10)				(10,12)			
	α_1	α_2	α_3	MNW	α_1	α_2	α_3	MNW	α_1	α_2	α_3	MNW	α_1	α_2	α_3	MNW
%EF	74	74	74	58	48	48	48	17	10	10	10	1	1	1	1	0
%EFX$_0$	97	97	97	88	96	96	96	88	88	86	88	78	72	82	83	80
%EFX	100	100	100	92	97	97	98	91	98	96	93	85	87	95	92	84
%EF1	100	100	100	100	100	100	100	100	100	100	100	100	100	100	100	100
%Pareto	100	100	100	100	73	76	72	100	64	66	67	100	51	62	64	100
%EF+PO	74	74	74	58	33	34	32	17	5	6	5	1	1	1	1	0
Time(s)	ϵ	ϵ	ϵ	ϵ	0.02	0.02	0.02	0.01	0.1	0.4	0.1	0.04	0.5	2.5	0.7	0.07

6 Conclusion

In this paper, we introduced a new fairness concept following the idea of minimizing envy. More particularly, we used an OWA to express fairness in the distribution of envy between agents. This generalizes several approaches using various definitions of degree of envies, which can be captured by adequate weight vector. In practice, we put a special focus on the egalitarian variant (minimizing the highest envy), the utilitarian variant (minimizing the sum of envies), and the compromise consisting of using the fair vector of decreasing weights. After implementing a MIP to compute min OWA allocations, we unveil several connections between the min OWA allocation and other famous fairness measures. In particular, we compare our approach with the alternative relaxations consisting of seeking "envy-freeness up to some/any good". Some of our conclusions show that these approaches correspond to very different perspectives: we show in particular that no algorithm minimizing a fair OWA can ever guarantee to return an EF1 (and thus nor EFX) allocation. This is however balanced by the fact that it never occured in our experiments. Indeed, even in the very few cases for which the min OWA allocation was not EF1 we easily found a weight for which it was the case. This raises the question of choosing the appropriate weight vector for example by elicitating it. We left that question open for now. Indeed, we also ran some experiments to test the performances of our method and compared it with other allocation protocols. The results are extremely encouraging. Our min OWA approaches do very well (in particular regarding the likelihood to return an EFX allocation, which may be somewhat paradoxical given our previous remarks) in terms of fairness, both on real Spliddit instances and randomly generated ones. In comparison, Nash social welfare –despite its guarantee to return an EF1 allocation– is dominated on that respect, as well as on the likelihood to return an EF and Pareto optimal allocation.

References

1. Aleksandrov, M., Ge, C., Walsh, T.: Fair division minimizing inequality. In: Moura Oliveira, P., Novais, P., Reis, L.P. (eds.) EPIA 2019. LNCS (LNAI), vol. 11805, pp. 593–605. Springer, Cham (2019). https://doi.org/10.1007/978-3-030-30244-3_49
2. Amanatidis, G., Birmpas, G., Filos-Ratsikas, A., Hollender, A., Voudouris, A.A.: Maximum nash welfare and other stories about EFX. In: Bessiere, C. (ed.) Proceedings of the Twenty-Ninth International Joint Conference on Artificial Intelligence, IJCAI 2020, pp. 24–30. ijcai.org, Yokohama (2020)
3. Amanatidis, G., Birmpas, G., Markakis, V.: Comparing approximate relaxations of envy-freeness. In: Proceedings of the Twenty-Seventh International Joint Conference on Artificial Intelligence, IJCAI 2018, pp. 42–48. Stockholm, Sweden (2018)
4. Bansal, N., Sviridenko, M.: The Santa Claus problem. In: Proceedings of the Thirty-Eighth Annual ACM Symposium on Theory of Computing, STOC '06, pp. 31–40. ACM, New York (2006)
5. Bouveret, S., Lemaître, M.: Characterizing conflicts in fair division of indivisible goods using a scale of criteria. Auton. Agents Multi-Agent Syst. 30(2), 259–290 (2015). https://doi.org/10.1007/s10458-015-9287-3
6. Budish, E.: The combinatorial assignment problem: approximate competitive equilibrium from equal incomes. J. Polit. Econ. 119(6), 1061–1103 (2011)
7. Caragiannis, I., Kurokawa, D., Moulin, H., Procaccia, A.D., Shah, N., Wang, J.: The unreasonable fairness of Maximum Nash Welfare. In: Proceedings of the 2016 ACM Conference on Economics and Computation, EC '16, pp. 305–322. ACM, New York, NY, USA (2016)
8. Chaudhury, B.R., Garg, J., Mehlhorn, K.: EFX exists for three agents. In: Proceedings of the 21st ACM Conference on Economics and Computation, EC '20, pp. 1–19. Association for Computing Machinery, New York (2020)

9. Chevaleyre, Y., Endriss, U., Maudet, N.: Distributed fair allocation of indivisible goods. Artif. Intell. **242**, 1–22 (2017)
10. Dickerson, J.P., Goldman, J., Karp, J., Procaccia, A.D., Sandholm, T.: The computational rise and fall of fairness. In: Proceedings of the 28th AAAI Conference on Artificial Intelligence (AAAI-14), pp. 1405–1411. AAAI Press, Québec City (2014)
11. Endriss, U.: Reduction of economic inequality in combinatorial domains. In: Gini, M.L., Shehory, O., Ito, T., Jonker, C.M. (eds.) International conference on Autonomous Agents and Multi-Agent Systems, AAMAS '13, Saint Paul, MN, USA, 6–10 May 2013, pp. 175–182. IFAAMAS (2013)
12. Foley, D.K.: Resource allocation and the public sector. Yale Econ. Essays **7**(1), 45–98 (1967)
13. Goldman, J., Procaccia, A.D.: Spliddit: unleashing fair division algorithms. SIGecom Exch. **13**(2), 41–46 (2015)
14. Heinen, T., Nguyen, N.-T., Rothe, J.: Fairness and rank-weighted utilitarianism in resource allocation. In: Walsh, T. (ed.) ADT 2015. LNCS (LNAI), vol. 9346, pp. 521–536. Springer, Cham (2015). https://doi.org/10.1007/978-3-319-23114-3_31
15. Kyropoulou, M., Suksompong, W., Voudouris, A.A.: Almost envy-freeness in group resource allocation. Theor. Comput. Sci. **841**, 110–123 (2020)
16. Lesca, J., Minoux, M., Perny, P.: The fair OWA one-to-one assignment problem: NP-Hardness and polynomial time special cases. Algorithmica **81**(1), 98–123 (2019)
17. Lipton, R., Markakis, E., Mossel, E., Saberi, A.: On approximately fair allocations of divisible goods. In: Proceedings of the 5th ACM Conference on Electronic Commerce (EC-04), pp. 125–131. ACM, New York (2004)
18. Nguyen, T.T., Rothe, J.: How to decrease the degree of envy in allocations of indivisible goods. In: Perny, P., Pirlot, M., Tsoukiàs, A. (eds.) ADT 2013. LNCS (LNAI), vol. 8176, pp. 271–284. Springer, Heidelberg (2013). https://doi.org/10.1007/978-3-642-41575-3_21
19. Ogryczak, W., Śliwiński, T.: On solving linear programs with the ordered weighted averaging objective. Eur. J. Oper. Res. **148**, 80–91 (2003)
20. Perny, P., Spanjaard, O.: An axiomatic approach to robustness in search problems with multiple scenarios. In: Meek, C., Kjærulff, U. (eds.) UAI '03, Proceedings of the 19th Conference in Uncertainty in Artificial Intelligence, Acapulco, Mexico, 7–10 August 2003, pp. 469–476. Morgan Kaufmann (2003)
21. Plaut, B., Roughgarden, T.: Almost envy-freeness with general valuations. In: Proceedings of the Twenty-Ninth Annual ACM-SIAM Symposium on Discrete Algorithms, SODA '18, pp. 2584–2603. USA (2018)
22. Rawls, J.: A Theory of Justice. Harvard University Press, Cambridge (1971)
23. Schneckenburger, S., Dorn, B., Endriss, U.: The Atkinson inequality index in multiagent resource allocation. In: Proceedings of the 16th Conference on Autonomous Agents and MultiAgent Systems, AAMAS 2017, São Paulo, Brazil, pp. 272–280. ACM (2017)
24. Weymark, J.A.: Generalized Gini inequality indices. Math. Soc. Sci. **1**(4), 409–430 (1981)
25. Yager, R.R.: On ordered weighted averaging aggregation operators in multicriteria decision making. IEEE Trans. Syst. Man. Cybern. **18**, 183–190 (1988)

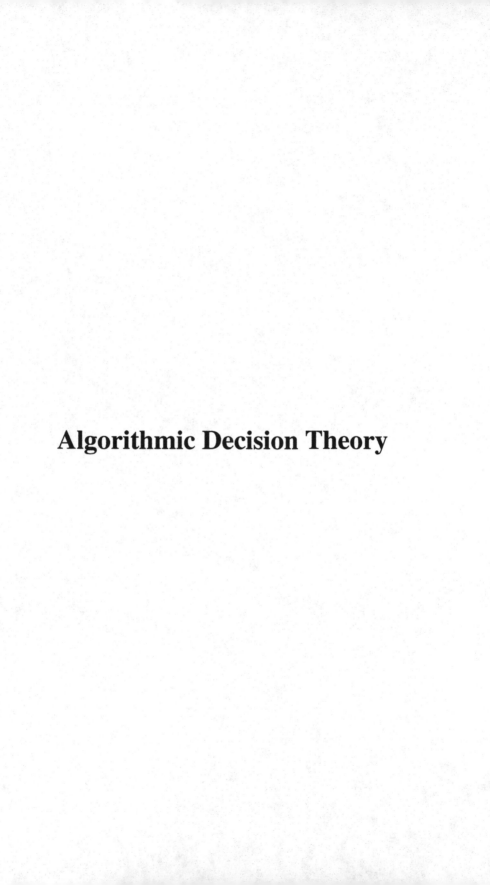

Algorithmic Decision Theory

Interactive Optimization of Submodular Functions Under Matroid Constraints

Nawal Benabbou, Cassandre Leroy, Thibaut Lust[✉], and Patrice Perny

Sorbonne Université, CNRS, LIP6, 75005 Paris, France
{nawal.benabbou,cassandre.leroy,thibaut.lust,patrice.perny}@lip6.fr

Abstract. Various practical optimization problems can be formalized as the search of an optimal independent set in a matroid. When the set function to be optimized is additive, this problem can be exactly solved using a greedy algorithm. However, in some situations, the set function is not exactly known and must be elicited before or during the optimization process. Moreover, the set function is not always additive due to possible interactions between the elements of the set. Here we consider the problem of maximizing a submodular set function under a matroid constraint. We propose two interactive approaches aiming at interweaving the elicitation of the submodular set function with the construction of an optimal independent subset subject to a matroid constraint. The first one is based on a greedy algorithm and the other is based on local search. These algorithms are tested on practical problems involving a matroid structure and a submodular function to be maximized.

Keywords: Submodular function · Matroid · Preference elicitation · Greedy search · Local search

1 Introduction

In many problems studied in combinatorial optimization, admissible solutions are defined as subsets of a ground set satisfying a structural property. A set function representing the utility or the cost of any subset is generally used to model preferences and the selection problem consists in determining an admissible subset having the maximal utility or the minimal cost. In particular, the optimization of a set function under a matroid constraint has received much attention since the seminal work of Edmonds [5]. This problem has multiple applications in various contexts such as recruitment, committee election, combinatorial auctions, scheduling, resource allocation, facility location and sensor placement, just to give a few examples. Various algorithms are now available to solve this problem either to optimality or approximately, for specific classes of set functions, see e.g., [4,10–12,14,15,17].

When the set function is additive (i.e., the value of any set is defined as the sum of the values of its elements), it is well known, after Edmonds [5], that the problem can be efficiently solved by a greedy algorithm. However, preferences

D. Fotakis and D. Ríos Insua (Eds.): ADT 2021, LNAI 13023, pp. 307–322, 2021.
https://doi.org/10.1007/978-3-030-87756-9_20

are not always representable by additive functions due to possible interactions among elements. In decision theory, the additivity of utilities is often relaxed and submodular utility functions are frequently used to guarantee a principle of diminishing returns [1,9,17]. This principle states that adding an element to a smaller set has more value than adding it to a larger set, formally the set function w should satisfy the following property: $w(X \cup \{i\}) - w(X) \geq w(Y \cup \{i\}) - w(Y)$ whenever $X \subseteq Y$ and $i \notin Y$. This is known to be equivalent to submodularity of function w defined by: $w(X \cup Y) + w(X \cap Y) \leq w(X) + w(Y)$ for all X, Y. Various set functions naturally considered in practical problems appear to be both submodular and monotonic with respect to set inclusion. Let us mention, for example, the budgeted-additive set function defined by $w(X) = \min\{\sum_{i \in X} w_i, B\}$, but also the coverage measure defined by $w(X) = |\bigcup_{i \in X} E_i|$ (where E_i is the list of elements covered by i) and satisfaction measures of the form $w(X) = \sum_{i \in I} p_i \max_{j \in X} \{u_{ij}\}$ used in the facility location problem (where I denotes a set of clients and u_{ij} the utility of location j for client i).

Although the problem of minimization of submodular functions is known to be polynomially solvable [14], the maximization of a submodular function is hard in general because it includes max-cut as special case. Approximate greedy and local search algorithms have been proposed for the maximization problem and some interesting worst case bounds on the quality of the approximations returned are known, see, e.g., [4,12,15,17]. In this paper we stay on this problem of finding a global maximum of a general submodular and monotonic set function under a matroid constraint, but in a different perspective. We propose an active learning approach aiming to iteratively collect preference information over sets to progressively infer new preference statements over other sets until an optimal subset can be determined. More precisely, our approach interleaves preference queries with some optimization steps of a combinatorial algorithm aiming to construct an optimal set. In particular, we propose an interactive greedy algorithm and an interactive local search algorithm combining preference elicitation and search to determine an optimal (or near-optimal) subset. The approach proposed in the paper extends to non-additive submodular functions a recent approach proposed in [2] for additive set functions.

The paper is organized as follows: in Sect. 2 we recall some background on matroids and regret-based incremental preference elicitation. Then in Sect. 3 we propose a near-admissible interactive greedy search algorithm for submodular optimization. In Sect. 4 we introduce an interactive local search algorithm based on improving sequences of element swaps for the same problem. In Sect. 5, both algorithms are tested on optimization problems involving different submodular set functions and different matroid constraints. We analyse and compare their performance and obtain various possible tradeoffs between the number of preference queries and the quality of the final solution returned.

2   Background

In this work, we consider the problem of finding a maximum weight independent set in a matroid. A matroid $\mathcal{M}$ is a pair $(E, \mathcal{I})$ where E is a set of size n

(called the *ground set*) and $\mathcal{I} \subseteq 2^E$ is a non-empty collection of sets (called the *independent sets*) such that, for all $X, Y \in 2^E$, the following properties hold:

(A_1): $(Y \in \mathcal{I}$ and $X \subseteq Y) \Rightarrow X \in \mathcal{I}$.
(A_2): $(X \in \mathcal{I}$ and $Y \in \mathcal{I}$ and $|Y| > |X|) \Rightarrow \exists e \in Y \setminus X$ s.t. $X \cup \{e\} \in \mathcal{I}$.

Axiom A_1 is sometimes called the *hereditary property* (or the *downward-closed property*) whereas A_2 is known as the *augmentation property* (or the *independent set exchange property*). Axiom A_2 implies that all maximal independent sets (w.r.t set inclusion) have the same cardinality. A maximal independent set is called a *basis* of the matroid, and the set of all bases will be denoted by $\mathcal{B}$ in the sequel. The cardinality of a basis is called the *rank* of the matroid and it will be denoted by $r(\mathcal{M})$ in the sequel. In this paper, a special focus will be given to the *uniform matroid*, which is defined by $\mathcal{I} = \{X \subseteq E : |X| \leq k\}$ for a given positive integer $k \leq n$. In the numerical tests, we will also consider the *partition matroid* which is defined by a collection $\mathcal{D} = \{D_1, \ldots, D_q\}$ of q disjoints subsets of E, a positive integer $d_i \leq |D_i|$ for all $i \in \{1, \ldots, q\}$ and $\mathcal{I} = \{X \subseteq \cup_{i=1}^q D_i : \forall i \in \{1, \ldots, q\}, |X \cap D_i| \leq d_i\}$.

The problem of finding a maximum weight independent set in a matroid can be defined as follows: given a matroid $\mathcal{M} = (E, \mathcal{I})$, we want to compute $\max_{X \in \mathcal{I}} w(X)$ where w is a positive set function defined on 2^E measuring the weight (or utility) of any subset of E. Here we assume that $w(\emptyset) = 0$, w is submodular (i.e., $w(X \cup Y) + w(X \cap Y) \geq w(X) + w(Y)$ for all $X, Y \subseteq E$) and w is monotonic with respect to set inclusion (i.e., $w(X) \leq w(Y)$ for all $X \subset Y \subseteq E$). Note that the latter assumption implies that we can focus on the bases of the matroid when searching for an optimal independent subset.

In this paper, we assume that w is a set function representing the subjective preferences of a Decision Maker (DM): for any two sets $X, Y \in 2^E$, X is preferred to Y if and only if $w(X) \geq w(Y)$. Hence finding a maximum weight basis amounts to determining an optimal basis according to the DM's preferences. Moreover, we assume that w is initially not known. Instead, we are given a (possibly empty) set $\mathcal{P}$ of pairs $(X, Y) \in \mathcal{I} \times \mathcal{I}$ such that X is known to be preferred to Y by the DM. Such preference data can be obtained by asking comparison queries to the DM (i.e., by asking the DM to compare two subsets and state which one is preferred). Let W be the *uncertainty set* implicitly defined as the set of all functions w that are compatible with $\mathcal{P}$, i.e., such that $w(X) \geq w(Y)$ for all $(X, Y) \in \mathcal{P}$. The problem is now to determine the most promising basis under preference imprecision. To this end, we consider the minimax regret decision criterion which is commonly used to make robust recommendations under preference imprecision in various decision contexts. The minimax regret (MMR) can be defined using pairwise max regrets (PMR) and max regrets (MR) as follows:

Definition 1. For any collection of sets $\mathcal{S} \subseteq 2^E$ and for any two sets $X, Y \in \mathcal{S}$:
$PMR(X, Y, W) = \max_{w \in W}\{w(Y) - w(X)\}$
$MR(X, \mathcal{S}, W) = \max_{Y \in \mathcal{S}} PMR(X, Y, W)$
$MMR(\mathcal{S}, W) = \min_{X \in \mathcal{S}} MR(X, \mathcal{S}, W)$

Thus $PMR(X, Y, W)$ is the worst-case loss when choosing X instead of Y. $MR(X, S, W)$ is the worst-case loss incurred when selecting X instead of any other set $Y \in S$. The set $\arg\min_{X \in S} MR(X, S, W)$ is the set of all optimal sets according to the minimax regret decision criterion. By definition, recommending any of these optimal sets allows to minimize the worst-case loss. Moreover, if $MMR(S, W) = 0$, then we know that these sets are necessarily optimal according to the DM's preferences.

Note that, depending on the available preference statements, the MMR value (representing the worst-case loss) might still be at an unacceptable level for the DM. As the MMR value can only decrease when adding new preference statements in $\mathcal{P}$, the minimax regret decision criterion can be used within an incremental elicitation process that progressively asks preference queries to the DM until the MMR value drops below a given *tolerance* threshold $\delta \geq 0$ (representing the maximum allowable gap to optimality) [3]. At that time, recommending any optimal basis for the minimax regret criterion ensures that the loss incurred by not choosing the preferred basis is bounded above by that threshold. This approach is sometimes referred to as *regret-based incremental elicitation* in the literature. Note that if we set $\delta = 0$, then the returned basis is necessarily optimal according to the DM's preferences. However, using $\delta > 0$ allows to reduce the number of generated preference queries in practice.

For matroid optimization problems, computing the MMR value at every step of the elicitation procedure may induce prohibitive computation times as it may require to compute the pairwise max regrets for all pairs of distinct bases in $\mathcal{B}$. Therefore, we propose instead to combine search and regret-based incremental elicitation to reduce both computation times and number of queries. More precisely, preference queries are generated during the search so as to progressively reduce the set W until being able to determine a (near-)optimal basis.

3   An Interactive Greedy Algorithm

For problems where w is exactly observable, good approximate solutions can be constructed using the following simple greedy algorithm: starting from $X = \emptyset$, the idea is to select an element $e \in E \setminus X$ that maximizes the marginal contribution to X, i.e.,

$$\Delta(e|X) = w(X \cup \{e\}) - w(X) \tag{1}$$

without loosing the independence property. The algorithm stops when no more element can be added to X (set X is a basis at the end of the procedure). For monotonic submodular set functions, this greedy algorithm has an approximation ratio of $(1 - \frac{1}{e}) \approx 0.63$ for the uniform matroid and an approximation ratio of $\frac{1}{2}$ in the general case [6,12]. For problems where the set function w is imprecisely known, we propose an interactive version of the greedy algorithm that generates preference queries only when it is necessary to discriminate between some elements. More precisely, queries are generated only when the available preference data is not sufficient to identify an element that could be added to set X so as to ensure that the returned basis is a good approximate solution with provable guarantees. We implement this idea by computing minimax regrets on sets $\mathcal{S} = \{X \cup \{e\} : e \in E \backslash X \text{ s.t. } X \cup \{e\} \in \mathcal{I}\}$, asking preference queries at step i until $MMR(\mathcal{S}, W)$ drops below a given threshold $\delta_i \geq 0$, where δ_i is a fraction of the tolerance threshold δ such that $\sum_{i=1}^{r(\mathcal{M})} \delta_i = \delta$ (see Algorithm 1).

Algorithm 1: Interactive Greedy Algorithm

---

1   $X \leftarrow \emptyset$;
2   $E_c \leftarrow E$;
3   **for** $i = 1 \ldots r(\mathcal{M})$ **do**
4      $\mathcal{S} \leftarrow \{X \cup \{e\} : e \in E_c\}$;
5   **while** $MMR(\mathcal{S}, W) > \delta_i$ **do**
6   Ask the DM to compare two elements of $\mathcal{S}$;
7   Update W according to the DM's answer;
8   **end**
9   Select $e \in E_c$ such that $\mathrm{MR}(X \cup \{e\}, \mathcal{S}, W) \leq \delta_i$ and move e from E_c to X;
10   Remove from E_c all elements e such that $X \cup \{e\} \notin \mathcal{I}$;
11 **end**
12 **return** X;

Note that Algorithm 1 generates no more than a polynomial number of queries. At every step, the number of queries is indeed bounded above by $|E|^2$ as comparison queries are generated until $MMR(\mathcal{S}, W) \leq \delta_i$, where $\mathcal{S} \subseteq \{X \cup \{e\} : e \in E\}$ (in the worst-case scenario, the DM is asked to compare all the elements of $\mathcal{S}$). Hence the number of steps of the while loop is also polynomial. Note however that the implementation of Algorithm 1 may differ significantly from one application context to another. In particular, checking whether $X \cup \{e\} \in \mathcal{I}$ can be more or less complex depending on the matroid under consideration. For example, when considering the uniform and partition matroids, the independence tests (line 10) can be performed in polynomial time. A second source of complexity is the computation of MMR values, which can be more or less simple depending on the assumptions made on w. An interesting option is to focus on parametric functions that are linear in their parameters (e.g., a linear combination of spline functions, or a linear multiattribute utility, or an ordered weighted average of criterion values). In that case, regret optimization can be performed in polynomial time using linear programming. Moreover, defining w by a parametric function enables to reduce the number of queries in practice, since any preference statement of type $w(X) \geq w(X')$ translates into a constraint on the parameter space, reducing possible preferences over other subsets.

We now provide theoretical guarantees on the quality of the returned solution. Before considering the general case, let us focus on the uniform matroid.

Proposition 1. *Let W_f be the final set W when Algorithm 1 stops. For the uniform matroid, Algorithm 1 is guaranteed to return a basis X such that:*

$$\forall w \in W_f, \; w(X) \geq \left(1 - \frac{1}{e}\right) w(X^*) - \delta, \; \text{where } X^* \in \arg\max_{Y \in \mathcal{I}} w(Y).$$

Proof. Let $w \in W_f$ and let $X^* \in \arg\max_{Y \in \mathcal{I}} w(Y)$. We want to prove that $w(X) \geq \left(1 - \frac{1}{e}\right) w(X^*) - \delta$ holds. Let e_i, $i \in \{1, \ldots, r(\mathcal{M})\}$, be the ith element inserted in X during the execution of Algorithm 1. Let X_i be the set X at the end of the ith iteration step (i.e., $X_i = \{e_1, \ldots, e_i\}$). Let W_i (resp. $\mathcal{S}_i$) denote the uncertainty set W (resp. the set $\mathcal{S}$) at the end of the ith iteration step. Let e_i^*, $i \in \{1, \ldots, r(\mathcal{M})\}$, denote the ith element of X^* in an arbitrary order.

For any step $i \in \{1, \ldots, r(\mathcal{M})\}$, we have $MR(X_{i-1} \cup \{e_i\}, \mathcal{S}_i, W_i) \leq \delta_i$ due to line 9. Since $w \in W_f \subseteq W_i$, we know that $w(X_{i-1} \cup \{e\}) - w(X_{i-1} \cup \{e_i\}) \leq \delta_i$ for all $e \in E_c$, where $E_c = E \backslash X_{i-1}$ for the uniform matroid (see lines 2 and 10). Then, from Eq. (1), we can derive $\Delta(e|X_{i-1}) - \Delta(e_i|X_{i-1}) \leq \delta_i$ for all $e \in E \backslash X_{i-1}$. Note that the last inequality also holds for all $e \in X_{i-1}$ as $\Delta(e|X_{i-1}) = 0$. Hence, for any step $i \in \{1, \ldots, r(\mathcal{M})\}$, we have:

$$\Delta(e|X_{i-1}) - \Delta(e_i|X_{i-1}) \leq \delta_i, \; \forall e \in E \tag{2}$$

Then we obtain:

$$w(X^*) \leq w(X_{i-1} \cup X^*) \; \text{(since w is monotonic)}$$

$$= w(X_{i-1}) + \sum_{j=1}^{r(\mathcal{M})} \left(w(X_{i-1} \cup \{e_1^*, \ldots, e_j^*\}) - w(X_{i-1} \cup \{e_1^*, \ldots, e_{j-1}^*\}) \right)$$

$$= w(X_{i-1}) + \sum_{j=1}^{r(\mathcal{M})} \Delta(e_j^*|X_{i-1} \cup \{e_1^*, e_2^*, \ldots, e_{j-1}^*\}) \; \text{(by Equation (1))}$$

$$\leq w(X_{i-1}) + \sum_{j=1}^{r(\mathcal{M})} \Delta(e_j^*|X_{i-1}) \; \text{(since w is submodular)}$$

$$\leq w(X_{i-1}) + \sum_{j=1}^{r(\mathcal{M})} \left(\Delta(e_i|X_{i-1}) + \delta_i \right) \; \text{(by Equation (2))}$$

$$= w(X_{i-1}) + r(\mathcal{M}) \times \left(\Delta(e_i|X_{i-1}) + \delta_i \right)$$

From the last inequality, we can derive:

$$\frac{1}{r(\mathcal{M})} \left(w(X^*) - w(X_{i-1}) \right) - \delta_i \leq \Delta(e_i|X_{i-1})$$

which can be rewritten as follows:

$$\frac{1}{r(\mathcal{M})} \left(w(X^*) - w(X_{i-1}) \right) - \delta_i \leq w(X^*) - w(X_{i-1}) - \left(w(X^*) - w(X_i) \right)$$

since $X_i = X_{i-1} \cup \{e_i\}$. Therefore we have $\frac{\Pi_{i-1}}{r(\mathcal{M})} - \delta_i \leq \Pi_{i-1} - \Pi_i$ or equivalently:

$$\Pi_i \leq \left(1 - \frac{1}{r(\mathcal{M})}\right)\Pi_{i-1} + \delta_i$$

where Π_i is simply defined by $\Pi_i = w(X^*) - w(X_i)$ for all $i \in \{0, \ldots, r(\mathcal{M})\}$. By recursively applying this inequality, we obtain:

$$\Pi_{r(\mathcal{M})} \leq \left(1 - \frac{1}{r(\mathcal{M})}\right)^{r(\mathcal{M})} \times \Pi_0 + \sum_{i=1}^{r(\mathcal{M})} \delta_i \left(1 - \frac{1}{r(\mathcal{M})}\right)^{r(\mathcal{M})-i}$$

Then, since $\Pi_0 = w(X^*)$ and $\Pi_{r(\mathcal{M})} = w(X^*) - w(X)$, we obtain:

$$w(X^*) - w(X) \leq \left(1 - \frac{1}{r(\mathcal{M})}\right)^{r(\mathcal{M})} \times w(X^*) + \sum_{i=1}^{r(\mathcal{M})} \delta_i \left(1 - \frac{1}{r(\mathcal{M})}\right)^{r(\mathcal{M})-i}$$

or equivalently:

$$w(X) \geq \left(1 - (1 - \frac{1}{r(\mathcal{M})})^{r(\mathcal{M})}\right)w(X^*) - \sum_{i=1}^{r(\mathcal{M})} \delta_i \left(1 - \frac{1}{r(\mathcal{M})}\right)^{r(\mathcal{M})-i}$$

Finally, using $1 - x \leq e^{-x}$ for all $x \in \mathbb{R}$, and $1 - \frac{1}{x} \leq 1$ for all $x \in \mathbb{R}_+^*$, we obtain:

$$w(X) \geq \left(1 - \frac{1}{e}\right)w(X^*) - \sum_{i=1}^{r(\mathcal{M})} \delta_i = \left(1 - \frac{1}{e}\right)w(X^*) - \delta$$

$\square$

Note that Proposition 1 cannot be extended to the case of general matroid, as inequalities of type $\Delta(e_i|X_{i-1}) + \delta_i \geq \Delta(e_j^*|X_{i-1})$ may not hold anymore ($E_c \neq E \backslash X_i$ in the general case). We now establish a more general result.

Proposition 2. *Let W_f be the final set W when Algorithm 1 stops. Algorithm 1 is guaranteed to return a basis X such that:*

$$\forall w \in W_f, \, w(X) \geq \frac{1}{2}(w(X^*) - \delta), \quad \text{where } X^* \in \arg\max_{Y \in \mathcal{I}} w(Y).$$

Proof. Let $w \in W_f$ and let $X^* \in \arg\max_{Y \in \mathcal{I}} w(Y)$. We want to prove that $w(X) \geq \frac{1}{2}(w(X^*) - \delta)$ holds. Let e_i, $i \in \{1, \ldots, r(\mathcal{M})\}$, be the ith element inserted in X during the execution of Algorithm 1. Let X_i be the set X at the end of the ith iteration step (i.e., $X_i = \{e_1, \ldots, e_i\}$). Let W_i (resp. S_i) denote the uncertainty set W (resp. the set S) at the end of the ith iteration step.

Due to a well-known multiple exchange theorem [7], there exists a one-to-one correspondence $\sigma : X \to X^*$ such that $B_i = (X \backslash \{e_i\}) \cup \{\sigma(e_i)\}$ is a basis of the matroid for every element $e_i \in X$. Then we can derive $X_{i-1} \cup \{\sigma(e_i)\} \in \mathcal{I}$ from $X_{i-1} \cup \{\sigma(e_i)\} \subseteq B_i$ (using Axiom A_1), and therefore we necessarily have

$X_{i-1} \cup \{\sigma(e_i)\} \in \mathcal{S}_i$ at step i. Since $MR(X_{i-1} \cup \{e_i\}, \mathcal{S}_i, W_i) \leq \delta_i$ (line 9), we obtain $w(X_{i-1} \cup \{\sigma(e_i)\}) - w(X_{i-1} \cup \{e_i\}) \leq \delta_i$, which can be rewritten:

$$\Delta(\sigma(e_i)|X_{i-1}) - \Delta(e_i|X_{i-1}) \leq \delta_i \qquad (3)$$

Then we obtain:

$w(X^*) \leq w(X \cup X^*)$ (since w is monotonic)

$$= w(X) + \sum_{i=1}^{r(\mathcal{M})} \left(w(X \cup \{\sigma(e_1), \ldots, \sigma(e_i)\}) - w(X \cup \{\sigma(e_1), \ldots, \sigma(e_{i-1})\}) \right)$$

$$= w(X) + \sum_{i=1}^{r(\mathcal{M})} \Delta(\sigma(e_i)|X \cup \{\sigma(e_1), \ldots, \sigma(e_{i-1})\}) \text{ (by Equation (1))}$$

$$\leq w(X) + \sum_{i=1}^{r(\mathcal{M})} \Delta(\sigma(e_i)|X_{i-1}) \text{ (since } w \text{ is submodular and } X_{i-1} \subseteq X)$$

$$\leq w(X) + \sum_{i=1}^{r(\mathcal{M})} \left(\Delta(e_i|X_{i-1}) + \delta_i \right) \text{ (by Equation (3))}$$

$$= 2w(X) + \sum_{i=1}^{r(\mathcal{M})} \delta_i \text{ (by Equation (1))}$$

$$= 2w(X) + \delta \text{ (which establishes the result)} \qquad \square$$

Example 1. We now present an execution of our algorithm. Consider an instance of the maximum coverage problem over a uniform matroid with a set $V = \{v_1, \ldots, v_q\}$, $q = 10$, and a family of $n = 8$ subsets $E = \{S_1, \ldots, S_n\}$ defined by (Table 1):

Table 1. Subsets used in Example 1.

S_1	S_2	S_3	S_4	S_5	S_6	S_7	S_8
v_3	v_1	v_6	v_2	v_7	v_6	v_2	v_1
v_4	v_3	v_{10}	v_8	v_9	v_7	v_8	v_3
v_5					v_{10}		v_5

A feasible solution is a collection of subsets $X \subseteq E$ such that $|X| \leq k$ (here we set $k = 2$), and the goal is to identify a feasible solution X maximizing $w(X)$ for a given set function w defined on 2^E. Here we assume that w is defined by:

$$w(X) = \sum_{v \in \bigcup_{S \in X} S} u(v) \qquad (4)$$

where $u(v) \geq 0$ is the utility of element $v \in V$. In that case, it can be proved that w is monotone and submodular [8]. We further assume that all elements $v \in V$ are evaluated with respect to 3 criteria (denoted by u_1, u_2, and u_3), and their evaluations are given in Table 2. Then, the utility of any element $v \in V$ is:

$$u(v) = \sum_{i=1}^{3} \lambda_i u_i(v) \tag{5}$$

where $\lambda = (\lambda_1, \lambda_2, \lambda_3) \in \mathbb{R}_+$ represents the value system of the DM.

Table 2. Performance vectors attached to elements in Example 1.

	v_1	v_2	v_3	v_4	v_5	v_6	v_7	v_8	v_9	v_{10}
u_1	4	2	2	3	7	6	8	7	7	1
u_2	5	7	1	2	3	1	5	1	9	1
u_3	4	5	3	7	2	5	3	8	4	4

In this example, we assume that the DM's preferences can be represented by the set function w^* defined by the hidden parameter $\lambda^* = (0.2, 0.5, 0.3)$. Here we start the execution with no preference data, and therefore we have to consider all weighting vectors λ in the set $\Lambda = \{\lambda \in [0,1]^3 : \sum_{i=1}^{3} \lambda_i = 1\}$, which implicitly defines the uncertainty set W using Eqs. (4–5). In Fig. 1, Λ is represented by triangle ABC in the space (λ_1, λ_2), λ_3 being implicitly defined by $\lambda_3 = 1 - \lambda_1 - \lambda_2$. Now, let us execute Algorithm 1 with $\delta = 0$. Note that only two iteration steps are needed as the rank of the uniform matroid is equal to k.

First Iteration Step: We have $X = \emptyset$ and $E_c = E$, and therefore $S = E$. Since $MMR(S, W) = 6 > 0$, the DM is asked to compare two elements of S, say S_5 and S_7. Since we have $w^*(\{S_5\}) = 12.1 \geq 9.7 = w^*(\{S_7\})$, the answer is: "subset S_5 is better than subset S_7". Then W is updated by imposing the constraint $w(\{S_5\}) \geq w(\{S_7\})$ which amounts to restricting Λ by imposing $\lambda_2 \geq \frac{1}{2} - \lambda_1$. Now Λ is represented by the polyhedron BCDE in Fig. 2. Since $MMR(S, W) = 2.5$, the DM is asked to compare two subsets, say S_5 and S_6. Since we have $w^*(\{S_5\}) = 12.1 \geq w^*(\{S_6\}) = 10.1$, the DM answers: "subset S_5 is better than subset S_6". Then, W is updated by imposing the constraint $w(\{S_5\}) \geq w(\{S_6\})$, which amounts to further restricting Λ by imposing $\lambda_2 \geq \frac{5}{12} - \frac{5}{12}\lambda_1$. Now Λ is represented by the polyhedron BCFE in Fig. 3. We have $MMR(S, W) = MR(\{S_5\}, S, W) = 0$, and therefore S_5 is added to X.

Second Iteration Step: We have $X = \{S_5\}$ and $E_c = E \setminus \{S_5\}$, and therefore $S = \{\{S_5\} \cup \{S\} : S \in E_c\}$. Since $MMR(S, W) = 1.5$, we ask the DM to compare two elements of S, say $\{S_5, S_8\}$ and $\{S_5, S_7\}$. The DM prefers the former option as $w^*(\{S_5, S_8\}) = 21.9 \geq w(\{S_5, S_7\}) = 21.8$. The uncertainty set W is therefore updated by imposing $w(\{S_5, S_8\}) \geq w(\{S_5, S_7\})$, i.e., $\lambda_2 \geq 1 - \frac{8}{3}\lambda_1$. Now Λ is represented by the triangle BGC in Fig. 4. Since we have $MMR(\mathcal{S}, W) = MR(\{S_5, S_8\}, \mathcal{S}, W) = 0$, then subset S_8 is added to X.

As $|X| = k = 2$, the algorithm stops and returns $X = \{S_5, S_8\}$ which is the optimal solution for this instance. This shows that we are able to make good recommendations without knowing λ^* precisely (here only 3 queries are needed).

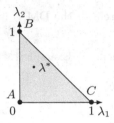

Fig. 1. Initial set Λ.

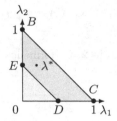

Fig. 2. Λ after 1 query.

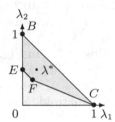

Fig. 3. Λ after 2 queries.

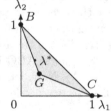

Fig. 4. Λ after 3 queries.

4   An Interactive Local Search

In this section, we consider another efficient way of constructing a good approximate solution to matroid optimization problems with monotonic submodular functions. More precisely, we focus on the following simple local search approach: starting from an arbitrary basis X, the idea is to replace one element $e \in X$ by an element $e' \in E \backslash X$ such that $X \cup \{e'\} \backslash \{e\}$ belongs to $\mathcal{I}$ and is better than X. This simple exchange principle can be iterated until reaching a local optimum. When w is exactly observable, the local search algorithm has an approximation ratio of $1/2$, even in the special case of the uniform matroid [6]. When w is not known, the local search algorithm can be combined with a preference elicitation method which collects preference data only when it is necessary to identify improving swaps. To implement this idea, we propose Algorithm 2 where N_X is the neighborhood of basis X (i.e., the set of bases that differ from X by exactly one element). The procedure ComputeInitialBasis called in line 1 can be any heuristic providing a good starting solution (see the numerical tests).

Algorithm 2: The Interactive Local Search Algorithm

---

1 $X \leftarrow$ ComputeInitialBasis($\mathcal{M}$);
2 improve $\leftarrow$ **true**;
3 **while** *improve* **do**
4 $\quad$ $N_X \leftarrow \{X' \in \mathcal{B} : |\{X \setminus X'\} \cup \{X' \setminus X\}| = 2\}$;
5 $\quad$ $\mathcal{S} \leftarrow N_X \cup \{X\}$;
6 $\quad$ **while** $MMR(\mathcal{S}, W) > \delta/r(\mathcal{M})$ **do**
7 $\quad\quad$ Ask the DM to compare two elements of $\mathcal{S}$;
8 $\quad\quad$ Update W according to the DM's answer;
9 $\quad$ **end**
10 $\quad$ **if** $MR(X, \mathcal{S}, W) \leq \delta/r(\mathcal{M})$ **then**
11 $\quad\quad$ improve $\leftarrow$ **false**
12 $\quad$ **else**
13 $\quad\quad$ $X \leftarrow$ RandomSelect($\arg \min_{X' \in N_X} MR(X', \mathcal{S}, W)$)
14 $\quad$ **end**
15 **end**
16 **return** X;

The following proposition shows that the basis returned by Algorithm 2 is a good approximate solution.

Proposition 3. *Let W_f be the final set W when Algorithm 2 stops. Algorithm 2 is guaranteed to return a basis X such that:*

$$\forall w \in W_f, \ w(X) \geq \frac{1}{2}(w(X^*) - \delta), \ \text{where } X^* \in \arg \max_{Y \in \mathcal{I}} w(Y).$$

Proof. Let $w \in W_f$ and let $X^* \in \arg\max_{Y \in \mathcal{I}} w(Y)$. We want to prove that $w(X) \geq \frac{1}{2}(w(X^*) - \delta)$ holds. Let e_i, $i \in \{1, \ldots, r(\mathcal{M})\}$, denote the ith element of X in an arbitrary order. Let X_i be the set defined by $X_i = \{e_1, \ldots, e_i\}$. Due to the multiple exchange theorem, there exists a one-to-one correspondence $\sigma : X \to X^*$ such that $B_i = (X \setminus \{e_i\}) \cup \{\sigma(e_i)\}$ is a basis of the matroid for every element $e_i \in X$. Note that $B_i \in N_X$ (the neighborhood of X) for all $i \in \{1, \ldots, r(\mathcal{M})\}$ since B_i differs from X by exactly one element. Moreover, we have $MR(X, N_X, W) \leq \delta/r(\mathcal{M})$ at the end of the execution (due to line 10). Therefore, $w(B_i) - w(X) \leq \delta/r(\mathcal{M})$ holds by definition of max regrets, which can be rewritten:

$$\Delta(\sigma(e_i)|X \setminus \{e_i\}) - \Delta(e_i|X \setminus \{e_i\}) \leq \frac{\delta}{r(\mathcal{M})} \tag{6}$$

using Eq. (1). Then, we obtain:

$$w(X^*) \leq w(X) + \sum_{e \in X^*} \Delta(e|X) \text{ (by submodularity, see [12] for a proof)}$$

$$= w(X) + \sum_{i=1}^{r(\mathcal{M})} \Delta(\sigma(e_i)|X)$$

$$\leq w(X) + \sum_{i=1}^{r(\mathcal{M})} \Delta(\sigma(e_i)|X \setminus \{e_i\}) \text{ (by submodularity)}$$

$$\leq w(X) + \sum_{i=1}^{r(\mathcal{M})} \left(\Delta(e_i|X \setminus \{e_i\}) + \frac{\delta}{r(\mathcal{M})}\right) \text{ (by Equation (6))}$$

$$\leq w(X) + \sum_{i=1}^{r(\mathcal{M})} \left(\Delta(e_i|X_{i-1}) + \frac{\delta}{r(\mathcal{M})}\right) \text{ (since } X_{i-1} \subseteq X \setminus \{e_i\})$$

$$= 2w(X) + \delta \text{ (which establishes the result)}$$

$\square$

Contrary to the greedy algorithm, we cannot prove that the local search algorithm generates a polynomial number of queries and ends after a polynomial number of iterations. More precisely, when $\delta \neq 0$, some cycles of type $(X_1, \ldots, X_t)$ with $X_{i+1} \in N_{X_i}$ and $X_1 = X_t$ can even occur. Fortunately, when a cycle is detected, it can be easily broken by iteratively dividing δ by two (while still guaranteeing the near-optimality of the returned basis). Despite these poor theoretical properties, we will see in the experimental section that the local search algorithm achieves good results in practice.

5   Experimental Results

In this section, we report the results obtained by our algorithms on two problems: the maximum coverage problem and the collective selection of items. Two matroid constraints have been considered for each problem: the uniform matroid and the partition matroid. The algorithms are evaluated through three performance indicators: number of queries, computation times (given in seconds) and empirical error, expressed as a percentage from the optimal solution. For the local search algorithm, we also report the number of iteration steps (NbI) performed by the algorithm. In our tests, two tolerance thresholds have been used: $\delta = 0$ and $\delta = 20\%$ of the initial maximum regret (to reduce the number of preference queries). Results are averaged over 30 runs. All the results have been obtained with a program written in C++ and tested on an Intel Core i7-9700, 3.00 GHz with 15,5 GB of RAM. Pairwise max regret optimizations were performed by CPLEX (https://www.ibm.com/analytics/cplex-optimizer).

To generate preference queries during the execution of our algorithms, we use the well-known query selection strategy called the Current Solution Strategy (CSS) [3] which consists in asking the DM to compare a solution X minimizing the max regret to one of its best challengers arbitrary chosen in the set $\arg\max_Y PMR(X, Y, W)$.

5.1 The Maximum Coverage Problem

Here we consider instances of the maximum coverage problem with a set $V = \{v_1, \ldots, v_q\}$ of $q = 100$ elements, and a family E of $n = 80$ subsets of V. The family of subsets are generated as suggested in [13]. The utility of an element $v \in V$ is defined by a weighted sum $u^\lambda(v) = \sum_{i=1}^p \lambda_i u_i(v)$ where u_i is the evaluation of v on criterion $i \in \{1, \ldots, p\}$. Utilities are randomly generated within $[1, 10]$ and three values of p are considered: $p = 4, 6$, and 8. The DM's preferences are then represented by a submodular monotone set function w defined by:

$$w(X) = \sum_{v \in \bigcup_{S \in X} S} u^\lambda(v)$$

for any $X \subseteq E$. Here we assume that λ is initially unknown. Answers to queries are simulated using a hidden vector λ randomly generated before running the algorithms. For the uniform matroid, we focus on subsets of size at most $k = 16$, i.e., $\mathcal{I} = \{X \subset E : |X| \leq 16\}$. For the partition matroid, set E is randomly partitioned into $q = 4$ sets $\mathcal{D} = \{D_1, \ldots, D_q\}$, and at most $d_i = 4$ elements can be selected for all $i \in \{1, \ldots, q\}$, i.e., $\mathcal{I} = \{X \subseteq E : \forall i \in \{1, \ldots, 4\}, |X \cap D_i| \leq 4\}$. The results are given in Table 3 and Table 4 respectively. For our local search algorithm, we consider two implementations

Table 3. Results obtained for the maximum coverage problem with uniform matroid.

		Greedy			Local search (random start)				Local search (greedy start)			
δ	p	Time(s)	Queries	Error(%)	Time(s)	Queries	Error(%)	NbI	Time(s)	Queries	Error(%)	NbI
0	4	4.4	19.4	1.1	0.2	17.2	0.9	12.6	0.2	12.1	0.5	2.9
	6	6.9	36.2	1.3	0.8	32.4	1.0	12.6	0.3	17.1	0.4	2.5
	8	9.9	48.3	1.1	1.4	42.9	0.8	12.5	0.5	28.5	0.3	3.0
0.2	4	2.8	7.2	1.3	0.1	7.6	6.1	7.7	0.1	9.5	0.5	2.9
	6	3.9	13.7	1.5	0.3	13.1	4.6	8.9	0.3	13.7	0.4	2.4
	8	5.0	17.3	1.3	0.6	19.5	4.0	9.4	0.4	21.8	0.3	2.9

Table 4. Results obtained for the maximum coverage problem with partition matroid.

		Greedy			Local search (random start)				Local search (greedy start)			
δ	p	Time(s)	Queries	Error(%)	Time(s)	Queries	Error(%)	NbI	Time(s)	Queries	Error(%)	NbI
0	4	3.8	20.5	4.2	0.2	16.2	4.1	11.1	0.1	9.0	2.2	2.8
	6	5.4	33.4	3.5	0.6	25.5	4.5	10.7	0.2	13.4	1.8	2.7
	8	6.8	44.3	4.1	0.9	35.9	4.3	10.6	0.3	17.7	2.2	3.0
0.2	4	2.4	7.4	4.3	0.1	7.5	6.8	7.7	0.1	7.2	2.3	2.7
	6	3.1	12.4	3.9	0.2	10.6	7.2	7.6	0.2	10.3	1.9	2.7
	8	4.1	17.8	4.5	0.4	15.9	5.9	8.8	0.3	14.6	2.2	3.0

of the procedure `ComputeInitialBasis`: we generate a basis at random (random start), or we use the standard greedy algorithm with the uniform weighting vector $\lambda = (1/p, \ldots, 1/p)$ (greedy start).

For $\delta = 0$, we observe that the interactive greedy algorithm is outperformed by the interactive local search procedures: the interactive greedy algorithm is about 10 times slower on average and asks more preference queries. Moreover, we observe that the local search performs better when considering the greedy start heuristic instead of the random start heuristic. We also observe that using $\delta = 0.2$ allows to significantly reduce the number of queries, without increasing the error too much (except for the local search with a random starting point). Finally, we observe that our algorithms perform better on the uniform matroid than on the partition matroid which is a little more complex.

5.2   The Collective Subset Selection Problem

In the collective subset selection problem, we are given a set A of m agents, and a set $E = \{e_1, \ldots, e_n\}$ of n items. Every agent $a \in A$ gives a score $s_a(e) \geq 0$ to each item $e \in E$, and the utility that agent a derives from a set $X \subseteq E$ is defined by an ordered weighted average (OWA) [18]. More precisely, for any $X = \{x_1, \ldots, x_\ell\} \subseteq E$ of size $\ell \leq n$, the utility of agent a is defined by:

$$u_a^\lambda(X) = \sum_{i=1}^{\ell} \lambda_i s_a(x_{(i)})$$

where $(\cdot)$ is a permutation of $\{1, \ldots, \ell\}$ sorting the elements of X is non-increasing order (i.e., $s_a(x_{(1)}) \geq \ldots \geq s_a(x_{(\ell)})$), and $\lambda = (\lambda_1, \ldots, \lambda_n) \in [0, 1]^n$ is a non-increasing normalized vector. Here the set function w is simply defined by:

$$w(X) = \sum_{a \in A} u_a^\lambda(X)$$

Note that function w is a submodular as functions u_a^λ, $a \in A$, are submodular whenever vector λ is non-increasing (see Skowron $et\ al.$ [16]). Here also we assume that λ is initially unknown, and answers to queries are simulated using a hidden weighting vector. We consider instances with $m = 50$ agents and $n = 50$ items, and scores are randomly generated within in $[1, 100]$. For the uniform matroid, two values of k have been tested: $k = 5$, and $k = 10$. For the partition matroid, E is randomly partitioned into 4 sets, and at most $d/4$ items can be selected in each set; we consider two values of d ($d = 8$ and $d = 16$). The results are given in Table 5 and Table 6 respectively. For this problem, we observe that our interactive greedy algorithm outperforms our interactive local search procedure. With $\delta = 0.2$, the greedy algorithm is very efficient: the error is less than 0.2% and the number of queries does not exceed 7.

Table 5. Collective selection of items problem under uniform matroid constraint.

		Greedy			Local search (random start)				Local search (greedy start)			
δ	d	Time(s)	Queries	Error(%)	Time(s)	Queries	Error(%)	NbI	Time(s)	Queries	Error(%)	NbI
0	5	0.9	8.2	0.2	3.1	14.4	0.1	5.8	1.6	11.4	0.0	3.1
	10	2.7	25.1	0.1	34.4	45.6	0.0	9.3	31.2	35.5	0.0	4.3
0.2	5	0.6	3.0	0.2	0.8	6.8	0.5	5.0	1.1	6.9	0.1	2.9
	10	1.2	4.3	0.1	12.8	16.6	0.6	7.5	17.7	18.5	0.0	4.1

Table 6. Collective selection of items problem under partition matroid constraint.

		Greedy			Local search (random start)				Local search (greedy start)			
δ	d	Time(s)	Queries	Error(%)	Time(s)	Queries	Error(%)	NbI	Time(s)	Queries	Error(%)	NbI
0	8	1.3	14.0	0.2	2.2	24.0	0.0	8.1	1.5	17.7	0.0	3.8
	16	3.1	38.9	0.1	21.5	56.9	0.0	12.0	0.4	34.0	0.0	4.1
0.2	8	0.8	4.2	0.2	1.8	18.5	0.1	8.1	1.2	13.4	0.1	3.7
	16	1.4	6.3	0.1	16.0	43.4	0.2	11.4	6.9	25.7	0.0	4.0

6 Conclusion

We have proposed two interactive algorithms (greedy and local search) combining the elicitation of a submodular utility function and the determination of the optimal independent subset in a weighted matroid. Both algorithms admit performance guarantees on the quality of the returned basis. The tradeoff between the quality of solutions and the number of preference queries used in the process can be controlled by the parameter δ used to define admissible max regrets. Our approach has been tested on two specific problems, but many others could be solved with similar performances due to the generality of matroids. Note also that a counterpart of this interactive approach could be proposed for the optimization of supermodular functions.

References

1. Ahmed, S., Atamtürk, A.: Maximizing a class of submodular utility functions. Math. Program. **128**(1), 149–169 (2011)
2. Benabbou, N., Leroy, C., Lust, T., Perny, P.: Combining preference elicitation with local search and greedy search for matroid optimization. In: 35th AAAI Conference on Artificial Intelligence (AAAI'21) (2021)
3. Boutilier, C., Patrascu, R., Poupart, P., Schuurmans, D.: Constraint-based optimization and utility elicitation using the minimax decision criterion. Artif. Intell **170**(8–9), 686–713 (2006)
4. Calinescu, G., Chekuri, C., Pal, M., Vondrák, J.: Maximizing a monotone submodular function subject to a matroid constraint. SIAM J. Comput. **40**(6), 1740–1766 (2011)
5. Edmonds, J.: Matroids and the greedy algorithm. Math. Program. **1**(1), 127–136 (1971)

 6. Fisher, M.L., Nemhauser, G.L., Wolsey, L.A.: An analysis of approximations for maximizing submodular set functions-II. In: Balinski, M.L., Hoffman, A.J. (eds.) Polyhedral Combinatorics. Mathematical Programming Studies, vol. 8, pp. 73–87. Springer, Heidelberg (1978). https://doi.org/10.1007/BFb0121195
 7. Greene, C., Magnanti, T.L.: Some abstract pivot algorithms. SIAM J. Appl. Math. **29**(3), 530–539 (1975)
 8. Hochbaum, D.S., Pathria, A.: Analysis of the greedy approach in problems of maximum k-coverage. Naval Res. Logist. (NRL) **45**(6), 615–627 (1998)
 9. Lehmann, B., Lehmann, D., Nisan, N.: Combinatorial auctions with decreasing marginal utilities. Games Econ. Behav. **55**(2), 270–296 (2006)
10. Minoux, M.: Accelerated greedy algorithms for maximizing submodular set functions. In: Stoer, J. (ed.) Optimization Techniques. Lecture Notes in Control and Information Sciences, vol. 2, pp. 234–243. Springer, Heidelberg (1978). https://doi.org/10.1007/BFb0006528
11. Nemhauser, G.L., Wolsey, L.A.: Maximizing submodular set functions: formulations and analysis of algorithms. In: North-Holland Mathematics Studies, vol. 59, pp. 279–301. Elsevier, Amsterdam (1981)
12. Nemhauser, G.L., Wolsey, L.A., Fisher, M.L.: An analysis of approximations for maximizing submodular set functions-I. Math. Program. **14**(1), 265–294 (1978)
13. Resende, M.: Computing approximate solutions of the maximum covering problem with GRASP. J. Heurist. **4**(2), 161–177 (1998)
14. Schrijver, A.: A combinatorial algorithm minimizing submodular functions in strongly polynomial time. J. Comb. Theory Ser. B **80**(2), 346–355 (2000)
15. Skowron, P.: FPT approximation schemes for maximizing submodular functions. Inf. Comput. **257**, 65–78 (2017)
16. Skowron, P., Faliszewski, P., Lang, J.: Finding a collective set of items: from proportional multirepresentation to group recommendation. Artif. Intell. **241**, 191–216 (2016)
17. Vondrák, J.: Optimal approximation for the submodular welfare problem in the value oracle model. In: Proceedings of the Fortieth Annual ACM Symposium on Theory of Computing, pp. 67–74 (2008)
18. Yager, R.R.: On ordered weighted averaging aggregation operators in multicriteria decisionmaking. IEEE Trans. Syst. Man Cybern. **18**(1), 183–190 (1988)

Necessary and Possible Interaction in a 2-Maxitive Sugeno Integral Model

Paul Alain Kaldjob Kaldjob[✉], Brice Mayag, and Denis Bouyssou

University Paris-Dauphine, University PSL, CNRS, LAMSADE, 75016 Paris, France
{paul-alain.kaldjob-kaldjob,brice.mayag,denis.bouyssou}@dauphine.fr

Abstract. This paper proposes and studies the notion of interaction between two criteria in a 2-maxitive Sugeno integral model. Within the framework of binary alternatives, we give a necessary and sufficient condition for preferential information on binary alternatives to be representable by a 2-maxitive Sugeno integral model. Using this condition, we show that it is always possible to choose a numerical representation, for which all the interaction indices are strictly positive. Outside the framework of binary alternatives, by introducing binary variables, we propose a MILP allowing to test whether an ordinal preference information is representable by a 2-maxitive Sugeno integral model and whether the interpretation of the interaction indices is ambiguous or not. We illustrate our results with examples.

Keywords: Binary alternatives · Sugeno integral model · Interaction indices · 2-maxitive capacity

1 Introduction

The Sugeno integral was introduced in [19] and it is an aggregation function in the ordinal approach to decision making. There are numerous applications of the Sugeno integral in decision making [6,10], and the problem of identification of fuzzy measures, based on which fuzzy integrals are defined, has attracted substantial attention [1,7,9].

In the context of a 2-additive Choquet integral model, the interaction index between two criteria coincides with the Möbius mass of this pair [16]. By analogy, in the context of a 2-maxitive Sugeno integral, in this paper we propose a definition of the interaction index between two criteria.

In [16], we find necessary and sufficient conditions for a preferential information on a set of binary alternatives to be represented by a 2-additive Choquet integral model. We give a similar result with the 2-maxitive Sugeno integral model.

In [15], it is proven that in the framework of binary alternatives, if the preferential information contains no indifference, and if it is representable by a 2-additive Choquet integral model, then it is always possible to represent these

© Springer Nature Switzerland AG 2021
D. Fotakis and D. Ríos Insua (Eds.): ADT 2021, LNAI 13023, pp. 323–337, 2021.
https://doi.org/10.1007/978-3-030-87756-9_21

preferences by a strict positive interaction. We obtain a similar result with the 2-maxitive Sugeno integral.

Outside the framework of binary alternatives, in [16] we find a simple LP allowing us to test whether ordinal preference information is representable by a 2-additive Choquet integral model and whether the interpretation of interaction indices is or not ambiguous. Outside the framework of binary alternatives, using the linearization of the min and max functions, we propose a MILP allowing to test whether an ordinal preference information is representable by a 2-maxitive Sugeno integral model and whether the interpretation of the interaction indices is or not ambivalent.

This paper is organized as follows. After having recalled in Sect. 2 some basic elements on the model of the Sugeno integral in MCDM, in Sect. 3, we use a classic example to argue that the usual interpretation of interaction indices is not always convincing. In Sect. 4, we give our main results. Outside the framework of binary alternatives, by introduicing binary variables, we propose a MILP model allowing to test whether an ordinal preference information is representable by a 2-maxitive Sugeno integral model and whether the interaction is necessary or not. We illustrate our results with an example in Sect. 6, and we end by a conclusion.

2   Notations and Definitions

2.1   The Framework

Let $N = \{1, 2, \ldots, n\}$ be a set of n criteria and L_i the evaluation scale for the criterion $i \in N$. We denote by 0_i (resp. 1_i) the smallest (resp. biggest) element of L_i. An alternative is a vector $x = (x_1, \cdots, x_n) \in L_1 \times L_2 \times \cdots \times L_n$ where x_i is the ordinal evaluation of the alternative with respect to the criterion $i \in N$. The criteria are recoded numerically using, for all $i \in N$, a function u_i from L_i into $\mathbb{R}$.

2.2   Sugeno Integral

The Sugeno integral [5,6,19] is an aggregation function known in MCDM. It is based on the notion of capacity [3,18] defined as a function μ from the powerset 2^N into L such that:

- $\mu(\emptyset) = 0$,
- $\mu(N) = 1$,
- $\forall S, T \in 2^N, [S \subseteq T \Longrightarrow \mu(S) \leq \mu(T)]$ (monotonicity).

For an alternative $x = (x_1, \cdots, x_n) \in L_1 \times L_2 \times \cdots \times L_n$, the expression of the Sugeno integral w.r.t. a capacity μ is given by:

$$S_\mu\big(u(x)\big) = \bigvee_{i=1}^{n} \left(u_{\sigma(i)}(x_{\sigma(i)}) \wedge \mu(N_{\sigma(i)}) \right) \tag{1}$$

where $u(x) = (u_1(x_1), \cdots, u_n(x_n))$, σ is a permutation on N such that: $N_{\sigma(i)} = \{\sigma(i), \cdots, \sigma(n)\}$ and $u_{\sigma(1)}(x_{\sigma(1)}) \leq u_{\sigma(2)}(x_{\sigma(2)}) \leq \cdots \leq u_{\sigma(n)}(x_{\sigma(n)})$.

Remark 1. The Sugeno integral is equivalent to the following expression (see [14])

$$S_\mu\big(u(x)\big) = \bigvee_{A \subseteq N} \left(\big(\bigwedge_{i \in A} u_i(x_i) \big) \wedge \mu(A) \right). \tag{2}$$

The ordinal Möbius transform [12] m^μ of μ is defined by:

$$m^\mu(S) = \begin{cases} \mu(S) & \text{if } \mu(S) > \bigvee_{T \subsetneq S} \mu(T) \\ 0 & \text{otherwise} \end{cases} \tag{3}$$

The definition of a capacity generally requires $2^n - 2$ coefficients which are the values of μ for all subsets of N. When n is large, this determination becomes difficult. This is why the concept of k-maxitive capacity, where k is an integer between 1 and n, has been introduced in order to reduce the number of parameters of μ to be determined.

Definition 1. *A capacity μ is said to be k-maxitive [2,17] if we have*

$$\mu(S) = \bigvee_{\substack{T \subseteq S \\ |T| \leq k}} \mu(T) \quad \text{for all } S \subseteq N. \tag{4}$$

Remark 2.

- k-maxitive capacities are thus completely determined by their values on the sets with at most k elements.
- A Sugeno integral defined with respect to a k-maxitive capacity is also said to be k-maxitive.
- It is not difficult to see from (4) that for a k-maxitive Sugeno integral S_μ, $m^\mu(S) = 0$ for all $S \subseteq N$ such that $|S| > k$, and thus S_μ can be expressed as a supremum of terms with at most k variables.

The following proposition gives a simplified expression of the k-maxitive Sugeno integral

Proposition 1. *If a capacity μ is $k-$maxitive, then we have:*

$$S_\mu\big(u(x)\big) = \bigvee_{\substack{A \subseteq N \\ |A| \leq k}} \left(\big(\bigwedge_{i \in A} u_i(x_i) \big) \wedge \mu(A) \right) \tag{5}$$

Proof. Let us assume that a capacity μ is k-maxitive. We then have
$$\mu(A) = \bigvee_{\substack{B \subseteq A \\ |B| \leq k}} \mu(B) \quad \text{for } |A| \geq k + 1.$$
Since $\bigwedge_{i \in A} u_i(x_i) \geq \bigwedge_{i \in B} u_i(x_i)$ when $A \subseteq B$, so

$$\bigvee_{\substack{A \subseteq N \\ |A| \leq k}} \left(\left(\bigwedge_{i \in A} u_i(x_i) \right) \wedge \mu(A) \right) \geq \bigvee_{\substack{A \subseteq N \\ |A| \geq k+1}} \left(\left(\bigwedge_{i \in A} u_i(x_i) \right) \wedge \mu(A) \right).$$

Thus we have:

$$S_\mu(x) = \bigvee_{A \subseteq N} \left(\left(\bigwedge_{i \in A} u_i(x_i) \right) \wedge \mu(A) \right)$$

$$= \left[\bigvee_{\substack{A \subseteq \\ |A| \leq k}} \left(\left(\bigwedge_{i \in A} u_i(x_i) \right) \wedge \mu(A) \right) \right] \vee \left[\bigvee_{\substack{A \subseteq N \\ |A| \geq k+1}} \left(\left(\bigwedge_{i \in A} u_i(x_i) \right) \wedge \mu(A) \right) \right]$$

$$= \bigvee_{\substack{A \subseteq N \\ |A| \leq k}} \left(\left(\bigwedge_{i \in A} u_i(x_i) \right) \wedge \mu(A) \right)$$

$\square$

Remark 3. For $k = 1$, the Sugeno integral simplifies in the form of a prioritized maximum [4]:

$$S_\mu(u(x)) = \bigvee_{i \in N} \left(u_i(x_i) \wedge \mu_i \right) \tag{6}$$

Remark 4. For $k = 2$, the 2-maxitive Sugeno integral is given by:

$$S_\mu(u(x)) = \left[\bigvee_{i \in N} \left(u_i(x_i) \wedge \mu_i \right) \right] \vee \left[\bigvee_{i,j \in N} \left(u_i(x_i) \wedge u_j(x_j) \wedge \mu_{ij} \right) \right] \tag{7}$$

In the context of a 2-additive Choquet integral, the interaction index between two criteria coincides with their Möbius transform [16]. By analogy, in the framework of a 2-maxitive Sugeno integral, we propose below a definition of the interaction index between two criteria. Indeed, we propose that $I_{ij}^\mu = m_{ij}^\mu$ for all $i, j \in N$.

Definition 2. *The interaction index w.r.t. a 2-maxitive capacity μ is defined by:*

$$I_{ij}^\mu = \begin{cases} \mu_{ij} & \text{if } \mu_{ij} > \mu_i \vee \mu_j \\ 0 & \text{otherwise} \end{cases} \tag{8}$$

Remark 5. Given a capacity μ, it is usual to interpret the interaction as follows:

– If $I_{ij}^\mu > 0$, then we say that criterion i and j are complementary(or in positive synergy) w.r.t μ.
– If $I_{ij}^\mu = 0$, then we say that criterion i and j are independent w.r.t μ.

Our definition for the 2-maxitive case implies that negative interactions are not taken into account.

3   A Motivating Example

We consider the following example inspired by [11]. Four students are evaluated on three subjects Mathematics (M), Statistics (S) and Language skills (L).

All marks are taken from the same scale, from 0 to 1. The evaluations of these students are given by the Table 1.

Table 1. Evaluations students

	1:Mathematics(M)	2:Language(L)	3: Statistics(S)
a	0.3	0.25	0.6
b	0.3	0.6	0.25
c	0.7	0.25	0.6
d	0.7	0.6	0.25

To select the best students, the Dean of the faculty expresses his/her preferences where the notation $x \, P \, y$ means x is strictly preferred to y. For a student bad in Mathematics, Language is more important that Statistics, so that

$$a \, P \, b, \tag{9}$$

for a student good in Mathematics, Statistics is more important that Language, so that

$$d \, P \, c. \tag{10}$$

It is not possible to model the two preferences $a \, P \, b$ and $d \, P \, c$ by an 1-maxitive Sugeno integral model. Indeed we have:

$$S_\mu(a) = (u_M(0.3) \wedge \mu_1) \vee (u_L(0.25) \wedge \mu_2) \vee (u_S(0.6) \wedge \mu_3)$$
$$S_\mu(b) = (u_M(0.3) \wedge \mu_1) \vee (u_L(0.6) \wedge \mu_2) \vee (u_S(0.25) \wedge \mu_3)$$
$$S_\mu(c) = (u_M(0.7) \wedge \mu_1) \vee (u_L(0.25) \wedge \mu_2) \vee (u_S(0.6) \wedge \mu_3)$$
$$S_\mu(d) = (u_M(0.7) \wedge \mu_1) \vee (u_L(0.6) \wedge \mu_2) \vee (u_S(0.25) \wedge \mu_3)$$

We then have: $(u_M(0.7) \wedge \mu_1) \vee S_\mu(a) = (u_M(0.3) \wedge \mu_1) \vee S_\mu(c)$ and
$$(u_M(0.7) \wedge \mu_1) \vee S_\mu(b) = (u_M(0.3) \wedge \mu_1) \vee S_\mu(d).$$

Therefore we have: $a \, P \, b \Longrightarrow S_\mu(a) > S_\mu(b),$
$$\Longrightarrow (u_M(0.7) \wedge \mu_1) \vee S_\mu(a) \geq (u_M(0.7) \wedge \mu_1) \vee S_\mu(b),$$
$$\Longrightarrow (u_M(0.3) \wedge \mu_1) \vee S_\mu(c) \geq (u_M(0.3) \wedge \mu_1) \vee S_\mu(d),$$
$$\Longrightarrow S_\mu(c) \geq S_\mu(d),$$
$$\Longrightarrow \text{not}(d \, P \, c). \text{ Contradiction with preference } d \, P \, c.$$

Let us assume that the scale of evaluation $[0, 1]$ corresponds to the utility function associated to each subject, i.e., $u_M(0.3) = 0.3, u_M(0.7) = 0.7, u_L(0.25) = 0.25, u_L(0.6) = 0.6, u_S(0.25) = 0.25$ and $u_S(0.6) = 0.6$. In this case, the preferences $a \, P \, b$ and $d \, P \, c$, are now representable by a 2-maxitive integral w.r.t. any capacity given in Table 2 below. We chose eight capacities compatible with these preferences (Cap. for short in Table 2) in order to illustrate the fact that the sign of an interaction index is strongly dependent upon to the chosen capacity.

In this illustration, the interpretation of the interaction between two criteria is not easy. For instance, w.r.t. the 2-maxitive capacity, the interaction between Mathematics and Statistics, I_{MS}^{μ}, could be strictly positive (Cap. 2, Cap. 7, Cap. 8) or null (Cap. 1, Cap. 3, Cap. 4, Cap. 5, Cap. 6). Thus, from the preferences given by the DM, it is not obvious whether the subjects Mathematics and Statistics are complementary or independent. This conclusion is still valid concerning the interaction I_{LS}^{μ} between Language and Statistics. Indeed, this interaction can be strictly positive (see Cap. 2, Cap. 3, Cap. 4, Cap. 6 and Cap. 7), or null (see Cap. 1 and Cap. 5).

Table 2. A set of eight 2-maxitive capacities compatible with the preferences $a\,P\,b$ and $d\,P\,c$.

	Cap. 1	Cap. 2	Cap. 3	Cap. 4	Cap. 5	Cap. 6	Cap. 7	Cap. 8
μ_M	0	0	0.3	0.5	0.3	0	0	0
μ_L	0	0	0	0	0	0	0	0
μ_S	0.5	0.4	0.5	0.5	0.5	0.5	0.49	0.49
μ_{ML}	1	1	0.6	0.6	1	0.6	0.6	1
μ_{MS}	0.5	0.5	0.5	0.5	0.5	0.5	0.5	0.5
μ_{LS}	0.5	0.5	1	1	0.5	1	1	0.49
$S_\mu(a)$	0.5	0.4	0.5	0.5	0.5	0.5	0.49	0.49
$S_\mu(b)$	0.3	0.3	0.3	0.3	0.3	0.3	0.3	0.3
$S_\mu(c)$	0.5	0.5	0.5	0.5	0.5	0.5	0.5	0.5
$S_\mu(d)$	0.6	0.6	0.6	0.6	0.6	0.6	0.6	0.6
I_{ML}^{μ}	1	1	0.6	0.6	1	0.6	0.6	1
I_{MS}^{μ}	0	0.5	0	0	0	0	0.5	0.5
I_{LS}^{μ}	0	0.5	1	1	0	1	1	0

Depending on the numerical representation μ, the interaction index can be null or strictly positive, this led to the definition of the notion of necessary and possible interaction, as introduced in [13,15] for the Choquet integral model.

4   Necessary and Possible Interaction

In the sequel, we will suppose that the DM is able to compare a number of alternatives in terms of strict preference (P) or indifference (I). The idea is to ask to the DM its preferences by comparing some elements of X. We then obtain the binary relations P and I defined as follows.

Definition 3. *An **ordinal preference information** $\{P,I\}$ on X is given by:*

$P = \{(x, y) \in X \times X : DM \ strictly \ prefers \ x \ to \ y\}$,
$I = \{(x, y) \in X \times X : DM \ is \ indifferent \ between \ x \ and \ y\}$.

We say that $\{P, I\}$ is representable by a 2-maxitive Sugeno integral model, if there exists a 2-maxitive capacity μ such that: for all $x, y \in X$,

$$(x, y) \in P \Longrightarrow S_\mu(u(x)) > S_\mu(u(y))$$
$$(x, y) \in I \Longrightarrow S_\mu(u(x)) = S_\mu(u(y)).$$

The set of all 2-maxitive capacities used to represent the preference information at hand will be denoted $S_{2\text{-max}}(P, I)$. When there is no ambiguity on the underlying preference information, we will simply write $S_{2\text{-max}}$.

The following definition of necessary and possible interactions will be central in the rest of this text.

Definition 4. *Let $i, j \in N$ be two distinct criteria, We say that:*

1. *there exists a possible complementarity (resp. independence) between i and j if there exists a capacity $\mu \in S_{2\text{-max}}$ such that $I_{ij}^\mu > 0$ (resp. $I_{ij}^\mu = 0$);*
2. *there exists a necessary complementarity (resp. independence) between i and j if $I_{ij}^\mu > 0$ (resp. $I_{ij}^\mu = 0$) for all capacity $\mu \in S_{2\text{-max}}$.*

Remark 6. Let $i, j \in N$ be two distinct criteria.

- If there exists a necessary complementarity (resp. independence) between i and j, then there exists a possible complementary (resp. independence) between i and j.
- If there is no necessary complementarity (resp. independence) between i and j, then there exists a possible independence (resp. synergy) between i and j.

5   Necessary and Possible Interaction with Binary Alternatives

5.1   Framework

We assume that the DM is able to identify on each criterion $i \in N$ two reference levels 1_i and 0_i:

- the level 0_i in L_i is considered as a neutral level and we set $u_i(0_i) = 0$;
- the level 1_i in L_i is considered as a good level and we set $u_i(1_i) = 1$.

For a subset $S \subseteq N$ we define the alternative $a_S = (1_S; 0_{-S})$ such that $a_i = 1_i$ if $i \in S$ and $a_i = 0_i$ otherwise. Our work is based on the set $\mathcal{B}$ that we can find in [16] and is defined as follows.

Definition 5. *The set of **binary alternatives** is defined by*
$\mathcal{B} = \{0_N, (1_i, 0_{N-i}), (1_{ij}, 0_{N-ij}) : i, j \in N, i \neq j\}$

where

- $0_N = (1_\emptyset, 0_N) =: a_0$ is an alternative considered neutral on all criteria.
- $(1_i, 0_{N-i}) =: a_i$ is an alternative considered satisfactory on criterion i and neutral on the other criteria.
- $(1_{ij}, 0_{N-ij}) =: a_{ij}$ is an alternative considered satisfactory on criteria i and j and neutral on the other criteria.

Remark 7. For any 2-maxitive capacity μ, we have:

$$S_\mu(u(a_0)) = 0; \quad S_\mu(u(a_i)) = \mu_i; \quad S_\mu(u(a_{ij})) = \mu_{ij}.$$

We add to this ordinal information a relation M modeling the relation of monotonicity between binary alternatives, and allowing us to ensure the satisfaction of the monotonicity conditions $\mu_i \geq 0$ and $\mu_{ij} \geq \mu_i$ for a capacity μ. For $(x, y) \in \{(a_i, a_0) : i \in N\} \cup \{(a_{ij}, a_i) : i, j \in N, i \neq j\}$, xMy if not $(x(P \cup I)y)$.

5.2   Results

The following proposition gives a necessary and sufficient condition for an ordinal preference information on $\mathcal{B}$ containing no indifference to be representable by a 2-maxitive integral model.

Proposition 2. *Let $\{P, I\}$ be an ordinal preference information on $\mathcal{B}$ such that $I = \emptyset$. Then $\{P, I\}$ is representable by a 2-maxitive Sugeno integral if and only if the binary relation $P \cup M$ contains no strict cycle.*

Proof.   **Necessity.** Suppose that the ordinal preference information $\{P, I\}$ on $\mathcal{B}$ is representable by a Sugeno integral. So there exists a capacity $\mu \in S_{2\text{-max}}$ such that $\{P, I\}$ is representable by S_μ.

   If $P \cup M$ contains a strict cycle, then there exists $x_0, x_1, \ldots, x_r$ on $\mathcal{B}$ such that $x_0 (P \cup M) x_1 (P \cup M) \ldots (P \cup M) x_r (P \cup M) x_0$ and there exists two elements $x_i, x_{i+1} \in \{x_0, x_1, \ldots, x_r\}$ such that $x_i P x_{i+1}$. Since $\{P, I\}$ is representable by S_μ, therefore $S_\mu(u(x_0)) \geq \ldots \geq S_\mu(u(x_i)) > S_\mu(u(x_{i+1})) \geq \ldots \geq S_\mu(u(x_0))$, then $S_\mu(u(x_0)) > S_\mu(u(x_0))$, contradiction.

Sufficiency. Assume that $(P \cup M)$ contains no strict cycle. The proof of sufficiency consists to extend the relation P to a total order on $\mathcal{B}$. This latter can be viewed as a partition of $\mathcal{B}$ elaborated by computing a topological sorting on $(P \cup M)$ detailed in Section 5.2. of [16].

   Then there exists $\{\mathcal{B}_0, \mathcal{B}_1, \ldots, \mathcal{B}_m\}$ a partition of $\mathcal{B}$, builds by using a suitable topological sorting on $(P \cup M)$ [8].

   We construct a partition $\{\mathcal{B}_0, \mathcal{B}_1, \ldots, \mathcal{B}_m\}$ as follows:

$\mathcal{B}_0 = \{x \in \mathcal{B} : \forall y \in \mathcal{B}, \text{not}[x(P \cup M)y]\}$,
$\mathcal{B}_1 = \{x \in \mathcal{B} \setminus \mathcal{B}_0 : \forall y \in \mathcal{B} \setminus \mathcal{B}_0, \text{not}[x(P \cup M)y]\}$,
$\mathcal{B}_i = \{x \in \mathcal{B} \setminus (\mathcal{B}_0 \cup \ldots \cup \mathcal{B}_{i-1}) : \forall y \in \mathcal{B} \setminus (\mathcal{B}_0 \cup \ldots \cup \mathcal{B}_{i-1}), \text{not}[x(P \cup M)y]\}$, for all $i = 1, 2, \ldots, m$.

Let us define the mapping $\phi : \mathcal{B} \longrightarrow \mathcal{P}(N)$, $f : \mathcal{P}(N) \longrightarrow \mathbb{R}$, $\mu : 2^N \longrightarrow [0,1]$ as follows: $\phi(a_S) = S$ for all $S \subseteq N$, $f(\phi(x)) = \ell$ for all $\ell \in \{0,1,\ldots,m\}$, $\forall x \in \mathcal{B}_\ell$,

$$\mu_\emptyset = 0, \ \mu_i = \frac{f_i}{\alpha}, \ \mu_{ij} = \frac{f_{ij}}{\alpha}, \ \mu(S) = \bigvee_{i,j \in S} \mu_{ij}, \ \forall i,j \in N, \ \forall S \subseteq N, \text{ where } f_i =$$

$f(\phi(a_i))$, $f_{ij} = f(\phi(a_{ij}))$ and $\alpha = \displaystyle\bigvee_{i,j \in N} \mu_{ij}$

The capacity μ, defined like this is 2-maxitive by construction and the ordinal information $\{P, I\}$ is then representable by a 2-maxitive Sugeno integral model S_μ. □

Given the ordinal preference information $\{P, I\}$ on $\mathcal{B}$, under the previous conditions, the following proposition shows that, it is always possible to choose in $S_{2\text{-max}}(P, I)$, a capacity allowing all the interaction indices strictly positive. This result shows that positive synergy interaction is always possible for all pairs of criteria in a 2-maxitive Sugeno integral model if the ordinal information does not contain indifference. This condition is the same as that obtained in the case of the 2-additive Choquet integral [15].

Proposition 3. *Let $\{P, I\}$ be an ordinal preference information on $\mathcal{B}$ such that $I = \emptyset$. Suppose that this information can be represented by the 2-maxitive Sugeno integral model. Then there exists a possible positive synergy between all pairs of criteria.*

Proof. The partition $\{\mathcal{B}_0, \ldots, \mathcal{B}_m\}$ of $\mathcal{B}$ and the capacity μ are built as in the proof of Proposition 2: $\phi(a_S) = S$ for all $S \subseteq N$,

Let be $i, j \in N$, there exist $p, q, s \in \{1, \ldots, m\}$ such that $a_{ij} \in \mathcal{B}_p$, $a_i \in \mathcal{B}_q$, $a_j \in \mathcal{B}_s$ with $p > q > 0$ and $p > s > 0$

The capacity μ, defined like this is 2-maxitive by construction and in Proposition 2 we have proved that S_μ represent $\{P, I\}$.

Moreover we have $f_{ij} = p$, $f_i = q$, $f_j = s$ with $p > q$ and $p > s$, therefore $p > q \vee s$, i.e., $\mu_{ij} > \mu_i \vee \mu_j$, then $I_{ij}^\mu = \mu_{ij} > 0$. Hence, we proved that, if $I = \emptyset$ then there exists a capacity μ such that $i, j \in N$, $I_{ij}^\mu > 0$, i.e., there exists a possible positive synergy between pair of criteria $\{i, j\}$. Hence, there is no necessary independence between criteria i and j. □

The following example illustrates the two previous results in this section.

Example 1. $N = \{1, 2, 3\}$, $P = \{(a_{23}, a_1), (a_{12}, a_{23})\}$.

The ordinal preference information $\{P, I\}$ contains no indifference and the binary relation $(P \cup M)$ contains no strict cycle, so $\{P, I\}$ is representable by a 2-maxitive Sugeno integral model. A suitable topological sorting on $(P \cup M)$ is given by: $\mathcal{B}_0 = \{a_0\}$; $\mathcal{B}_1 = \{a_1, a_2, a_3\}$; $\mathcal{B}_2 = \{a_{13}, a_{23}\}$; $\mathcal{B}_3 = \{a_{12}\}$. The preference information $\{P, I\}$ is representable by the capacity μ given by Table 3. We can see that $I_{ij}^\mu > 0$, $\forall i, j \in N$.:

Table 3. A capacity μ and the corresponding interaction indices.

S	$\emptyset$	$\{1\}$	$\{2\}$	$\{3\}$	$\{1,3\}$	$\{2,3\}$	$\{1,2\}$	$\{1,2,3\}$
$\mu(S)$	0	1/3	1/3	1/3	2/3	2/3	1	1
$I^S_{\mu(S)}$	–	–	–	–	2/3	2/3	1	–

6   A MILP Testing Necessary Interactions

In this section, we relax the hypothesis that we only ask a preference information on binary alternatives since considering only binary alternatives is restrictive. Given two criteria i and j, we elaborate a MILP to test in two steps if a preference information on the set of alternatives is representable by a 2-maxitive Sugeno integral model. Then, in the third step, we test the existence of a necessary null or positive interaction between i and j. In the next subsection, we show how to linearize the min and max functions so as to obtain a MILP.

6.1   Linearization of min and max Functions

Given n real numbers $x_1, x_2, \cdots, x_n$, we have:

- $m = \min(x_1, x_2, \cdots, x_n) \iff m \leq x_i;\ m \geq x_i - A\delta_i;\ \delta_i \in \{0,1\}\ \forall i = 1, 2, \cdots, n;\ \delta_1 + \delta_2 + \cdots + \delta_n = n - 1$ and A is a "big" positive constant arbitrarily chosen.
- $M = \max(x_1, x_2, \cdots, x_n) \iff M \geq x_i;\ M \leq x_i + B\delta_i;\ \delta_i \in \{0,1\}$ $\forall i = 1, 2, \cdots, n;\ \delta_1 + \delta_2 + \cdots + \delta_n = n - 1$ and B is a "big" positive constant arbitrarily chosen.

  This transformation of the min and max functions into linear constraints allows us to transform the following program into a MILP.

6.2   Algorithm

Step 1. The following MILP $(MIPL_1)$ models each preference of $\{P, I\}$ by introducing two nonnegative slack variables α^+_{xy} and α^-_{xy} in the corresponding constraints (Eqs. (11) and (12)). Equation (13) (resp. (14)) ensures the normalization (resp. monotonicity) of capacity μ. Equation (15) reflects 2-maxitivity condition. The objective function Z_1 minimizes all the nonnegative variables introduced in (11) and (12).

$$\text{Minimize } Z_1 = \sum_{(x,y) \in P \cup I} (\alpha_{xy}^+ + \alpha_{xy}^-) \qquad (MILP_1)$$

Subject to

$$S_\mu(u(x)) - S_\mu(u(y)) + \alpha_{xy}^+ - \alpha_{xy}^- \geq \varepsilon \ \forall x,y \in X \text{ such that } x\,P\,y \quad (11)$$

$$S_\mu(u(x)) - S_\mu(u(y)) + \alpha_{xy}^+ - \alpha_{xy}^- = 0 \ \forall x,y \in X \text{ such that } x\,I\,y \quad (12)$$

$$\alpha_{xy}^+ \geq 0, \ \alpha_{xy}^- \geq 0 \ \forall x,y \in X \text{ such that } x(P \cup I)y$$

$$\varepsilon \geq 0$$

$$\mu(N) = 1 \qquad (13)$$

$$\mu_i \geq 0, \ \mu_{ij} \geq \mu_i, \ \mu_{ij} \geq \mu_j, \text{ for all } i,j \in N. \qquad (14)$$

$$\mu(S) = \bigvee_{i,j \in S} \mu_{ij} \forall S \subseteq N, \ |S| \geq 3 \qquad (15)$$

The MILP $(MIPL_1)$ is always feasible due to the introduction of the non-negative variables α_{xy}^+ and α_{xy}^-. There are two possible cases:

1. If the optimal solution of $(MIPL_1)$ is $Z_1^* = 0$, then we can conclude that, depending on the sign of the variable ε, the preference information $\{P,I\}$ may be representable by a 2-maxitive Sugeno integral model. The next step of the procedure, Step 2 hereafter, will confirm or not this possibility.
2. If the optimal solution of $(MIPL_1)$ is $Z_1^* > 0$, then there is no 2-maxitive Sugeno integral model compatible with $\{P,I\}$.

Step 2. Here, the MILP $(MIPL_2)$ ensures the existence of a 2-maxitive Sugeno integral model compatible with $\{P,I\}$, when the optimal solution of $(MIPL_1)$ is $Z_1^* = 0$. Compared to the previous MILP, in this formulation, we only removed the nonnegative variables α_{xy}^+ and α_{xy}^- (or put them equal to zero) and change the objective function by maximizing the value of the variable ε, in order to satisfy the strict preference relation.

$$\text{Maximize } Z_2 = \varepsilon \qquad (MILP_2)$$

Subject to

$$S_\mu(u(x)) - S_\mu(u(y)) \geq \varepsilon \ \forall x,y \in X \text{ such that } x\,P\,y \qquad (16)$$

$$S_\mu(u(x)) - S_\mu(u(y)) = 0 \ \forall x,y \in X \text{ such that } x\,I\,y \qquad (17)$$

$$\varepsilon \geq 0$$

$$\mu(N) = 1 \qquad (18)$$

$$\mu_i \geq 0, \ \mu_{ij} \geq \mu_i, \ \mu_{ij} \geq \mu_j \text{ for all } i,j \in N. \qquad (19)$$

$$\mu(S) = \bigvee_{i,j \in S} \mu_{ij} \ \forall S \subsetneq N, \ |S| \geq 3 \qquad (20)$$

Notice that $(MIPL_2)$ is solved only if $Z_1^* = 0$. Hence, the linear program $(MIPL_2)$ is always feasible and it does not have an unbounded solution (it is not restrictive to suppose that $S_\mu(u(x)) \in [0,1]; \forall x \in X$). Hence, we have one of the following two cases:

1. If $(MIPL_2)$ is feasible with optimal solution $Z_2^* = 0$, then there is no 2-maxitive Sugeno integral model compatible with $\{P, I\}$.
2. If the optimal solution of is $(MIPL_2)$ is $Z_2^* > 0$, then ordinal information $\{P, I\}$ is representable by a 2-maxitive Sugeno integral model.

Step 3. At this step, we suppose that the preference information $\{P, I\}$ is representable by a 2-maxitive Sugeno integral model, i.e., $Z_2^* > 0$. In order to know if the interaction between i and j is necessarily null (resp. positive) w.r.t. the provided preference information. At $(MILP_2)$, we add the contraint $I_{ij}^\mu > 0$ (resp. $I_{ij}^\mu = 0$) and we obtain the MILP denoted by $MIPL_{NN}^{ij}$ (resp. $MIPL_{NP}^{ij}$). After a resolution of the MILP, we have one of the following three possible conclusions:

1. If $MIPL_{NN}^{ij}$ (resp. $MIPL_{NP}^{ij}$) is not feasible, then there is a necessary positive (resp. null) interaction between i and j. Indeed, as the program $(MIPL_2)$ is feasible with an optimal solution, the contradiction about the representation of $\{P, I\}$ only comes from the introduction of the constraint $I_{ij}^\mu > 0$ (resp. $I_{ij}^\mu = 0$).
2. If $MIPL_{NN}^{ij}$ (resp. $MIPL_{NP}^{ij}$) is feasible and the optimal solution $Z_3^* = 0$, then the contraint $S_\mu(u(x)) - S_\mu(u(y)) \geq \varepsilon \ \forall x, y \in X$ such that $x \ P \ y$ is satisfied with $\varepsilon = 0$, i.e., it is not possible to model strict preference by adding the constraint $I_{ij}^\mu > 0$ (resp. $I_{ij}^\mu = 0$) in $MIPL_{NN}^{ij}$ (resp. $MIPL_{NP}^{ij}$). Therefore, we can conclude that there is a necessary positive (resp. null) interaction between i and j.
3. If $MIPL_{NN}^{ij}$ (resp. $MIPL_{NP}^{ij}$) is feasible and the optimal solution $Z_3^* > 0$, then there is no necessary null (resp. positive) interaction between i and j.

6.3 Example

We consider the preferences given by the DM in the classic example given by Table 1. We proved in Sect. 3 that these preferences are representable by a 2-maxitive Sugeno integral. The following $MIPL_{MS}^{NN}$ corresponding to the test of the existence of a necessary null interaction between the Mathematics (1) and Statistics (3):

Maximize $Z_3 = \varepsilon$

Inputs of example
$a_1 = 0.3; a_2 = 0.25; a_3 = 0.6; b_1 = 0.3; b_2 = 0.6; b_3 = 0.25; c_1 = 0.7;$
$c_2 = 0.25; c_3 = 0.6; d_1 = 0.7; d_2 = 0.6; d_3 = 0.25;$
$S_\mu(a) \geq S_\mu(b) + \varepsilon; S_\mu(d) \geq S_\mu(c) + \varepsilon; \varepsilon \geq 0.1$

Constraints related of linearization of $S_\mu(x) = \max(\alpha_{x1}, \alpha_{x2},$
$\alpha_{x3}, \alpha_{x12}, \alpha_{x13}, \alpha_{x23})$ **with the introduction of binary variables**
$\delta_1^x, \delta_2^x, \delta_3^x, \delta_{12}^x, \delta_{13}^x, \delta_{23}^x$, **where** $x \in \{a, b, c, d\}$.
$S_\mu(x) \geq \alpha_{x1}; S_\mu(x) \geq \alpha_{x2}; S_\mu(x) \geq \alpha_{x3}; S_\mu(x) \geq \alpha_{x12}; S_\mu(x) \geq \alpha_{x13};$
$S_\mu(x) \geq \alpha_{x23}; S_\mu(x) \leq \alpha_{x1} + 500\delta_1^x; S_\mu(x) \leq \alpha_{x2} + 500\delta_2^x;$

$S_\mu(x) \leq \alpha_{x3} + 500\delta_3^x$; $S_\mu(x) \leq \alpha_{x12} + 500\delta_{12}^x$; $S_\mu(x) \leq \alpha_{x13} + 500\delta_{13}^x$; $S_\mu(x) \leq \alpha_{x23} + 500\delta_{23}^x$; $\delta_1^x + \delta_2^x + \delta_3^x + \delta_{12}^x + \delta_{13}^x + \delta_{23}^x = 5$; $\delta_1^x, \delta_2^x, \delta_3^x, \delta_{12}^x, \delta_{13}^x, \delta_{23}^x \in \{0,1\}$.

Constraints related of linearization of $\alpha_{xi} = \min(x_i, \mu_i)$ **with the introduction of binary variables** $\delta_{i1}^x, \delta_{i2}^x$, **where** $x \in \{a,b,c,d\}$ **and** $i \in \{1,2,3\}$.

$\alpha_{xi} \leq x_i$; $\alpha_{xi} \leq \mu_i$; $\alpha_{xi} \geq x_i - 500\delta_{i1}^x$; $\alpha_{xi} \geq \mu_i - 500\delta_{i2}^x$; $\delta_{i1}^x + \delta_{i2}^x = 1$; $\delta_{i1}^x, \delta_{i2}^x \in \{0,1\}$.

Constraints related of linearization of $\alpha_{xij} = \min(x_i, x_j, \mu_{ij})$ **with the introduction of binary variables** $\delta_{ij1}^x, \delta_{ij2}^x, \delta_{ij3}^x$, **where** $x \in \{a,b,c,d\}$ **and** $i, j \in \{1,2,3\}$, $i \neq j$.

$\alpha_{xij} \leq x_i$; $\alpha_{xij} \leq x_j$; $\alpha_{xij} \leq \mu_{ij}$; $\alpha_{xij} \geq x_i - 500\delta_{ij1}^x$; $\alpha_{xij} \geq x_j - 500\delta_{ij2}^x$; $\alpha_{xij} \geq \mu_{ij} - 500\delta_{ij3}^x$; $\delta_{ij1}^x + \delta_{ij2}^x + \delta_{ij3}^x = 2$; $\delta_{ij1}^x, \delta_{ij2}^x, \delta_{ij3}^x \in \{0,1\}$.

2-maxitivity constraints

$\mu_{12} \geq \mu_1$; $\mu_{12} \geq \mu_2$; $\mu_{13} \geq \mu_1$; $\mu_{13} \geq \mu_3$; $\mu_{23} \geq \mu_2$; $\mu_{23} \geq \mu_3$; $\mu_{123} = \max(\mu_{12}, \mu_{13}, \mu_{23})$.

Contraint of normalization $\mu_{123} = 1$.

Constraints related of linearization of $\mu_{123} = \max(\mu_{12}, \mu_{13}, \mu_{23})$ **with the introduction of binary variables** $\delta^1, \delta^2, \delta^3$.

$\mu_{123} \geq \mu_{12}$; $\mu_{123} \geq \mu_{13}$; $\mu_{123} \geq \mu_{23}$; $\mu_{123} \leq \mu_{12} + 500\delta^1$; $\mu_{123} \leq \mu_{13} + 500\delta^2$; $\mu_{123} \leq \mu_{23} + 500\delta^3$; $\delta^1 + \delta^2 + \delta^3 = 2$; $\delta^1, \delta^2, \delta^3 \in \{0,1\}$.

Constraints related of linearization of $I_{13}^\mu > 0$

$\mu_{13} \geq \mu_1 + \varepsilon$; $\mu_{13} \geq \mu_3 + \varepsilon$; $\varepsilon \geq 0.1$.

The results obtained by solving $MIPL_{NN}^{MS}$ are given in Table 4. We can conclude that the interaction between Mathematics and Statistics is not necessarily null, because the optimal solution of the program $MIPL_{NN}^{MS}$ is $Z_3^* = 0.1 > 0$. Besides, we have $S_\mu(a) = 0.4$, $S_\mu(b) = 0.3$, $S_\mu(c) = 0.5$ and $S_\mu(d) = 0.6$.

Table 4. Results of $MIPL_{NN}^{MS}$ testing necessary null interaction between Mathematics and Statistics

	$Z_3 = \varepsilon$	1	2	3	$\{1,2\}$	$\{1,3\}$	$\{2,3\}$	$\{1,2,3\}$
Optimal solution Z_3^*	0.1	–	–	–	–	–	–	
Capacity μ	–		0	0	0.4	0.6	0.5	1	1
Interaction index I_{ij}^μ	–		–	–	–	0.6	**0.5**	1	–

7 Conclusion

This article proposes and studies the notion of interaction between two criteria in a 2-maxitive Sugeno integral model. We make a restriction in the case where the DM gives preference information on a set of finite number of alternatives. The 2-maxitive capacity that is elicited in such a setting is not unique. The interpretation of the interaction effects between two criteria requires some caution. Indeed, we have give some examples in which the sign of the interaction index depends upon the arbitrary choice of a capacity within the set of all 2-maxitive capacities compatible with the preference information. Only necessary interactions are robust since their sign and, hence, interpretation, does not vary within the set of all representing capacities.

Within the framework of binary alternatives, our first result gives a necessary and sufficient condition for an ordinal preference information containing no indifference to be representable by a 2-maxitive Sugeno integral model. This result is similar to that obtained in our paper on the general Choquet integral model (see Proposition 1 on [13]).

Under the conditions of our first result, in the framework of binary alternatives, if the ordinal preference information contains no indifference, our second result shows that it is always possible to represent it by a 2-maxitive Sugeno integral model which all interaction indices between two criteria are strictly positive. This result is similar to that obtained in paper on the general Choquet integral model (see Proposition 2 on [13]).

Outside the framework of binary alternatives, using the linearization of the min and max functions, we propose a MILP allowing to test whether the interpretation of the interaction indices is ambivalent or not.

The subject of this paper offer several avenues for future research.

In fact, this paper proposes an interaction index for the 2-maxitive Sugeno integral, it would be interesting to propose others for the k-maxitive Sugeno integral, with $k \geq 3$.

The notion of interaction would deserve further study. In particular, it would be interesting to have a definition that would not depend on a particular aggregation technique or on a particular index.

It would finally be interesting to study the case of bipolar scales. We are already investigating some of these research avenues.

References

1. Beliakov, G.: Construction of aggregation functions from data using linear programming. Fuzzy Sets Syst. **160**, 65–75 (2009)
2. Calvo, T., de Baets, B.: Aggregation operators defined by k-order additive/maxitive fuzzy measures. Int. J. Uncertainty Fuzziness Knowl.-Based Syst. **06**(06), 533–550 (1998)
3. Choquet, G.: Theory of capacities. Annales de l'Institut Fourier **5**, 131–295 (1954)
4. Couceiro, M., Dubois, D., Fargier, H., Grabisch, M., Prade, H., Rico, A.: New directions in ordinal evaluation: sugeno integrals and beyond. In: Doumpos, M., Figueira, J.R., Greco, S., Zopounidis, C. (eds.) New Perspectives in Multiple Criteria Decision Making. MCDM, pp. 177–228. Springer, Cham (2019). https://doi.org/10.1007/978-3-030-11482-4_7
5. Couceiro, M., Dubois, D., Prade, H., Waldhauser, T.: Decision-making with Sugeno integrals. Order **33**(3), 517–535 (2016)

6. Dubois, D., Marichal, J.-L., Prade, H., Roubens, M., Sabbadin, R.: The use of the discrete Sugeno integral in decision-making: a survey. Int. J. Uncertainty Fuzziness Knowl.-Based Syst. **9**, 539–561 (2001)
7. Sanchez, T.C., Beliakov, G., Sola, H.B.: A Practical Guide to Averaging Functions. Springer, Heidelberg (2016). https://doi.org/10.1007/978-3-319-24753-3
8. Gondran, M., Minoux, M.: Graphes et algorithmes, 3rd edn. Eyrolles, Paris (1995)
9. Grabisch, M.: A new algorithm for identifying fuzzy measures and its application to pattern recognition. In: Proceedings of 1995 IEEE International Conference on Fuzzy Systems., vol. 1, pp. 145–150 (1995)
10. Grabisch, M.: The application of fuzzy integrals in multicriteria decision making. Fuzzy Sets Syst. **89**, 445–456 (1995)
11. Grabisch, M., Labreuche, C.: Fuzzy measures and integrals in MCDA. In: Greco, S., Ehrgott, M., Figueira, J.R. (eds.) Multiple Criteria Decision Analysis. ISORMS, vol. 233, pp. 553–603. Springer, New York (2016). https://doi.org/10.1007/978-1-4939-3094-4_14
12. Grabisch, M., Marichal, J.-L., Mesiar, R., Pap, E.: Aggregation Functions (Encyclopedia of Mathematics and Its Applications), 1st edn. Cambridge University Press, Cambridge (2009)
13. Kaldjob Kaldjob, P.A., Mayag, B., Bouyssou, D.: Necessary and possible interaction between criteria in a general choquet integral model. In: Lesot, M.J., et al. (eds.) IPMU 2020. CCIS, vol. 1238, pp. 457–466. Springer, Cham (2020). https://doi.org/10.1007/978-3-030-50143-3_36
14. Marichal, J.-L.: An axiomatic approach of the discrete Sugeno integral as a tool to aggregate interacting criteria in a qualitative framework. IEEE Trans. Fuzzy Syst. **9**(1), 164–172 (2001)
15. Mayag, B., Bouyssou, D.: Necessary and possible interaction between criteria in a 2-additive Choquet integral model. Eur. J. Oper. Res. **283**, 308–320 (2019)
16. Mayag, B., Grabisch, M., Labreuche, C.: A representation of preferences by the Choquet integral with respect to a 2-additive capacity. Theory Decis. **71**(3), 297–324 (2011)
17. Mesiar, R.: Generalizations of k-order additive discrete fuzzy measures. Fuzzy Sets Syst. **102**(3), 423–428 (1999)
18. Pignon, J.P., Labreuche, C.: A methodological approach for operational and technical experimentation based evaluation of systems of systems architectures. In: International Conference on Software & Systems Engineering and their Applications (ICSSEA), Paris, France, 4–6 December 2007 (2007)
19. Sugeno, M.: Theory of fuzzy integrals and its applications. PhD thesis, Tokyo Institute of Technology (1974)

Coalition Formation

Democratic Forking: Choosing Sides with Social Choice

Ben Abramowitz[1(✉)], Edith Elkind[2], Davide Grossi[3,4], Ehud Shapiro[5], and Nimrod Talmon[6]

[1] Rensselaer Polytechnic Institute, Troy, NY 12180, USA
abramb@rpi.edu
[2] University of Oxford, Oxford, UK
[3] University of Groningen, Groningen, The Netherlands
d.grossi@rug.nl
[4] University of Amsterdam, Amsterdam, The Netherlands
[5] Weizmann Institute of Science, Rehovot, Israel
ehud.shaprio@weizmann.ac.il
[6] Ben-Gurion University, Be'er Sheva, Israel
talmonn@bgu.ac.il

Abstract. Any community in which membership is voluntary may eventually break apart, or fork. For example, forks may occur in political parties, business partnerships, social groups, and cryptocurrencies. Forking may be the product of informal social processes or the organized action of an aggrieved minority or an oppressive majority. The aim of this paper is to provide a social choice framework in which agents can report preferences not only over a set of alternatives, but also over the possible forks that may occur in the face of disagreement. We study the resulting social choice setting, concentrating on stability issues, preference elicitation and strategy-proofness.

Keywords: Forking · Blockchain · Group activity selection

1 Introduction

Collective decisions can produce conflict when no outcome is acceptable to all agents involved. Tensions may arise while options are being considered and discussed, but they may also appear or worsen once a decision has been made, particularly if preference strengths have not been accounted for or if some agents are disenfranchised entirely.

In many situations there is an implicit recourse for the frustrated and the downtrodden—they can leave. An agent, or a set of agents, may leave a community if they are sufficiently dissatisfied with the outcome of a decision, or a bundle of decisions. When group cohesiveness is valued, it is sensible for decision makers to consider how each potential decision might effect camaraderie. This is commonly done informally through discussion when group sizes are small

© Springer Nature Switzerland AG 2021
D. Fotakis and D. Ríos Insua (Eds.): ADT 2021, LNAI 13023, pp. 341–356, 2021.
https://doi.org/10.1007/978-3-030-87756-9_22

enough to deliberate, and through polling when communities are large. Here we propose a formal mechanism to take such considerations into account.

In typical voting scenarios, agents express preferences over alternatives and a single alternative is elected, which must then be universally accepted as the outcome by all agents in the community. In our setting, agents do not have to accept a particular alternative, and can fork instead. Consequently, the set of possible outcomes for every decision is not just the set of alternatives; rather, an outcome is a partition of the agents into one or more coalitions, with each coalition selecting an alternative that appeals to its members.

By designing voting rules that account for forking preferences, we empower aggrieved minorities to leverage the threat of leaving in order to pressure the majority into concessions. Importantly, we enable minorities to coordinate with minimal overhead, by eliciting additional information during the voting process. In our model, voters individually indicate conditions under which they prefer to leave, and the mechanism then identifies a group that can benefit from forking, thereby eliminating the need for campaigning or coordination among the disgruntled minority.

The value of stability is inherent in digital and analogue communities alike. In most current blockchain protocols (i.e., proof-of-work and proof-of-stake) a fork can only be initiated by a majority or a powerful minority. When a fork does occur, enacted by only a subset of the agents, all agents must determine what "side" of the fork they want to be on.[1] In this sense, the forks we study are also relevant to version control systems such as Git, where anyone can initiate a fork. In our setting this is an edge case where a voter is willing to split from the group by themselves and others may follow, but we also allow groups of voters to fork together when none of them may be willing to do so on their own.

Since forks can be tumultuous, we wish to design democratic systems that enable communities to efficiently find states that are stable, in the sense that no further forks will occur. To this end, we put forward a formal model that approaches this challenge from the perspective of computational social choice.

Related Work. Our paper is positioned at the interface of the research on blockchain technology and computational social choice.

Blockchain. A blockchain is a replicated data structure designed to guarantee the integrity of data (e.g., monetary transactions in cryptocurrency applications such as Bitcoin [11]) and computations on them, combined with consensus protocols, which allow peers to agree on their content (e.g., who has been paid) and to ensure that no double spending of currency has occurred (see [12,17] for recent overviews). By now, blockchain is an established technology, and cryptocurrency applications are attracting considerable attention [13,14].

---

[1] In blockchain forks, currency owners may be able to run both protocols, but miners must choose how to allocate their computing resources, and programmers must choose how to allocate their personal time.

When the community of a specific blockchain protocol—such as the Bitcoin protocol—is not satisfied with it, it may break into several subcommunities. The community typically consists of developers, who build the software, miners, who operate the protocol, and users, e.g., account holders. When some of the developers of the current protocol decide to modify it, they create an alternative branch that obeys their new protocol, and if it attracts a sufficient number of miners and users, the result is a so-called *hard fork*. Several such hard forks have been documented, including among the most influential cryptocurrencies. Bitcoin (cf. [18]), despite its relatively short history, has already undergone seven hard forks. At the moment, a key feature of these hard forks is that they happen through an informal social process, and, crucially, in ways that are completely exogenous to the protocol underpinning the blockchain. In blockchain terminology, they are said to happen 'off-chain'. This points to a lack of governance in most current blockchain systems, for better or worse.

Against this backdrop there have been attempts at incorporating protocol amendment procedures within blockchain protocols themselves (so-called 'on-chain' governance, cf. [1]); more generally, the issue of governance is attracting increasing attention [2,15]. We are not aware, however, of research that approaches forking as a social choice problem, and aims for an algorithmic solution. We lay the foundations for this approach here.

Computational Social Choice. Social choice theory studies preference aggregation methods for various settings [3]. Our social choice setting is closely related to assignment problems, as the result of a fork is that each agent is "assigned" to a community. In this context, we mention works on judgement aggregation [10], [3, ch. 17] (which, formally, can capture assignments as well) and on partition aggregation [4]. To the best of our knowledge, the specific social choice setting we consider is novel. In a broader context, we mention work on stability in coalition formation games [5], [3, ch. 15] as well as the recent paper on deliberative majorities [9], which studies coalition formation in a general voting setting. Our model is also related to the group activity selection problem with (increasing) ordinal preferences (o-GASP) [6] (see also [7,8]), where our notion of stability corresponds to core stability in o-GASP. However, due to our focus on strategy-proofness, and the fact that our setting does not admit a no-choice option (void activity in o-GASP), most of the existing results for o-GASP are not directly relevant to our study, so we chose not to use the o-GASP formalism.

Version Control. Forking is not limited to the cryptocurrency setting; in particular, forking is relevant to projects of open-source code, in which a community jointly writes a piece of code and may experience different opinions regarding the code that is being written. Indeed, there is some work on using social choice mechanisms (and, in particular, liquid democracy) for revision control systems [16]. Others have been studying the phenomena of forking in open source projects; see, e.g., the work of Zhou et al. [19].

Outline and Contributions. We describe a formal model of social choice for community forking, in which agents report their preferences over alternatives relative to possible forks. Throughout the paper, we focus on the setting where the number of available alternatives is two; towards the end of the paper, we discuss the challenges in extending our approach to three or more alternatives. The paper is structured as follows. Section 2 describes the formal model. Section 3 examines whether stable solutions always exist and whether they can be found efficiently in terms of computation and elicitation. In Sect. 4, we consider strategic agent behavior. In Sect. 5 we discuss the extension of our framework to more than two alternatives. We conclude in Sect. 6. The main contributions of our paper are as follows:

- We devise a polynomial time algorithm (Algorithm 1) for our setting that finds a stable assignment for a very broad and natural domain restriction.
- We propose a modification of Algorithm 1 (Algorithm 2) that allows for efficient iterative preference elicitation.
- We prove an impossibility result (Theorem 3), showing that there is no algorithm that is strategyproof for profiles with more than one stable assignment.
- We establish that Algorithm 1 is strategyproof for profiles that admit a unique stable solution; the impossibility result mentioned above then implies that it is optimal in that sense.

2 Formal Model

Setting. We have a set of agents $V = \{v_1, \ldots, v_n\}$. This community will vote on a set of two alternatives $\{A, B\}$ (say, cryptocurrency protocols or locations). However, unlike in most voting scenarios, the agents are not all bound to accept the same winning alternative. Agents have the ability to *fork*, or forge a new community centered around the "losing" alternative. Ultimately, either all of the agents will remain in a single community or they will split into two communities that have accepted opposite alternatives.

Agent Preferences. Agents care about what alternative their community adopts and how many people are in their community, but not the identities of the other agents in their community. We can represent agent preferences as total orders over the possible tuples (S, j), where $S \in \{A, B\}$ is the alternative to which they are assigned and $j \in [1, n]$ is the number of agents in their community (including themselves and $j - 1$ other agents). We denote the set of all such tuples by $\mathcal{S}$ The preference relation $(A, j) \succ_i (B, k)$ means that agent v_i would prefer to be in a community of size j that accepts alternative A rather than a community of size k that accepts alternative B. Agent preferences are *monotonic* in the size of their community, so given a fixed alternative, they would always prefer to be in a larger community. Formally, for each agent $v_i \in V$ we have $(S, j) \succ_i (S, k)$ for all $1 \leq k < j \leq n$ and $S \in \{A, B\}$. We denote the set of all monotonic total orders over $\mathcal{S}$ by $\mathcal{T}$. For $S \in \{A, B\}$, let V_S^* denote the set of agents who prefer (S, n) to (S', n). We will overload notation and use v_i to represent both an agent $v_i \in V$ and their preference ordering $v_i \in \mathcal{T}$. In a similar fashion, V is the set of agents and also the preference *profile*, or collection of the voters' total orders, $V \in \mathcal{T}^n$. We refer to the pair $(V, \{A, B\})$, with $V \in \mathcal{T}^n$, as a *forking problem*.

Example 1. Suppose we have $n = 3$ agents, $V = \{v_1, v_2, v_3\}$. Consider the preferences of a single agent v_i. By monotonicity, $(A, 3) \succ_i (A, 2) \succ_i (A, 1)$ and $(B, 3) \succ_i (B, 2) \succ_i (B, 1)$ must hold for all agents $v_i \in V$. However, these two orders may be interleaved differently for different agents.

Assignments. We refer to the community that accepts alternative A (resp., B) as community A (resp., B). An assignment $f : V \to \{A, B\}$ assigns agents to one of the two communities, and we denote by $f(v_i) \in \{A, B\}$ the community into which agent v_i is placed. The set $\mathcal{F}$ is the set of all 2^n possible assignments, or partitions, of the agents. Voters' preferences over S induce preferences over assignments in $\mathcal{F}$: a voter v_i prefers an assignment f to an assignment g if $(f(v_i), |f^{-1}(f(v_i))|) \succ_i (g(v_i), |g^{-1}(g(v_i))|)$. Given a forking problem $(V, \{A, B\})$ a voting rule $R : \mathcal{T}^n \to \mathcal{F}$ selects an assignment $R(V) = f \in \mathcal{F}$. We let $a = |f^{-1}(A)|$ be the size of community A, and similarly for $b = |f^{-1}(B)|$.

3   Stability

Our primary goal is to construct stable assignments. An assignment is stable if no subset of agents has an incentive to move simultaneously to a new community.

Definition 1 (k-Stability). *An assignment $f : V \to \{A, B\}$ is stable if there is no assignment $f' : V \to \{A, B\}$ such that each voter v_i with $f'(v_i) \neq f(v_i)$ prefers f' over f.*

A voting rule R is *stable* if it returns a stable assignment whenever one exists.

Example 2. Consider two agents, $V = \{v_1, v_2\}$, where $v_1 : (A, 2) \succ_1 (A, 1) \succ_1 (B, 2) \succ_1 (B, 1)$ and $v_2 : (B, 2) \succ_2 (B, 1) \succ_2 (A, 2) \succ_2 (A, 1)$. Each agent would prefer to be alone at their preferred alternative to being together with the other agent at their less preferred alternative. Thus, the only stable assignment f has $f(v_1) = A$ and $f(v_2) = B$.

3.1   Finding Stable Solutions

When preferences are monotonic, there must be at least one stable assignment, and it can be computed in polynomial time.

Theorem 1. *There is a polynomial time assignment rule (Algorithm 1) that finds a stable assignment for any monotonic profile.*

Algorithm 1. General Stable Assignment Rule

$V_A = V$, $V_B = \emptyset$, $a \leftarrow |V_A|$, $b \leftarrow |V_B|$
while true do
   $k \leftarrow \max\{j : 0 \leq j \leq a, |\{v_i \in V : (B, b + j) \succ_i (A, a)\}| \geq j\}$
   if $k = 0$ **then**
      return $\{V_A, V_B\}$
   else
      Let $X = \{v_i \in V : (B, b + k) \succ_i (A, a)\}$
      $V_B \leftarrow V_B \cup X$, $V_A \leftarrow V_A \setminus X$
      $a \leftarrow |V_A|$, $b \leftarrow |V_B|$

Algorithm 2. General Stable Assignment Rule with Iterative Elicitation

$V_A = V$, $V_B = \emptyset$, $a \leftarrow |V_A|$, $b \leftarrow |V_B|$
while true do
   Ask each agent v_i in V_A for the smallest value $j \in [0, a]$ such that $(B, b+j) \succ_i (A, a)$
   $k \leftarrow \min\{j : j \in [0, a], |\{v_i \in V : (B, b + j) \succ_i (A, a)\}| \geq j\}$
   if $k = 0$ **then**
      return $\{V_A, V_B\}$
   else
      Let $X = \{v_i \in V : (B, b + k) \succ_i (A, a)\}$
      $V_B \leftarrow V_B \cup X$, $V_A \leftarrow V_A \setminus X$
      $a \leftarrow |V_A|$, $b \leftarrow |V_B|$

Proof. Consider Algorithm 1. Let $a = |V_A|$ and $b = |V_B|$. Initially, we place all agents in V_A, so $a = n$ and $b = 0$. If this assignment is not stable, then there exists a subset X of some $k > 0$ agents that all prefer (B, k) to (A, n). We move all these agents to V_B. Monotonicity implies that moving additional agents from V_A to V_B will never cause agents in V_B to want to move back to V_A; thus, as long as we are not in a stable state, there must be a subset of agents at V_A who would prefer to move together to V_B. As long as such a set of agents exists, we continue to move them over together. This procedure halts in at most n steps, and when it halts, the result must be stable, as there is no subset of agents who will move together. A naive implementation of the algorithm loops at most n times, and each computation of the set X of agents to move takes $O(n^2)$ time. □

3.2 Elicitation

Algorithm 1 does not use all of the information in agents' preferences. An iterative version of the algorithm can ask only for the information it needs. Instead of assuming that the total orders of all agents are given explicitly in the input, we place all of the agents at A and at each iteration we ask the a remaining agents at A for the minimum value j such that if j agents could be moved to B, they would now prefer the new community $(B, b + j)$ over their current community (A, a). Once an agent has been moved to B there is no need to ask them for any more information. In Algorithm 2 we repeatedly query agents about the conditions under which they are willing to leave their current community. Agents indicate their preferences with a single integer that says how many agents would have to move with them for them to prefer leaving over the status quo.

Ideally, we would like to only have to query each agent a small number of times. If agents' preferences are structured, it becomes possible to compute stable assignments with little information. To capture this intuition, we introduce the concept of non-critically-interleaving preferences.

Definition 2 (Non-critically-interleaving). *A preference is* non-critically-interleaving *if it is monotonic and* $(A, j) \succ (B, n) \succ (B, n - j) \succ (A, j - 1)$ *or* $(B, j) \succ (A, n) \succ (A, n - j) \succ (B, j - 1)$ *for some* $j \in [1, n]$. *A profile is* non-critically-interleaving *if it contains only non-critically-interleaving preferences.*

When preferences are non-critically-interleaving, we only need to ask each agent whether they prefer (A, n) or (B, n), and the minimum value of j such that they would rather be at their preferred alternative in a coalition of size j than at the other alternative in a coalition of size n. From this information the relevant part of the preference order of each agent can be inferred, and so Algorithm 2 will compute a stable assignment.

3.3 Uniqueness

While at least one stable assignment must exist for all monotonic profiles (Theorem 1), it is not necessarily unique.

Example 3. Let $V = \{v_1, v_2, v_3, v_4\}$ be a set of four agents with preferences that contain the following prefixes, respectively:

- $v_1 : (B, 4) \succ_1 (B, 3) \succ_1 (A, 4) \succ_1 (B, 2) \succ_1 (A, 3) \succ_1 \cdots$
- $v_2 : (B, 4) \succ_2 (B, 3) \succ_2 (B, 2) \succ_2 (A, 4) \succ_2 (B, 1) \succ_2 \cdots$
- $v_3 : (A, 4) \succ_3 (A, 3) \succ_3 (A, 2) \succ_3 (B, 4) \succ_3 (A, 1) \succ_3 \cdots$
- $v_4 : (A, 4) \succ_4 (A, 3) \succ_4 (B, 4) \succ_4 (A, 2) \succ_4 (B, 3) \succ_4 \cdots$

Regardless of how the remainder of the preference profile is filled, as long as monotonicity is maintained, there are at least three stable assignments: (1) all agents at A; (2) all agents at B; or (3) v_1 and v_2 at B and v_3 and v_4 at A.

We would like to identify conditions under which a profile admits a unique stable assignment. One extreme case is when preferences are non-interleaving.

Definition 3 (Non-interleaving). *A preference order is* non-interleaving *if it is monotonic and either* $(A, 1) \succ (B, n)$ *or* $(B, 1) \succ (A, n)$. *A profile is* non-interleaving *if it contains only non-interleaving preference orders.*

The profile in Example 2 is an instance of a non-interleaving profile. If an agent's preference is non-interleaving, then their choice of community is independent of the other agents: they would rather be alone at their preferred alternative than with everyone else at the other alternative. Thus, their preference is described by a single bit of information: it suffices to know whether they are in V_A^* or in V_B^*. Non-interleaving preferences can be viewed as a degenerate case of non-critically-interleaving preferences when $j = 1$.

Observation 1. *When preferences are non-interleaving, there is a unique stable assignment.*

Proof. The only stable assignment assigns to all agents in V_A^* to A and all agents in V_B^* to B. Otherwise, an agent assigned to the opposite community will wish to move, even if on their own. □

Non-interleaving preferences can be generalized to domains of preferences that guarantee unique stable assignments. Informally, we say that an agent is k-loyal to an alternative S if they prefer to be at S with k other agents to being at the other alternative in a coalition of size n.

Definition 4 (k-Loyalty). *An agent $v_i \in V$ is k-loyal to alternative S, $k \in [n]$, if $v_i \in V_S^*$ and $(S, k) \succ_i (S', n)$ for $S' \neq S$.*

When all agents are sufficiently loyal to their preferred alternatives, there is a unique stable assignment.

Proposition 1. *Suppose there exist some k_1, k_2 such that $k_1 \leq |V_A^*|$, $k_2 \leq |V_B^*|$, every agent in V_A^* is k_1-loyal, and every agent in V_B^* is k_2-loyal. Then there is a unique stable assignment.*

Proof. By construction, any stable assignment must have all agents in V_A^* assigned to A, because otherwise those assigned to B would prefer to move together to A, forming a coalition of size $|V_A^*| \geq k_1$ at A. Symmetrically, any stable assignment must have all agents in V_B^* assigned to B, as otherwise those assigned to A would prefer to move together to B, forming a coalition of size at least $|V_B^*| \geq k_2$ at B. □

Proposition 1 holds because all agents must necessarily be assigned to their preferred alternative. We now examine a sub-domain of non-critically-interleaving preferences in which there is always a unique stable assignment, but not all agents are necessarily assigned to their preferred alternative.

Proposition 2. *Suppose agents' preferences are non-critically-interleaving. Let*

$$V_A' = \arg\max_{U \subseteq V} |\{v_i \in U : (A, |U|) \succ (B, n)\}| \,,$$

$$V_B' = \arg\max_{U \subseteq V} |\{v_i \in U : (B, |U|) \succ (A, n)\}| \,.$$

If none of the agents in $V_A^* \setminus V_A'$ are $(n - |V_B'|)$-loyal and none of the agents in $V_B^* \setminus V_B'$ are $(n - |V_A'|)$-loyal, then there is a unique stable assignment.

Proof. Note first that, by monotonicity, the set arg max in the definition of V'_A and V'_B is a singleton, so V'_A and V'_B are well-defined. As with Proposition 1, for any assignment to be stable it must assign all agents in V'_A to A and those in V'_B to B. For the remaining agents, they must necessarily be assigned to the opposite alternative, because there cannot be enough agents at their most preferred alternative for them to stay there.   □

The maximal class of profiles for which there is a unique stable solution is still more general than those we describe above. We can use Algorithm 1 to characterize the set of profiles that admit a unique stable assignment. Let R_A be the assignment rule given by Algorithm 1, and let R_B be the complementary assignment rule that starts with all agents at B and iteratively moves them to A in the same manner.

Theorem 2. *Algorithm 1 (R_A) and the reverse assignment rule (R_B) return the same assignment if and only if the profile admits a unique stable assignment.*

Proof. If the stable assignment is unique, then both R_A and R_B must return this assignment. We now show that if R_A and R_B return the same stable assignment, then it must be the unique stable assignment. Let V_A^1 and V_B^1 be the communities in the stable assignment $f_1 = R_A(V)$. Let V_A^2 and V_B^2 be the communities according to a different stable assignment f_2. By monotonicity and the properties of Algorithm 1 we have $V_A^2 \subsetneq V_A^2$. Consider the set of agents $V_A^2 \setminus V_A^1$, and in particular, the agent(s) in this set that were the first to be moved to B by R_A. At the beginning of the iteration in which they were moved, the number of agents at A had to be at least $|V_A^2|$ (before moving). This contradicts the claim that f_2 is stable, as there are agents in V_A^2 preferring to move together to B. □

3.4   Cohesiveness

Not all stable assignments may be equally attractive. In real life, forking comes at a cost, such as the need to replicate infrastructure and to carve out or abandon intellectual property or goodwill, as well as the social and emotional cost of separation. The cost of forking within our framework is implicit in the preferences of the agents. In line with the monotonicity of preferences, it is natural that the community may want to avoid forks when possible. When it is desirable to avoid forking, we prefer stable assignments that place all agents at the same alternative over those that fork. We call these non-forking assignments. A profile is said to be *cohesive* if it admits at least one non-forking stable assignment; otherwise, we say that a profile is *forking*. The profile in Example 3 is cohesive, although it also permits a forked stable assignment.

The assignment in Example 2 is stable, but the input profile is forking, because no stable assignment exists with all agents in one community. For a profile to be cohesive there must be at least one alternative (w.l.o.g, A) such that for all $j \leq n$, there are fewer than j agents who prefer (B, j) over (A, n). The following example shows a cohesive profile with no stable forked assignments.

Example 4. Let $V = \{v_1, v_2\}$, where $(A, 2) \succ_1 (B, 2) \succ_1 (A, 1) \succ_1 (B, 1)$ and $(B, 2) \succ_2 (A, 2) \succ_2 (B, 1) \succ_2 (A, 1)$. Each agent would prefer to be together with the other agent at their less preferred alternative rather than alone at their preferred alternative. This profile is cohesive, and only admits stable assignments that are non-forking.

4   Strategyproofness

So far we have considered the existence and the possibility of efficiently computing stable assignments when agents report their preferences truthfully. Another important question is whether there exist strategyproof stable assignment rules, i.e., rules that output stable assignments and do not incentivize the agents to misreport their true preferences.

Definition 5 (Strategyproofness). *A rule R is* strategyproof *over domain $D \subseteq T^n$ if for all profiles $V \in D$ and assignments $f = R(V)$, there is no agent $v_i \in V$ that can unilaterally change her preference order to v_i', creating a new profile V' such that she prefers $f'(v_i)$ over $f(v_i)$, where $f' = R(V')$.*

Similarly, a rule is k-strategyproof in our setting if no subset of agents of size k can simultaneously report false preferences to yield an assignment they all prefer. Naturally, k-strategyproofness implies $(k - 1)$-strategyproofness.

Definition 6 (k-Strategyproofness). *A rule R is k-strategyproof over domain $D \subseteq T^n$ if for all profiles $V \in D$ and assignments $f = R(V)$, there is no subset of agents $U \subseteq V$ of size $|U| \leq k$ that can simultaneously change their preferences, creating a new profile V' such that each agent $v_i \in U$ prefers $f'(v_i)$ over $f(v_i)$, where $f' = R(V')$.*

For the domain of all monotonic profiles, no strategyproof stable rules exist. This can be seen from Example 4. In this example there are two stable assignments, one creating $(A, 2)$ and the other creating $(B, 2)$. Suppose the agents are both assigned to B. If v_1 were to change their reported preferences to $(A, 2) \succ_1 (A, 1) \succ_1 (B, 2) \succ_1 (B, 1)$, then $(A, 2)$ would become the only stable assignment for the new profile, which v_1 clearly prefers over $(B, 2)$. This profile is symmetric, so if the agents were to be assigned to $(A, 2)$ (by some tie-breaking mechanism) then v_2 has the opportunity to be strategic. In fact, no strategyproof stable assignment rule can exist for any domain containing a profile that admits two or more stable assignments.

Theorem 3. *No assignment rule can be strategyproof over a domain that includes a profile that admits more than one stable assignment.*

Proof. Suppose profile V permits at least two stable assignments, and our assignment rule R picks one of them, $f_1 = R(V)$. Let f_2 be the closest stable assignment to f_1, in the sense that there is no other assignment f_3 such that $|f_3^{-1}(A)|$ is between $|f_1^{-1}(A)|$ and $|f_2^{-1}(A)|$, and consequently no $|f_3^{-1}(B)|$

between $|f_1^{-1}(B)|$ and $|f_2^{-1}(B)|$ (indeed, $f^{-1}(A) + f^{-1}(B) = n$ for any f). For brevity, let $V_A^1 = f_1^{-1}(A)$, $V_A^2 = f_2^{-1}(A)$, $V_B^1 = f_1^{-1}(B)$, $V_B^2 = f_2^{-1}(B)$. Assume that $|V_A^1| > |V_A^2|$ and $|V_B^2| > |V_B^1|$; later we will see that this assumption is without loss of generality.

From monotonicity, we know that $V_A^2 \subset V_A^1$ and $V_B^1 \subset V_B^2$. Consider an agent $v_i \in V_A^1$ with $(B, |V_B^2|) \succ_i (A, |V_A^1|) \succ_i (B, |V_B^1| + 1)$. At least one such agent must exist because otherwise f_2 could not be stable, as all the agents in $V_A^1 \cap V_B^2$ would prefer to move together to A rather than stay in $(B, |V_B^2|)$. If v_i commits to B by falsely reporting that they prefer $(B, 1)$ to (A, n), then they must be assigned to B (as otherwise the assignment would not be stable). By construction, however, no intermediate stable state can exist between f_1 and f_2, so agents will prefer to move from V_A^1 to B until it is of size $|V_B^2|$, which is what v_i preferred. By symmetry, if our rule R had picked assignment f_2 instead of f_1 then we would have the same result; thus, our assumption that $|V_A^1| > |V_A^2|$ and $|V_B^2| > |V_B^1|$ is indeed without loss of generality. $\square$

The above theorem means that the domain consisting of those profiles that admit a unique stable assignment is the maximal domain for which a stable assignment rule can be strategyproof. Next, we show that, whenever a profile admits a unique stable assignment, Algorithm 1 is strategyproof.

Lemma 1. *Algorithm 1 is strategyproof over the domain of all profiles that admit a unique stable assignment.*

Proof. Consider a run of Algorithm 1 in which it assigns some agent v_i to B and outputs the assignment f. First, observe that v_i cannot misreport their preferences so that they would end up in a larger community at B that they prefer. If the agents assigned to A do not move at any iteration, then v_i moving at an earlier to later iteration has no effect on them. And, due to monotonicity, v_i staying at A would not further entice anyone to move to B.

Second, we want to show that agent v_i cannot manipulate the outcome so that it will be assigned to a larger community at A that it prefers. Suppose that at the beginning of the iteration when v_i is moved from A to B, the size of the community at A is a. The agents who were moved from A to B at an iteration before the iteration at which v_i is moved will be assigned to B regardless of what v_i reports. In general, agents moved to B together at one iteration must end up at B regardless of the preferences of those moved to B at later iterations and those who stay at A. As a consequence, v_i can never induce an assignment with a community at A of size greater than a. Since v_i was moved at the iteration when the size of A was a, it must be to a community B that they prefer over (A, a). Therefore no agent $v_i \in B$ can deviate profitably from their true preferences.

It remains to show that no agent assigned to A can benefit from strategic behavior. By symmetry, if there is a unique stable assignment, then if Algorithm 1 starts with all agents at A and moves them in batches to B, or starts at B and moves them in batches to A, then it must return the same assignment. We can therefore use the same argument as above for agents assigned to A according to the algorithm that initializes A and B in the opposite way. $\square$

The result extends to n-strategyproofness, or group-strategyproofness, since, if we consider any coalition of agents assigned to B by Algorithm 1, and look at only those who were moved first (in the same iteration as one another, but before everyone else in the coalition who was moved), then they have no incentive to misreport their preferences for the same reason as the agent in the proof of Lemma 1. By combining this with Theorem 3, we arrive at our main result.

Theorem 4. *Algorithm 1 is group-strategyproof over the domain of all profiles that admit a unique stable assignment.*

Just how common are profiles that admit strategyproof stable assignment rules? One specific domain restriction that implies group-strategyproofness of Algorithm 1 is the domain of non-interleaving profiles (recall that, for this domain restriction, placing all agents at their preferred alternative is stable). Non-interleaving preferences are indeed very extreme, in that agents ignore each other completely. However, if we relax this extreme constraint on preferences even the slightest bit, we can lose strategyproofness.

Definition 7 (Minimally-interleaving). *A preference order is minimally-interleaving if it is monotonic and* $(A, 2) \succ (B, n) \succ (A, 1) \succ (B, n - 1)$ *or* $(B, 2) \succ (A, n) \succ (B, 1) \succ (A, n - 1)$. *A profile is a* minimally-interleaving *if it contains only non-interleaving and minimally-interleaving preferences.*

Minimally-interleaving preferences can be interpreted as just barely extending non-interleaving preferences to allow that agents may be willing to go with their less preferred alternative if they would otherwise be alone with their more preferred alternative. Note that the minimally-interleaving domain is still a severely restricted domain. In particular, it allows each agent to specify only one of four possible orders. However, it turns out that if we allow just minimal interleaving, then there is no assignment rule that is both stable and strategyproof.

Observation 2. *There is no assignment rule that is both stable and strategyproof for all minimally-interleaving preference profiles.*

Proof. Let R be a stable assignment rule and consider the profile from Example 4: $v_1 : (A, 2) \succ_1 (B, 2) \succ_1 (A, 1) \succ_1 (B, 1)$; $v_2 : (B, 2) \succ_2 (A, 2) \succ_2 (B, 1) \succ_2 (A, 1)$. Note that, indeed, this profile is minimally-interleaving. Observe that the only stable assignments are the two that place both agents in the same community. Since there are two stable assignments, Theorem 3 implies that R cannot be strategyproof for this profile. For illustrative purposes, consider the case in which v_1 votes strategically by reporting $v_1' : (A, 2) \succ_1 (A, 1) \succ_1 (B, 2) \succ_1 (B, 1)$. The only stable assignment places both agents at A, creating $(A, 2)$, which v_1 prefers to $(B, 2)$. So if R placed both agents at $(B, 2)$, it cannot be strategyproof.   □

We say that a profile is k-*interleaving* if it may contain preference orders in which $(S', n) \succ \ldots \succ (S, n) \succ (S', k)$, but not in which $(S', n) \succ \ldots \succ (S, n) \succ$

$(S', k+1)$, where $S' \neq S$. Hence, non-interleaving preferences are equivalent to 0-interleaving; minimally-interleaving preferences are the same as 1-interleaving; and n-interleaving is the domain of all monotonic preferences. Naturally, the set of all k-interleaving preferences encompasses all $(k-1)$-interleaving preferences, so no strategyproof stable assignment rule can exist for $k \geq 1$. While non-interleaving is a sufficient condition for strategyproofness, the next example demonstrates that it is not a necessary condition.

Example 5. Consider two agents, $V = \{v_1, v_2\}$, where $v_1 : (A, 2) \succ_1 (A, 1) \succ_1 (B, 2) \succ_1 (B, 1)$ and $v_2 : (B, 2) \succ_2 (A, 2) \succ_2 (B, 1) \succ_2 (A, 1)$. Any stable assignment must have v_1 at A, independent of the preferences of v_2. Agent v_2 would prefer to be at A with v_1 to being alone at B, so the only stable assignment has both agents at A, and neither agent has an incentive to be strategic.

Notice that our interleaving conditions apply to the preference order of each agent individually. Example 5 suggests that we should instead consider restrictions on the profile as a whole. While we know that the necessary and sufficient conditions for stable strategyproof assignment rules to exist is that there be a unique stable assignment, characterizing the profiles for which this occurs is an interesting challenge.

5 Forking with More Than Two Alternatives

So far, we focused on the case of two alternatives. We conclude the paper with two observations about the general case: (1) stable assignments are no longer guaranteed to exist; (2) deciding whether an assignment is stable is NP-complete.

Proposition 3. *There exist monotonic profiles with no stable assignment.*

Proof. Consider the problem with three agents $V = \{v_1, v_2, v_3\}$, three alternatives $\{A, B, C\}$, and a following profile:

- $v_1 : \cdots \succ_1 (B, 2) \succ_1 (A, 2) \succ_1 (A, 1) \succ_1 (B, 1) \succ_1 (C, 3) \succ_1 \cdots$
- $v_2 : \cdots \succ_2 (C, 2) \succ_2 (B, 2) \succ_2 (B, 1) \succ_2 (C, 1) \succ_2 (A, 3) \succ_2 \cdots$
- $v_3 : \cdots \succ_3 (A, 2) \succ_3 (C, 2) \succ_3 (C, 1) \succ_3 (A, 1) \succ_3 (B, 3) \succ_3 \cdots$

Assume for contradiction that this profile admits a stable assignment f. As v_2 prefers $(B, 1)$ to $(A, 3)$, the assignment f cannot assign v_2 to A. By considering v_1 and v_3, we conclude that $|f^{-1}(s)| < 3$ for every $S \in \{A, B, C\}$. Suppose $f(v_1) = A$, $f(v_3) = A$. Then we have $f(v_2) = B$ because $(B, 1) \succ_2 (C, 1)$. But in this case, v_1 would prefer to move to B since $(B, 2) \succ_1 (A, 2)$. By the same reasoning $|f^{-1}(S)| \neq 2$ for each $S \in \{A, B, C\}$. The only remaining option is to have one voter at each alternative. Let v be the voter at A. If $v = v_1$, then v_1 prefers to move to B and if $v = v_3$ then v_3 would prefer to move to C. Finally, if $v = v_2$ then v_1 prefers to join v_2 at A. □

From a complexity-theoretic perspective, it is then natural to ask if there are efficient algorithms for (a) checking whether a given assignment is stable, and (b) deciding if a given profile admits a stable assignment. It turns out that, while the answer to the first question is 'yes', the answer to the second question is likely to be 'no'.

Proposition 4. *We can decide in polynomial time whether a given assignment for a forking problem is stable.*

Proof. Note first that if an assignment f is not stable, then this can be witnessed by a deviation in which all deviating agents move to the same alternative (say, A). Indeed, the agents who deviate from f by moving to A would find this move beneficial even if other agents did not move (in particular, due to monotonicity, they benefit from other agents not moving away from A). Thus, to decide if a given assignment f is stable, it suffices to consider deviations that can be described by a pair (S, n_S), where S is an alternative and $n_S > f^{-1}(S)$. For each such pair, we need to check if there are $n_S - f^{-1}(S)$ agents who are currently not assigned to S, but prefer (S, n_S) to their current circumstances.   □

Proposition 5. *Deciding whether a forking problem admits a stable assignment is NP-complete.*

Proof (Sketch). By Proposition 4, our problem is in NP. For hardness we adapt the reduction argument of Darmann [6, Theorem 3], establishing NP-hardness for the core stability problem in o-GASP with increasing preferences. That construction makes use of so-called void activities, which are available in o-GASP, but not in forking problems. In the profile constructed for [6, Theorem 3], the occurrence of void activities in each agent's preference needs to be replaced by $(S^*, 1)$, where S^* denotes the top alternative in the agent's preference.   □

Our hardness reduction produces an instance where the number of alternatives is linear in the number of voters. The complexity of finding a stable assignment for a fixed number of alternatives (e.g., $m = 3$) remains open.

6   Conclusions and Future Work

In the real world, communities sometimes fracture, or fork. This can generally be seen as a consequence of the decisions the community has made. If agents associate freely, with the ever-present option of leaving, then we can account for this possibility within collective decision making procedures. This enables minorities to threaten a fork in protest against the tyranny of the majority while giving the majority an opportunity to concede to prevent a fork. Such a forking process also facilitates the emergence of new communities, as it may be easier to sprout a community from an existing one rather than to build one from scratch.

We have shown that, while it may not be difficult to find stable partitions of a set of agents, constructing strategyproof rules is only possible in restricted domains. While the necessary and sufficient conditions for strategyproofness remain an interesting open question, we have identified a range of circumstances that are sufficient for strategyproofness. Lastly, we have shown that efficient preference elicitation is possible and desirable.

The social choice setting we considered is, to the best of our knowledge, novel and our work has only made the first steps towards its analysis. Several directions for future research present themselves: (1) first, as mentioned above, settling the question about the domain restrictions that are necessary and sufficient for the existence of stable and strategy-proof assignment rules is a priority; (2) second, natural generalizations of the setting we propose will be worth investigating— e.g., settings with several alternatives (similarly to how, e.g., large miners can be present in several forks), or settings in which the identities of the agents matter (as agents may wish to fork with other specific agents); (3) third, studying mechanisms for the converse problem, in which several communities could merge into a new one; and (4) fourth, enabling a majority to remove troublesome or faulty agents (e.g. Sybils) by forcing a fork.

Acknowledgements. Ehud Shapiro is the Incumbent of The Harry Weinrebe Professorial Chair of Computer Science and Biology. We thank the generous support of the Braginsky Center for the Interface between Science and the Humanities. Nimrod Talmon was supported by the Israel Science Foundation (ISF; Grant No. 630/19). Ben Abramowitz was supported in part by NSF award CCF-1527497.

References

1. Allombert, V., Bourgoin, M., Tesson, J.: Introduction to the Tezos blockchain. In: Proceedings of HPCS '19, pp. 1–10 (2019)
2. Beck, R., Müller-Bloch, C., King, J.L.: Governance in the blockchain economy: a framework and research agenda. J. Assoc. Inf. Syst. **19**(10), 1 (2018)
3. Brandt, F., Conitzer, V., Endriss, U., Lang, J., Procaccia, A.D.: Handbook of Computational Social Choice. Cambridge University Press, Cambridge (2016)
4. Bulteau, L., Jain, P., Talmon, N.: Partition aggregation for budgeting. In: Proceedings of M-PREF '20 (at ECAI '20) (2020)
5. Chalkiadakis, G., Elkind, E., Wooldridge, M.: Computational Aspects of Cooperative Game Theory. Morgan & Claypool Publishers, San Rafael (2011)
6. Darmann, A.: Group activity selection from ordinal preferences. In: Walsh, T. (ed.) ADT 2015. LNCS (LNAI), vol. 9346, pp. 35–51. Springer, Cham (2015). https://doi.org/10.1007/978-3-319-23114-3_3
7. Darmann, A.: A social choice approach to ordinal group activity selection. Math. Soc. Sci. **93**, 57–66 (2018)
8. Darmann, A., Elkind, E., Kurz, S., Lang, J., Schauer, J., Woeginger, G.: Group activity selection problem with approval preferences. Int. J. Game Theory **47**(3), 767–796 (2018)
9. Elkind, E., Grossi, D., Shapiro, E., Talmon, N.: United for change: deliberative coalition formation to change the status quo. In: Proceedings of AAAI '21 (2021)
10. Grossi, D., Pigozzi, G.: Judgment Aggregation: A Primer. Morgan & Claypool Publishers, San Rafael (2014)

11. Nakamoto, S.: Bitcoin: a peer-to-peer electronic cash system. Technical report, Manubot (2019)
12. Narayanan, A., Bonneau, J., Felten, E., Miller, A., Goldfeder, S.: Bitcoin and Cryptocurrency Technologies. Princeton University Press, Princeton (2016)
13. Ng, D., Griffin, P.: The wider impact of a national cryptocurrency. In: Global Policy, p. 1 (2018)
14. Phillip, A., Chan, J.S.K., Peiris, S.: A new look at cryptocurrencies. Econ. Lett. **163**, 6–9 (2018)
15. Reijers, W., et al.: Now the code runs itself: On-chain and off-chain governance of blockchain technologies. In: Topoi, pp. 1–11 (2018)
16. Swierczek, B.: Democratic file revision control with liquidfeedback. Liquid Demo. J. (2021)
17. Wattenhofer, R.: Distributed Ledger Technology: The Science of the Blockchain. Createspace Independent Publishing Platform, Scotts Valley (2017)
18. Webb, N.: A fork in the blockchain: income tax and the bitcoin/bitcoin cash hard fork. North Carolina J. Law Technol. **19**(4), 283 (2018)
19. Zhou, S., Vasilescu, B., Kästner, C.: How has forking changed in the last 20 years? a study of hard forks on github. In: Proceedings of ICSE '20, pp. 445–456. IEEE (2020)

Hedonic Diversity Games Revisited

Andreas Darmann[✉]

Department of Operations and Information Systems,
University of Graz, Graz, Austria
andreas.darmann@uni-graz.at

Abstract. A hedonic diversity game (HDG) is a coalition formation problem, where the set of agents is partitioned into two types of agents (say red and blue agents), and each agent has preferences over the relative number (fraction) of agents of her own type in her coalition. In a dichotomous hedonic diversity game (DHDG) each agent partitions the set of possible fractions into a set of approved and a set of disapproved fractions. The solution concepts for these games considered in the literature so far are concerned with stability notions such as core and Nash stability. We add to the existing literature by providing NP-completeness results for the decision problems whether a DHDG admits (i) a Nash stable outcome and (ii) a strictly core stable outcome respectively, in restricted settings with only two (and three, respectively) approved fractions per agent. In addition, applying approval and Borda scores from voting theory we aim at outcomes that maximize social welfare (i.e., the sum of scores) in (dichotomous) hedonic diversity games. In that context we provide an NP-completeness result for HDGs under the use of Borda scores. For DHDGs with approval scores, we draw the sharp separation line between polynomially solvable and NP-complete cases with respect to the number of approved fractions per agent.

Keywords: Coalition formation · Stable outcomes · Approval and Borda scores · Computational complexity

1 Introduction

A hedonic diversity game (HDG) is a coalition formation problem, where the set of agents is partitioned into two types (say red and blue agents), and each agent has preferences over the relative number (fraction) of agents of her own type in her coalition. An outcome of a HDG is a partition of the agents into disjoint coalitions, i.e., subsets of agents. As a particular example, consider the situation in which two institutes of the same university merge into one large institute. While, for their next project, some researchers want to collaborate intensively with the members of the formerly other institute (who they do not know very well yet)—and hence prefer a high fraction of researchers of the formerly other institute in their coalition—some (more shy) researchers might prefer to work alone or together with the colleagues of their original institute, and hence prefer a

© Springer Nature Switzerland AG 2021
D. Fotakis and D. Ríos Insua (Eds.): ADT 2021, LNAI 13023, pp. 357–372, 2021.
https://doi.org/10.1007/978-3-030-87756-9_23

low fraction. Applications of similar flavor are interdisciplinary collaboration and group formation among exchange and local students (see Bredereck et al. [11]). Other examples of a HDG include Bakers and Millers games (Aziz et al. [1] and Bredereck et al. [11]), where each red agent (baker) would like to be in a coalition with an as large as possible fraction of blue agents (millers) and vice versa.

A number of solution concepts have been considered for hedonic diversity games that are concerned with stability against individual or group deviations, including the notions of Nash stable, individually stable, and core stable outcomes (see Bredereck et al. [11] and Boehmer and Elkind [7]). The particular special case of a dichotomous hedonic diversity game (DHDG) arises when the agents' preferences are dichotomous, i.e., each agent i partitions the set of fractions of possible coalitions containing i into a set of approved fractions and a set of disapproved fractions respectively. In this setting it is known that deciding whether a Nash stable outcome exists is NP-complete even when each agent approves of at most 4 fractions, while in any HDG an individually stable outcome always exists and can be found in polynomial time (Boehmer and Elkind [7]). To the best of our knowledge, however, the complexity of deciding whether a HDG, and, in particular, a DHDG, admits a strictly core stable outcome has not been considered yet.

In this work, we add to the existing literature by showing that deciding whether a DHDG admits a Nash stable outcome remains NP-complete even if each agent approves of at most 2 fractions and one type of agents does not approve of coalitions that contain only agents of the same type. In addition, we show that deciding whether a DHDG admits a strictly core stable outcome is NP-complete, even when each agent approves of at most three fractions and one type of agents does not approve of coalitions that contain only agents of the same type.

Finally, we expand the set of solution concepts considered in the literature by considering the concept of social welfare. Using scores from voting theory— in particular, approval and Borda scores—the goal would be an outcome that maximizes social welfare (i.e., the sum of scores). In that respect, we prove that in a DHDG in which each agent approves of exactly one fraction (or at most one fraction) an outcome that maximizes social welfare (i.e., total approval score) can be found in polynomial time. In contrast, the problem of deciding whether a DHDG admits an outcome with total approval score exceeding some given integer is NP-complete, even when each agent approves of at most two fractions. Leaving the setting of a DHDG, we show that in a hedonic diversity game in which each agents' preferences are strict orders over the possible fractions, the problem of deciding whether there is an outcome with total Borda score exceeding some given integer is NP-complete.

Related Work

Hedonic diversity games were introduced by Bredereck et al. [11]. The main solution concepts considered in Bredereck et al. [11]—Nash stability, individual stability, and core stability—stem from game theory and are, in particular, adopted from the hedonic game literature. They remark that, even when all agents have single-peaked preferences, a Nash stable outcome does not always exist. On the positive side, they prove that in that case of single-peaked preferences an individually stable outcome always exists and provide a polynomial time algorithm that computes such an outcome. In addition, it is shown that deciding whether a HDG admits a non-empty core is NP-complete, and that even with single-peaked preferences the core might be empty. Boehmer and Elkind [7] focus on Nash stability and individual stability in hedonic diversity games. They prove that in fact any HDG admits an individually stable outcome and provide a polynomial time algorithm for computing such an outcome. On the other hand, they prove NP-completeness for the decision problem whether a HDG admits a Nash stable outcome, and show that this holds even in the setting of a DHDG when each agent approves of four fractions only. As mentioned above, we improve upon that result by showing that hardness also holds in DHDGs with each agent approving of only two fractions and one type of agents approving only of coalitions containing at least one agent of the other type. In addition, it is known that every DHDG admits a core stable outcome (this follows from a more general result for dichotomous hedonic games (Peters [18])), and such an outcome can be found in polynomial time (Boehmer [6]). However, in this work we prove that deciding whether a DHDG admits a strictly core stable outcome is NP-complete even in a restricted setting.

In *hedonic games*, originally introduced by Drèze and Greenberg [14], a given set of agents needs to be partitioned into coalitions (subgroups of agents), where each agent has preferences over the members of her coalition; see Aziz and Savani [2] for a survey. The computational complexity of stable coalition formation has been well-studied, e.g., by Bogomolnaia and Jackson [8], Ballester [3], and Peters [18]. In particular, HDGs have a certain vicinity to *anonymous hedonic games* (see, e.g., Bogomolnaia and Jackson [8]) and *fractional hedonic games* (see, e.g., Aziz et al. [1]).

In anonymous hedonic games, the agents do not care about the identity of the other agents in their coalition, but have preferences over the size of the coalitions only. Stable outcomes—for stability concepts such as Nash, individual, and (strict) core stability—in hedonic games, and in particular anonymous hedonic games, have been well-studied from a computational viewpoint (see, e.g., Ballester [3] and Bogomolnaia and Jackson [8]). In fractional hedonic games, each agent associates a numerical value with each other agent, and, for agent i, the value of her coalition is the average value of the other agents' values in i's coalition. The computational complexity involved in finding Nash, individually, or (strictly) core stable outcomes (or deciding whether such an outcome exists) in fractional hedonic games has been studied, for instance, by Bilò et al. [5], Brandl et al. [10], and Aziz et al. [1]. In contrast to these kinds of games, in HDGs we are concerned with two types of agents who have preferences over the fraction of agents of her own type. Bakers and Millers games, however, can be formulated as fractional hedonic games; it is known that the strict core (and thus the core) of a Bakers and Millers game is always non-empty, and a finest partition in the strict core can be found in linear time (Aziz et al. [1]).

Finally, we point out that scores from voting theory (see Brams and Fishburn [9] for a survey) such as approval and Borda scores have also been applied outside of their classical framework to evaluate outcomes, for instance, in fair division problems (see, e.g., Baumeister et al. [4] and Darmann and Schauer [13]), in combinatorial optimization problems like the traveling salesperson problem (Klamler and Pferschy [17]), or in the group activity selection problem (Darmann [12]).

This paper is structured as follows. In Sect. 2 we present the preliminaries, i.e., we formally introduce the model of a hedonic diversity game and the solution concepts considered in this work. Sections 3 and 4 are concerned with dichotomous hedonic diversity games: in Sect. 3 we present our computational complexity results for Nash stability and strict core stability; in Sect. 4 we focus on outcomes maximizing the total number of approvals, and draw the sharp separation line between polynomially solvable and NP-complete cases with respect to the number of approved fractions per agent. In Sect. 5 we turn to hedonic diversity games with strict preferences and prove that it is NP-complete to decide whether such a hedonic diversity game admits an outcome with total Borda score exceeding some given threshold.

2   Preliminaries

A *hedonic diversity game* $G = (R, B, (\succsim_i)_{i \in R \cup B})$ consists of two disjoint sets R, B of agents—the agents in R are called red agents, the agents in B are called blue agents—and we set $N = R \cup B$. Each agent $i \in N$ specifies a weak order $\succsim_i$ (with indifference part $\sim_i$ and strict preference part $\succ_i$) over the set Θ of all fractions of red agents in some subset of N containing agent i. Hence, for a red agent we have $\Theta = \{\frac{r}{r+b} \mid r \in \{1, \dots |R|\}, b \in \{1, \dots, |B|\}\} \cup \{1\}$, and for a blue agent we have $\Theta = \{\frac{r}{r+b} \mid r \in \{1, \dots |R|\}, b \in \{1, \dots, |B|\}\} \cup \{0\}$; observe that the cardinality of Θ is the same for a red and a blue agent.

A subset $C \subseteq N$ is called coalition, and $\mathcal{C}_i$ denotes the set of all coalitions containing agent $i \in N$. We interpret $\succsim_i$ as the preferences of agent i over all possible fractions of red agents in some coalition containing her. For coalition C, we denote the fraction of red agents in C by $\theta_R(C)$. A coalition is *mixed* if it contains both blue and red agents, otherwise it is *pure*. A *purely red (blue) coalition* consists of red (blue) agents only. An *outcome* π is a partition of $R \cup B$ into disjoint coalitions. For outcome π, let $\pi(i)$ denote the coalition containing agent i; conversely, we write $C \in \pi$ if $\pi(i) = C$ holds for some agent i. Abusing notation, for $C, D \in \mathcal{C}_i$ we have $C \succsim_i D$ iff $\theta_R(C) \succsim_i \theta_R(D)$ holds; we say that agent i strictly prefers coalition C over coalition D, $C \succ_i D$, iff $\theta_R(C) \succ_i \theta_R(D)$ holds.

In a *dichotomous hedonic diversity game (DHDG)* $G = (R, B, (A_i)_{i \in R \cup B})$, each agent i specifies a set A_i of approved fractions in Θ; agent i is indifferent between all fractions in A_i (i.e., for $\theta, \bar{\theta} \in A_i$ we have $\theta \sim_i \bar{\theta}$), strictly prefers any $\theta \in A_i$ to any $\bar{\theta} \notin A_i$, and is indifferent between all fractions not contained in A_i.

Solution Concepts

We will consider two kinds of solution concepts: On the one hand, we take into account the game-theoretic notions of stability against individual and group deviations, where we focus on Nash stability and strict core stability; on the other hand, we apply approval scores and Borda scores from voting theory to our setting in order to determine outcomes that maximize (utilitarian) social welfare, i.e., the total sum of scores.

Stability Notions. Nash stable outcomes require that no agent can make herself better off by forming a singleton coalition or by deviating towards some other coalition. Formally, an outcome π of a hedonic diversity game is *Nash stable*, if there is no agent i with $S \cup \{i\} \succ_i \pi(i)$ for some $S \in \pi \cup \{\emptyset\}$. In a DHDG, an outcome π is hence Nash stable if there is no agent i with $\theta_R(\pi(i)) \notin A_i$ but $\theta_R(S \cup \{i\}) \in A_i$ for some $S \in \pi \cup \{\emptyset\}$.

Strictly core stable outcomes require that there is no group of agents S such that, by forming a deviating coalition, at least one member of S is better off while no member of S changes for the worse. This can be formalized as follows. A coalition $S \subseteq N$ *weakly blocks* an outcome π of N if for every agent $i \in S$ we have $S \succsim_i \pi(i)$, and for some $i \in S$ we have $S \succ_i \pi(i)$. An outcome π is said to be *strictly core stable* (or in the *strict core*) if there is no weakly blocking coalition for π.

Social Welfare. *The score* of outcome π for agent i, $sc_\pi(i)$, is a non-negative integer assigned to $\pi(i)$. Under given scores, the social welfare of outcome π, $SW(\pi)$, is the sum of the scores over all agents: $SW(\pi) = \sum_{i \in N} sc_\pi(i)$. We consider the following two kinds of scores.

In a DHDG, the *approval score* of outcome π for agent i is 1 if $\theta_R(\pi(i)) \in A_i$ and 0 otherwise. In a DHDG, using approval scores the social welfare $SW(\pi)$ (or total approval score) of outcome π is hence the number of agents $i \in N$ for which $\theta_R(\pi(i)) \in A_i$ holds.

Given a hedonic diversity game $G = (R, B, (\succ_i)_{i \in R \cup B})$ with strict preferences $\succ_i$ over the set Θ, the *Borda score* of outcome π for agent i is given by $sc_\pi(i) = |\{\theta \in \Theta \mid \theta_R(\pi(i)) \succ_i \theta\}|$.

3 DHDG: Nash Stability and the Strict Core

In this section, we provide NP-completeness results in restricted settings with a small number of approvals per agent for the decision problems whether a DHDG admits a Nash stable outcome and a strictly core stable outcome respectively.

3.1 Nash Stability: NP-completeness for 2 Approvals per Agent

Our first result states that deciding whether a DHDG admits a Nash stable outcome is computationally intractable, even when restricted to instances with at most two approvals per agent and one type of agents approving of mixed coalitions only.

Theorem 1. *The problem of deciding whether a dichotomous hedonic diversity game $G = (R, B, (A_i)_{i \in R \cup B})$ admits a Nash stable outcome is NP-complete, even when (i) each agent approves of at most two fractions and (ii) none of the red agents approves of a purely red coalition.*

Proof. We provide a reduction from EXACT COVER BY 3-SETS (X3C). An instance of X3C is a pair $(X, \mathcal{Y})$, where $X = \{1, \ldots, 3q\}$ and $\mathcal{Y} = \{Y_1, \ldots, Y_p\}$ is a collection of 3-element subsets (3-sets) of X; it is a "yes"-instance iff X can be covered by exactly q sets from $\mathcal{Y}$. We assume that every element of X appears in exactly three sets in $\mathcal{Y}$; X3C is known to be NP-complete even under this restriction [15]. Observe that the restriction implies $p = 3q$, which allows us to omit q. Given such a restricted instance of X3C, we construct an instance $G = (R, B, (A_i)_{i \in R \cup B})$ of a dichotomous hedonic diversity game as follows. We set $R = \{\hat{r}_{i,j} \mid i \in \{1, \ldots, 3p\} \setminus \{2p\}, 1 \leq j \leq 3p + 1\} \cup \{r_{k,t} \mid 1 \leq k \leq p, 1 \leq t \leq 3\}$ and $B = \{b_{k,t} \mid 1 \leq k \leq p, 1 \leq t \leq 3\}$. For $x_k \in X$ let $Y_{k_1}, Y_{k_2}, Y_{k_3}$ denote the three sets of $\mathcal{Y}$ that contain x_k. We identify $x_k \in X$ with the agents $b_{k,t}$ and $r_{k,t}$, and set Y_{k_t} with the fraction $\frac{4+3k_t}{7+3k_t}$.

The agents' approved fractions are as follows. For each k,

- blue agent $b_{k,t}$'s set of approved fractions is $\{0, \frac{4+3k_t}{7+3k_t}\}$, $t \in \{1, 2, 3\}$,
- red agent $r_{k,1}$'s set of approved fractions is $\{\frac{5+3k_2}{8+3k_2}, \frac{5+3k_3}{8+3k_3}\}$,
- red agent $r_{k,2}$'s set of approved fractions is $\{\frac{5+3k_1}{8+3k_1}, \frac{5+3k_3}{8+3k_3}\}$, and
- red agent $r_{k,3}$'s set of approved fractions is $\{\frac{5+3k_1}{8+3k_1}, \frac{5+3k_2}{8+3k_2}\}$.

Finally,

- each red agent $\hat{r}_{i,j}$ approves of $\frac{1}{i+1}$ exclusively.

Observe that

- each fraction $\theta = \frac{4+3k_i}{7+3k_i}$—induced by the 3-set Y_{k_i}—is approved by exactly three blue agents, since for each element $x_k \in Y_{k_i}$ exactly one blue agent approves of θ;
- each fraction $\tilde{\theta} = \frac{5+3k_i}{8+3k_i}$ is approved by exactly six red agents, because for each element $x_k \in Y_{k_i}$ exactly two red agents approve of $\tilde{\theta}$.

We show that $(X, \mathcal{Y})$ admits an exact cover by 3-sets from $\mathcal{Y}$ iff G has a Nash stable outcome.

Assume $(X, \mathcal{Y})$ admits an exact cover, say $Z \subset \mathcal{Y}$, by 3-sets from $\mathcal{Y}$. We derive the following partition of the agents in G:

– For each set $Y_k \in Z$ form a coalition made up of the three blue agents approving of $\frac{4+3k}{7+3k}$ together with the six red agents approving of $\frac{5+3k}{8+3k}$ plus exactly $(4+3k) - 6$ arbitrarily chosen red agents $\hat{r}_{i,j}$. The remaining $2p$ blue agents form the purely blue coalition S. The remaining red agents form singleton coalitions each.

Observe that each blue agent approves of its coalition's fraction, hence no such agent has an incentive to deviate. No red agent approves of 1 or $\frac{1}{2p+1}$, hence no red agent wants to deviate towards a purely red coalition or towards S. In addition, for any choice of $k, \ell \in \mathbb{N}$ we have $\frac{(4+3k)+1}{(7+3k)+1} \neq \frac{1}{\ell}$ because otherwise $\ell = \frac{8+3k}{5+3k} = 1 + \frac{3}{5+3k}$ in contradiction with $\ell, k \in \mathbb{N}$. Therefore, no red agent $\hat{r}_{i,j}$ has an incentive to deviate towards a mixed coalition. Finally, for any coalition of fraction $\frac{4+3k}{7+3k}$, by construction the coalition contains all the six agents $r_{k,t}$ approving of $\frac{5+3k}{8+3k}$. Therefore, none of the agents $r_{k,t}$ has an incentive to deviate towards a mixed coalition either. Thus, the partition is Nash stable.

On the other hand, let π be a Nash stable outcome. *Let C be a mixed coalition in π. Coalition C must contain exactly the three blue agents approving of its fraction:* C cannot contain a blue agent not approving of its fraction since she would otherwise wish to form a singleton coalition instead. Also, each mixed coalition requires at least three blue agents, and in case C contains more than three blue agents at least one of them wishes to form a singleton coalition instead because for each fraction $\theta = \frac{4+3k}{7+3k}$ there are exactly three blue agents approving of θ.

Now, we show that π cannot contain a purely blue coalition of size $s \neq 2p$, $s \geq 1$. Assume the opposite and let S be such a purely blue coalition of size $s \neq 2p$, $s \geq 1$. Observe that there are exactly $(3p+1)$ red agents $\hat{r}_{s,j}$ approving of $\frac{1}{s+1}$ for any choice of $s \neq 2p$. Each agent $\hat{r}_{s,j}$ hence prefers $S \cup \{\hat{r}_{s,j}\}$ over its current coalition—and hence has an incentive to deviate—unless she is already in a coalition of fraction $\frac{1}{s+1}$. However, it is impossible that each such agent $\hat{r}_{s,j}$ is in a coalition of fraction $\frac{1}{s+1}$ since this would require $(3p+1) \cdot s > 3p$ blue agents. Thus, π cannot contain a purely blue coalition of size $s \neq 2p$.

Hence, *there must be at least p blue agents which are engaged in some mixed coalitions.* Since each blue agent needs to approve of the fraction of the mixed coalition C she is part of, C must be of fraction $\theta = \frac{4+3k}{7+3k}$ for some k. Also, recall that fraction $\theta = \frac{4+3k}{7+3k}$ is induced by set $Y_k = \{x_u, x_v, x_w\}$ in $\mathcal{Y}$. Coalition C is thus made up of

1. exactly the three blue agents approving of θ, i.e., three agents $b_{u,h_u}, b_{v,h_v}, b_{w,h_w}$ for some choices of $h_u, h_v, h_w \in \{1,2,3\}$, and
2. exactly $4+3k$ red agents including all the six red agents who approve of $\frac{5+3k}{8+3k}$
   —say $r_{u,t_u}, r_{u,\bar{t}_u} r_{v,t_v}, r_{v,\bar{t}_v} r_{w,t_w}, r_{w,\bar{t}_w}$ for some choices of $t_u, \bar{t}_u, t_v, \bar{t}_v, t_w, \bar{t}_w$—since otherwise π is not Nash stable.

Note that any two mixed coalitions must have different fractions since each mixed coalition must have exactly three blue agents, all of which approving of its fraction, and by construction for each fraction there are exactly three such agents.

In addition, observe that due to Point 2. above, for each u *at most one* of $b_{u,1}, b_{u,2}, b_{u,3}$ can be contained in a mixed coalition: the fact that r_{u,t_u} and $r_{u,\bar{t}_u}$ are contained in mixed coalition C with $\theta = \frac{4+3k}{7+3k}$ implies that $r_{u,\tilde{t}_u}$ with $\tilde{t}_u \notin \{t_u, \bar{t}_u\}$ who approves of $\frac{5+3\ell}{8+3\ell}$, $\ell \neq k$, cannot be contained in a mixed coalition $D \neq C$ with fraction $\frac{4+3\ell}{7+3\ell}$ because (i) one of $\{r_{u,t_u}, r_{u,\bar{t}_u}\}$ approves of $\frac{5+3\ell}{8+3\ell}$, and (ii) D would need to contain all the six red agents approving of $\frac{5+3\ell}{8+3\ell}$.

Since at least p blue agents need to be engaged in some mixed coalition it follows that for each u *exactly one* of $b_{u,1}, b_{u,2}, b_{u,3}$ is contained in some mixed coalition. Due to 1. that coalition C has to contain also the two other blue agents approving of its fraction. As a consequence, the collection Z of sets Y_k for which π contains a coalition of size $\frac{4+3k}{7+3k}$ forms an exact cover by 3-sets in instance $(X, \mathcal{Y})$.   □

3.2 Strict Core Stability: NP-completeness for 3 Approvals per Agent

We now turn to strictly core stable outcomes and show that the decision problem whether a DHDG admits such an outcome is computationally hard even in a restricted setting with only three approvals per agent.

Theorem 2. *The problem of deciding whether a dichotomous hedonic diversity game $G = (R, B, (A_i)_{i \in R \cup B})$ admits a strictly core stable outcome is NP-complete, even when (i) each agent approves of at most three fractions and (ii) none of the blue agents approves of a purely blue coalition.*

Proof. We provide a reduction from the restricted NP-complete version of Exact Cover by 3-Sets (X3C) used in the proof of Theorem 1. Given such an instance $(X, \mathcal{Y})$ of X3C, where $X = \{1, \dots, 3q\}$ and $\mathcal{Y} = \{Y_1, \dots, Y_p\}$ is a collection of 3-element subsets of X such that every element of X appears in exactly three sets in $\mathcal{Y}$, we construct an instance $G = (R, B, (A_i)_{i \in R \cup B})$ of a dichotomous hedonic diversity game. Recall that we have $p = 3q$. We set $R = \{r_k \mid 1 \leq k \leq p\} \cup \{\hat{r}_{k,j} \mid 1 \leq k \leq p, \ 1 \leq j \leq 3k - 2\}$ and $B = \{b_k \mid 1 \leq k \leq p\}$. For $x_k \in X$ let $Y_{k_1}, Y_{k_2}, Y_{k_3}$ denote the three sets of $\mathcal{Y}$ that contain x_k. We identify $x_k \in X$ with the agents b_k and r_k, and we associate set $Y_i \in \mathcal{Y}$ with the fraction $\frac{1+3i}{4+3i}$. The agents' approvals are as follows:

- for each k, agent b_k's and agent r_k's set of approved fractions is $\{\frac{1+3k_t}{4+3k_t} \mid 1 \leq t \leq 3\}$, and
- for each k and j, agent $\hat{r}_{k,j}$'s set of approved fractions is $\{1, \frac{1+3k}{4+3k}\}$.

Observe that by construction each fraction $\frac{1+3i}{4+3i}$, $1 \leq i \leq p$, is approved by exactly three blue agents. We now show that $(X, \mathcal{Y})$ admits an exact cover by 3-sets from $\mathcal{Y}$ iff G admits a non-empty strict core.

Assume that in instance $(X, \mathcal{Y})$ there is an exact cover Z by 3-sets. We construct partition π of N as follows. For each set $Y_i \in Z$ let coalition $C_i = \{b_k, r_k \mid x_k \in Y_i\} \cup \{\hat{r}_{i,j} \mid 1 \leq j \leq 3i - 2\}$, and let each $\hat{r}_{k,j}$ with $Y_k \notin Z$ form a singleton coalition. Each of the agents in a singleton coalition approves of its fraction. Observe that C_i contains exactly three blue agents and $(3 + 3i - 2)$ red agents. The fraction of coalition C_i is hence $\frac{1+3i}{4+3i}$ which, due to $x_k \in Y_i$, is approved by all of its agents. Note that by the fact that Z is an exact cover each of the agents r_k, b_k is in exactly one mixed coalition. Therefore, each agent is engaged in some coalition and approves of its fraction. Thus, partition π is strictly core stable.

On the other hand, assume that there is a strictly core stable outcome π. For the sake of contradiction, assume that at least one blue agent b_k is in a coalition with a fraction she disapproves of. Let $Y_i \in \mathcal{Y}$ denote one of the three sets that contain element x_k. As above, form the coalition $C_i = \{b_\ell, r_\ell \mid x_\ell \in Y_i\} \cup \{\hat{r}_{i,j} \mid 1 \leq j \leq 3i - 2\}$ with fraction $\frac{1+3i}{4+3i}$ which is approved by all members of C_i. Since $b_k \in C_i$ holds we can conclude that C_i weakly blocks π, in contradiction with the assumption that π is strictly core stable. Therewith, each blue agent must be in a coalition with a fraction θ she approves of. Note that by construction (for each blue agent, the denominator of each approved fraction exceeds the nominator by three), this requires that all the three agents approving of θ must be in the same coalition. Thus, the set $Z = \{Y_k \in \mathcal{Y} \mid \exists S \in \pi \text{ with } \theta_R(S) = \frac{1+3k}{4+3k}\}$ forms an exact cover by 3-sets in $(X, \mathcal{Y})$. $\square$

4   Maximizing Social Welfare: A Dichotomy for DHDG

Apart from stability notions, from a social choice perspective an outcome that maximizes social welfare is of interest. In this section, we consider DHDGs and use approval scores to measure the social welfare induced by an outcome. We first show that an outcome that maximizes social welfare, i.e., total approval score, can be found in polynomial time when each agent approves of exactly (or at most) one fraction. However, we then prove that the corresponding decision problem turns NP-complete already as soon as agents may approve of two fractions. Therewith we draw the sharp separation line between polynomially solvable and NP-complete cases with respect to the number of approved fractions per agent.

We introduce some additional notation. For set $N' \subseteq N$ of agents and fraction θ, let $R_\theta(N')$ and $B_\theta(N')$ denote the set of red and blue agents in N' approving of θ respectively. Let $\#r(N')$ and $\#b(N')$ denote the number of red and blue agents in set N' respectively.

Theorem 3. *In a dichotomous hedonic diversity game* $G = (R, B, (A_i)_{i \in R \cup B})$ *with approval scores in which each agent approves of exactly one fraction, an outcome that maximizes social welfare can be found in polynomial time.*

Proof. We will reduce a dichotomous hedonic diversity game with a single approval per agent to a two-constraint knapsack problem[1]. An instance of the two-constraint knapsack problem consists of a set J of items, where each item

---

[1] Also known as two-dimensional knapsack problem (the latter notion, however, often refers to the geometric variant of that problem).

$j \in J$ is associated with a profit p_j, a weight w_j and a volume v_j; the goal is to select a subset $J^* \subseteq J$ of items of maximum total profit $p^* = \sum_{j \in J^*} p_j$ such that the total weight does not exceed a given weight bound W and the total volume does not exceed a given volume bound V (i.e., $\sum_{j \in J^*} w_j \leq W$ and $\sum_{j \in J^*} v_j \leq V$). By dynamic programming, the maximum profit in an instance of the two-constraint knapsack problem can be determined in $\mathcal{O}(nWV)$ time, determining both the optimal profit and the profit maximizing set of items can be done in $\mathcal{O}(n^2WV)$ time (see Ch. 9.3.2 of [16]).

Given a dichotomous hedonic diversity game $G = (R, B, (A_i)_{i \in R \cup B})$ with exactly one approval per agent, we construct an instance $\mathcal{I}$ of the two-constraint knapsack problem. W.l.o.g., we assume that the fractions in G cannot be reduced anymore, i.e., the nominator and denominator of each fraction θ in G are coprime.

For each fraction $\theta = \frac{r_\theta}{r_\theta + b_\theta}$ (with $0 = \frac{0}{1}$ and $1 = \frac{1}{1+0}$) approved of by at least one agent we first partition the set of agents approving of θ into sets $S_{\theta,i}$ and introduce the items for instance $\mathcal{I}$ on basis of these sets.

In order to construct the sets, the idea is that while there are red or blue agents approving of θ, add them to $S_{\theta,1}$ as long as it contains less than r_θ red agents (b_θ blue agents); then continue with $S_{\theta,2}$, etc. We proceed as follows:

- let $q_\theta = \max\{c, d \mid |R_\theta(N)| = r_\theta \cdot c + e, |B_\theta(N)| = b_\theta \cdot d + f$ for some $e < r_\theta$, $f < b_\theta\}$;
- for each fraction θ construct the sets $S_{\theta,i}$, $1 \leq i \leq q_0$, containing agents of $R_\theta(N)$ and $B_\theta(N)$ exclusively, such that
  - each such set contains at most r_θ red agents and b_θ blue agents, and
  - for red agent $r \in R_\theta(N)$ we have $r \in S_{\theta,i+1}$ iff $S_{\theta,i}$ contains r_θ agents (for blue agent b of $B_\theta(N)$ we have $b \in S_{\theta,i+1}$ iff $S_{\theta,i}$ contains b_θ agents).

We say that set $S_{\theta,i}$ is *full*, if it contains exactly r_θ red agents and b_θ blue agents.

Observe that each agent of N is contained in exactly one of the sets $S_{\theta,i}$ (hence we have at most n such sets), and each agent in set $S_{\theta,i}$ approves of θ.

In order to construct instance $\mathcal{I}$ of the two-constraint knapsack problem, for each set $S_{\theta,i}$ we introduce an item $\theta^{(i)}$ with profit $p_{\theta^{(i)}} = |S_{\theta,i}|$, weight r_θ and volume b_θ, and set $W = |R|$, $V = |B|$.

We show that there is solution of $\mathcal{I}$ with profit $\geq p^*$ iff there is an outcome π for G with $SW \geq p^*$.

"$\Rightarrow$": A solution ($=$ set of items) J^* of total profit ℓ in $\mathcal{I}$ induces an outcome of social welfare $\geq \ell$ in G as follows. Consider the set $S^* = \{S_{\theta,i} \mid \theta^{(i)} \in J^*\}$, i.e., S^* is the set of sets $S_{\theta,i}$ corresponding to the items in J^*. We construct outcome π of our DHDG in two steps.

First, for all sets $S_{\theta,i} \in S^*$ which are full, define the coalition $C_{\theta,i} = S_{\theta,i}$. Next, for a non-full set $S_{\theta,i} \in S^*$, the set contains $|R \cap S_{\theta,i}| < r_\theta$ red agents or $|B \cap S_{\theta,i}| < b_\theta$ blue agents; however, the weight and volume of the corresponding item $\theta^{(i)}$ are r_θ and b_θ, respectively. Hence, for each non-full set $S_{\theta,i} \in S^*$ the weight contribution of $\theta^{(i)}$ exceeds the number of red agents in $S_{\theta,i}$ by $r_\theta - |R \cap S_{\theta,i}|$ and the volume contribution of $\theta^{(i)}$ exceeds the number of blue agents in $S_{\theta,i}$ by $b_\theta - |B \cap S_{\theta,i}|$. Together with the choice of $W = |R|$ (and $V = |B|$ respectively) it follows that there must be at least

$$\sum_{S_{\theta,i} \in S^* : |S_{\theta,i}| < r_\theta + b_\theta} r_\theta - |R \cap S_{\theta,i}|$$

red agents and at least

$$\sum_{S_{\theta,i} \in S^* : |S_{\theta,i}| < r_\theta + b_\theta} b_\theta - |B \cap S_{\theta,i}|$$

blue agents in N that are not contained in some set of S^*. Therefore, for all sets $S_{\theta,i} \in S^*$ which are not full we are able to construct a coalition $C_{\theta,i}$ of fraction θ by "filling up" $S_{\theta,i}$ with such red and blue agents—i.e., create $C_{\theta,i}$ by adding to $S_{\theta,i}$ red and blue agents of $N \setminus \bigcup_{S_{\theta,i} \in S^*} S_{\theta,i}$ until it contains exactly r_θ red and b_θ blue agents.

Now let π be the outcome made up of the coalitions $C_{\theta,i}$ for $S_{\theta,i} \in S^*$ plus coalition D containing all remaining agents. Observe that, for each $C_{\theta,i} \in \pi$, $|S_{\theta,i}|$ denotes the number of agents in $C_{\theta,i}$ that approve of its fraction θ. In addition, recall that by definition $p_{\theta(i)} = |S_{\theta,i}|$. Thus, for outcome π we have $SW(\pi) \geq \sum_{S_{\theta,i} \in S^*} |S_{\theta,i}| = \sum_{\theta(i) \in J^*} p_{\theta(i)} = \ell$.

"$\Leftarrow$": Assume there is an outcome π of $SW(\pi) = \ell \geq p^*$. W.l.o.g. we assume that each coalition in π cannot be split into smaller coalitions of the same fraction, i.e., for each coalition C and $\tilde{C} \subset C$ it holds that $\theta_R(C) \neq \theta_R(\tilde{C})$.

Let S be the set of coalitions $C \in \pi$ in which all agents approve its fraction $\theta_R(C)$, and let S' be the set of coalitions C' for which at least one agent disapproves of $\theta_R(C')$. Let N' be the set of agents engaged in some coalition $C' \in S'$, and let $N_a' \subseteq N'$ be the set of agents of N' who approve of its coalition fraction, and $N_d' \subseteq N'$ be the set of agents of N' who disapprove of its coalition fraction. From π we construct a new partition π' as follows:

- for the agents N_a' build, for all θ' such that $\theta' = \theta_R(C')$ for at least one set $C' \in S'$, the sets $S_{\theta',i}$ as described in the construction of instance I;
- fill up the sets $C' = S_{\theta',i}$ with agents in N_d' as follows (i.e., add exactly $(r_{\theta'} - |R_{\theta'}(N') \cap S_{\theta',i}|)$ red and $(b_{\theta'} - |B_{\theta'}(N') \cap S_{\theta',i}|)$ blue agents from N_d' to $S_{\theta',i}$):
  - as long as there is a set $C' = S_{\theta',i}$ with $\#r(C') < r_{\theta'}$ (resp. $\#b(C') < b_{\theta'}$), and a red (resp. blue) agent from N_d' who approves of θ', add that agent to the set $S_{\theta',j}$ with the smallest index j among such sets;
  - after that, as long as there is a set $C' = S_{\theta',i}$ with $\#r(C') < r_{\theta'}$ (resp. $\#b(C') < b_{\theta'}$), add an arbitrary red (resp. blue) agent from N_d' to C';
- the remaining agents of N_d' form a single coalition D.

Observe that compared with π, for each $\theta' = \theta_R(C')$ such that $C' \in S'$ the number of sets $S_{\theta',i}$ does not exceed the number of coalitions of fraction θ' in π. Hence, the number of agents from N_d' added to the sets $S_{\theta',i}$ in order to achieve the required fraction θ' is in fact sufficient since π is a feasible partition.

As a consequence, for each θ such there is coalition C in partition π with $\theta_R(C) = \theta$, there are at most as many coalitions of fraction θ in π' as in π. Also, observe that an agent who approves of her coalition's fraction in π also approves of her coalition's fraction in π'. Thus, the number of agents engaged in some coalitions in $\pi' \setminus D$ who approve of their coalition's fraction is at least ℓ. However, by construction of π', there is a one-to-one correspondence between the set of coalitions of $\pi' \setminus D$ and a subset J' of items in $\mathcal{I}$. Therewith, for each coalition $S \in \pi'$, $S \neq D$, there must be an item in instance $\mathcal{I}$ with the profit corresponding to the number of agents approving of $\theta_R(S)$. Also, by the choice of the items' weights and volume, J' is a feasible solution for instance $\mathcal{I}$. Therewith, $\mathcal{I}$ admits solution J' with profit $\geq \ell$.

Finally, observe that the optimal profit in two-constraint knapsack problem can be determined in $\mathcal{O}(nWV) = \mathcal{O}(n^3)$ time, together with backtracking of the solution this can be done in $\mathcal{O}(n^4)$ time. $\square$

On the negative side, as soon as agents approve of up to two fractions the problem of deciding whether a DHDG with approval scores admits an outcome with social welfare exceeding some given integer becomes computationally difficult.

Theorem 4. *Given integer ℓ, the problem of deciding whether a dichotomous hedonic diversity game $G = (R, B, (A_i)_{i \in R \cup B})$ with approval scores admits an outcome with $SW \geq \ell$ is NP-complete, even when (i) each agent approves of at most two fractions and (ii) each blue agent approves of only one fraction.*

Proof. Again we reduce from the NP-complete variant of EXACT COVER BY 3-SETS (X3C) restricted to instances $(X, \mathcal{Y})$ with $X = \{1, \ldots, 3q\}$ and $\mathcal{Y} = \{Y_1, \ldots, Y_p\}$ such that every element of X appears in exactly three sets in $\mathcal{Y}$. Let $\mathcal{I}$ be such a restricted instance of X3C, and recall that $p = 3q$ holds. We construct an instance $G = (R, B, (A_i)_{i \in R \cup B})$ of a dichotomous hedonic diversity game as follows. We set $R = \{\hat{r}_i \mid i \in \{1, \ldots, \frac{p}{3}(p^2 + p - 6)\}\} \cup \{r_{k,t} \mid 1 \leq k \leq p, 1 \leq t \leq 3\}$, and $B = \{b_{k,t} \mid 1 \leq k \leq p, 1 \leq t \leq 3\}$. For $x_k \in X$ the three sets containing x_k are denoted by $Y_{k_1}, Y_{k_2}, Y_{k_3}$. The agents' approvals are given by:

- blue agent $b_{k,t}$'s approved fraction is $\frac{p^2+k_t}{p^2+k_t+3}$, $t \in \{1, 2, 3\}$,
- red agent $r_{k,1}$'s set of approved fractions is $\{\frac{p^2+k_2}{p^2+k_2+3}, \frac{p^2+k_3}{p^2+k_3+3}\}$,
- red agent $r_{k,2}$'s set of approved fractions is $\{\frac{p^2+k_1}{p^2+k_1+3}, \frac{p^2+k_3}{p^2+k_3+3}\}$,
- red agent $r_{k,3}$'s set of approved fractions is $\{\frac{p^2+k_1}{p^2+k_1+3}, \frac{p^2+k_2}{p^2+k_2+3}\}$, and
- each red agent $\hat{r}_i$ approves of $\frac{1}{1+3p}$ exclusively.

Observe that each fraction $\theta = \frac{p^2+k_t}{p^2+k_t+3}$—induced by set Y_{k_t}—is approved by exactly three blue and six red agents. We now argue that $(X, \mathcal{Y})$ admits an exact cover by 3-sets from $\mathcal{Y}$ iff G admits an outcome π with $SW(\pi) \geq 3p$.

Assume there is an exact cover Z. Recall that Z contains exactly $\frac{p}{3}$ sets of $\mathcal{Y}$. Observe that for any $x_k \in X$ exactly one of $Y_{k_1}, Y_{k_2}, Y_{k_3}$ is in Z. To construct partition π in G,

- for each set $Y_{k_t} \in Z$, form a coalition C made up of the three blue and six red agents approving of $\theta = \frac{p^2+k_t}{p^2+k_t+3}$ together with $(p^2 + k_t - 6)$ arbitrarily chosen agents of $\hat{r}_i$;
- the remaining agents form singleton coalitions each.

Due to the fact that Z is an exact cover by 3-sets partition π is well-defined. Since we are concerned with exactly $\frac{p}{3}$ mixed coalitions, the number of agents of $\hat{r}_i$ engaged in a mixed coalition is at most $\frac{p}{3}(p^2 + p - 6)$, and hence partition π is feasible. Each mixed coalition contains 9 agents who approve of its fraction, which yields a total social welfare of $SW(\pi) = 3p$.

On the other hand, assume there is an outcome π with $SW(\pi) \geq 3p$ in G. No agent approves of being in a pure coalition, so only mixed coalitions can contribute a positive value to the social welfare of π. In order to do so, a mixed coalition C must be either of fraction $\frac{1}{1+3p}$ or of fraction $\frac{p^2+j}{p^2+j+3}$ for some $1 \leq j \leq p$. In the former case C contains all blue agents and hence must be the only mixed coalition, implying $SW(\pi) \leq 1$ in contradiction with our assumption. In the latter case, C requires at least $(p^2 + j)$ red agents. Given that G contains exactly $\frac{p}{3}(p^2 + p - 6) + 3p = \frac{p}{3}(p^2 + p + 3)$ red agents, at most $\frac{p}{3}$ such coalitions can exist, because otherwise at least $(\frac{p}{3} + 1)(p^2 + 1) = \frac{p}{3}(p^2 + 3p + 1) + 1$ red agents would be required. Due to the fact that any fraction $\frac{p^2+j}{p^2+j+3}$ is approved by exactly 9 agents, this means that we must have exactly $\frac{p}{3}$ such coalitions and each of them must contain all the 9 agents approving of its fraction. For each k this means that for at most one—and by the fact that we have $\frac{p}{3}$ such coalitions this means for exactly one—$t \in \{1, 2, 3\}$ there is a coalition of fraction $\frac{p^2+k_t}{p^2+k_t+3}$ in π; hence, for each k exactly one of $b_{k,1}, b_{k,2}, b_{k,3}$ is engaged in such a coalition, and that coalition contains all the three blue agents approving of its fraction. Therewith, the collection of sets $Z = \{Y_{k_t} \mid \exists C \in \pi : \theta_R(C) = \frac{p^2+k_t}{p^2+k_t+3}\}$ forms an exact cover by 3-sets in $\mathcal{I}$. $\square$

5 Maximizing Social Welfare in HDG Under Borda Scores

We now leave the setting of DHDGs and consider the case in which each agents' preferences are given by means of a strict order over the possible coalition fractions. It turns out that in such a scenario under the use of Borda scores maximizing social welfare is computationally hard.

Theorem 5. *Given integer ℓ, the problem of deciding whether a hedonic diversity game $G = (R, B, (\succ_i)_{i \in R \cup B})$, with strict order $\succ_i$ over Θ for $i \in N$, under Borda scores admits an outcome with $SW \geq \ell$ is NP-complete.*

Proof. We reduce from the NP-complete variant of EXACT COVER BY 3-SETS (X3C) restricted to instances $(X, \mathcal{Y})$ with $X = \{1, \ldots, 3q\}$ and $\mathcal{Y} = \{Y_1, \ldots, Y_p\}$ such that every element of X appears in exactly three sets in $\mathcal{Y}$. Let $\mathcal{I}$ be such a restricted instance of X3C (recall that $p = 3q$ holds). From $\mathcal{I}$ we derive instance

$G = (R, B, (\succ_i)_{i \in R \cup B})$ of a hedonic diversity game as described below. The set of agents is made up of the sets $R = \{r_i \mid i \in \{1, \ldots, p^5\}\}$ and $B = \{b_k \mid 1 \le k \le p\}$. Again, for $x_k \in X$ we denote the three sets containing x_k by $Y_{k_1}, Y_{k_2}, Y_{k_3}$. Agent $b_k \in B$ represents element $x_k \in X$, and we associate fraction $\frac{j+3}{j+6}$ with set $Y_j \in \mathcal{Y}$. The agents' rankings—up to the respective position where fraction $\frac{1}{|R|+|B|}$ is ranked—are given in Table 1. Let $T = |\Theta| - 1$, i.e., T is the maximum possible Borda score for a single agent.

Table 1. Rankings of agents b_k and r_i up to fraction $\frac{1}{|R|+|B|}$ (used in the proof of Theorem 5)

Borda score	agent b_k's ranking		Borda score	agent r_i's ranking
T	$\frac{k_1+3}{k_1+6}$		T	$\frac{4}{7}$
$T-1$	$\frac{k_2+3}{k_2+6}$		$T-1$	$\frac{5}{8}$
$T-2$	$\frac{k_3+3}{k_3+6}$		$\vdots$	$\vdots$
$T-3$	$\frac{|R|}{|R|+|B|}$		$T-p+1$	$\frac{p+3}{p+6}$
$T-4$	$\frac{|R|-1}{|R|-1+|B|}$		$T-p$	1
$T-5$	$\frac{|R|-2}{|R|-2+|B|}$		$T-p-1$	$\frac{|R|}{|R|+|B|}$
$\vdots$	$\vdots$		$T-p-2$	$\frac{|R|-1}{|R|-1+|B|}$
$T-2-|R|$	$\frac{1}{1+|B|}$		$T-p-3$	$\frac{|R|-2}{|R|-2+|B|}$
			$\vdots$	$\vdots$
			$T-p-|R|$	$\frac{1}{1+|B|}$

We claim that $\mathcal{I}$ is a "yes"-instance of X3C iff G admits an outcome π with total Borda score $SW(\pi) \ge \ell = (T-2)p + (T-p)p^5$.

"$\Rightarrow$": Let Z be an exact cover by 3-sets in instance $\mathcal{I}$. Consider partition π which

- for each $Y_j \in Z$ forms a coalition C_j made up of the three blue agents who have $\frac{j+3}{j+6}$ among their top 3 ranked fractions together with $(j+3)$ arbitrarily chosen red agents,
- and assigns the remaining agents (who, by the fact that Z is an exact cover by 3-sets, must all be red agents) to the single coalition D.

By the fact that Z is an exact cover by 3-sets each blue agent is in a coalition with a fraction she ranks first, second, or third. Each red agent is in a coalition of fraction 1 or $\frac{j+3}{j+6}$ for some $1 \le j \le p$. Thus, we have $SW(\pi) \ge (T-2)p + (T-p)p^5 = \ell$.

"$\Leftarrow$": Let π be an outcome with $SW(\pi) \ge \ell$. Note that any outcome in which all blue agents are engaged in the same coalition yields a total Borda score of at most $(T-3)p + (T-p-1)p^5 < \ell$. Hence, any outcome meeting the desired bound splits the set of blue agents into at least two coalitions.

Assume there is a blue agent b_k who is not in a coalition with fraction ranked among her top 3 fractions. Since the blue agents are not in a single coalition, the maximum possible Borda score for that agent is $sc_\pi(i) = T - 2 - |R| - 1 = T - 3 - p^5$. For the remaining $p - 1$ blue agents the maximum Borda score is T. Next, observe that any coalition's fraction among the first p ranked fractions of agents r_i corresponds to $\frac{k+3}{k+6}$ for some $1 \le k \le p$, and hence the number of blue agents required in such a coalition is a multiple of 3. Since there are only p blue agents in total, there are less than $\frac{p}{3}(p+3)$ red agents involved in coalitions of fraction $\frac{k+3}{k+6}$ for some k. Thus the largest possible total Borda score contributed by all red agents is $\frac{p}{3}(p+3)T + (p^5 - \frac{p}{3}(p+3))(T-p)$. Thus,

$$
\begin{aligned}
SW(\pi) &\le T - 3 - p^5 + T(p-1) + \tfrac{p}{3}(p+3)T + (p^5 - \tfrac{p}{3}(p+3))(T-p) \\
&= T(p + \tfrac{p}{3}(p+3) + p^5 - \tfrac{p}{3}(p+3)) - p^6 + \tfrac{p^3}{3} + p^2 - 3 - p^5 \\
&= T(p^5 + p) - p^6 - p^5 + \tfrac{p^3}{3} + p^2 - 3 \\
&< T(p^5 + p) - p^6 - 2p \\
&= \ell
\end{aligned}
$$

in contradiction with our assumption.

As a consequence, each blue agent's coalition must have a fraction ranked among her top 3 fractions. Since each such fraction is among the top 3 fractions of exactly three blue agents, each respective coalition requires exactly 3 blue agents (because such a coalition requires a multiple of 3 agents). Therewith, the collection of sets $Z = \{Y_j \mid \exists C \in \pi : \theta_R(C) = \frac{j+3}{j+6}\}$ forms an exact cover by 3-sets in $\mathcal{I}$. □

6 Conclusion

We have provided several computational complexity results for hedonic diversity games with respect to two kinds of solution concepts: stability notions that stem from game theory on the one hand, and the concept of (maximum) social welfare that origins from social choice theory on the other. Concerning the latter, we have taken into account two prominent types of scores from voting theory, namely approval scores and Borda scores. Some interesting questions, however, are still open. In particular, what is the computational complexity involved in deciding whether a dichotomous hedonic diversity game admits a Nash stable outcome in the case of exactly one approval per agent? How hard is it to decide whether a dichotomous hedonic diversity game admits a strictly core stable outcome when each agent approves of exactly one or at most two fractions? And more generally, which (additional) plausible domain restrictions allow for a polynomial time computation of outcomes that are stable or maximize social welfare?

Acknowledgments. The author would like to thank Edith Elkind for useful discussion and is grateful for the valuable comments provided by the reviewers.

References

1. Aziz, H., Brandl, F., Brandt, F., Harrenstein, P., Olsen, M., Peters, D.: Fractional hedonic games. ACM Trans. Econ. Comput. **7**(2), 6:1–6:29 (2019)
2. Aziz, H., Savani, R.: Hedonic games. In: Brandt, F., Conitzer, V., Endriss, U., Lang, J., Procaccia, A.D. (eds.), Handbook of Computational Social Choice, chapter 15. Cambridge University Press (2016)
3. Ballester, C.: NP-completeness in hedonic games. Games Econ. Behav. **49**, 1–30 (2004)
4. Baumeister, D., et al.: Axiomatic and computational aspects of scoring allocation rules for indivisible goods. In: Proceedings of the 5th International Workshop on Computational Social Choice (COMSOC 2014), pp. 1–22 (2014)
5. Bilò, V., Fanelli, A., Flammini, M., Monaco, G., Moscardelli, L.: Nash stability in fractional hedonic games. In: Liu, T.-Y., Qi, Q., Ye, Y. (eds.) WINE 2014. LNCS, vol. 8877, pp. 486–491. Springer, Cham (2014). https://doi.org/10.1007/978-3-319-13129-0_44
6. Boehmer, N.: Algorithmic analysis of hedonic games with diversity preferences. Master's thesis, University of Oxford (2019)
7. Boehmer, N., Elkind, E.: Individual-based stability in hedonic diversity games. In: Proceedings of the 34th AAAI Conference on Artificial Intelligence (AAAI 2020), pp. 1822–1829 (2020)
8. Bogomolnaia, A., Jackson, M.: The stability of hedonic coalition structures. Games Econ. Behav. **38**, 201–230 (2002)
9. Brams, S.J., Fishburn, P.C.: Voting procedures. In: Arrow, K.J., Sen, A.K., Suzumura, K. (eds.), Handbook of Social Choice and Welfare, vol. 1, pp. 173–236 (2002)
10. Brandl, F., Brandt, F., Strobel, M.: Fractional hedonic games: individual and group stability. In: Proceedings of the 14th International Conference on Autonomous Agents and Multiagent Systems (AAMAS 2015), pp. 1219–1227 (2015)
11. Bredereck, R., Elkind, E., Igarashi, A.: Hedonic diversity games. In: Proceedings of the 18th International Conference on Autonomous Agents and Multi-Agent Systems (AAMAS 2019), pp. 565–573 (2019)
12. Darmann, A.: A social choice approach to ordinal group activity selection. Math. Soc. Sci. **93**(C), 57–66 (2018)
13. Darmann, A., Schauer, J.: Maximizing Nash product social welfare in allocating indivisible goods. Eur. J. Oper. Res. **247**(2), 548–559 (2015)
14. Drèze, J., Greenberg, J.: Hedonic coalitions: optimality and stability. Econometrica **48**(4), 987–1003 (1980)
15. Gonzalez, T.F.: Clustering to minimize the maximum intercluster distance. Theoret. Comput. Sci. **38**, 293–306 (1985)
16. Kellerer, H., Pferschy, U., Pisinger, D.: Knapsack Problems. Springer, Berlin (2004). https://doi.org/10.1007/978-3-540-24777-7
17. Klamler, C., Pferschy, U.: The traveling group problem. Soc. Choice Welfare **29**(3), 429–452 (2007)
18. Peters, D.: Complexity of hedonic games with dichotomous preferences. In: Proceedings of the 30th AAAI Conference on Artificial Intelligence (AAAI 2016), pp. 579–585 (2016)

Stable Matchings

Multi-agent Reinforcement Learning for Decentralized Stable Matching

Kshitija Taywade[✉], Judy Goldsmith, and Brent Harrison

University of Kentucky, Lexington, KY, USA
kshitija.taywade@uky.edu, {goldsmit,harrison}@cs.uky.edu

Abstract. In the real world, people/entities usually find matches independently and autonomously, such as finding jobs, partners, roommates, etc. It is possible that this search for matches starts with no initial knowledge of the environment. We propose the use of a multi-agent reinforcement learning (MARL) paradigm for a spatially formulated decentralized two-sided matching market with independent and autonomous agents. Having autonomous agents acting independently makes our environment very dynamic and uncertain. Moreover, agents lack the knowledge of preferences of other agents and have to explore the environment and interact with other agents to discover their own preferences through noisy rewards. We think such a setting better approximates the real world and we study the usefulness of our MARL approach for it. Along with conventional stable matching case where agents have strictly ordered preferences, we check the applicability of our approach for stable matching with incomplete lists and ties. We investigate our results for stability, level of instability (for unstable results), and fairness. Our MARL approach mostly yields stable and fair outcomes.

Keywords: Stable matching · Multi-agent reinforcement learning · Decentralized system

1 Introduction

Matching markets are prevalent in the real world, for example, matching of students to colleges, doctors to hospitals, employees to employers, men and women, etc. A two-sided market consists of two disjoint sets of agents. In a two-sided stable matching problem, each participant has preferences over the participants on the other side. A matching is stable if it does not contain a blocking pair. A blocking pair is formed if two agents from disjoint sets prefer each other rather than their current partner. Although, most of the prior literature focuses on centralized algorithms where the entire set of preferences is known to some central agency, having such a central clearinghouse is not always feasible. Therefore, we consider a decentralized matching market with independent and autonomous agents.

There have been several decentralized matching methods proposed in recent years. However, many of them assume that the agents have knowledge of one

© Springer Nature Switzerland AG 2021
D. Fotakis and D. Ríos Insua (Eds.): ADT 2021, LNAI 13023, pp. 375–389, 2021.
https://doi.org/10.1007/978-3-030-87756-9_24

another's preferences and can easily approach/contact each other, i.e., negligible search friction. In reality, it takes time to meet a partner and to learn the value of said partnership. Furthermore, there is seldom a scope for knowing the preferences of other agents. Also, it is a crucial task to locate and approach a potential match, either by navigating physically or virtually. Several decentralized matching markets, such as worker-employer markets and buyer-seller trading markets, consist of locations at which matching agents may meet, be it physically or online. The level of information, search cost, medium of interaction, and commitment laws can vary across markets. Nonetheless, these are the important features of decentralized markets. Some research works study the impact of these features on the final outcomes for certain types of markets [5,22,23]. To better represent these features, we propose a generalized matching problem in which agents are placed in a grid world environment and must learn to navigate it in order to form matches. We see this as a generalized case for matching problems. While it contains the features described above, it does not conform to the standards set by any individual market type.

There are multiple factors involved in deciding a preferred match in real-world situations, and having a score for each match is more expressive. Thus, we consider weighted preferences for a stable matching problem, which is discussed in [8,13,24]. The weighted preference is used as the utility value (or *reward*) for being in the match. These scores reflect the underlying preference order. Agents are initially unaware of others' preferences as well as of their own. In many matching markets, knowledge acquisition is important: in labor markets, employers interview workers; in matching markets, men and women date; and in real estate markets, buyers attend open houses. We have taken this into account, so an agent gets to know a noisy version of its utility for a match only after being part of it. Noise represents uncertainty in the value of a partnership, e.g., the uncertain nature of human behavior in relationships.

Finding a long-term match in this scenario is quite a complex task. Therefore, we propose multi-agent reinforcement learning (MARL) as an alternative paradigm where agents must learn how to find a match based on their experiences interacting with others. We equip each agent with their own reinforcement learning (RL) module. We use *SARSA*, a model-free, online RL algorithm. Instead of a common reward signal, each agent has a separate intrinsic reward signal. Therefore, we model this problem as a stochastic/Markov game [17], which is useful in modeling multi-agent decentralized control where the reward function is separate for each agent. Agents learn to operate in the environment with the goal of increasing expected total reward, by getting into a long-term, stable, or close-to-stable and fair match. We impose search cost as a small negative reward (−1) for each step whenever an agent is not in a match.

We investigate the applicability of the MARL approach to the conventional *stable matching (SM)* problem, as well as its extensions, such as *stable matchings with incomplete lists (SMI)*, where agents are allowed to declare one or more partners unacceptable [9], and *stable matching with ties (SMT)*, where agents have the same preference for more than one agent [9,12]. Moreover, we study both the cases of symmetric and asymmetric preferences of agents towards each other. Stability is one of the main measures in our investigation. We check whether our method yields stable results and then to which stable matching the method will converge if there are multiple stable matchings. As we have a dynamic decentralized system with agents having incomplete information, it is hard to guarantee stability for every instance. For unstable outcomes, we check instability with three measures: the degree of instability calculates the number of blocking agents, i.e., those agents who are part of blocking pairs [27]; the ratio of instability gives the proportion of blocking pairs out of all possible pairs [6]; and maximum dissatisfaction, which is the maximum difference between an agent's current utility and their obtainable utility by being part of the blocking pair. Overall, we found that many of our outcomes are stable, or if not, they are close-to-stable. Also, it is easy to get stable outcomes for instances with symmetric preferences and harder for the asymmetric ones.

It is important for the outcome to be fair to all the agents, as the goal of the agents is to increase only their own happiness. Therefore, we use three measures of fairness: set-equality cost, regret cost, and egalitarian cost [9]. We compare the fairness of our results to those of bidirectional local search, a centralized approach, and two decentralized approaches: Hoepman's algorithm [11], and a decentralized algorithm by Comola and Fafchamps. Note that these algorithms solve the much easier, non-spatial problems, usually with the assumptions of complete information on the part of agents. Nonetheless, our approach performs competitively in terms of fairness. Lastly, similar to [5], we check the proportion of overall median stable matchings, as well as individual median matchings in our results, which are other important measures of fairness.

2   Related Work

Reinforcement Learning has not been used for the decentralized two-sided stable matching problem, but researchers have applied both RL and Deep RL mechanisms to solve coalition formation problems. This is closest to our work as matching problems are a special case of coalition formation problems. In [1], Bachrach et al. proposed a framework for training agents to negotiate and form teams using deep reinforcement learning. They have also formulated the problem spatially. Bayesian reinforcement learning has also been used for coalition formation problems [2,18]. Unlike most of the work in MARL approaches for coalition formation and task allocation, our agents cannot communicate with each other (although they can observe other agents in the same cell). Nonetheless, their utilities get affected by the actions of other agents.

Researchers have studied several decentralized matching markets, and have proposed frameworks for modeling them and techniques for solving them, and have also analyzed different factors that affect the results [4,5,10,20–23,29]. Most of these works focus on job markets. Echenique and Yariv, in their study of one-to-one matching markets, proposed a decentralized approach for which stable outcomes are prevalent, but unlike our formulation of a problem, agents have complete information of everyone's preferences [5]. Unlike our work, none of these works have formulated the problem spatially, and also, they have used different methods than RL. Some distributed algorithms for weighted matching include algorithms that are distributed in terms of agents acting on their own either synchronously or asynchronously [11,15,33]. The crucial assumption in these works is that agents already know their preferences over the members of the opposite set and can directly contact other agents to propose matches.

Most of the decentralized methods mentioned here allow agents to make matching offers and accept/reject such offers. While in our approach, an agent shows interest in pairing with an agent from the other set that is present at the same location, by selecting a relevant action. The agent's state space represents those agents from the opposite set that are present at the same cell location and also which ones among them are interested in pairing. Agents get matched only when both the agents select an action for pairing with each other. While this may seem similar to making, accepting, or rejecting offers, it is not exactly the same.

3   Preliminaries

Our two-sided stable matching problem consists of n agents divided equally into two disjoint sets S_1 and S_2. These agents are placed randomly on the grid with dimensions $H \times L$. We investigate if agents can learn good matching policies in a decentralized, spatial setting.

Definition 1. *In the classical two-sided **stable matching problem (SM)**, each agent has a strict preference order p over the members of the other set. Given matching M, the pair (i,j) with an agent $i \in S_1$ and an agent $j \in S_2$ is a blocking pair for M, if i prefers j and j prefers i to their respective partners in M. A matching is said to be stable if it does not contain any blocking pairs.*

The preferences are expressed as weights and hence referred to as weighted preferences. A weight/score represents the true utility value an agent may receive by being in a particular match. These weights still correspond to a strict preference order for each agent. An agent only gets to know the utility from a match when it is in that particular match. Even then, it only receives a noisy utility value for that match rather than the true, underlying utility value. It can be formally written as: for $i \in S_1$ and $j \in S_2$, agent i receives the utility $U_{ij} \cdot C$ for being in a match with j, where C is the noise, sampled from a normal distribution with mean $\mu = 1$ and standard deviation $\sigma = 0.1$ and U_{ij} is the true utility value that agent i can get from a match with agent j. Agents still have a strict preference order p over the agents on the other side. U_{ij} is picked uniformly from range $[k,l] \in \mathbb{Z}$, while maintaining the strict preference order.

We also consider following two extensions of the SM problem.

Definition 2. *The **stable matching problem with incomplete preference lists (SMI)** may have incomplete preference lists for those involved. In this* *case, the members of the opposite set who are unacceptable to an agent simply do not appear in their preference list [9].*

As we have a score based formulation, an agent has negative scores for unacceptable agents of the other set.

Definition 3. *Some agents may be indifferent (i.e., have the same utility) between two or more members of the opposite set. This is called the **stable matching problem with ties (SMT)** [9,12].*

We consider two types of preferences among agents: symmetric and asymmetric.

For $i \in S_1$ and $j \in S_2$, let $p_i(j)$ ($p_j(i)$, respectively) denote the position of i in j's preference list (the position of j in i's preference list, respectively). In **symmetric preferences**, $p_i(j) = p_j(i)$ (in our case, $U_{ij} = U_{ji}$ as well), which is not guaranteed to be the case in **asymmetric preferences**. With asymmetric preferences (similar to random preferences in literature), there can be many different stable matchings in a market. However, in the case of symmetric preferences, there can be only one stable matching where each agent gets their best choice. Our environment is dynamic and uncertain, and also, due to the narrow difference between noisy utilities, it can be hard for agents to discriminate between their choices efficiently. This can cause unstable outcomes, especially for asymmetric preferences. Therefore, if a stable outcome does not emerge, then we investigate the nature of instability with the following three measures.

Definition 4. *The **degree of instability (DoI)** of the matching is the number of blocking agents, i.e., the agents that belong to some blocking pair [27].*

Eriksson and Häggström pointed out that, rather than only looking for the number of blocking agents, it can also be helpful to look at the number of blocking partners of an agent, as it gives insight into how likely the agent will exploit instability [6]. Their notion of instability is defined as follows.

Definition 5. *For any matching M under preference structure $P^{(m)}$ on a set of m agents, let $B_P^{(m)}(M)$ denote the number of blocking pairs. Let $\hat{B}_P^{(m)}(M)$ denote the proportion of blocking pairs: $\hat{B}_P^{(m)}(M) = B_P^{(m)}(M) \,/\, m^2$ [6].*

While Eriksson and Häggström call this measure the 'instability' of the matching M, we call it the **ratio of instability (RoI)**. We also use a third measure, **maximum dissatisfaction (MD)**. It is inspired by the notion of α-stability in [24] which is specific to SM with weighted preferences.

Definition 6. *In matching M, for every blocking agent x, let y be their current match and v be their partner in some blocking pair, then*

$$MD(M) = \max_{(x,v)}\{U_{xv} - U_{xy}\}.$$

Increase in this number may lead to exploitation of instability by agents in the market. Stability in the outcomes does not guarantee fairness. We consider three measures of fairness to check the quality of matchings as given in [9].

Definition 7. *The **regret cost**,* $r(M) = \max\limits_{(i,j) \in M} max\{p_i(j), p_j(i)\}.$

Definition 8. *The **egalitarian cost**,* $c(M) = \sum\limits_{(i,j) \in M} p_i(j) + \sum\limits_{(i,j) \in M} p_j(i).$

Definition 9. *The **set-equality cost**,* $d(M) = \sum\limits_{(i,j) \in M} p_i(j) - \sum\limits_{(i,j) \in M} p_j(i).$

Lower values for these measures indicate better quality of the matchings. Especially, low regret cost and set-equality cost indicate fairness among agents. It is well known that the Gale-Shapley algorithm provides a matching that is optimal for only one side, over all possible stable matchings. Thus, one notion of fairness is to consider the median of the set of stable matchings, so as to privilege neither set over the other. Thus, similar to [5], we check whether the final matchings are median stable matchings (MSM), and overall, what proportion of individual matches are median matches (MM). The well-known median property is first discovered by Conway [9]. A median matching exists whenever there is an odd number of stable outcomes. It is the matching that is in the middle of the two sides' orders of preference. Thus, the median stable matching represents some sense of fairness as it balances the interests of both sides.

Definition 10. *Let P be a preference profile with the set of stable matchings* $S(P)$. *If* $K = |S(P)|$ *is odd, the **median stable matching (MSM)** is a matching* $M \in S(P)$ *such that for all agents* $a \in S_1 \cup S_2$, $M(a)$ *occupies the* $\frac{K+1}{2}$ *th place in a's preference among the agents in* $\{M'(a)|M' \in S(P)\}$. $M(a)$ *is a's median partner among a's stable-matching partners [5]. While MSM is for overall matching, **median match (MM)** refers to individual matches between the agents, i.e., an individual agent being matched to its median stable match partner.*

Multi-agent Reinforcement Learning

As mentioned earlier, we propose a multi-agent reinforcement learning (MARL) approach that enables each agent to learn independently to find a good match for itself. A reinforcement learning agent learns by interacting with its environment. The agent perceives the state of the environment and takes an action, which causes the environment to transition into a new state at each time step. The agent receives a reward reflecting the quality of each transition. The agent's goal is to maximize the expected cumulative reward over time [30]. In our system, although agents learn independently and separately, their actions affect the environment and in turn affect the learning process of other agents as well. As agents receive separate intrinsic rewards, we modeled our problem as a Markov game. Stochastic/Markov games [17] are used to model multi-agent decentralized control where the reward function is separate for each agent, as each agent works only towards maximizing its own total reward.

A *Markov game* with n players specifies how the state of an environment changes as the result of the joint actions of n players. The game has a finite set of states S. The observation function $O : S \times \{1, \ldots, n\} \to R_d$ specifies a d-dimensional view of the state space for each player. We write $O_i = \{o_i | s \in S, o_i = O(s, i)\}$ to denote the observation space of player i. From each state, players take actions from the set $\{A_1, \ldots, A_n\}$ (one per player). The state changes as a result of the joint action $\langle a_1, ..., a_n \rangle \in \langle A_1, ..., A_n \rangle$, according to a stochastic transition function $T : S \times A_1 \times ... \times A_n \to \Delta(S)$, where $\Delta(S)$ denotes the set of probability distributions over S. Each player receives an individual reward defined as $r_i : S \times A_1 \times ... \times A_n \to \mathbb{R}$ for player i. In our multi-agent reinforcement learning approach, each agent learns independently, through its own experience, a behavior policy $\pi_i: O_i \to \Delta(A_i)$ (denoted $\pi(a_i|o_i)$) based on its observation o_i and reward r_i. Each agent's goal is to find policy π_i which maximizes a long term discounted reward [30].

4   Method

We propose a MARL approach for decentralized two-sided stable matching problems that are formulated spatially on a grid. For each agent, the starting location is picked uniformly randomly from the grid cells. As agents go through episodic training, they start in this same cell location in each episode and explore the environment. Agents must first find each other before they can potentially form matches. This approximates the spatial reality of meeting with individuals (at, e.g., bars or parties) or organizations (at, e.g., job fairs).

We believe that finding a partner for oneself is an independent task, where agents do not necessarily need to compete or even co-operate. Agents only need to learn to find a suitable partner. Each agent independently learns a policy using the RL algorithm, SARSA [28,30], with a multi-layer perceptron as a function approximator to learn the set of Q-values. An agent's learning is independent of other agents' learning as all the agents have separate learning modules (neural networks). We use SARSA because it is an on-policy algorithm in which agents improve on the current policy. Unlike off-policy algorithms like deep Q-learning where agents' behavior while learning can be erratic due to inconsistencies in the policy, on-policy algorithms follow the same policy and improve on it, which is useful when the agent's exploratory behavior matters. In real-world matching markets, there is a value to the path of finding a final match. SARSA is also a model-free algorithm, so that agents directly learn policies, without having to learn the model.

While exploring, agents cannot perceive any part of the environment other than their cell location. If an agent encounters another agent from the opposite set in the same cell and both the agents show interest in matching with each other, at the same time step, then they get matched. As agents can only view their current grid cell, agents can only match with one another if they are in the same cell. As long as agents are matched, they receive a noisy reward as a utility value at each time step. This noise is sampled from the normal distribution and the true utility value is multiplied by this noise. Note that our environment is deterministic. We now describe the agents' observation space, action space, and reward function.

Observation Space: An observation O_i for an agent i at time step t, let's say $O_i[t]$, consists of three one-hot vectors. The first one represents an agent's position on the grid, the second vector represents which members of the opposite set are present in the current cell, and third one shows if any of those agents are interested in forming a match. The size of O_i is $R \times C + 2 \cdot m$, where R and C are the number of rows and columns in the grid and m is the total number of members of the opposite set. The size of the first hot vector is equal to the total number of grid cells, and the size of the second and third vector is equal to the size of the opposite set. Thus, an agent initially starts out knowing only the dimensions of the grid and the total number of agents in the opposite set.

Action Space: There are two types of actions available to an agent: navigating the grid and expressing an interest in matching with an agent from the opposite set. The action space is of size $m + 4$, where m is the size of the opposite set and each member has an action associated with it for showing an interest in matching with that member. There are 4 additional actions for navigating the grid by moving up, down, left, and right. There is no specific action for staying in the same grid cell because whenever an agent is interested in forming a match with another agent, it automatically stays in the same cell. When two collocated agents show an interest in forming a match with one another, then the match is considered to be formed. Note that once a match is formed, the agents must continue to express interest in each other at each time step in order to maintain the match. If at some point, one ceases to express interest, the match is dissolved.

Reward Function: We have a noisy reward function described as: (1) -1 reward for not being in a match. (2) The immediate reward received by an agent i for a matching of agents i and j is $R_{ij} = U_{ij} \cdot C$, where C is the noise, sampled from a normal distribution with mean $\mu = 1$ and standard deviation $\sigma = 0.1$ and U_{ij} is the true utility value that agent i can get from a match with agent j.

Agents have prior knowledge of the grid size and the total number of agents in the opposite set because of the way the states are constructed. However, they completely lack the knowledge of the weighted preferences/utility values of other agents. Furthermore, agents only get to know their own utility for an agent on the other side when they get into a match with it, and that utility value is noisy. In our setup, individuals may choose to be in a match until someone better comes along or may choose to leave a match in order to explore further and look for someone better. Thus, a time step in which all agents are paired is not necessarily *stable*, because agents may break off a partnership to explore, or another, more appealing agent may be willing to partner with them.

5   Experiments

In this section, we present the ways we tested our approach on stable matching problems. Our main focus is on investigating the applicability of our MARL approach. Along with the classical stable matching case (SM), we examine how MARL performs on variations such as stable matching with incomplete lists (SMI) and ties (SMT). We consider two types of preferences among agents: symmetric and asymmetric. As mentioned earlier, agents have weighted preferences over agents on the other side. For an agent $i \in S_1$, it can be seen as the utility value U_{ij} that it gets while in a match with agent $j \in S_2$. In the case of SM and SMT problems, these weights are generated from a uniform random distribution in the range $[1, 10]$; for SMI, the weights are generated from a uniform random distribution in the range $[-10, 10]$ (negative weights indicate how much one agent dislikes the other). For SM and SMI problems, the instances where agents have weights reflecting the strictly ordered preferences are chosen for the experiments. This constraint is removed while choosing SMT instances.

As we formulate the problems on a grid, we investigate results for increasingly complex environments. This complexity is in terms of grid size and the number of agents. We use grid sizes 3×3, 4×4, and 5×5 in combination with 8, 10, 12, and 14 agents as follows: (1) Grid: 3×3 ; Agents: 8; (2) Grid: 4×4 ; Agents: 8, 10, 12, 14; (3) Grid: 5×5 ; Agents: 8, 10, 12, 14. We do not place more than 8 agents on a 3×3 grid to keep a reasonable density of the population. We chose grid sizes such that agents find other agents easily accessible. This is motivated by real-world places like bars, parties, job fairs, etc. We think that our choices of grid size and number of agents are sufficient to get the essence of realistic situations. Starting cell locations of agents are chosen uniformly randomly from the grid cells, and agents are placed back to these locations at the start of each episode. We run experiments for every possible combination of matching problem variation (SM, SMI, and SMT), preference type (symmetric and asymmetric), grid size, and total agents. We implement 10 different instances of each of these combinations. Each instance is generated by assigning weights between the agents uniformly randomly while still maintaining the preference order if needed.

Parameter Settings: Each agent independently learns a policy using SARSA [28, 30] with a multi-layer perceptron as a function approximator to learn a set of Q-values. Each network consists of 2 hidden layers with 50 and 25 hidden units, respectively. We trained models using the Adam optimizer [16] with learning rate 10^{-4} to minimize TD-control loss. We used discount factor, $\gamma = 0.9$. We have combined SARSA with experience replay for better results. The use of experience replay along with SARSA has been proposed by Zhao et al. [34]. As SARSA is an on-policy algorithm, we only used data from recent (last 10) episodes in our experience replay buffer, which increased our performance over not using a replay buffer. The number of training episodes and steps varies based on grid-size and the total number of agents in an instance. The number of steps per episode varies between 300–700 and training can take between 100k to 400k episodes to converge. When there are multiple suitable matches available in the environment for an agent, a proper exploration strategy is needed to find the best among them. Therefore, we used exploration rate with non-linear decay, such that it is high in the beginning but decays later (with a minimum exploration rate, $\epsilon = 0.05$). Learning rate and discount factor are fine tuned as the outcomes are slightly sensitive to these hyper-parameters; however, results are robust to the changes in other hyper-parameters.

We investigate stability and fairness of the outcomes. Roth hypothesized that the success of a centralized labor market depends on whether the matchmaking mechanism generates a stable matching [26]. Although we have decentralized matching market, we think that stability is still an important measure of the success. For the *SM* problem, stable matchings always exist, and for the *SMI* and *SMT* problems, at least a weakly stable matching exists [14]. In weak stability, a blocking pair is defined as (i, j) such that $M(i) \neq j, j \succ_i M(i)$, and $i \succ_j M(j)$ [14]. Note that in *SMI* instances, agents can end up without a partner as incomplete lists make some potential matches unacceptable.

As we have a dynamic and uncertain environment and agents with incomplete knowledge, there is a scope for the rise of instability. Economic experiments on decentralized matching markets with incomplete information [19,31] have yielded outcomes with considerable instability. We use three more measures to study instability: the degree of instability (DoI), the ratio of instability (RoI), and maximum dissatisfaction (MD) (details in Preliminaries). Stable or close-to-stable solutions do not guarantee fairness, specifically for asymmetric preference cases. As agents are independent and autonomous, we need to check the efficacy of our approach from an individual agent's point of view. Therefore, we use three fairness measures: set-equality cost, regret cost, and egalitarian cost. Additionally, we check the proportion of both median stable matchings as well as individual median matches. We compare our results with both centralized and decentralized algorithms. The comparison baselines are detailed below.

Bidirectional Local Search Algorithm (BLS). [32] is a centralized local search algorithm for stable matching with set-equality. It uses the Gale-Shapley algorithm [7] to compute S_1-optimal and S_2-optimal stable matchings and executes bi-directional search from those matchings until the search frontiers meet.

Hoepman's Algorithm (HA). [11] is a variant of the sequential greedy algorithm [25] which computes a weighted matching at most a factor of two away from the maximum. It is a distributed algorithm in which agents asynchronously message each other.

Decentralized Algorithm by Comola and Fafchamps (D-CF). [3] is designed to compute a matching in a decentralized market with deferred acceptance. Deferred acceptance means an agent can be paired with several other partners in the process of reaching their final match. This algorithm includes a sequence of rounds in which agents take turns in making proposals to other agents, who can accept or reject them. While Comola and Fafchamps focused on many-to-many matching, the method can be easily adapted for one-to-one matching.

Note that not only BLS but also HA and D-CF are non-spatial algorithms where agents already have knowledge of every other agent present in the system. This gives them a significant advantage over the agents in our system, both because the agents know whom they prefer and because they have instantaneous contact, rather than having to wander around in a grid world. Both of the decentralized algorithms use randomness while forming their final matching, giving different results each time. Therefore, we run each instance 5 times and compare to the average of those runs. We also run our MARL approach 5 times for each instance. We discovered that if a stable outcome is found, the same one is found consistently, but if not, then the outcomes vary.

6   Results and Discussion

We evaluate the results for stability, as well as the level of instability for unstable outcomes. We use three measures to evaluate instability: the degree of instability (DoI), the ratio of instability (RoI), and maximum dissatisfaction (MD). As fairness in the outcomes is also important, we use three fairness measures: set-equality cost, regret cost, and egalitarian cost. In addition to this, we check what percent of the stable matchings are median stable matchings, as well as what percent of the individual matches are median matches. Results for SM and SMT problems with symmetric preferences and for SMI problem with both symmetric and asymmetric preferences are straightforward, therefore, are mentioned in the text. However, the results for SM and SMT problems with asymmetric preferences needed more analysis. We elaborate on the results of SM problem with asymmetric preferences in Tables 1, 2, 3, as we think that this is the most relevant and adverse case. Due to lack of space, we omitted the similar analysis of the results for the SMT-asymmetric case; however, those results are very similar to the ones presented for the SM-asymmetric case.

Table 1. For SM (asymmetric) case, MARL results on stability (%), instability measures ($Avg \pm Std$) and median matches (%).

Grid	3×3	4×4				5×5			
Agents	8	8	10	12	14	8	10	12	14
Stability(%)	100	92.0	82.0	68.0	56.0	80.0	74.0	54.0	46.0
DoI	0	2 ± 0.0	2 ± 0.0	3.3 ± 1.3	3.2 ± 1.7	2 ± 0.0	2.5 ± 1.1	2.8 ± 0.9	3.5 ± 1.6
RoI	0	0.04 ± 0.0	0.04 ± 0.0	0.06 ± 0.01	0.07 ± 0.02	0.04 ± 0.0	0.04 ± 0.01	0.05 ± 0.02	0.07 ± 0.03
MD	0	1.75 ± 0.9	2.89 ± 1.7	3.13 ± 1.9	3.89 ± 1.8	2.33 ± 1.5	2.77 ± 1.4	3.25 ± 1.9	4.44 ± 2.3
MM(%)	83.1	73.2	65.3	63.4	52.4	75.0	67.1	58.9	48.7

Many of our outcomes are stable or close-to-stable. For SM problem with symmetric preferences, there is only one possible stable matching, and all the outcomes converge to that. However, for asymmetric preferences, more than one stable matching is possible. The instances with symmetric preferences converge faster than the asymmetric ones. The instances of SM and SMT with asymmetric preferences take longer to converge, with lower rates of convergence to stability. The results of SM and SMT are similar. Additionally, for SM asymmetric instances, we have observed that the agents disliked by everyone in the opposite set (low utility associated with them by everyone) find it difficult to get a long-term match. Similarly, unsurprisingly, the most-liked agent (high utility associated with them by everyone) easily settles with its ideal match. We also noticed that the noise in utilities adversely affects convergence to stable outcomes.

When it comes to SMI, our results are always stable. The number of agents that are matched is the maximum possible. This is important because when agents have incomplete lists (negative utilities for matches), it is hard to get a match for everyone, even though it is easier for some agents to find stable partner due to fewer choices. Here, the final outcome always has the lowest regret cost. Importantly, between the agents in the matched pair, there can be an agent having zero utility towards its match, while the other agent still has positive utility for the same match. As the agent with positive utility tries to get in a match, having noise in the reward causes the agent with zero utility to stick to the match. Note that this does not happen when both the agents in the match have zero utility for the match, as neither of them tries to stick with the match.

Table 2. For SM (asymmetric) case, comparison of set-equality cost and regret cost in ($Avg \pm Std$) format; results for 8 agents on 3×3 grid not included due to limited space.

N	Set-equality Cost; $d(M)$					Regret Cost; $r(M)$				
	MARL (4×4)	MARL (5×5)	BLS	HA	D-CF	MARL (4×4)	MARL (5×5)	BLS	HA	D-CF
8	3.1 ± 2.4	3.9 ± 2.5	2.9 ± 2.1	2.6 ± 1.8	3.1 ± 1.7	3.6 ± 0.8	3.5 ± 0.8	3.5 ± 0.8	3.7 ± 0.7	3.5 ± 0.8
10	2.9 ± 2.1	3.2 ± 2.8	3 ± 2.8	3.5 ± 1.6	4 ± 2.7	4.3 ± 0.8	4.1 ± 0.7	4 ± 0.8	4.6 ± 0.5	4.2 ± 0.9
12	4.6 ± 3.5	6.6 ± 3.8	5 ± 4.2	4 ± 4.2	4 ± 4.2	5.4 ± 0.8	5.3 ± 0.9	5.1 ± 0.9	5.3 ± 1.1	5.1 ± 0.9
14	7 ± 9.2	7.4 ± 7.7	7.5 ± 4.9	7.2 ± 4.4	7.5 ± 4.9	6.7 ± 0.7	5.9 ± 1.3	5.7 ± 1.1	5.6 ± 1.0	5.7 ± 1.1

From Table 1, which elaborates on the results of the SM-asymmetric case, we can see that the curse of dimensionality in how the number of agents affects stability. Although the grid size also affects stability, its impact is much less. Both of these factors affect the convergence rate as well: more complex environments take longer to converge. The environment with 8 agents on a 3×3 grid is the easiest one for training agents, and 100% of the outcomes are stable, while the one with 14 agents on a 5×5 grid is the hardest to train and the stability of the final outcomes declined significantly to 46%. Nonetheless, we can also see from the measures of instability that the outcomes are close-to-stable. Note that in Table 1 the values associated with these measures are averaged over only unstable outcomes. The average number of blocking agents (DoI) is low in all cases. We also checked the proportion of blocking pairs (RoI), as the greater this number, the more likely that blocking agents will discover and exploit the instability at some point [6]. Our approach does well for this measure. This follows the suggestion by Eriksson and Häggström that if agents increase the search effort rather than picking random partners, then we can expect outcomes to have a very small proportion of blocking pairs [6].

Table 3. For SM (asymmetric) case, comparison of egalitarian cost in $(Avg \pm Std)$ format.

N	Egalitarian Cost; $c(M)$				
	MARL (4×4)	MARL (5×5)	BLS	HA	D-CF
8	15.3 ± 2.8	15.1 ± 2.3	15.5 ± 2.7	16.6 ± 2.8	15.5 ± 2.7
10	20.3 ± 2.7	20.4 ± 3.4	19.8 ± 2.8	25.1 ± 3.7	20.4 ± 2.9
12	31 ± 5.9	28.4 ± 4.6	27.6 ± 3.4	32.6 ± 5.1	27.8 ± 3.3
14	41.4 ± 6.1	39.4 ± 5.6	34.9 ± 3.5	41.2 ± 6.9	34.9 ± 3.5

Furthermore, we look at the maximum dissatisfaction (MD) that an agent can have for an outcome, as great dissatisfaction may also lead to exploiting instability. This number is also low, which assures that there is a low likelihood of blocking agents exploiting unstable outcomes in the market. We think that the dynamic and uncertain environment, incomplete information, noisy utilities, and the narrow differences in the utilities between matches found for an individual over different episodes are potential reasons behind the emergence of instability in the outcomes. Especially in the case of asymmetric preferences, it is unlikely that an agent's ideal partner also best prefers that agent.

In Tables 2 and 3, we compare fairness in the outcomes with three other algorithms. Here, we can see that MARL performs competitively, and there is no significant difference between the fairness results. The regret cost of MARL is slightly, but not significantly, higher for all the types of instances. Hoepman's algorithm (HA) and the decentralized algorithm by Comola and Fafchamps (D-CF) are decentralized approaches. While D-CF always produces stable outcomes, that may not be the case with HA. Our approach performed better than HA in almost all the cases and very similarly to the D-CF algorithm. Again, our approach performs well despite being implemented on a fundamentally more complex formulation of the problem than the ones for HA and D-CF. Further, when our outcomes are stable, they usually match with those found by BLS. It shows that despite the decentralization, our MARL approach is capable of producing outcomes as good as those found by a central agency. This is further supported by the fact that the good proportion of individual matches are median matches (shown in Table 1). Also, approximately half of the stable matchings are median stable matchings. We think that the fairness is achieved because agents are self-interested and independent, and the stability is achieved as agents learn to find their best viable matches.

The learned policies include agents moving to a fixed location from their starting point and getting into a match corresponding to the final outcome. It is possible that more than one pair is formed at the same location, but it is rare. The location where agents in a pair move to form the match is not necessarily the mid-point of the distance between starting points of two agents, nor is it guaranteed to be close to either starting point. Centralized algorithms do not work on a grid; they produce matchings but not the learned policies. This shows that the real-world entities can benefit from using our MARL approach to learn to efficiently navigate the environment in finding and maintaining the good match.

7   Conclusion and Future Work

We have shown that the MARL paradigm can be successfully used for decentralized stable matching problems that are formulated spatially in a dynamic and uncertain environment, with independent and autonomous agents having minimum initial knowledge. Our MARL approach is also applicable for variations such as SM with incomplete lists and ties. Agents tend to be happy with their final matches, as outcomes are stable or close-to-stable and fair for everyone. Even with unstable outcomes, agents are less likely to exploit instability. In future work, we plan to work on bigger instances, environments where agents can arbitrarily enter and exit the matching market, and investigate environments where a few agents can learn while others have a fixed policy.

References

1. Bachrach, Y., et al.: Negotiating team formation using deep reinforcement learning (2018)
2. Chalkiadakis, G., Boutilier, C.: Bayesian reinforcement learning for coalition formation under uncertainty. In: Proceeding of AAMAS 2004, pp. 1090–1097 (2004)
3. Comola, M., Fafchamps, M.: An experimental study on decentralized networked markets. J. Econ. Behav. Organ. **145**, 567–591 (2018)
4. Diamantoudi, E., Miyagawa, E., Xue, L.: Decentralized matching: the role of commitment. Games Econ. Behav. **92**, 1–17 (2015)
5. Echenique, F., Yariv, L.: An experimental study of decentralized matching (2012)
6. Eriksson, K., Häggström, O.: Instability of matchings in decentralized markets with various preference structures. Int. J. Game Theor. **36**(3–4), 409–420 (2008)
7. Gale, D., Shapley, L.S.: College admissions and the stability of marriage. Am. Math. Monthly **69**(1), 9–15 (1962)
8. Gusfield, D.: Three fast algorithms for four problems in stable marriage. SIAM J. Comput. **16**(1), 111–128 (1987)
9. Gusfield, D., Irving, R.W.: The Stable Marriage Problem: Structure and Algorithms. MIT Press, Cambridge (1989)
10. Haeringer, G., Wooders, M.: Decentralized job matching. Int. J. Game Theor. **40**(1), 1–28 (2011)
11. Hoepman, J.H.: Simple distributed weighted matchings. arXiv cs/0410047 (2004)
12. Irving, R.W.: Stable marriage and indifference. Discrete Appl. Math. **48**(3), 261–272 (1994)
13. Irving, R.W., Leather, P., Gusfield, D.: An efficient algorithm for the optimal stable marriage. J. ACM (JACM) **34**(3), 532–543 (1987)
14. Iwama, K., Miyazaki, S.: A survey of the stable marriage problem and its variants. In: International Conference on Informatics Education and Research for Knowledge-Circulating Society, pp. 131–136. IEEE Computer Society (January 2008)

15. Khan, A., et al.: Efficient approximation algorithms for weighted b-matching. SIAM J. Sci. Comput. **38**(5), S593–S619 (2016)
16. Kingma, D.P., Ba, J.: Adam: a method for stochastic optimization. arXiv preprint arXiv:1412.6980 (2014)
17. Littman, M.L.: Markov games as a framework for multi-agent reinforcement learning. In: Machine Learning Proceedings 1994, pp. 157–163. Elsevier (1994)
18. Matthews, T., Ramchurn, S.D., Chalkiadakis, G.: Competing with humans at fantasy football: Team formation in large partially-observable domains. In: Twenty-Sixth AAAI Conference on Artificial Intelligence, aaai.org (2012)
19. Niederle, M., Roth, A.E.: Making markets thick: How norms governing exploding offers affect market performance. preprint (2006)
20. Niederle, M., Yariv, L.: Matching through decentralized markets. Stanford University, Discussion Paper (2007)
21. Niederle, M., Yariv, L.: Decentralized matching with aligned preferences. Technical report, National Bureau of Economic Research (2009)
22. Pais, J., Pintér, A., Veszteg, R.F.: Decentralized matching markets: a laboratory experiment (2012)
23. Pais, J., Pintér, Á., Veszteg, R.F.: Decentralized matching markets with (out) frictions: a laboratory experiment. Exp. Econ. 1–28 (2017)
24. Pini, M.S., Rossi, F., Venable, K.B., Walsh, T.: Stability and optimality in matching problems with weighted preferences. In: Filipe, J., Fred, A. (eds.) ICAART 2011. CCIS, vol. 271, pp. 319–333. Springer, Heidelberg (2013). https://doi.org/10.1007/978-3-642-29966-7_21
25. Preis, R.: Linear time 1/2-approximation algorithm for maximum weighted matching in general graphs. In: Meinel, C., Tison, S. (eds.) STACS 1999. LNCS, vol. 1563, pp. 259–269. Springer, Heidelberg (1999). https://doi.org/10.1007/3-540-49116-3_24
26. Roth, A.E.: A natural experiment in the organization of entry-level labor markets: regional markets for new physicians and surgeons in the United Kingdom. Am. Econ. Rev. 415–440 (1991)
27. Roth, A.E., Xing, X.: Turnaround time and bottlenecks in market clearing: decentralized matching in the market for clinical psychologists. J. Political Econ. **105**(2), 284–329 (1997)
28. Rummery, G.A., Niranjan, M.: On-Line Q-Learning Using Connectionist Systems, vol. 37. University of Cambridge, Department of Engineering England (1994)
29. Satterthwaite, M., Shneyerov, A.: Dynamic matching, two-sided incomplete information, and participation costs: existence and convergence to perfect competition. Econometrica **75**(1), 155–200 (2007)
30. Sutton, R.S., Barto, A.G.: Reinforcement Learning: An Introduction. MIT Press, Cambridge (2018)
31. Ünver, M.U.: On the survival of some unstable two-sided matching mechanisms. Int. J. Game Theor. **33**(2), 239–254 (2005)
32. Viet, H.H., Trang, L.H., Lee, S., Chung, T.: A bidirectional local search for the stable marriage problem. In: 2016 International Conference on Advanced Computing and Applications (ACOMP), pp. 18–24. ieeexplore.ieee.org (November 2016)
33. Wattenhofer, M., Wattenhofer, R.: Distributed weighted matching. In: Guerraoui, R. (ed.) DISC 2004. LNCS, vol. 3274, pp. 335–348. Springer, Heidelberg (2004). https://doi.org/10.1007/978-3-540-30186-8_24
34. Zhao, D., Wang, H., Shao, K., Zhu, Y.: Deep reinforcement learning with experience replay based on SARSA. In: 2016 IEEE Symposium Series on Computational Intelligence (SSCI), pp. 1–6. IEEE (2016)

Lazy Gale-Shapley for Many-to-One Matching with Partial Information

Taiki Todo, Ryoji Wada, Kentaro Yahiro, and Makoto Yokoo[✉]

Kyushu University, Fukuoka, Japan
{todo,yokoo}@inf.kyushu-u.ac.jp, {r-wada,yahiro}@agentinf.kyushu-u.ac.jp

Abstract. In the literature of two-sided matching, each agent is assumed to have a complete preference. In practice, however, each agent initially has only partial information and needs to refine it by costly actions (interviews). For one-to-one matching with partial information, the student-proposing Lazy Gale-Shapley policy (LGS) minimizes the number of interviews when colleges have identical partial preferences. This paper extends LGS to a significantly more practical many-to-one setting, in which a college can accept multiple students up to its quota. Our extended LGS uses a student hierarchy and its performance (in terms of the required number of interviews) depends on the choice of this hierarchy. We prove that when colleges' partial preferences satisfy a condition called compatibility, we can obtain an optimal hierarchy that minimizes the number of interviews in polynomial-time. Furthermore, we propose a heuristic method to obtain a reasonable hierarchy when compatibility fails. We experimentally confirm that compatibility is actually much weaker than being identical, i.e., when the partial preferences of each college are obtained by adding noise to an ideal true preference, our requirement is much more robust against such noise. We also experimentally confirm that our heuristic method obtains a reasonable hierarchy to reduce the number of required interviews.

Keywords: Two-sided matching · Many-to-one matching · Gale-shapley algorithm · Partial information · Partial preferences

1 Introduction

Two-sided matching deals with finding an appropriate matching between two types of agents (i.e., students and colleges). This topic has attracted much attention both from economics and computer science [3,10,19]. In traditional settings, agents are assumed to have complete information in advance about all the agents of the other type. However, such an assumption may not hold in practice. For instance, in a school choice program, a student may initially lack sufficient information to distinguish among several colleges. *Matching with partial information* captures and formalizes such cases. An agent is initially endowed with partial knowledge of her underlying strict preference and refines it through actions:

D. Fotakis and D. Ríos Insua (Eds.): ADT 2021, LNAI 13023, pp. 390–405, 2021.
https://doi.org/10.1007/978-3-030-87756-9_25

costly interviews [17,23]. Recent works tackle this problem from various viewpoints, e.g., combining interviews and queries to elicit preferences [6], restricting available partial information [7], considering other variants of stability [11,22], considering cases where each student's quality is unknown and estimating it by colleges' private signals [2,5], or examining communication complexity when partial preferences are clarified by simple queries [9,20].

The *Lazy Gale-Shapley* policy (LGS) [23] is an extension of the Gale-Shapley algorithm (GS) [8] for one-to-one matching with partial information. LGS obtains the student-optimal matching by performing interviews. When all colleges have identical partial preferences (i.e., the *Identical Equivalence Class* condition (IEC) is satisfied), LGS minimizes the number of interviews among all the policies that obtain the student-optimal matching.

This paper extends the results obtained by Rastegari et al. [23] to many-to-one matching, which has many real-life applications, including school choice [16] and hospital-residency matching [13]. To the best of our knowledge, no discussion exists on extending LGS to many-to-one matching. Our extended LGS is guaranteed to obtain the student-optimal matching. It uses a student hierarchy; its performance (in terms of the required number of interviews) depends on the choice of this hierarchy.

Our first main contribution is to show that we can obtain an optimal hierarchy that minimizes the number of interviews when colleges' partial preferences satisfy a condition called *compatibility*. This condition is more generally applicable than being identical. We quantitatively show that our compatibility condition is fairly robust against preference diversity, while IEC is likely to fail immediately. Compatibility is efficiently verified by a graph-based algorithm where students are vertices and the colleges' preferences are edges. Our second main contribution is to develop a heuristic, greedy method to obtain a reasonable hierarchy when compatibility fails. We experimentally evaluate how well our heuristic method obtains a reasonable hierarchy to reduce the number of required interviews.

2   Model

In this section, we introduce the model of a many-to-one matching problem with partial information. The model and properties introduced in this section are fairly standard in two-sided matching literature (e.g., see [24] for a comprehensive survey), except that we assume each agent initially has only partial information on her preference. Most of the concepts and representations related to partial information are based on [23]. More specifically, an instance of a many-to-one matching problem with partial information is given as: $(S, C, p_S, p_C, \succ_S, \succ_C, q_C)$.

- $S = \{s_1, \ldots, s_n\}$ is a set of n students.
- $C = \{c_1, \ldots, c_m\}$ is a set of m colleges.
- $p_S = (p_s)_{s \in S}$ is a profile of the partial preferences of the students, where each p_s is the partial preference of student s. More specifically, p_s partitions $C \cup \{\varnothing\}$ into finite equivalence classes $(p_s^1, p_s^2, \ldots)$. Each $p_s^k \subseteq C \cup \{\varnothing\}$, $\bigcup_{k=1}^{|p_s|} p_s^k =$

$C \cup \{\varnothing\}$, and for any $k \neq k'$, p_s^k and $p_s^{k'}$ are disjoint. p_s represents a strict preference order among equivalence classes, as described later.

- $p_C = (p_c)_{c \in C}$ is a profile of the partial preferences of the colleges, where each p_c is the partial preference of college c over $S \cup \{\varnothing\}$. As with p_s, p_c partitions $C \cup \{\varnothing\}$ into finite equivalence classes $(p_c^1, p_c^2, \ldots)$. We say p_C satisfies *Identical Equivalence Class* condition (IEC), if p_c is identical for all $c \in C$.

- $\succ_S = (\succ_s)_{s \in S}$ is a profile of the underlying preferences of the students, where each $\succ_s$ is the underlying strict preference of student s over $C \cup \{\varnothing\}$, which must be *consistent* with p_s. We call $\succ_s$ consistent with p_s if for any $c \in p_s^k$ and $c' \in p_s^{k'}$ such that $k < k'$ holds, $c \succ_s c'$ holds.

- $\succ_C = (\succ_c)_{c \in C}$ is a profile of the underlying preferences of the colleges, where each $\succ_c$ is the underlying strict preference of c over $S \cup \{\varnothing\}$, which must be consistent with p_c. Consistency is defined analogously to p_s and $\succ_s$.

- $q_C = (q_c)_{c \in C} \in \mathbb{N}_{\geq 0}^m$ is a profile of the colleges' quotas.

Student s prefers college c to college $c' (\neq c)$ under p_s if $c \in p_s^k$ and $c' \in p_s^{k'}$ such that $k < k'$. Note that c and c' may be $\varnothing$. p_c's meaning is defined analogously. Student s is acceptable to college c if $s \succ_c \varnothing$ holds. College c is acceptable to student s if $c \succ_s \varnothing$ holds. Similarly, student s (or college c) is unacceptable to college c (or student s) if s (or c) is not acceptable to c (or s).

Initially, s is not sure about her preference over colleges in the same equivalence class. In particular, if $c, \varnothing \in p_s^k$, she is initially not sure whether c is acceptable for her. Matching $\mu \subseteq S \times (C \cup \{\varnothing\})$ is an assignment of the students to colleges such that each student is assigned to at most one college. $\mu(s) \in C \cup \{\varnothing\}$ denotes the college to which s is matched, and $\mu(c) \subseteq S$ denotes the set of students assigned to c. We assume $\mu(s) = c$ if and only if $s \in \mu(c)$ holds. $\mu(s) = \varnothing$ denotes that s is not assigned to any college. Matching μ is student-feasible if, for each $s \in S$, either $\mu(s) = \varnothing$ or $\mu(s) \succ_s \varnothing$ holds. Matching μ is college-feasible if, for each $c \in C$, $|\mu(c)| \leq q_c$ holds, and $s \succ_c \varnothing$ holds for each $s \in \mu(c)$. A matching is feasible if it is student and college feasible.

Stability and student-optimality are two important desiderata [24].

Definition 1 (Stability). *For matching μ, pair (s, c) is a blocking pair if $c \succ_s \mu(s)$, and either (i) $|\mu(c)| < q_c$ and $s \succ_c \varnothing$ hold, or (ii) $s' \in \mu(c)$ exists such that $s \succ_c s'$ holds. A matching is stable if it is feasible and has no blocking pair.*

Definition 2 (Student-Optimality). *Student s weakly prefers matching μ to matching μ' if $\mu(s) \succ_s \mu'(s)$ or $\mu(s) = \mu'(s)$. A matching is student-optimal if it is stable and weakly preferred by all students to any other stable matching.*

The existence of the unique student-optimal matching is guaranteed since it is obtained by GS [24]. To achieve the student-optimal matching, we need more detailed information on the preferences of both the students and the colleges. One way is to do *interviews* [23]. Here we formally define the interview process. An interview between student s and college c is represented as $(s : c)$. Through it, student s learns more about college c and vice versa. If two colleges, c and

c', belong to the same equivalence class of s, she knows her preference between them only after both $(s : c)$ and $(s : c')$ are held.[1]

An information state represents part of an underlying preference revealed by a sequence of interviews.

Definition 3 (Information State). *Information state $\mathcal{I}_s$ of student s is the strict order of the interviewed colleges (as well as $\varnothing$). Information state $\mathcal{I}_c$ of college c is analogously defined.*

Information state $\mathcal{I}_s$ (resp., $\mathcal{I}_c$) *refines* partial preference p_s, i.e., $\mathcal{I}_s$ is consistent with p_s. Next we introduce a policy and its desirable characteristics.

Definition 4 (Policy). *A policy is a procedure that conducts a sequence of interviews and returns a matching for given S, C, p_S, p_C, and q_C. A policy is* sound *if it returns the student-optimal matching under any underlying preferences $\succ_S$ and $\succ_C$, which are consistent with p_S and p_C.*

Note that a sound policy may allow a student and a college to be matched without any interview. In practice, this may be inappropriate. For example, a firm might be required to examine an applicant's qualifications before hiring her. We theoretically and practically pursue this notion by *diligence*.

Definition 5 (Diligence). *A policy is* diligent *if it is sound, and for obtained matching μ, $\mu(s) = c$ holds only if an interview was conducted between s and c.*

For diligent policies, we compare the number of interviews and define a dominance relation as follows.

Definition 6 (Very Weak Dominance). *Policy f very weakly dominates another diligent policy g if f conducts no more interviews than g for any underlying preferences.[2] A policy is* very weakly dominant *if it is diligent, and very weakly dominates any other diligent policy.*

3   Extension of Lazy Gale-Shapley

We formally describe our extension of LGS [23] for many-to-one matching. The policy uses a hierarchy of students $o = (o^1, o^2, \ldots)$, which partitions S into finite equivalence classes $o^1, o^2, \ldots$, such that $o^k \subseteq S$ for each k, $\bigcup_{k=1}^{|o|} o^k = S$, and o^k and $o^{k'}$ are disjoint $(k \neq k')$. During LGS process, student s can be *forbidden* by college c, meaning that s cannot apply to c. Each college initially forbids students who are ranked strictly lower than $\varnothing$. Also, c is *unchecked* for s if $(s : c)$ has

---

[1] Intuitively, we can assume that a student and a college have an underlying cardinal preference, e.g., $v_s(c)$ represents the utility of s to be assigned to c, which is normalized by $v_s(\varnothing) = 0$. She prefers c to c' if $v_s(c) > v_s(c')$ holds. Through interviews, the students/colleges learn the utility of the colleges/students.

[2] For two policies that behave exactly the same, they very weakly dominate each other. We call this definition "very weak dominance" since it is weaker than the standard notion of weak dominance.

not been conducted yet, c is acceptable to s, and s is not forbidden by c. As standard GS, it keeps on determining the tentative assignment of students. The tentative assignment is finalized when the policy terminates.

For a given hierarchy $o = (o^1, o^2, \ldots)$, as well as S, C, p_S, p_C, and q_C, LGS runs as follows.

Policy 1 (Lazy Gale-Shapley for Many-to-One Matching).

Initialization: *Set μ to an empty assignment. For each $s \in S$, add $\varnothing$ to $\mathcal{I}_s$. For each $c \in C$, add $\varnothing$ to $\mathcal{I}_c$. Set k to 1.*

Stage $k\, (\geq 1)$: *As long as a student remains in $o^1 \cup \ldots \cup o^k$, whose assignment is not determined yet, do the following; otherwise, go to Stage $k + 1$ (when o^{k+1} does not exist, return μ and terminate).*

Step 1: *Choose student $s \in o^1 \cup \ldots \cup o^k$ whose assignment is not determined yet.*

Step 2: *If an acceptable college in $\mathcal{I}_s$ exists, by which s is not forbidden, go to Step 3. Otherwise, choose the top remaining equivalence class p_s^ℓ that contains an unchecked college. Conduct interview $(s : c)$ for each unchecked college $c \in p_s^\ell$. Update $\mathcal{I}_s$ and $\mathcal{I}_c$ accordingly. If s is unacceptable to c, let c forbid s. If no such equivalence class remains, set $\mu(s)$ to $\varnothing$ and repeat Stage k.*

Step 3: *Choose the most preferred, acceptable college c in $\mathcal{I}_s$ from which s is not forbidden. Set $\mu(s)$ to c, $\mu(c)$ to $\mu(c) \cup \{s\}$.*

Step 4: *If $|\mu(c)| > q_c$, choose student s', who is ranked the lowest in $\mu(c)$ based on $\mathcal{I}_c$. Unassign s' from c, and let c forbid s'.*

Step 5: *If $|\mu(c)| = q_c$, let s' be the student who is ranked the lowest in $\mu(c)$ according to $\mathcal{I}_c$. For each student s'', who is ranked strictly lower than s' either in $\mathcal{I}_c$ or p_c, let c forbid s''. Repeat Stage k.*

If we assume IEC is satisfied, i.e., all colleges have identical partial preferences over students, we can use this unique preference for student hierarchy o. By further assuming that each college c can accept at most one student, i.e., for each college c, $q_c = 1$ holds, Policy 1 becomes equivalent to LGS for one-to-one matching presented in [23]. The differences are that we do not assume IEC is (always) satisfied (thus Policy 1 can be very weakly dominant even when IEC fails), and each college can tentatively accept at most q_c students (Step 4).

The following example shows how LGS works.

Example 1. Consider four students, $S = \{s_1, s_2, s_3, s_4\}$, and three colleges, $C = \{c_1, c_2, c_3\}$, whose quotas are $q_{c_1} = q_{c_2} = 1$ and $q_{c_3} = 2$. Assume that LGS uses hierarchy $o = (\{s_1, s_3\}, \{s_2\}, \{s_4\})$. Partial preferences p_S and p_C are:

$$p_{s_1} : (\{c_1, c_3\}, \{\varnothing\}, \{c_2\}), \quad p_{c_1} : (\{s_1, s_3\}, \{s_4, \varnothing\}, \{s_2\}),$$
$$p_{s_2} : (\{c_1, c_3\}, \{c_2, \varnothing\}), \quad\;\; p_{c_2} : (\{s_1\}, \{s_2, \varnothing\}, \{s_4\}, \{s_3\}),$$
$$p_{s_3} : (\{c_1, c_2, c_3, \varnothing\}), \quad\;\;\; p_{c_3} : (\{s_3\}, \{s_2\}, \{s_4, \varnothing\}, \{s_1\}),$$
$$p_{s_4} : (\{c_2, c_3\}, \{c_1, \varnothing\}).$$

Before running LGS, college c_1 forbids student s_2 who is ranked strictly lower than $\varnothing$. Similarly, college c_2 forbids s_3 and s_4, and c_3 forbids s_1. First, from

$o^1 = \{s_1, s_3\}$, one student is chosen. Assume s_1 is chosen at Step 1. Then the interviews to her most preferred unchecked colleges are conducted. In this case, since s_1 is already forbidden by c_3, interview $(s_1 : c_1)$ is conducted. The information states are: $\mathcal{I}_{s_1} : (c_1, \varnothing)$, $\mathcal{I}_{c_1} : (s_1, \varnothing)$. Student s_1 is assigned to her most preferred and not yet forbidden college in $\mathcal{I}_{s_1}$, i.e., c_1. Then $\mu = \{(s_1, c_1)\}$. Since $q_{c_1} = |\mu(c_1)|$, c_1 forbids less preferred student s_4 (Step 5).

Next student $s_3 \in o^1$ is chosen, who was initially forbidden by c_2. Two interviews $(s_3 : c_1)$ and $(s_3 : c_3)$ are conducted here. Assuming $c_1 \succ_{s_3} \varnothing \succ_{s_3} c_3$ and $s_3 \succ_{c_1} s_1$, the information states change: $\mathcal{I}_{s_3} : (c_1, \varnothing, c_3)$, $\mathcal{I}_{c_1} : (s_3, s_1, \varnothing)$, and $\mathcal{I}_{c_3} : (s_3, \varnothing)$. Student s_3 is assigned to c_1. Since $q_{c_1} > |\mu(c_1)|$, s_1, who is ranked lowest in $\mu(c_1)$, is unassigned and forbidden by c_1 (Step 4). Student s_1 is chosen again in Step 1. Since she has neither unchecked nor acceptable colleges, $\mu(s_1)$ is set to $\varnothing$. Thus, $\mu = \{(s_1, \varnothing), (s_3, c_1)\}$.

Next student $s_2 \in o^2$ is chosen. Since she is forbidden by c_1, interview $(s_2 : c_3)$ is conducted. The information states change: $\mathcal{I}_{s_2} : (c_3, \varnothing)$ and $\mathcal{I}_{c_3} : (s_3, s_2, \varnothing)$. Student s_2 is assigned to c_3. Thus, $\mu = \{(s_1, \varnothing), (s_2, c_3), (s_3, c_1)\}$. Finally, $s_4 \in o^3$ is chosen and interview $(s_4 : c_3)$ is conducted. The information states change: $\mathcal{I}_{s_4} : (c_3, \varnothing)$ and $\mathcal{I}_{c_3} : (s_3, s_2, s_4, \varnothing)$. Student s_4 is assigned to c_3. As a result, $\mu = \{(s_1, \varnothing), (s_2, c_3), (s_3, c_1), (s_4, c_3)\}$. Now the assignment of every student is fixed; LGS returns μ.

We show that LGS inherits the following fundamental desiderata from one-to-one settings.

Theorem 1. *LGS is diligent, and runs in polynomial-time for many-to-one matching.*

Proof. Student s applies to the most preferred acceptable college in $\mathcal{I}_s$ by which she is not forbidden. Student s is forbidden by college c only when she has no chance to be assigned. Thus, the matching obtained by LGS and GS must be identical; LGS is sound. Also, student s is matched to college c only after interview $(s : c)$ has been conducted. Thus, LGS is diligent. Once student s is forbidden from college c, she will never apply to c. Thus, Step 1 in Policy 1 is executed at most $O(nm)$ times. The number of interviews is at most $O(nm)$. Thus, LGS runs in polynomial-time. □

4   Very Weak Dominance of LGS

In this section, we first introduce a condition on hierarchy o called *prudence*, such that LGS becomes a very weakly dominant policy with o. Then we clarify a sufficient condition on colleges' partial preferences called *compatibility* such that a prudent hierarchy exists.

Assume interview $(s : c)$ is conducted. Afterwards, if it turns out that c can forbid s based on the current assignment regardless of the outcome of $(s : c)$, we say interview $(s : c)$ becomes *irrelevant*. For example, assume c prefers s to s' in p_c, where $q_c = 1$. When s is assigned to c, interview $(s' : c)$ becomes irrelevant since c forbids s' even though interview $(s' : c)$ is not conducted. On the other hand, if s and s' are in the same equivalence class in p_c, $(s' : c)$ will not be irrelevant since c can forbid s' only after conducting $(s' : c)$.

Definition 7 (Prudent Hierarchy). *Given p_C and q_C, o is prudent if, under LGS with o, the following conditions hold.*

(i) The assignment of $s \in o^k$ does not change after the assignment of all the students in $o^1, \ldots, o^k$ is determined. More specifically, s will not be rejected due to student in $o^{k'}$ such that $k < k'$ and stays with the college to the end.

(ii) The assignment of $s \in o^k$ never makes any interview irrelevant, which is conducted by student s' in $o^1, \ldots, o^k$.

Condition (i) means that, when $s' \in o^{k'} (k < k')$ applies to college c that accepts s, c never rejects s due to s'. Thus, in each stage k in Policy 1, no student in $o^1, \ldots, o^{k-1}$ is selected (again) in Step 1. Condition (ii) means that no interview performed by LGS becomes irrelevant. The next lemma shows that o's prudence is a sufficient condition for LGS with o to be very weakly dominant.

Lemma 1. *If o is prudent, then LGS with o is a very weakly dominant policy.*

Proof. By contradiction, assume another diligent policy g conducts strictly fewer interviews than LGS with prudent hierarchy o. Since g is sound, obtained matching μ must be identical in g and LGS. Interview $(s : c)$ exists which is conducted in LGS and not in g. Assume $s \in o^k$. Since g is diligent, $\mu(s) = c' (\neq c)$ holds, and $(s : c')$ is also conducted in LGS. Two cases are possible: (a) $c \succ_s c'$ and (b) $c' \succ_s c$.

For case (a), since μ is stable, $|\mu(c)| = q_c$ and for each $s' \in \mu(c)$, $s' \succ_c s$ must hold. Since o is prudent, each $s' \in \mu(c)$ must be in $o^1, o^2, \ldots, o^k$ (otherwise, $s \in o^k$ is rejected due to some student $s' \in o^{k'}$ s.t. $k < k'$). Let us assume $s \in p_c^\ell$. Then, since for each $s' \in \mu(c)$, $s' \succ_c s$ holds, $s' \in p_c^{\ell'} (\ell' \le \ell)$ holds. Assume $S' = p_c^\ell \cap \mu(c)$ is non-empty. Then, for each $s' \in S'$, without interview $(s : c)$, we cannot determine whether $s' \succ_c s$ holds. Thus, we cannot guarantee that g always obtains a stable matching. Thus, S' must be empty. Next, assume $S'' = o^k \cap \mu(c)$ is non-empty. Then, when all students in S'' are assigned to c, interview $(s : c)$ becomes irrelevant, even though $s \in o^k$. This violates the fact that o is prudent. Thus, S'' must be empty. Then, for each $s' \in \mu(c)$, s' must be in $o^1, o^2, \ldots, o^{k-1}$. Then, c must forbid s in LGS when the assignments of the students in $o^1, \ldots, o^{k-1}$ are determined. This contradicts our assumption that LGS conducts interview $(s : c)$.

For case (b), the fact that interview $(s : c)$ is conducted while s is assigned to c' in LGS means that c and c' belong to the same equivalence class in p_s. Then without interview $(s : c)$, we cannot determine whether $c' \succ_s c$ holds. Thus, g does not always obtain a stable matching; if $c \succ_s c'$ holds and s can be assigned to c, μ is no longer stable.   □

Next we clarify the condition where a prudent hierarchy exists by introducing the following classification of students.

Definition 8 (Possibly/Certainly Acceptable Students). *For college* $c \in C$ *such that* $\varnothing \in p_c^\ell$, *we call the students in* $S_c = \bigcup_{k=1}^{\ell} p_c^k \setminus \{\varnothing\}$ *possibly acceptable. We call the students in* $S \setminus S_c$ *certainly unacceptable.*

For college $c \in C$ *such that* $|\bigcup_{k=1}^{\ell'} p_c^k \setminus \{\varnothing\}| \leq q_c < |\bigcup_{i=k}^{\ell'+1} p_c^k \setminus \{\varnothing\}|$ *holds,*[3] *we call the students in* $\hat{S}_c = \{s \in S_c \mid s \in p_c^k, 1 \leq k \leq \ell'\}$ *certainly acceptable.*

Possibly acceptable students S_c include all acceptable students $\{s \in S \mid s \succ_c \varnothing\}$. Furthermore, S_c includes all students in p_c^ℓ, where $\varnothing \in p_c^\ell$. College c might realize that such a student is actually unacceptable after an interview. A possibly acceptable student can be assigned to c based on the preferences/applications of the other students, but a certainly unacceptable student has no chance to be assigned to c (i.e., she is also unacceptable). A certainly acceptable student will not be rejected if she applies to c (except for the case that $s \in p_c^\ell$, and after interview $(s : c)$, $\varnothing \succ_c s$ holds), regardless of the preferences/applications of the other students.

With this classification, we introduce *compatibility*.

Definition 9 (Compatible Preferences). p_C *and* q_C *are compatible if a hierarchy of students* $o = (o^1, o^2, \ldots)$ *exists such that for each pair of students* s, s', *for each college* c, *one of the following conditions holds:*

(i) the order of s and s' is identical in o and p_c. More specifically, (a) if $s, s' \in o^k$, then s, s' are in the same equivalence class in p_c, (b) if $s \in o^k$ and $s' \in o^{k'}$ such that $k < k'$, c prefers s to s' in p_c, and (c) if $s \in o^k$ and $s' \in o^{k'}$ such that $k > k'$, c prefers s' to s in p_c.
(ii) $s, s' \in \hat{S}_c$, or
(iii) either s or $s' \notin S_c$.

When p_C and q_C are compatible, we call o, which satisfies the above conditions, a compatible hierarchy.

We require q_C to define a compatible hierarchy since the definitions of S_c and $\hat{S}_c$ depend on q_c. Informally, compatible preferences require colleges to have the same partial structure among any pair of students, except when both are certainly acceptable ((ii) in Definition 9) or when at least one of them is certainly unacceptable ((iii) in Definition 9). Here we relax the restrictive IEC condition, i.e., colleges' partial preferences must be identical, by fully exploiting the fact that a college can accept multiple students up to its quota, and some students are certainly unacceptable in the many-to-one matching extension. When colleges' partial preferences are identical, we can use that partial preference as compatible hierarchy o. Thus, our compatibility condition is a strict generalization of IEC.

The following lemma shows that compatibility is a sufficient condition where a prudent hierarchy exists.

Lemma 2. *If p_C and q_C are compatible and o is a compatible hierarchy, then o is prudent.*

---

[3] We assume $\ell' = 0$ when $|p_c^1 \setminus \{\varnothing\}| > q_c$ and $\hat{S}_c = \{\}$.

Proof. By contradiction, assume that o is compatible but not prudent. Two cases are possible: (a) student $s \in o^k$ is rejected by c by assigning student $s' \in o^{k'}$ to c, s.t. $k < k'$, and (b) by the assignment of $s \in o^k$ to c, interview $(s' : c)$, s.t. s' in $o^{k'}$ and $k' \leq k$, becomes irrelevant.

In case (a), $s' \succ_c s$ holds. Then either c prefers s' to s in p_c or both s and s' are in the same equivalence class. Since s is rejected by c after assigning s', s cannot be certainly acceptable. Thus, $s \in S_c \setminus \hat{S}_c$ holds. Also, $s' \in S_c$ must hold. Thus, in either case, $k' \leq k$ must hold assuming o is a compatible hierarchy. This is a contradiction.

In case (b), college c must (at least) weakly prefer s to s' in p_c. However, if s and s' are in the same equivalence class, s' cannot be forbidden by c without an interview. Thus, c must strictly prefer s to s' in p_c. Also, since s' is forbidden from c, s' cannot be certainly acceptable. Thus, $s' \in S_c \setminus \hat{S}_c$ holds. Also, $s \in S_c$ must hold. Since o is a compatible hierarchy, $k < k'$ must hold. This is a contradiction. $\qquad\square$

We can check whether p_C and q_C are compatible and obtain a compatible hierarchy o when they are compatible by creating *compatibility graph* $G(p_C, q_C)$, described below, and applying the following procedure to it.

Definition 10 (Compatibility Graph). *Given p_C and q_C, their compatibility graph is weighted digraph $G(p_C, q_C) = (S, E)$ defined as follows:*

- *S is a set of vertices (identical to the set of students).*
- *If for any pair $s \in S_c$ and $s' \in S_c \setminus \hat{S}_c$ such that $s \neq s'$ and c prefers s to s' in p_c for some $c \in C$, edge $(s, s') \in E$ with weight -1 exists.*
- *For any pair $s, s' \in S_c \setminus \hat{S}_c$ such that $s \neq s'$ and s and s' are in the same equivalence class in p_c for some c (and no edge (s, s') with weight -1 exists), edge $(s, s') \in E$ with weight 0 exists.*

As described later in Lemma 3, the fact that $G(p_C, q_C)$ has no negative cycle (i.e., a cycle that includes at least one negative edge) is a sufficient and necessary condition that p_C and q_C are compatible. Therefore, we can verify the compatibility and obtain the compatible hierarchy when no negative cycle exists by implementing the following procedure.

Procedure 1.

1. *Divide vertices in $G(p_C, q_C)$ into connected components $V_1, V_2, \ldots V_p$ by 0 edges. We say V_k is a connected component by 0 edges if for any $i, j \in V_k$, a path exists between i and j consisting of 0 edges (here, we ignore the edge direction of 0 edges).*
2. *If any -1 edge exists within a connected component, terminate the procedure with failure.*
3. *Otherwise, construct a component graph, i.e., $V_1, V_2, \ldots, V_p$ are vertices, and a directed edge from V_k to V_ℓ exists if there exists -1 edge (i, j) and $i \in V_k$ and $j \in V_\ell$.*

4. For the obtained component graph, we first identify a set of vertices $\mathcal{V}^*$, each of which has no incoming -1 edge. If no such vertex exists, then the original compatibility graph has a negative cycle. Thus, we terminate the procedure with failure. Otherwise, we set o^1 to $\bigcup_{V_k \in \mathcal{V}^*} V_k$. Remove each $V_k \in \mathcal{V}^*$ and its connecting edges from it, and repeat the same procedure to obtain o^2, and so on.

Example 2. Let us illustrate how Procedure 1 works. Consider the setting of Example 1. Certainly/possibly acceptable students are: $\hat{S}_{c_1} = \{\}$, $S_{c_1} = \{s_1, s_3, s_4\}$, $\hat{S}_{c_2} = \{s_1\}$, $S_{c_2} = \{s_1, s_2\}$, $\hat{S}_{c_3} = \{s_2, s_3\}$, and $S_{c_3} = \{s_2, s_3, s_4\}$. We first divide this graph into connected components by 0 edges $\{s_1, s_3\}$, $\{s_2\}$, $\{s_4\}$. There is no -1 edge within a connected component. The component graph has no cycle. The vertex for $\{s_1, s_3\}$ has no incoming edge. Thus, $o^1 = \{s_1, s_3\}$. After removing this vertex and connecting edges, the vertex for $\{s_2\}$ has no incoming edge. Thus, $o^2 = \{s_2\}$. After removing this vertex and connecting edges, there is only one vertex for s_4. Thus, $o^3 = \{s_4\}$. Procedure 1 terminates successfully and obtains $o = (\{s_1, s_3\}, \{s_2\}, \{s_4\})$. It is easy to verify that p_C and q_C are compatible. In Example 1, this hierarchy is used; LGS is very weakly dominant.

The following lemma shows that having no negative cycle is a sufficient and necessary condition for compatibility (thus, Procedure 1 is correct).

Lemma 3. p_C and q_C are compatible if and only if their compatibility graph has no negative cycle.

Proof. For the "if" part, when the compatibility graph has no negative cycle, Procedure 1 returns hierarchy o. We show that such o is a compatible hierarchy. By contradiction, assume o is not a compatible hierarchy, i.e., for some c and some pair (s, s'), the order of s and s' in o is different from their order in p_c, either s or s' is not in $\hat{S}_c$, and both s and s' are in S_c. W.l.o.g., assume c weakly prefers s to s' in p_c. There are two cases: (1) s and s' are in the same equivalence class in p_c, while they are in different equivalence classes in o, or (2) c prefers s to s' in p_c, where $s \in o^k$, $s' \in o^{k'}$, and $k \geq k'$.

In case (1), due to p_c, there must be edges (s, s') and (s', s) with weight 0; there cannot be a negative weight edge between s and s' since we assume no negative cycle exists. Then, s and s' must be in the same connected component. Thus, they must be in the same equivalence class in o. This is a contradiction.

In case (2), due to p_c, edge (s, s') with weight -1 exists in $G(p_C, q_C)$. Thus, the connected component that includes s must be chosen earlier than that includes s'. Then, if $s \in o^k$ and $s' \in o^{k'}$, $k < k'$ must hold. This is a contradiction.

For the "only if" part, assume by contradiction that a negative cycle exists in $G(p_C, q_C)$. Since p_C and q_C are compatible, there exists a compatible hierarchy o. Assume edge (s, s') exists in $G(p_C, q_C)$ and $s \in o^k$ and $s' \in o^{k'}$. When its weight is -1, college c exists such that $s \in S_c$, $s' \in S_c \setminus \hat{S}_c$, and c prefers s to s' in p_c. Since o is a compatible hierarchy, $k < k'$ must hold. When its weight is 0, college c exists such that $s, s' \in S_c \setminus \hat{S}_c$ and s and s' are in the same equivalence class in p_c. Since o is a compatible hierarchy, $k = k'$ holds. Assume a negative cycle of students $s_1, s_2, \ldots, s_\ell, s_1$ exists. Let $o(s)$ denote the index of the equivalence class to which s belongs in o. Then $o(s_1) \leq o(s_2) \leq \ldots \leq o(s_\ell) \leq o(s_1)$ holds and at least one inequality must be strict. This is a contradiction.   □

From Theorem 1 and Lemmas 1 and 2, we immediately obtain Theorem 2. Theorem 3 shows the time complexity for checking compatibility.

Theorem 2. *Assume p_C and q_C are compatible, and o is a compatible hierarchy. Then LGS with o is a very weakly dominant policy for any p_S.*

Theorem 3. *For given p_C and q_C, checking their compatibility (and obtaining a compatible hierarchy when they are compatible) can be done in $O(mn^2)$.*

Proof. From Lemma 3, Procedure 1 can check the compatibility of p_C and q_C, and obtain a compatible hierarchy when they are compatible. Building compatibility graph $G(p_C, q_C)$ requires $O(mn^2)$, i.e., for each pair of vertices/students, we need to check the partial preference of each college. In Procedure 1, we can identify connected components by performing depth-first search, which requires $O(|S|+|E|)$ time. Then, checking the existence of any -1 edge within a connected component requires $O(|E|)$ time. Finally, Step. 4, which is a slight modification of a topological sort algorithm [12], requires $O(|S| + |E|)$ time. Thus, time complexity of Procedure 1 in total is $O(n^2)$.   □

5   When Compatibility Fails

We can apply LGS with any hierarchy and obtain the student-optimal matching even when colleges' partial preferences do not satisfy compatibility, as shown by Theorem 1. However, the required number of interviews can vary according to the chosen hierarchy. It is desirable to choose a hierarchy that is as *close to* the colleges' partial preferences as possible. Now, we need to define a criterion to measure the closeness among preferences/hierarchies. One plausible criterion is the distance metric used in Kemeny's voting scheme [14], which is introduced by Kemeny [14]. It is the unique voting scheme that satisfies several natural axiomatic conditions [27]. In Kemeny's voting scheme, we choose a preference that minimizes the total Kendall tau distance (the number of pairwise disagreements between two students) [15].

Unfortunately, using Kemeny's voting scheme is computationally too expensive; a decision version of Kemeny's voting scheme is NP-complete [4]. Thus, we propose the following heuristic, greedy procedure to find a reasonable hierarchy. Its time complexity is $O(n^2)$, assuming $G(p_C, q_C)$ is already built.

Procedure 2.

1. $k \leftarrow 1$, $G \leftarrow G(p_C, q_C)$.
2. If a vertex exists that has no incoming -1 edge in G, then choose such a vertex s (if a vertex also has some 0 edge between removed vertices, choose such a vertex first). Otherwise, choose vertex s using a certain criterion.
3. $o^k \leftarrow \{s\}$, $k \leftarrow k + 1$, remove s and its connecting edges from G, and go to 2.

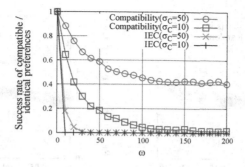

Fig. 1. Success rate of compatibility/IEC by varying ω

Procedure 2 obtains hierarchy o, which is actually a total order (i.e., $|o^k| = 1$ for each k). When p_C and q_C are compatible, there always exists a vertex that has no incoming -1 edge. Then, the obtained hierarchy is essentially equivalent to a compatible hierarchy, where the order among students in the same equivalence class is determined arbitrarily, while the order among students in different equivalence classes are preserved. For the criterion choosing a vertex, we adopt the following heuristic method.

– For each -1 edge, we count its *support*, i.e., the number of colleges that require it. We choose vertex s with the fewest total supports of its incoming -1 edges.

6   Experimental Evaluation

All the experiments presented in this section were executed on a Windows computer with 32 GB memory and Intel CPU Core i7-8700K with Java 14, Python 3.6, and R 3.3.0. We set $n = 400$, $m = 10$, and $q_c = 40$ for every college c and created $1,000$ problem instances for each parameter setting. Let σ_C denote the average size of the equivalence classes in p_c; we split n students into n/σ_C equivalence classes. As σ_C increases, the information held initially by each college becomes more coarse-grained. σ_S is defined analogously.

First, we quantitatively compared the *generality* of our compatibility condition against IEC. We randomly generated colleges' partial preferences p_C as follows and checked whether the compatibility condition and IEC are satisfied. Initially, all colleges have the same underlying strict order among students. Then for each college, we performed ω swaps of contiguous students in this order. Every swap was randomly chosen. Then we inserted acceptability threshold $\varnothing$ in the $q_c(1 + \psi)$-th position in the strict student order. By splitting the order for each college c into n/σ_C pieces, we obtained p_c. More specifically, let us assume n/σ_C, i.e., there are ℓ partitions. Then we randomly choose $\ell - 1$ different natural numbers from $\{1, 2, \ldots, n - 1\}$ and obtain $x_1, x_2, \ldots, x_{\ell-1}$ (sorted in the increasing order). Then, the first equivalence class has x_1 students, the second equivalence

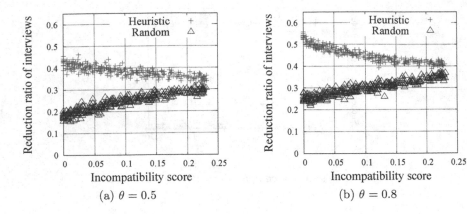

Fig. 2. Reduction ratios and incompatibility scores

class has $x_2 - x_1$ students, and so on. The way of splitting is common to all colleges, i.e., for all c, c', and ℓ, $|p_c^\ell| = |p_{c'}^\ell|$ holds. We plotted the rate that obtained partial preferences remain compatible or satisfying IEC (which we call success rate) by varying σ_C and ω (Fig. 1), while ψ is fixed to 0.25.

As σ_C increases, the success rate also rises, since the partial preferences become more coarse-grained and the chance decreases that a swap matters. By increasing ω, the compatibility success rate gradually decreases. The compatibility condition is fairly robust compared to IEC. For example, after 30 swaps, the IEC success rate becomes zero, although the compatibility success rate remains over 65% when $\sigma_C = 50$.

Next we evaluated the ratio of interviews reduced in LGS by choosing an appropriate hierarchy. For each student s, we first generated her underlying strict preference $\succ_s$ based on the Mallows model [18,26] with spread parameter θ. Then we decomposed it into m/σ_S pieces to obtain p_s (the way for splitting is analogous to the method used for creating p_c). In the Mallows model, the students' underlying strict preferences become more similar as θ increases.

For each instance, we counted the upper-bound of the required interviews as follows. For each student s, who is matched to college $c \in p_s^k$ in the student-optimal matching, we assume (s, c') is performed for all $c' \in \cup_{1 \le \ell \le k} p_s^\ell$, except when s is certainly unacceptable for c'. LGS never performs interview (s, c'), where $c' \in p_s^{k'}$ and $k' > k$. For a problem instance, if LGS performs x interviews while the upper-bound is $\bar{x}$, then the reduction ratio is given as $(\bar{x} - x)/\bar{x}$.

To measure how the colleges' partial preferences are *incompatible*, we introduce the following *incompatibility score*. For each pair of students s and s', assume each college c votes for their relative order, i.e., s (or s') must be in a preceding equivalence class, they must be in the same equivalence class, or abstain (c does not care since conditions (ii) or (iii) holds in Definition 9). By assuming the majority wins, we can count the number of wasted votes (excluding the abstentions). We use a normalized count, i.e., the total number of wasted votes for all pairs and for all colleges divided by $\frac{n(n-1)m}{2}$.

We show the reduction ratio by varying ω from 0 to 500, where $\sigma_C = 4$, $\sigma_S = 2$, $\psi = 6.5$, $\theta = 0.5$ (Fig. 2 (a)), and $\theta = 0.8$ (Fig. 2 (b)). We show a scattered plot where the x-axis denotes the incompatibility score and the y-axis denotes the reduction ratio. We also plotted the result where hierarchy o is chosen at random. The reduction ratio of our heuristic method decreases as the incompatibility score becomes large. However, the slope is rather gentle; e.g., when the score is 0, the reduction ratio is about 0.43 for $\theta = 0.5$, while it remains 0.35 when the score is 0.23 in Fig. 2 (a). The reduction ratio of a random hierarchy increases as the incompatibility score increases, since no hierarchy is particularly good/bad when colleges' partial preferences are almost at random. Our heuristic method also obtains a much better hierarchy than a randomly generated one. Furthermore, the reduction ratio improves when θ is large, i.e., when students' preferences become more similar. This is because, as the competition among students becomes more severe, the chance increases that a college forbids a student without an interview. The upper-bound of the required interviews is larger when $\theta = 0.8$: about twice as many as the case where $\theta = 0.5$. Thus, we have more room for reduction.

From these results, we conclude that the hierarchy obtained by our heuristic method is reasonable since its reduction ratio is not too far from the case where a theoretically optimal hierarchy is available.

7   Concluding Remarks

Our extension of LGS is practically important since many-to-one matching has many real-life applications, where agents are not likely to have complete information in advance. Since GS plays a central role in the literature, clarifying its theoretical properties and the experimental performance of its extension under partial information is indispensable. Obvious future works include identifying a necessary and sufficient condition of colleges' preferences to guarantee the minimality of LGS, as well as examining other heuristic methods for obtaining a reasonable hierarchy. Understanding to what extent the complexity results presented in [23] carry over is also important. Scrutinizing the robustness against various manipulations would also be interesting [1,21,25].

References

1. Afacan, M.O.: Fictitious students creation incentives in school choice problems. Econ. Theor. 56(3), 493–514 (2014). https://doi.org/10.1007/s00199-014-0804-4
2. Ashlagi, I., Braverman, M., Kanoria, Y., Shi, P.: Clearing matching markets efficiently: informative signals and match recommendations. Manage. Sci. 66(5), 2163–2193 (2020)
3. Aziz, H., et al.: Stable matching with uncertain pairwise preferences. In: Proceedings of the 16th International Conference on Autonomous Agents and MultiAgent Systems (AAMAS-2017), pp. 344–352 (2017)

4. Bartholdi, J., Tovey, C.A., Trick, M.A.: Voting schemes for which it can be difficult to tell who won the election. Soc. Choice Welfare **6**, 157–165 (1989)
5. Chakraborty, A., Citanna, A., Ostrovsky, M.: Two-sided matching with interdependent values. J. Econ. Theor. **145**(1), 85–105 (2010)
6. Drummond, J., Boutilier, C.: Preference elicitation and interview minimization in stable matchings. In: Proceedings of the 28th AAAI Conference on Artificial Intelligence (AAAI-2014), pp. 645–653 (2014)
7. Echenique, F., Lee, S., Shum, M., Yenmez, M.B.: The revealed preference theory of stable and extremal stable matchings. Econometrica **81**(1), 153–171 (2013)
8. Gale, D., Shapley, L.S.: College admissions and the stability of marriage. Am. Math. Monthly **69**(1), 9–15 (1962)
9. Gonczarowski, Y.A., Nisan, N., Ostrovsky, R., Rosenbaum, W.: A stable marriage requires communication. Games Econ. Behav. **118**, 626–647 (2019)
10. Goto, M., Kojima, F., Kurata, R., Tamura, A., Yokoo, M.: Designing matching mechanisms under general distributional constraints. Am. Econ. J. Microeconomics **9**(2), 226–62 (2017)
11. Irving, R.W., Manlove, D.F., Scott, S.: The Hospitals/Residents problem with ties. In: SWAT 2000. LNCS, vol. 1851, pp. 259–271. Springer, Heidelberg (2000). https://doi.org/10.1007/3-540-44985-X_24
12. Kahn, A.B.: Topological sorting of large networks. Commun. ACM **5**(11), 558–562 (1962)
13. Kamada, Y., Kojima, F.: Efficient matching under distributional constraints: theory and applications. Am. Econ. Rev. **105**(1), 67–99 (2015)
14. Kemeny, J.G.: Mathematics without numbers. Daedalus **88**(4), 577–591 (1959)
15. Kendall, M.G.: A new measure of rank correlation. Biometrika **30**(1/2), 81–93 (1938)
16. Kurata, R., Hamada, N., Iwasaki, A., Yokoo, M.: Controlled school choice with soft bounds and overlapping types. J. Artif. Intell. Res. **58**, 153–184 (2017)
17. Lee, R.S., Schwarz, M.: Interviewing in two-sided matching markets. Technical report, National Bureau of Economic Research (2009)
18. Mallows, C.L.: Non-null ranking models. I. Biometrika **44**(1–2), 114–130 (1957)
19. Manlove, D.F.: Algorithmics of Matching Under Preferences. World Scientifc, Singapore (2013)
20. Ng, C., Hirschberg, D.S.: Lower bounds for the stable marriage problem and its variants. SIAM J. Comput. **19**(1), 71–77 (1990)
21. Pini, M.S., Rossi, F., Venable, K.B., Walsh, T.: Manipulation complexity and gender neutrality in stable marriage procedures. Auton. Agent. Multi-Agent Syst. **22**(1), 183–199 (2011)
22. Rastegari, B., Condon, A., Immorlica, N., Irving, R., Leyton-Brown, K.: Reasoning about optimal stable matchings under partial information. In: Proceedings of the 15th ACM Conference on Economics and Computation (EC-2014), pp. 431–448 (2014)
23. Rastegari, B., Condon, A., Immorlica, N., Leyton-Brown, K.: Two-sided matching with partial information. In: Proceedings of the 14th ACM Conference on Economics and Computation (EC-2013), pp. 733–750 (2013)
24. Roth, A.E., Sotomayor, M.A.O.: Two-Sided Matching: A Study in Game-Theoretic Modeling and Analysis. No. 18, Cambridge University Press, Cambridge (1992)
25. Teo, C.P., Sethuraman, J., Tan, W.P.: Gale-Shapley stable marriage problem revisited: strategic issues and applications. Manage. Sci. **47**(9), 1252–1267 (2001)

26. Tubbs, J.: Distance based binary matching. In: Page C., LePage R. (eds.) Computing Science and Statistics, pp. 548–550. Springer, New York (1992) https://doi.org/10.1007/978-1-4612-2856-1_97

27. Young, H.P., Levenglick, A.: A consistent extension of Condorcet's election principle. SIAM J. Appl. Math. **35**(2), 285–300 (1978)

Participatory Budgeting

Participatory Funding Coordination: Model, Axioms and Rules

Haris Aziz[✉] and Aditya Ganguly

UNSW Sydney, Sydney, Australia
{haris.aziz,a.ganguly}@unsw.edu.au

Abstract. We present a new model of collective decision making that captures important crowd-funding and donor coordination scenarios. In the setting, there is a set of projects (each with its own cost) and a set of agents (that have their budgets as well as preferences over the projects). An outcome is a set of projects that are funded along with the specific contributions made by the agents. For the model, we identify meaningful axioms that capture concerns including fairness, efficiency, and participation incentives. We then propose desirable rules for the model and study, which sets of axioms can be satisfied simultaneously. An experimental study indicates the relative performance of different rules as well as the price of enforcing fairness axioms.

Keywords: Social choice · Participatory budgeting · Fairness · Crowd-funding

1 Introduction

Consider a scenario in which a group of residents want to pitch in money to buy some common items for the house but not every item is of interest or use to everyone. Each of the items (e.g. TV, video game console, music system, etc.) has its price. The residents each have a maximum amount they can spend towards the common items. Residents would like to have as much money as possible used toward items that are useful to them. It is a scenario that is encountered regularly in shared houses or apartments.

As a second scenario, hundreds of donors want to fund charitable projects. Each of the projects (e.g. building a well, enabling a surgery, funding a scholarship, etc.) has a cost requirement. Donors have upper caps on their individual budgets and care about the amount of money that is used towards projects of which they approve. The question of how to coordinate the funding in a principled and effective way is a fundamental problem in crowdfunding and donor coordination. The model that we propose is especially suitable for coordinating donations from alumni at various universities.

Both of the settings above are coordination problems in which agents contribute money, and they have preferences over the social outcomes. A collective outcome specifies which projects are funded and how much agents are charged.

© Springer Nature Switzerland AG 2021
D. Fotakis and D. Ríos Insua (Eds.): ADT 2021, LNAI 13023, pp. 409–423, 2021.
https://doi.org/10.1007/978-3-030-87756-9_26

For these problems, we consider the following question. *What is a desirable and principled way of aggregating the preferences and financial contributions of the agents?*

Contributions. We propose a formal model that we refer to as *Participatory Funding Coordination (PFC)* that captures many important donor coordination scenarios. In this model, agents have an upper budget limit. The outcome for the problem is a set of projects that are funded and the respective monetary contributions of the agents for the funded projects. The utility of the agents is the amount of money used for projects that are approved by them. It reflects the approved investment from the perspective of an individual agent. We lay the groundwork for work on the model by formulating new axioms for the model. The logical relations between the axioms are established and the following question is studied: which sets of axioms are simultaneously achievable? We propose and study rules for the problems that are inspired by welfarist concerns but satisfy participation constraints. In addition to an axiomatic study of the rules, we also undertake an experimental comparison of the rules. The experiment sheds light on the impact that various fairness or participation constraints can have on the social welfare. This impact has been referred to as the price of fairness in other contexts. In particular, we investigate the effects of enforcing fairness properties on instances that model real-world applications of PFC, including crowdfunding.

2   Related Work

Our model generally falls under the umbrella of a collective decision making setting in which agents' donations and preferences are aggregated to make funding decisions. It is a concrete model within the broad agenda of achieving effective altruism [18–20].

The model we propose is related to the discrete participatory budgeting model [3–5,14,16,21]. In discrete participatory budgeting, agents do not make personal donations towards the projects. They only express preferences over which projects should be funded. We present several axioms that are only meaningful for our model and not for discrete participatory budgeting. Algorithms for discrete participatory budgeting cannot directly be applied to our setting because they do not take into account individual rationality type requirements.

Another related setting is multi-winner voting [13]. Multi-winner voting can be viewed as a restricted version of discrete participatory budgeting. The Participatory Funding Coordination (PFC) setting differs from multi-winner voting in some key respects: in our model, each project (winner) has an associated cost, and we select projects subject to a knapsack constraint as opposed to having a fixed number of winners.

Our PFC model relies on approval ballots in order to elicit agents' preferences. Dichotomous preferences have been considered in several important setting including committee voting [2,17] and discrete participatory budgeting [4,15].

Another related model that takes into account the contributions of agents was studied by Brandl et al. [8]. Just like in our model, an agent's utilities are based on how much money is spent on projects approved by the agent. However, their model does not have any costs and agents can spread their money over projects in any way. Our model has significant differences from the model of [7,8]: (1) in our setting, the projects are indivisible and have a minimum cost to complete; and (2) agents may not be charged the full amount of their budgets. The combination of these features leads to challenges in even defining simple individual rationality requirements. Furthermore, it creates difficulties in finding polynomial-time algorithms for some natural aggregation rules (utilitarian, egalitarian, Nash product, etc.). Our model is more appropriate for coordinating donations where projects have short-term deadlines and a target level of funding which must be reached for the project to be successfully completed. We show that the same welfarist rules that satisfy some desirable properties in the model [7,8], fail to do so in our model. Just as the work of Brandl et al. [7,8], Buterin et al. [9] consider donor coordination for the divisible model in which the projects do not have costs and agents do not have budget limits. They also assume quasi-linear utilities, whereas we model charitable donors who are not interested in profit but want their money being used as effectively as possible towards causes that matter to them.

The features of our PFC model enable the model to translate smoothly to a number of natural settings. Crowdfunding, in particular, is a scenario in which we would like to capitalise upon commonalities in donors' charitable preferences [11]. Furthermore, crowdfunding projects (e.g. building a well, funding a scholarship, etc.) often have provision points (see e.g. Agrawal et al. [1,10,12]), and it can be critical for these targets to be met. For example, a project to raise funds for a crowdfunding recipient to pay for a medical procedure would have to raise a minimum amount of money to be successful, otherwise all donations are effectively wasted.

Crowdfunding projects have been discussed in a broader context with various economic factors and incentive issues presented [1]. Bagnoli and Lipman [6] discuss additional fairness and economic considerations for the related topic of the division of public goods. The discrete model that we explore, where projects have finite caps, has the potential to coordinate donors and increase the effectiveness of a crowdfunding system.

3   Participatory Funding Coordination

A *Participatory Funding Coordination (PFC)* setting is a tuple (N, C, A, b, w) where N is the set of agents/voters, C is the set of projects (also generally referred to as candidates). The function $w : C \to \mathbb{R}^+$ specifies the cost $w(c)$ of each project $c \in C$. The function $b : N \to \mathbb{R}_{\geq 0}$ specifies the budget b_i of each agent $i \in C$. The budget b_i can be viewed as the maximum amount of money that agent i is willing to spend. For any set of agents $M \subseteq N$, we will denote $\sum_{i \in M} b_i$ by $b(M)$. The approval profile $A = (A_1, \ldots, A_n)$ specifies for each agent, her set of acceptable projects A_i. An *outcome* is a pair (S, x) where $S \subseteq C$ is the set of funded projects and x is a vector of payments that specify for each $i \in N$, the payment x_i that is charged from agent i. We will restrict our attention to feasible outcomes in which $x_i \leq b_i$ for all $i \in N$ and only those projects get financial contributions that receive their required amount. Also, note that the projects that are funded are only those that receive the entirety of their price in payments from the agents. For any given PFC instance, a mechanism F returns an outcome. We will denote the set of projects selected by F as F_C and the payments by F_x. For any outcome (S, x), since $x_i \leq b_i$, the money $b_i - x_i$ can either be kept by the agent i or it can be viewed as going into some common pool. The main focus of our problem is to fund a maximal set of projects while satisfying participation constraints.

We suppose that an agent's preferences are *approval-based*. For any set of funded projects S, any agent i's utility is

$$u_i(S) = \sum_{c \in S \cap A_i} w(c).$$

That is, an agent cares about how many dollars are *usefully* used on his/her approved projects. Our preferences domain is similar to the one used by Brandl et al. [8] who considered a continuous model in which projects do not have target costs. In their model, agents also care about how much money is used for their liked projects.

4   Axiom Design

In this section, we design axioms for outcomes of the PFC setting. We consider an outcome (S, x). For any axiom **Ax** for outcomes, we say that a mechanism satisfies **Ax** if it always returns an outcome that satisfies **Ax**.

We first present three axioms for our setting that are based on the principle of participation:

- **Minimal Return (MR)**: each agent's utility is at least as much as the money put in by the agent: $u_i(S) \geq x_i$. In other words, the societal decision is as good for each agent i as i's best use of the money x_i that she is asked to contribute. We will use this as a minimal condition for all feasible outcomes.
- **Implementability (IMP)** : There exists a payment function $y : N \times C \to \mathbb{R}_{\geq 0}$ such that $\sum_{c \in C} y(i, c) = x_i$ for all $i \in N$, $\sum_{i \in N} y(i, c) \in \{0, w(c)\}$ for all $c \in C$ and there exists no $i \in N$ and $c \notin A_i$ such that $y(i, c) > 0$. Here $y(i, c)$ represents the money paid by i to project c. IMP captures the requirement that an agent's contribution should only be used on projects that are approved by the agent.
- **Individual Rationality (IR):** the utility of an agent is at least as much as an agent can get by funding alone: $u_i(S) \geq \max_{S' \subseteq A_i, w(S') \leq b_i}(w(S'))$. Note that IR is easily achieved if the project costs are high enough: if for $i \in N$ and $c \in C$, $w(c) > b_i$, then every outcome is IR.

We note that MR is specified with respect to the amount x_i charged to the agent. It can be viewed as a participation property: an agent would only want to participate in the market if she gets at least as much utility as the money she spent. We will show IMP is stronger than MR. IMP can also be viewed as a fairness property: agents are made to coordinate but they only spend their money on the projects they like.

Remark 1. If (S, x) is an IMP outcome with associated payment function y, then for any subset of projects $S' \subseteq S$, there is an IMP outcome that funds only the set of projects S'. In particular, the payment function y' for one such implementable outcome is obtained by setting (for each agent i) $y'(i, c) = y(i, c)$ for all $c \in S'$ and $y'(i, c) = 0$ for all $c \in S \setminus S'$.

Next, we present axioms that are based on the idea of efficiency.

– **Exhaustive (EXH)**: An outcome (S, x) satisfies EXH if there exists no set of agents $N' \subseteq N$ and unfunded project $c \in C \setminus S$ such that $c \in \cap_{i \in N'} A_i$ with $w(c) \leq \sum_{i \in N'} (b_i - x_i)$. In words, agents in N' cannot pool in their unspent money and fund another project liked by all of them.
– **Pareto optimality (PO)-X**: An outcome (S, x) is Pareto optimal within the set of outcomes satisfying property X if there exists no outcome (S', x') satisfying X such that $u_i(S') \geq u_i(S)$ for all $i \in N$ and $u_i(S') > u_i(S)$ for some $i \in N$. Note that Pareto optimality is a property of the set of funded projects S irrespective of the payments.
    • PO is Pareto optimal among the set of all outcomes.
    • PO-IMP: Pareto optimal among the set of IMP outcomes.
    • PO-MR: Pareto optimal among the set of MR outcomes.
– **Payment constrained Pareto optimality (PO-Pay)**: An outcome is PO-Pay if it is not Pareto dominated by any outcome of at most the same price. Formally, there exists no (S', x') such that $\sum_{i \in N} x'_i \leq \sum_{i \in N} x_i$, $u_i(S') \geq u_i(S)$ for all $i \in N$ and $u_i(S') > u_i(S)$ for some $i \in N$.
– **Weak Payment constrained Pareto optimality (weak PO-Pay)**: An outcome is weakly PO-Pay if it is not Pareto dominated by any outcome that charges at most the same cost from each agent. Formally, there exists no (S', x') such that $x'_i \leq x_i$ and $u_i(S') \geq u_i(S)$ for all $i \in N$ and $u_i(S') > u_i(S)$ for some $i \in N$.

A concept that can be viewed in terms of participation, efficiency, and fairness is the adaptation of the principle of core stability for our setting.

– **Core stability (CORE)**: There exists no set of agents who can pool in their budget and each gets a strictly better outcome. In other words, an outcome (S, x) is CORE if for every subset of agents $N' \subseteq N$, for every subset of projects $C' \subseteq C$ such that $w(C') \leq \sum_{i \in N'} b_i$, the following holds for some agent $i \in N'$: $u_i(S) \geq w(C' \cap A_i)$.

We also describe a basic fairness axiom for outcomes and rules based on the idea of proportionality.

- **Proportionality (PROP)**: Suppose a set of agents $N' \subseteq N$ each have approval set that is *exactly* some (common) set of projects $C' \subseteq C$ such that $\sum_{i \in N'} b_i \geq w(C')$. In that case, all the projects in C' are selected.

Finally, we consider an axiom that is defined for mechanisms rather than outcomes. We say that a mechanism satisfies **strategyproofness** if there exists no instance under which some agent has an incentive to misreport her preference relation.

We conclude this section with some remarks on computation. The following proposition follows via a reduction from the Subset Sum problem.

Proposition 1. *Even for one agent, computing an IR, PO, PO-MR, or PO-IMP outcome is NP-hard.*

Note that IMP is a property of an outcome rather than a set of projects. We say that a set of projects S is IMP if there exists a feasible vector of charges to agents x such that the outcome (S, x) is IMP. The property IMP can be tested in polynomial time via reduction to network flows.

Proposition 2. *For a given set of projects S, checking whether there exists a vector of charges x such that (S, x) is implementable can be done in polynomial time.*

5   Axioms: Compatibility and Logical Relations

In this section, we study the compatibility and relations between the axioms formulated.

Remark 2. Note that IR and MR are incomparable. Any outcome in which every agent is not charged any money trivially satisfies MR. However, it will not satisfy IR if any agent could afford one of their approved projects by themselves. On the other hand, an IR outcome may not be MR. Consider a profile with one agent and one project. Say the agent has budget greater than the cost of the project, but does not approve of the project. Then, the outcome where the agent is forced to fund the project is IR but not MR.

Next, we point out that PO-Pay is equivalent to weak PO-Pay.

Proposition 3. *PO-Pay is equivalent to weak PO-Pay.*

Proof. Suppose an outcome (S, x) is not weakly PO-Pay. Then, it is trivially not PO-Pay. Now suppose (S, x) is not PO-Pay. Then, there exists another outcome (S', x') such that $\sum_{i \in N} x'_i \leq \sum_{i \in N} x_i$ and $u_i(S') \geq u_i(S)$ for all $i \in N$ and $u_i(S') > u_i(S)$ for some $i \in N$. Note that S' can be funded with total amount $\sum_{i \in N} x'_i$ irrespective of who paid what. So S' is still affordable if $x'_i \leq x_i$ for each agent i. □

The next proposition establishes further logical relations between the axioms.

Proposition 4. *The following logical relations hold between the properties.*

1. *IMP implies MR.*
2. *PO implies PO-Pay.*
3. *PO-X implies PO-Y if Y implies X.*
4. *PO-IMP implies EXH.*
5. *PO-IR implies EXH.*
6. *CORE implies IR.*
7. *The combination of PO-IMP and IMP imply PROP.*

Next, we show that MR is compatible with PO-Pay.

Proposition 5. *Suppose an outcome is MR and there is no other MR outcome that Pareto dominates it. Then, it is PO-Pay.*

Proof. Suppose the outcome (S, x) is MR and PO-MR. We claim that (S, x) is PO-Pay. Suppose it is not PO-Pay. Then there exists another outcome (S', x') such that $\sum_{i \in N} x'_i \leq \sum_{i \in N} x_i$, $u_i(S') \geq u_i(S)$ for all $i \in N$ and $u_i(S') > u_i(S)$ for some $i \in N$. Note that S' is affordable with total amount $\sum_{i \in N} x'_i$ irrespective of who paid what. So S' is still affordable if $x'_i \leq x_i$. Therefore, we can assume that $x'_i \leq x_i$ for all $i \in N$. Note that since S' Pareto dominates S and since (S, x) is MR, $u_i(S') \geq u_i(S) \geq x_i \geq x'_i$ for all $i \in N$. Hence (S', x') also satisfies MR. Since (S', x') is MR and since S' Pareto dominates S, it contradicts the fact that (S, x) is PO-MR. $\qquad \square$

Proposition 6. *There always exists an outcome that satisfies IMP, IR, PO-IMP and hence also MR and EXH.*

Proof. For each $i \in N$ compute $S_i = \arg\max_{S' \subseteq A_i, w(S') \leq b_i} w(S')$, i.e. a maximum total weight set of approved projects that has weight at most b_i. Then, observe that any outcome that funds all of $S = \bigcup_{i \in N} S_i$ is necessarily IR. Thus, in order to construct an IMP and IR outcome, we can construct a payment function $y : N \times C \to \mathbb{R}_{\geq 0}$ that funds S. For each project $c \in C$, let n_c denote the number of agents i with $c \in S_i$. Note that if $c \in S$, then $n_c \geq 1$. Then, for each $i \in N$ let $y(i, c) = \frac{w(c)}{n_c}$ if $c \in S_i \subseteq S$ and $y(i, c) = 0$ if $c \notin S_i$. It is then simple to check that each agent's total cost is affordable to them, each project in S is fully paid for, and each agent i only pays for projects in $S_i \subseteq A_i$ (i.e. projects that they approve of). Therefore, we can let $x = (\sum_{c \in C} y(1, c), \ldots, \sum_{c \in C} y(n, c))$ and see that the outcome (S, x) is IR and IMP.

Now, observe that if any IMP outcome (S', x') Pareto dominates (S, x), then it must be also be IR because the utility of each agent is at least as high as before. There can only be a finite number of Pareto improvements to (S, x) since the number of possible subsets of projects to be funded is finite, and Pareto dominance depends only on the projects funded, not the costs to the agents. Hence, there must exist such a Pareto improvement which is IR, IMP and PO-IMP. Finally, Proposition 4 gives that this outcome must be MR and EXH. $\qquad \square$

Note that PO-Pay and IMP are both satisfied by an empty outcome with zero charges. PO-IMP and IMP are easily satisfied by computing a PO outcome from the set of IMP outcomes. PO-Pay and PO-IMP are easily satisfied by computing a PO outcome which may not necessarily satisfy IMP.

Proposition 7. *There always exists an outcome that satisfies MR, IR, PO-MR and hence also EXH.*

Proof. Existence of an outcome that satisfies MR, IR, PO-MR: From the proof of Proposition 6, we know that an IMP and IR outcome always exists. Also, from Proposition 4, we know that every IMP outcome is MR, so there always exists an MR and IR outcome. Now suppose the outcome satisfying MR and IR does not satisfy PO-MR. Then there exists another outcome satisfying MR that Pareto dominates the original outcome, which is still IR. There cannot exist an infinite number of Pareto improvements because there are only finitely many possible subsets of projects that can be funded. Hence we can reach a PO-MR outcome that is also IR and MR. Proposition 4 gives that this outcome is EXH.
□

We note that if no agent can individually fund a project, then every outcome is IR. In crowdfunding settings in which projects have high costs, the IR requirement is often easily satisfied.

6 Aggregation Rules

In this section, we take a direct welfarist view to formalize rules that maximize some notion of welfare. We consider three notions of welfare: utilitarian, egalitarian, and Nash welfare; and we define the following rules.

- UTIL: define the utilitarian welfare derived from an outcome (S, x) as $\sum_{i \in N} u_i(S)$. Then, UTIL returns an outcome that maximises the utilitarian welfare.
- EGAL: given some outcome (S, x), write the sequence of agents' utilities from that outcome as a tuple $u(S) = (u_i(S))_{i \in N}$, where $u(S)$ is sorted in non-decreasing order. Then, EGAL returns an outcome (S, x) such that $u(S)$ is lexicographically maximal among the outcomes.
- NASH: maximises the Nash welfare derived from an output (S, x), i.e. $\prod_{i \in N} u_i(S)$.

Proposition 8. *UTIL, EGAL, and NASH satisfy PO and hence PO-MR, PO-IMP, PO-Pay, and EXH.*

One notes that the rules UTIL, EGAL, and NASH do not satisfy minimal guarantees such as MR. The reason is that an agent may donate her budget to a widely approved project even though she may not approve any of such projects.

Given that the existing aggregation rules do not provide us with guarantees that the outcomes they produce will satisfy our axioms, we can instead define rules that optimize social welfare within certain subsets of feasible outcomes. For a property X, we can define UTIL-X, EGAL-X, and NASH-X as rules that maximise the utilitarian, egalitarian and Nash welfare respectively among only those outcomes that satisfy property X.

Next, we analyse the properties satisfied by rules EGAL/UTIL/NASH constrained to the set of MR or IMP outcomes. In the continuous model introduced by Brandl et al. [8], there is no need to consider the rule NASH-IMP, as the NASH rule in the case where projects can be funded to an arbitrary degree (given there is sufficient budget) already satisfies IMP.

Before we study the axiomatic properties, we note that most meaningful axioms and rules are NP-hard to achieve or compute. The following result follows from Proposition 1.

Proposition 9. *Even for one agent, computing a UTIL, UTIL-MR, UTIL-IMP, EGAL, EGAL-MR, EGAL-IMP, NASH, NASH-MR, NASH-IMP outcome is NP-hard.*

Similarly, the next result follows from Proposition 5.

Proposition 10. *UTIL-MR, EGAL-MR, and NASH-MR satisfy PO-Pay.*

From Proposition 5, it follows that UTIL-MR, EGAL-MR, and NASH-MR satisfy PO-Pay. In contrast, we show that UTIL-IMP, EGAL-IMP, and NASH-IMP do not satisfy PO-Pay. In order to show this, we prove that it is possible in some instances for the set of jointly IMP and PO-IMP outcomes to be disjoint from the set of PO-Pay outcomes.

Proposition 11. *UTIL-IMP, EGAL-IMP and NASH-IMP do not satisfy PO-Pay. In fact it is possible that no IMP and PO-IMP outcome satisfies PO-Pay.*

Similarly, the following also holds.

Proposition 12. *EGAL, EGAL-MR and EGAL-IMP are not strategyproof.*

Table 1 shows the axioms that are satisfied by restricting the aggregation rules to optimising within the space of MR or IMP outcomes.

7   Experiment

In addition to the axiomatic study of the welfare-based rules, we undertake a simulation-based experiment to gauge the performance of different rules with respect to utilitarian and egalitarian welfare. Our study shows the impact of fairness axioms such as MR and IMP on welfare.

We generate random samples of profiles in order to simulate two potential real-world applications of PFC.

Table 1. Properties satisfied by UTIL-MR, EGAL-MR, NASH-MR, UTIL-IMP, EGAL-IMP and NASH-IMP.

	UTIL-MR	EGAL-MR	NASH-MR	UTIL-IMP	EGAL-IMP	NASH-IMP
MR	✓	✓	✓	✓	✓	✓
IMP	–	–	–	✓	✓	✓
PROP	–	–	–	✓	✓	✓
IR	–	–	–	–	–	–
PO	–	–	–	–	–	–
PO-MR	✓	✓	✓	–	–	–
PO-IMP	✓	✓	✓	✓	✓	✓
PO-Pay	✓	✓	✓	–	–	–
EXH	✓	✓	✓	✓	✓	✓
CORE	–	–	–	–	–	–
SP	–	–	–	–	–	–

1. Share-house setting: In this example, we can imagine a group of house-mates pooling their resources to fund communal items for their house. We operate under the following assumptions:
   - Number of agents from 3–6: this represents a reasonable number of house-mates in a share-house.
   - Number of projects from 5–12: projects may include buying items such as tables, chairs, sofas, televisions, lights, kitchen appliances, washing machines, dryers, etc.
   - Agent budgets are from 300–600 and project costs are from 50–1000. We base these costs on typical rent and furniture costs in Australia as well as costs of the above items in first and second-hand retailers. We expect that each agent brings some money to the communal budget, and would spend around one or two weeks' worth of rent on one-time communal expenses.
2. Crowdfunding setting: In this example, we imagine a relatively small number of expensive projects to be funded, and a large number of philanthropic donors, and make the following assumptions.
   - Number of agents from 20–50: A review of crowdfunding websites such as Kickstarter and GoFundMe shows that the most promoted projects are typically funded by thousands of donors, and smaller projects can attract tens of donors. For the purposes of our simulation, we use between 20–50 donors, which is still relatively large compared to the number of available projects.
   - Number of projects from 3–8: In crowdfunding, there are far more projects available than a donor actually sees. However, we can estimate that in a browsing session, a donor might view the top 3–8 promoted projects.
   - Agent budgets from 0–400 and project costs from 1000–10000: Projects in real-life crowdfunding can have vastly varying costs. For our simulation, we want for the agents with all their money combined to be able to afford

some, but not all of the available projects in order to create instances that are not trivially resolved by funding all or none of the projects.

The results of the experiments are shown in Figs. 1, 2, 3, 4, 5, 6, 7 and 8.

Imposing MR on a rule seems to have a significant impact on both utilitarian and egalitarian welfare on average. Of course, since IMP implies MR, we expect that imposing IMP as a constraint will have an even greater cost on welfare, but from our experiment, this cost is a relatively small increase on top of the

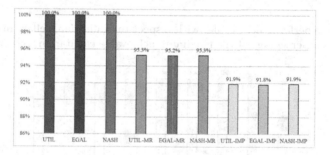

Fig. 1. Average performance of rules with respect to utilitarian welfare in share-house simulations as a percentage of the maximum achievable utilitarian welfare.

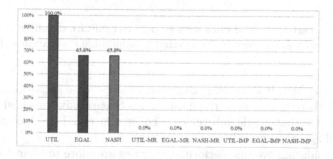

Fig. 2. Average performance of rules with respect to utilitarian welfare in crowdfunding simulations as a percentage of the maximum achievable utilitarian welfare.

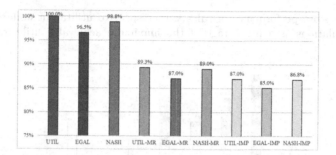

Fig. 3. Worst-case performance of rules with respect to utilitarian welfare in share-house simulations as a percentage of the maximum achievable utilitarian welfare.

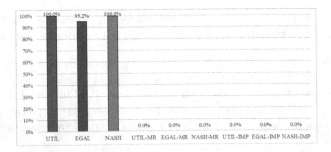

Fig. 4. Worst-case performance of rules with respect to utilitarian welfare in crowd-funding simulations as a percentage of the maximum achievable utilitarian welfare.

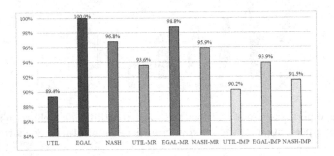

Fig. 5. Average performance of rules with respect to egalitarian welfare in share-house simulations as a percentage of the maximum achievable egalitarian welfare.

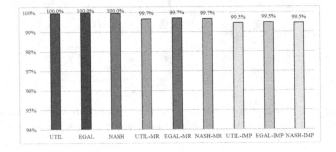

Fig. 6. Average performance of rules with respect to egalitarian welfare in crowdfunding simulations as a percentage of the maximum achievable egalitarian welfare.

cost of imposing MR. It is worth noting that in worst-case scenarios, it is always possible that there are no non-trivial outcomes that satisfy the constraints, and so there is a risk that a rule subject to a constraint could produce an outcome that gives all agents zero utility.

When considering average performance, rules are more resilient to the imposition of fairness constraints for instances that simulate crowdfunding scenarios compared to share-house scenarios. When the number of agents is high and

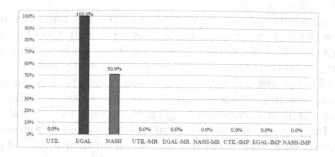

Fig. 7. Worst-case performance of rules with respect to egalitarian welfare in share-house simulations as a percentage of the maximum achievable egalitarian welfare.

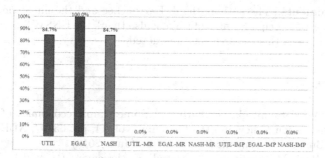

Fig. 8. Worst-case performance of rules with respect to egalitarian welfare in crowd-funding simulations as a percentage of the maximum achievable egalitarian welfare.

the number of projects is small, and project costs are high compared to agent budgets, it seems to be easier to achieve fairness properties.

We typically expect the NASH rule to be a compromise between UTIL and EGAL. This manifests in the results, where the performance losses for NASH with respect to utilitarian welfare are considerably less than those for EGAL. NASH loses considerably less with respect to egalitarian welfare than UTIL.

8 Conclusions

We proposed a concrete model for coordinating funding for projects. A formal approach is important to understand the fairness, participation, and efficiency requirements a system designer may pursue. We present a detailed taxonomy of such requirements and clarify their properties and relations. We also analyse natural welfarist rules both axiomatically and experimentally.

In practical applications of PFC, it is important to balance welfare demands with fairness conditions. Our experiment investigated the cost of fairness when imposing MR or IMP on UTIL, EGAL and NASH rules over instances that model crowdfunding and share-house scenarios. We find that imposing MR alone significantly reduces welfare on average, but imposing IMP as well produces a relatively small additional cost on welfare. The costs of imposing any fairness condition are much more pronounced on instances that model a share-house setting than a crowdfunding setting, suggesting that for a large number of agents and large project costs, fairness conditions are more easily met.

Our model is not just a rich setting to study collective decision making. We feel that the approaches considered in the paper go beyond academic study and can be incorporated in portals that aggregate funding for charitable projects. We envisage future work on online versions of the problem. We studied a utility model in which agents want as much money spent on their approved projects. It will be interesting to examine utility models in which agents care about which unapproved projects are funded or factor in the payments they have been made.

Acknowledgements. The authors thank Barton Lee and the anonymous reviewers of ADT 2021 for their helpful comments. They also thank the UNSW Taste of Research program under which this research was conducted.

References

1. Agrawal, A., Catalini, C., Goldfarb, A.: Some simple economics of crowdfunding. In: Innovation Policy and the Economy, vol. 14, pp. 63–97, NBER Chapters, National Bureau of Economic Research, Inc (June 2013)
2. Aziz, H., Brill, M., Conitzer, V., Elkind, E., Freeman, R., Walsh, T.: Justified representation in approval-based committee voting. Soc. Choice Welfare **48**(2), 461–485 (2017). https://doi.org/10.1007/s00355-016-1019-3
3. Aziz, H., Lee, B.E.: Proportionally representative participatory budgeting with ordinal preferences. In: Proceedings of the 35th AAAI Conference on Artificial Intelligence (AAAI) (2021)
4. Aziz, H., Lee, B.E., Talmon, N.: Proportionally representative participatory budgeting: axioms and algorithms. In: Proceedings of the 17th International Conference on Autonomous Agents and MultiAgent Systems, AAMAS 2018, Stockholm, Sweden, July 10–15, 2018, pp. 23–31 (2018)
5. Aziz, H., Shah, N.: Participatory budgeting: models and approaches. In: Rudas, T., Péli, G. (eds.) Pathways Between Social Science and Computational Social Science. CSS, pp. 215–236. Springer, Cham (2021). https://doi.org/10.1007/978-3-030-54936-7_10
6. Bagnoli, M., Lipman, B.L.: Provision of public goods: fully implementing the core through private contributions. Rev. Econ. Stud. **56**(4), 583–601 (1989)
7. Brandl, F., Brandt, F., Peters, D., Stricker, C., Suksompong, W.: Donor coordination: collective distribution of individual contributions. In: GAIW (Games, Agents and Incentives Workshops) (2019)
8. Brandl, F., Brandt, F., Peters, D., Stricker, C., Suksompong, W.: Funding public projects: a case for the nash product rule. Working paper (2020)
9. Buterin, V., Hitzig, Z., Weyl, E.G.: A flexible design for funding public goods. Manage. Sci. **65**(11), 5171–5187 (2019)
10. Chandra, P., Gujar, S., Narahari, Y.: Crowdfunding public projects with provision point: a prediction market approach. In: Proceedings of the 22nd European Conference on Artificial Intelligence (ECAI), pp. 778–786 (2016)

11. Corazzini, L., Cotton, C., Valbonesi, P.: Donor coordination in project funding: evidence from a threshold public goods experiment. J. Public Econ. **128**(1), 16–29 (2015)
12. Damle, S., Moti, M.H., Chandra, P., Gujar, S.: Civic crowdfunding for agents with negative valuations and agents with asymmetric beliefs. In: Proceedings of the 28th International Joint Conference on Artificial Intelligence (IJCAI), pp. 208–214 (2019)
13. Elkind, E., Faliszewski, P., Skowron, P., Slinko, A.: Properties of multiwinner voting rules. Soc. Choice Welfare **48**(3), 599–632 (2017). https://doi.org/10.1007/s00355-017-1026-z
14. Fain, B., Goel, A., Munagala, K.: The core of the participatory budgeting problem. In: Proceedings of the 12th International Conference on Web and Internet Economics (WINE 2016), pp. 384–399 (2016)
15. Fluschnik, T., Skowron, P., Triphaus, M., Wilker, K.: Fair knapsack. In: Proceedings of the 33rd AAAI Conference on Artificial Intelligence (AAAI) (2019)
16. Goel, A., Krishnaswamy, K., Sakshuwong, S., Aitamurto, T.: Knapsack voting for participatory budgeting. ACM Trans. Econ. Comput. (TEAC) **7**(2), 8:1–8:27 (2019). ISSN 2167–8375
17. Lackner, M., Skowron, P.: A quantitative analysis of multi-winner rules. In: Proceedings of the 28th International Joint Conference on Artificial Intelligence (IJCAI), pp. 407–413 (2019)
18. MacAskill, W.: Doing Good Better: How Effective Altruism Can Help You Make a Difference. Avery (2015)
19. MacAskill, W.: Essays Philos. Effective altruism: introduction **18**(1), 1–5 (2017)
20. Peters, D.: Economic Design for Effective Altruism, pp. 381–388 (2019). https://app.dimensions.ai/details/publication/pub.1122621382
21. Talmon, N., Faliszewski, P.: A framework for approval-based budgeting methods. In: Proceedings of the 33rd AAAI Conference on Artificial Intelligence (AAAI) (2019)

Complexity of Manipulative Interference in Participatory Budgeting

Dorothea Baumeister, Linus Boes[(⊠)], and Johanna Hillebrand

Heinrich-Heine-Universität Düsseldorf, Düsseldorf, Germany
{d.baumeister,linus.boes,johanna.hillebrand}@uni-duesseldorf.de

Abstract. A general framework for approval-based participatory budgeting has recently been introduced by Talmon and Faliszewski [17]. They use satisfaction functions to model the voters' agreement with a given outcome based on their approval ballots. We adopt two of their satisfaction functions and focus on two types of rules. That is, rules that maximize the overall voters' satisfaction and greedy rules that iteratively extend a partial budget by an item that maximizes the satisfaction in each incremental step. An important task in participatory budgeting is to study different forms of manipulative interference that may occur in practice. We investigate the computational complexity of different problems related to determining the outcome of a given rule and give a very general formulation of manipulative interference problems. A special focus is on problems dealing with a varying cost of the items and a varying budget limit. The results range from polynomial-time algorithms to completeness in different levels of the polynomial hierarchy.

Keywords: Participatory budgeting · Control · Computational complexity

1 Introduction

Participatory budgeting is often implemented as a mean of making democratic decisions. Thus, citizens can usually express their opinions on how a portion of a city's budget should be distributed in such a process, sometimes making new suggestions on which projects could be realized as well. The first implementation of participatory budgeting can be found in Porto Alegre (Brazil) in 1989 as an attempt by the Workers Party to break with traditionally authoritarian public policies (see Sintomer et al., [15]). Starting here the idea spread around the world, taking different forms and magnitudes regarding size, budget, and other factors. As described by Cabannes [5], integrating participatory budgeting into a city's form of government has yielded several positive effects. These range from an increased accountability of politicians, due to the collective will being rather visible, to directing larger parts of the city's budget towards education, health care, infrastructure, and childcare. The different stages of a participatory budgeting cycle are described by Aziz and Shah [3]. These stages include the

© Springer Nature Switzerland AG 2021
D. Fotakis and D. Ríos Insua (Eds.): ADT 2021, LNAI 13023, pp. 424–439, 2021.
https://doi.org/10.1007/978-3-030-87756-9_27

division into different districts, the determination of the total available budget, the emergence of project proposals, deliberation steps, and finally the voting stage. While Rey et al. [14] study multiple stages in one model, we will solely focus on the last step in this paper. Here, the citizens express their preferences regarding the projects they want the budget to be spent on. A participatory budgeting method then aggregates these votes in order to reach a decision about which projects will be realized. Here we assume that each project is either fully funded or not at all. So, in this last step we have a fixed set of projects, each associated with a cost, and we have an overall budget limit. The total cost of the funded projects must be within this limit. A crucial point is how the preferences of the voters are expressed, as a decision between expressiveness and compactness of the presentation has to be made. We follow a very simple approach by assuming approval ballots, where every voter chooses for every project whether she thinks that it should be funded or not. These individual votes are independent of the budget limit, i.e., a voter may approve a set of projects which could not be realized within the given budget limit (in contrast to, e.g., Goel et al. [8]). Regarding the aggregation of the approval ballots, we follow the approach of Talmon and Faliszewski [17]. As a first step, we define a satisfaction function that returns for each voter and each possible committee the satisfaction of said voter based on her approval ballot. Then, an (ir)resolute budgeting method chooses a (set of) winning projects. One method is to output bundles that maximize the sum of the voters' satisfaction while taking into account the budget limit. As we will see more detailed in Sect. 3, this may lead to winner determination problems of high complexity in some cases. A different approach commonly used in practice is a simple greedy approach. In each iteration, the set of winning projects is extended by the project that maximizes the satisfaction in each incremental step, again respecting the budget limit. In this case, winner determination is more straightforward but depends on some tie-breaking mechanism in every step. Combinations of different satisfaction functions and budgeting methods have been studied by Talmon and Faliszewski [17] with respect to their axiomatic properties, see Baumeister et al. [4] for an adaption to irresolute variants of these rules. Unfortunately, many of the desired axioms are not satisfied by the proposed methods, which opens the possibility of manipulative interference on participatory budgeting processes.

Due to the combinatorial structure in participatory budgeting (i.e., the set of implemented items may not exceed the available funds), there are new types of control to consider. The axioms proposed by Talmon and Faliszewski [17] focus on the way a budgeting method should react to certain changes of the parameters. If, for example, an item's costs are less than originally anticipated, this should not lead to the item becoming unfunded. This is a reasonable assumption, however budgeting methods using a cost-based satisfaction function do not satisfy it. A budgeting method violating this axiom could be vulnerable to control if a chair would be able to influence an item's cost, in order to either exclude it from the chosen budget or to ensure it being funded. Another axiom requests that an increase of the budget limit may not lead to an item being excluded from the winning budget. As this axiom is not satisfied by any of the budgeting methods in question, this leaves the possible vulnerability to control via a change of the budget limit in order to in- or exclude an item. In order to examine these possible vulnerabilities to control further, we initially investigate the computational complexity of determining a winner for the greedy and maximizing rule in Sect. 3. Then we provide a general definition of manipulative interference in Sect. 4 with a specific focus on problems where either the budget limit or the cost of a specific item may be manipulated.

Related Work. Regarding traditional election problems this refers to different variants of control. Here, an election chair alters the structure of the election to make some distinguished candidate win or to prevent some distinguished candidate from winning. There is a huge amount of literature studying different kinds of control problems in voting. For an overview, we refer to the book chapter by Faliszewski and Rothe [6]. Related work on participatory budgeting close to our assumptions (i.e., approval ballots, binary outcomes, and satisfaction functions) was studied by Jain et al. [9], who considered satisfaction functions under project interactions, and Rey et al. [13], who embedded the framework introduced by Talmon and Faliszewski [17] into the framework of judgment aggregation. Aziz et al. [2] considered aggregation using an axiomatic approach instead of predefined rules. A well studied special case of approval-based participatory budgeting are multiwinner elections, where we assume uniform cost for each candidate. Lackner and Skowron [12] compare a variety of rules, that also use approval-based satisfaction functions as a measure of the voters' agreement with a committee.

2 Preliminaries

For a formal study of the voting step in participatory budgeting we follow the approach of Talmon and Faliszewski [17].

Definition 1. *A budgeting scenario $E = (A, V, c, \ell)$ consists of a set $A = \{a_1, \ldots, a_m\}$ of m items, associated with a cost function $c : A \to \mathbb{N}_+$, and a set $V = \{v_1, \ldots, v_n\}$ of n voters, where each voter $v \in V$ has an associated ballot $A_v \subseteq A$ containing a set of preferred items, and a budget limit $\ell \in \mathbb{N}_+$.*

Without giving a formal definition, let $\mathcal{E}$ denote the set of all possible budgeting scenarios without fixing any of the parameters (apart from mentioned dependencies of parameters in the definition above). The goal in participatory budgeting is to select a subset B of the items, called budget, such that the total cost of B does not exceed the budget limit ℓ. Slightly abusing notation we write $c(B) = \sum_{a \in B} c(a)$ to denote the total cost of some budget $B \subseteq A$. Moreover, we call a budget feasible if $c(B) \leq \ell$ and denote the set of feasible budgets by $\mathcal{B}(E) = \{B \subseteq A \mid c(B) \leq \ell\}$. Feasibility is a hard constraint, but of course the budget should take the ballots of the voters into account. Therefore, we introduce satisfaction functions for the voters.

Definition 2. *The satisfaction of a voter $v \in V$ with a given budget $B \subseteq A$ is modeled by a **satisfaction function** $s : 2^A \times 2^A \to \mathbb{N}_0$. For simplicity, we define $B_v = A_v \cap B$ to be the set of items, which are both, approved by a voter v and in a given budget B. In this paper we consider the following satisfaction functions focussing on:*

- *$\mathbf{quantity}$: $s(A_v, B) = |B_v|$, the number of budgeted approved items, and*
- *$\mathbf{cost}$: $s(A_v, B) = c(B_v)$, sum of the cost of the budgeted approved items.*

Slightly abusing notation, we write $s(V, B) = \sum_{v \in V} s(A_v, B)$ to denote the overall satisfaction of the voters in V with a budget B. The presented satisfaction functions follow different intentions and model different application scenarios. The intuition for satisfaction by quantity is straightforward. The satisfaction of a voter correlates with the number of implemented projects she likes. For satisfaction by cost, we assume satisfaction correlates with the amount of funds that are spent on preferred projects. Now, in order to compute a set of winning budgets based on the voters' preferences, we define an irresolute budgeting method $\mathcal{R}$, which maps a budgeting scenario E to a set of feasible budgets. The rules we study use the underlying satisfaction functions we defined previously.

Definition 3. *Given a budgeting scenario $E = (A, V, c, \ell) \in \mathcal{E}$ and a satisfaction function s we define:*

- *$\mathbf{Max\ rules\ (m)}$: as $\mathcal{R}_s^m(E) = \arg \max_{B \in \mathcal{B}(E)} s(V, B)$, and*
- *$\mathbf{Greedy\ rules\ (g)}$: starting with $B = \emptyset$ iteratively extend B by $a \in A \setminus B$, maximizing $s(V, B \cup \{a\})$, until there is no item $a \in A \setminus B$ with $c(B \cup \{a\}) \leq \ell$. Finally, set $\mathcal{R}_s^g(E) = \{B\}$.*

The max rules return all budgets that maximize the sum of the voters' satisfaction according to the function s. This rule is irresolute since there may be several budgets satisfying this requirement. In contrast, the greedy rules work iteratively. In each step one item that maximizes the sum of the voters' satisfaction when added to the current budget, will be added. We assume that some tie-breaking mechanism is used in each round, such that exactly one item is added. This leads to a resolute rule, always returning a set containing a single budget, also referred to as the budget returned by the rule. Together with the two satisfaction functions defined above, we consider four different rules.

Example 1. Let $E = (A, V, c, \ell)$ be a budgeting scenario with $A = \{a_1, a_2, a_3, a_4\}$, $V = \{v, v'\}$ with $A_v = A$ and $A_{v'} = \{a_1\}$, $c(a_i) = i$, and $\ell = 7$. For the greedy rules we break ties in favor of the item with a higher index. We have $\mathcal{R}_{|B_v|}^m(E) = \{\{a_1, a_2, a_3\}, \{a_1, a_2, a_4\}\}$, as both bundles yield a satisfaction of four, while the only bundle with a higher satisfaction is $A \notin \mathcal{B}(E)$. Similarly, it holds that $\mathcal{R}_{c(B_v)}^m(E) = \{\{a_1, a_2, a_4\}\}$ with a satisfaction of eight. For the greedy rules we list the items in the order they are added, that is $\mathcal{R}_{|B_v|}^g = \{\{a_1, a_4, a_2\}\}$ (where a_3 is skipped in the third iteration due to feasibility), and $\mathcal{R}_{c(B_v)}^g = \{\{a_4, a_3\}\}$.

In Sect. 4, we will define different decision problems related to different kinds of manipulative interference and study them from a computational point of view. As an intermediate step, it is important to determine the complexity for winner-determination problems first. Of course, computing a winning bundle for the greedy rule is easy, as it is a rather simple algorithm, that tries to find a solution that is close to the one from the max rule. Yet, in the following section, we will see that depending on the satisfaction function, there is little hope for an efficient algorithm that returns at least one budget that maximizes the sum of the voters' satisfaction. Our results range from polynomial-time algorithms to completeness in the polynomial hierarchy. We refer the reader to the textbook by Arora and Barak [1] for further details on computational complexity. In the rest of the paper, we assume that the reader is familiar with the complexity classes NP, coNP, $\Delta_2^p = \text{P}^{\text{NP}}$, and $\Sigma_2^p = \text{NP}^{\text{NP}}$. Further, in this paper, for a decision problem X, let $\overline{\text{X}}$ denote its complement, and for $i, j \in \mathbb{N}_+$ with $i < j$ we denote $[i, j] = \{i, i + 1, \ldots, j\}$, $[i, i] = \{i\}$, $[j, i] = \emptyset$, and $[i] = [1, i]$.

3   Winner Determination

In this section, we investigate the computational complexity for a variety of winner-determination problems associated with the considered budgeting rules. We assume, that for any greedy rule, a tie-breaking is fixed priorly and applied every round, resulting in a single final budget. Assuming that the given satisfaction function s and the tie-breaking rule are efficiently computable, computing a winning budget for a greedy rule can be done in polynomial time, since in each round the number of possible budgets that has to be considered equals the number of actually non-funded items. Therefore, we study decision problems related to winner determination only for maximizing rules combined with some efficiently computable satisfaction function s. The first problem we study asks whether there is some feasible budget where the sum of the voters' satisfaction exceeds some given bound. Additionally, we focus on some desired budget B^*, and ask whether it is a winning budget.

$\mathcal{R}_s$-BUDGET SCORE ($\mathcal{R}_s$-SC)

Given: A budgeting scenario $E = (A, V, c, \ell) \in \mathcal{E}$ and some bound $t \in \mathbb{N}_0$.
Question: Is there a budget $B \in \mathcal{B}(E)$ with $s(V, B) \geq t$?

$\mathcal{R}_s$-WINNING BUDGET ($\mathcal{R}_s$-WB)

Given: A budgeting scenario $E = (A, V, c, \ell) \in \mathcal{E}$ and some desired budget $B^* \subseteq A$.
Question: Is $B^* \in \mathcal{R}_s(E)$?

Since the max rule we consider is irresolute, we also ask whether a given bundle is a subset of at least one, respectively every, winning budget. Formally the problem $\mathcal{R}_s^m$-POSSIBLY BUDGETED ($\mathcal{R}_s^m$-PB) has the same input as $\mathcal{R}_s^m$-WB, but the question is whether there is some $B \in \mathcal{R}_s^m(E)$ with $B^* \subseteq B$. Accordingly, we ask for the problem $\mathcal{R}_s^m$-NECESSARILY BUDGETED ($\mathcal{R}_s^m$-NB), whether $B^* \subseteq B$ for every budget $B \in \mathcal{R}_s^m(E)$. Now, we provide general upper bounds.

Lemma 1. *Consider $E \in \mathcal{E}$, an efficiently computable satisfaction function s, and $t^* = \max_{B \in \mathcal{B}(E)} s(V, B)$. For $\mathcal{R}_s^m$-SC being a member of complexity class $\mathcal{A}$, it holds that*

(i) $\mathcal{R}_s^m$-WB is in co$\mathcal{A}$, and
(ii) $\mathcal{R}_s^m$-PB and $\mathcal{R}_s^m$-NB are in $\mathrm{P}^{\mathcal{A}[O(\log(t^*))]}$, and
(iii) $\mathcal{R}_s^m$-SC $\in$ NP.

Proof. Consider a budgeting scenario E, a set of items $B^* \subseteq A$, and a satisfaction function s. For (i) we first may verify if B^* is feasible and compute $s(V, B^*)$ in polynomial time. Then to solve $\mathcal{R}_s^m$-WB we may decide in co$\mathcal{A}$ if every feasible budget has an overall satisfaction of less than $s(V, B^*) + 1$ by solving $\overline{\mathcal{R}_s^m\text{-SC}}$.

For (ii) we may compute the optimal score t^* of a winning budget by sending $O(\log(t^*))$ queries to an $\mathcal{A}$-oracle using binary search. To solve $(E, B^*) \in \mathcal{R}_s^m$-PB we construct another satisfaction function s' with $s'(V, B) = 2 \cdot s(V, B) + 1$ if $B^* \subseteq B$ and $s'(V, B) = 2 \cdot s(V, B)$ otherwise. To answer $(E, B^*) \in \mathcal{R}_s^m$-PB, we send a final query to our $\mathcal{A}$-oracle, asking whether $(E, 2t^* + 1) \in \mathcal{R}_{s'}$-SC is a yes-instance. We can use similar techniques to solve $(E, B^*) \in \mathcal{R}_s^m$-NB, by defining s', such that a bundle B with $B^* \not\subseteq B$ is assigned the slightly increased score. Then $(E, B^*) \in \mathcal{R}_s^m$-NB is a yes-instance if and only if $(E, 2t^* + 1) \in \mathcal{R}_{s'}$-SC is a no-instance. Overall we can query an $\mathcal{A}$-oracle $O(\log(t^*))$ times.

For (iii) recall that s is efficiently computable, so verifying that there is a budget B with $s(V, B) \geq t$ can be done in polynomial time. □

From the above lemma it follows that if $\mathcal{R}_s^m$-SC is efficiently computable for some satisfaction function s, then the other winner-determination problems are also in P. Another implication is, that $\mathcal{R}_s^m$-PB and $\mathcal{R}_s^m$-NB are in Δ_2^p in general, and in $\Theta_2^p = \mathrm{P}^{\mathrm{NP}[\log]}$ for satisfaction functions, where the satisfaction for a bundle is at most polynomial in the (binary encoded) size of the budgeting scenario. For the rules we consider, we will see that P and Δ_2^p are suitable upper bounds for $\mathcal{R}_s^m$-PB and $\mathcal{R}_s^m$-NB. Yet, there are satisfaction functions,[1] for which $\mathcal{R}_s$-SC is known to be NP-complete, but for $E = (A, V, c, \ell)$ and $B \subseteq A$, the score $s(V, B)$ is bounded by $|A| \cdot |V|$, yielding an upper bound of Θ_2^p (which is not necessarily tight).

We continue by establishing tight bounds for the winner determination problems. Talmon and Faliszewski [17] already showed, that $\mathcal{R}_{|B_v|}^m$-SC is solvable in polynomial time. Following Lemma 1 we can formulate the following corollary.

Corollary 1. $\mathcal{R}_{|B_v|}^m$-WB, $\mathcal{R}_{|B_v|}^m$-PB, *and* $\mathcal{R}_{|B_v|}^m$-NB *are in* P.

---

[1] For example the Chamberlin-Courant rule for approval ballots, studied by Skowron and Faliszewski [16].

Next, we establish lower bounds for $\mathcal{R}^m_{c(B_v)}$. Talmon and Faliszewski [17] showed NP-hardness for $\mathcal{R}^m_{c(B_v)}$-SC by reducing from the well known problem SUBSET SUM (see Garey and Johnson [7]).

Theorem 1. $\mathcal{R}^m_{c(B_v)}$-WB *is* coNP-*complete, and* $\mathcal{R}^m_{c(B_v)}$-PB *and* $\mathcal{R}^m_{c(B_v)}$-NB *are* Δ^p_2-*complete.*

Proof. We start by showing coNP-hardness for $\mathcal{R}^m_{c(B_v)}$-WB. We will reduce from $\overline{\text{SUBSET SUM}}$, where the input is a set of integers $N = \{n_1, \ldots, n_m\} \subseteq \mathbb{N}_+$ and a bound $n \in \mathbb{N}_+$, and the question is, whether there is no subset $S \subseteq N$ with $\sum_{i \in S} i = n$. We transform an arbitrary instance (N, n) to an instance of $\mathcal{R}^m_{c(B_v)}$-WB. Let $A = \{a_1, \ldots, a_m, b\}$, $V = \{v\}$ with $A_v = A$, $c(a_i) = 2n_i$ for each $i \in [1, m]$, and $c(b) = 2n - 1$, and $\ell = 2n$. Finally, we set $B^* = \{b\}$ and claim that $(E, B^*) \in \mathcal{R}^m_{c(B_v)}$-WB if and only if $(N, n) \in \overline{\text{SUBSET SUM}}$. In particular, B^* is a winning budget if there is no set of items, which adds up to a cost of $2n$. This is exactly the case if there is no $S \subseteq N$ which sums up to n.

To show Δ^p_2-completeness for the remaining problems, we will use the following Δ^p_2-complete problem, based on Krentel's results [11, Thm 2.1, Thm 3.3].[2]

EVEN SUBSETSUM (ESS)

Given: A finite set of integers $N \subset \mathbb{N}_+$ and a distinct integer $n \in \mathbb{N}_+$.
Question: Let $t = \sum_{i \in S} i$ be the largest possible value with $t \leq n$ over all $S \subseteq N$. Is $t \mod 2 \equiv 0$?

For an upper bound, $\mathcal{R}^m_{c(B_v)}$-PB and $\mathcal{R}^m_{c(B_v)}$-NB are in Δ^p_2 following Lemma 1. Since the overall satisfaction derived from a winning budget t^* depends on the cost, $\log(t^*)$ is polynomial in the instance size (but not logarithmical).

To show hardness, we reduce from ESS. Consider any ESS instance $\mathcal{I} = (N, n)$ with $N = \{n_1, \ldots, n_m\}$. For simplicity and without loss of generality assume that $n \geq n_i$ holds for every $i \in [m]$. Further, let $k = |n|$ denote the length of the binary representation of n. We construct a $\mathcal{R}^m_{c(B_v)}$-PB instance $\mathcal{I}' = (E, B^*)$, with $A = \{a_1, \ldots, a_m, b_1, \ldots, b_k\}$ and $V = \{v\}$ with $A_v = A$. For our cost function c, we interpret the cost $c(a)$ of some item $a \in A$ in its binary encoding. By construction, each cost $c(a)$ will be consisting of two different zones, which are k bit long (to prevent carries). The front zone will be used to verify if the maximum achievable cost is even and the end zone will be used to still respect our bound n. For each a_i we will simply set cost n_i in both zones, i.e., $c(a_i) = (2^{k+1} + 1) \cdot n_i$. For each b_i we will only use the front zone to set the i-th bit to one, i.e., $c(b_i) = 2^{i+k}$. We choose ℓ in a way that the first k bits are set to one and the last k bits are set to the binary representation of n, that is $\ell = 2^{k+1}(2^{k+1} - 1) + n$. Finally, we set $B^* = \{b_1\}$.

To prove equivalence, note that each winning budget in $\mathcal{I}'$ has a satisfaction of at least $\sum_{i=1}^{k} c(b_i) = 2^{k+1}(2^{k+1} - 1)$ (e.g., by budgeting all b_i) and at most ℓ (by definition). Also note, that the cost of any budget not containing any b_i,

---

[2]   Also known as a KNAPSACK variant in related literature (see Kellerer et al. [10]).

n is always either exceeded in both or neither zones simultaneously. Therefore, any winning budget in $\mathcal{I}'$ has an equivalent cost in the last zone to the largest possible value for $\mathcal{I}$, while the front zone can always be filled up bitwise by values b_i. Finally, note, that by construction b_1 is only part of a winning budget if and only if the optimal value adds up to an even value, so its corresponding bit in the front zone can be flipped to one by adding b_1. By construction $B^* = \{b_1\}$ is necessarily and thus possibly budgeted if and only if $\mathcal{I}$ is a yes instance.   □

4   Manipulative Interference

To study problems of manipulative interference in a generic way, we consider an alteration function f, that maps from a given budgeting scenario to a set of possible scenarios after the alteration of specified parameters (e.g., the cost function or the voters' ballots). Formally, we have $f : \mathcal{E} \to 2^{\mathcal{E}}$. We assume, that we can efficiently verify whether $E' \in f(E)$ holds. We distinguish between a constructive and a destructive variant of manipulative interference.

CONSTRUCTIVE-$\mathcal{R}_s$-MANIPULATIVE-INTERFERENCE (C-$\mathcal{R}_s$-MI)

Given: A budgeting scenario E, a set of items $B_\heartsuit$, an integer k, and an alteration function f.

Question: Is there a budgeting scenario $E' \in f(E)$, such that there is a winning budget $B \in \mathcal{R}_s(E')$ with $|B_\heartsuit \cap B| \geq k$?

For DESTRUCTIVE-$\mathcal{R}_s$-MANIPULATIVE-INTERFERENCE (D-$\mathcal{R}_s$-MI) the input remains the same, but now we ask whether there is a budgeting scenario $E' \in f(E)$, such that there is a winning budget $B \in \mathcal{R}_s(E')$ with $|B_\heartsuit \cap B| < k$. Both definitions are very general. In particular, we have a set $B_\heartsuit$ of distinguished items. A natural restriction is the focus on a single item with $|B_\heartsuit| = 1$. In the constructive case we ask, whether there is a winning budget that contains at least k of the preferred items. This again gives the freedom to choose between having at least one to having all items in the winning budget. Accordingly, in the destructive case we ask whether there is a winning budget where less than k of the distinguished items are included. By setting $k = 1$ we obtain the special case where we ask for a winning budget containing none of the items in $B_\heartsuit$. A more strict variant of constructive manipulative interference would be to require that all winning budgets contain at least k of the preferred items. Accordingly, in the destructive variant one could require that the condition holds for all winning budgets. In this paper, we will however focus on the above presented variants.

For a trivial upper bound, we may guess an altered budgeting scenario $E' \in f(E)$ and a budget $B \in \mathcal{B}(E')$ with $|B_\heartsuit \cap B| \geq k$, and verify whether $B \in \mathcal{R}_s(E')$ holds by querying an oracle to answer $(E', B) \in \mathcal{R}_s$-WB.

Lemma 2. *Fix some alteration function f such that $\mathcal{R}_s$-WB restricted to budgeting scenarios E' with $E' \in f(E)$ is in $\mathcal{A}$. Then*

(i) C-$\mathcal{R}_s$-MI and D-$\mathcal{R}_s$-MI restricted to f are in $\mathrm{NP}^{\mathcal{A}}$, and
(ii) $\mathcal{R}_s$-WB restricted to f is in $\mathcal{A}$.

Hence, any form of manipulative interference, like manipulation, bribery, or control in classical voting, is bound upwards by NP, for rules, where $\mathcal{R}_s$-WB can be solved efficiently, including all greedy rules. Following Lemma 1, an upper bound for all maximizing rules is Σ_2^p. For lower bounds, we investigate specific forms of control, as a subtype of manipulative interference, to determine how vulnerable the rules in question are to seemingly small changes of a given budgeting scenario. In particular, we study the impact of influencing the budget limit or an item's cost on the outcome. While initially putting their combinatorial budgeting methods forward, Talmon and Faliszewski [17] simultaneously proposed several axioms a budgeting method should satisfy. As these axioms are not satisfied by, in some cases any and in other cases several of the proposed rules, we derive ways in which to exploit these particular weaknesses in order to exert control over the results of the participatory budgeting process. We investigate tight bounds for these specific forms of control, by studying the complexity of $\mathcal{R}_s$-MI under respective alteration functions f. Table 1 summarizes our results.

Changing the Budget Limit. The first type of control we consider is by altering the budget limit, which originates from the axiom of limit monotonicity as defined by Talmon and Faliszewski [17]. The idea is that if the budget limit is increased, no previously budgeted item becomes unfunded. All budgeting rules we consider violate said axiom. Thus, we define a variant of manipulative interference capturing different possibilities of taking influence on the budget limit.

Definition 4. *Given* $L, H \in \mathbb{N}_+$ *with* $L \leq H$, *define an alteration function* $f_{L,H}$ *such that* $(A, V, c, d) \in f_{L,H}(E)$ *for every* $E = (A, V, c, \ell)$ *and* $d \in [L, H]$. *The restriction of manipulative interference to such alteration functions and* $k \leq |B_\heartsuit|$ *will be denoted by* $\mathcal{R}_s$-CONTROL-BY-SETTING-THE-BUDGET-LIMIT *($\mathcal{R}_s$-CSBL).*

In the constructive case C-$\mathcal{R}_s$-CSBL asks whether it is possible to increase or decrease the budget limit such that at least k of the desired items are contained in one winning budget. In the destructive variant D-$\mathcal{R}_s$-CSBL, asks whether it is possible to obtain a winning budget containing less than k of the distinguished items by increasing or decreasing the budget limit. Since the rules we consider here violate limit monotonicity, they are obviously vulnerable to this type of control. Now, we will show that for the max rules and the quantity based satisfaction functions both control problems are solvable in polynomial time.

Theorem 2. C-$\mathcal{R}_{|B_v|}^m$-CSBL *and* D-$\mathcal{R}_{|B_v|}^m$-CSBL *are in* P.

Proof. We start with the constructive variant, showing C-$\mathcal{R}_{|B_v|}^m$-CSBL $\in$ P. Let $B_\heartsuit \subseteq A$ be the set of items, from which we want to include at least k items, by a successful control, in at least one winning budget. We reduce the given instance $\mathcal{I} = (E, B_\heartsuit, k, f_{L,H})$ to $\mathcal{I}' = (E', B_\heartsuit, k, f_{L,H})$, by modifying the set of voters. For $w = |B_\heartsuit|$, we clone each voter $w+1$ times and add one additional voter v with $A_v = B_\heartsuit$, resulting in a set of voters V'. This enforces, that budgets containing more items from $B_\heartsuit$ yield a slightly higher satisfaction in case of ties. We set $E' = (A, V', c, \ell)$. It holds that $\mathcal{I} \in$ C-$\mathcal{R}^m_{|B_v|}$-CSBL $\Leftrightarrow \mathcal{I}' \in$ C-$\mathcal{R}^m_{|B_v|}$-CSBL, because for every $d \in [L, H]$ and any two budgets $B \in \mathcal{R}^m_{|B_v|}((A, V, c, d))$ and $B' \in \mathcal{R}^m_{|B_v|}((A, V', c, d))$ it holds that $|B \cap B_\heartsuit| \leq |B' \cap B_\heartsuit|$. Note, that for E' the maximum achievable satisfaction for any feasible budget in $\mathcal{B}(E')$ is at most $s(V', A) = (w + 1) \cdot s(V, A) + w$. We use dynamic programming as described by Talmon and Faliszewski [17], to determine the minimum cost of a budget with a satisfaction of exactly t for each $t \in [0, s(V', A)]$, for the budgeting scenario E'. We may compute those values and store them in a list $\mathcal{T}$. Formally, for every $t \in [0, s(V', A)]$, if there is no feasible budget with a satisfaction of exactly t, let $\mathcal{T}(t) = \infty$ and otherwise, let $\mathcal{T}(t) = \min_{B' \in \{B \in \mathcal{B}(E') | s(V', B) = t\}} c(B)$. Finally, we solve C-$\mathcal{R}^m_{|B_v|}$-CSBL by identifying if there is a value $d \in [L, H]$ we can set the budget limit to, such that there is a winning budget B with $|B \cap B_\heartsuit| \geq k$. We can search for d in a polynomial number of steps. First, we initialize to $d = H$ and determine the highest value t^* with $\mathcal{T}(t^*) \leq d$. We express this value as $t^* = (w + 1) \cdot t_1 + t_2$, such that $t_2 \in [0, w]$. If $t_2 \geq k$, a control can be executed by choosing $\ell = d$. Otherwise, we can decrease d to $d = \mathcal{T}(t^*) - 1$ and repeat until we either found d, or stop if $d < L$. Note, that this procedure stops after at most $s(V', A) < |A| \cdot |V'|$ steps.

To show, that D-$\mathcal{R}^m_{|B_v|}$-CSBL $\in$ P also holds, we can use the same algorithm. We deviate by slightly permutating the values for the function $\mathcal{T}$, such that bundles are preferred, that include less items from $B_\heartsuit$. In particular, for every $t_1 \in [0, w + 1]$ and $t_2 \in [0, w]$, we set $\mathcal{T}'((w+1) \cdot t_1 + t_2) = \mathcal{T}((w+1) \cdot t_1 + w - t_2)$. Again, we search for $d \in [L, H]$, starting at $d = H$, while our condition for identifying a yes-instance changes to $t_2 \geq w - k$. □

For C-$\mathcal{R}^m_{c(B_v)}$-CSBL and D-$\mathcal{R}^m_{c(B_v)}$-CSBL tight complexity bounds are still open. Following Lemma 2, both problems are in Σ^p_2 and following Theorem 1 they are Δ^p_2 hard. The latter follows easily, as problems of winner-determination can be reduced to control problems without altering the parameters at all.

A very general result holds for all additive satisfaction functions, i.e., for any function s with $s(A_v, B) = \sum_{a \in A_v} \sum_{b \in B} s(\{a\}, \{b\})$ for all $A_v, B \subseteq A$.

Theorem 3. *For additive satisfaction functions s it holds that C-$\mathcal{R}^g_s$-CSBL and D-$\mathcal{R}^g_s$-CSBL are in P.*

Proof. Since s is additive by assumption and not dependent on the budget limit $d \in [L, H]$, the processing order of a greedy rule $\mathcal{R}^g_s$ is determined prior execution, using a fixed linear tie-breaking scheme $\succ$ if necessary. Without loss of generality we assume, the set of items is labeled in this ordering. That is, for $A = \{a_1, \ldots, a_m\}$ we assume that for each $1 \leq i < j \leq m$ it holds that either $s(V, \{a_i\}) > s(V, \{a_j\})$ or $s(V, \{a_i\}) = s(V, \{a_j\})$ and $a_i \succ a_j$. Further, we denote $A_i = \{a_1, \ldots, a_i\}$ and $E_d = (A, V, c, d)$.

We use dynamic programming to compute all values for $d \in [L, H]$, such that we can include exactly $j \in [0, |B_\heartsuit|]$ items from $B_\heartsuit$, only using items from A_i for $i \in [0, m]$. We generate a $(|B_\heartsuit| + 1) \times (m + 1)$ table $\mathcal{T}$, where each column represents a processing step after investigating an item a_i and each row represents the number of items shared with $B_\heartsuit$ in a possible (partial) solution. More precisely, the leftmost column ($i = 0$) represents an initial state, column i represents partial solutions after processing the first i items A_i, and the values in the rightmost column ($i = m$) represent possible (full) solutions.

The intuition behind $\mathcal{T}(j, i)$ is, that the greedy rule has already executed its first i iterations for an unknown budget limit $d \in [L, H]$, such that j items from $B_\heartsuit$ have already been added to the (partial) budget B_i. As we might have added in items in some iterations, we assume that some of the budget has already been filled by the respective items cost $c(B_i)$. For $d \in [L, H]$ and every possible resulting partial budget $B_i = \mathcal{R}_s^g(E_d) \cap A_i$ containing j items from $B_\heartsuit$, we add $d - c(B_i) \in \mathcal{T}(j, i)$. Note that $\mathcal{T}(j, i)$ is empty, if there is no $d \in [L, H]$ such that $\mathcal{R}_s^g(E_d)$ contains exactly j items from $B_\heartsuit$ after the first i iterations, i.e., if there is no $d \in [L, H]$ with $|\mathcal{R}_s^g(E_d) \cap A_i \cap B_\heartsuit| = j$. In particular, $\mathcal{T}(j, i)$ contains every value, such that we can extend the cost $c(B_i)$ of a partial budget $B_i = \mathcal{R}_s^g(E_d)$ satisfying above conditions to retrieve the input value d. Additionally, we claim that each $\mathcal{T}(j, i)$ can be represented by two discrete intervals, such that we can encode the values for each cell efficiently (to be shown at the end of the proof).

We initialize every cell to $\mathcal{T}(j, i) = \emptyset$ for $j \in [0, |B_\heartsuit|]$ and $i \in [0, m]$, except for $\mathcal{T}(0, 0) = [L, H]$. Next, we populate $\mathcal{T}$ left-to-right and top-to-bottom, where any cell $\mathcal{T}(j, i)$ is used to extend $\mathcal{T}(j, i + 1)$ and $\mathcal{T}(j + 1, i + 1)$, i.e., we only populate to the right. By design, each cell might be populated from two different cells; in this case we consider the union of both values. We will explain in detail how to populate in the first iteration ($i = 1$) to generalize from there.

We start by investigating $\mathcal{T}(0, 0)$ and reduce our problem to smaller instances, where the decision on a_1 is already made and thus, we only need to consider $A \backslash A_1$ in following iterations. In particular, we study two main cases. In case $d < c(a_1)$ holds, then in the first iteration we cannot add a_1 to the bundle. Hence, in case $d \in [L, c(a_1) - 1]$, we can reduce to an instance, which considers only $A \backslash A_1$, i.e., we extend cell $(0, 1)$ by $\mathcal{T}(0, 1) = \mathcal{T}(0, 1) \cup [L, c(a_1) - 1]$. Otherwise, for $d \geq c(a_1)$, we certainly need to add a_1 to the budget in this iteration. Again, we can reduce this to an instance not considering a_1, by choosing the budget limit, such that $d \geq c(a_1)$ holds in any case. Instead of enforcing d to have a minimum value (of at least $c(a_1)$), we reduce by decreasing the respective values to choose from by $c(a_1)$. In case $a_1 \notin B_\heartsuit$, we set $\mathcal{T}(0, 1) = \mathcal{T}(0, 1) \cup [0, H - c(a_1)]$, otherwise we also increment j, i.e., $\mathcal{T}(1, 1) = \mathcal{T}(1, 1) \cup [0, H - c(a_1)]$.

More general, for some iteration, in which we investigate the cell (j, i), we again study two seperate cases. We split $\mathcal{T}(j, i)$ into two disjoint sets based on the respective items cost $c(a_i)$. That is, $X = \mathcal{T}(j, i) \cap [0, c(a_i) - 1]$ and $Y = \mathcal{T}(j, i) \cap [c(a_i), H]$. We extend $\mathcal{T}(j, i + 1)$ by X. Before extending a cell with values from Y, we shift all values of Y by $-c(a_i)$. Formally, that is $Y' = \{y - c(a_i) \mid y \in Y\}$. Finally, if $a_i \notin B_\heartsuit$, we extend $\mathcal{T}(j, i + 1)$ by Y', otherwise we extend $\mathcal{T}(j + 1, i + 1)$ by Y'.

After populating the table $\mathcal{T}$, there is a $d \in [L, H]$ with $|\mathcal{R}_s^g(E_d) \cap B_\heartsuit| = j$ if and only if $\mathcal{T}(j, m) \neq \emptyset$. Additionally, we can use backtracking on every value $d' \in \mathcal{T}(j, m)$, to compute a distinct value $d \in [L, H]$ with $|\mathcal{R}_s^g(E_d) \cap B_\heartsuit| = j$.

It is left to show, that each cell of the table can be stored efficiently. Therefore, we show, that each cell can be represented with at most two intervals $I_1, I_2 \subseteq [0, H]$ with $0 \in I_1$. Of course, this claim holds for $\mathcal{T}(0, 0) = [L, H]$ by assumption and for the remaining values in the leftmost column, as they are never populated. Next, we show that, the if the claim holds for previously populated cells, then it also holds after populating the next cell. We start with the first row. Consider some iteration, where we are investigating cell $\mathcal{T}(0, i)$. For simplicity, we imagine the (at most) two intervals in $\mathcal{T}(0, i)$ to occupy respective space on the larger interval $[0, H]$. We imagine this interval to be ordered left-to-right by ascending values. In any iteration investigating $\mathcal{T}(j, i) \subseteq [0, H]$, we split $\mathcal{T}(j, i)$ at $c(a_i)$. The left part (excluding $c(a_i)$) is added to $\mathcal{T}(0, i+1)$ without any shifting operation. If $a_i \in B_\heartsuit$, we use the values on the right (including $c(a_i)$) to populate $\mathcal{T}(1, i+1)$, which is not in the first row. Otherwise, we shift those values to the left by subtracting $c(a_i)$ and add them to $\mathcal{T}(0, i+1)$. If $c(a_i)$ did not intersect one of the intervals, the claim holds. If on the other hand $c(a_i)$ did intersect an interval, then the rightmost part is shifted to the left, such that the starting value is 0. By assumption in $\mathcal{T}(0, i+1)$ there are now two intervals starting with 0. Thus, those two intervals collapse to a single interval. For the remaining rows first note, that if we split and shift any interval $[0, x]$, the two resulting intervals both have a starting value of 0. Subsequently, the only way there is an interval $I \in \mathcal{T}(j, i)$ with $j > 0$ and $0 \notin I$, is that in some previous iteration i' a preferred item $a_{i'}$ was added, whose cost $c(a_{i'})$ did not intersect the right interval in $\mathcal{T}(j-1, i')$. In particular, the right interval was shifted to the left and added to $\mathcal{T}(j, i'+1)$. This especially means, that $\mathcal{T}(j - 1, i' + 1)$ can only hold the left interval, which is always sticking to 0 when using the operations of splitting and shifting. Overall, in each column there can be at most one interval I with $0 \notin I$.   □

Changing an Item's Cost. Another type of control is the alteration of a given item's cost. This is based on the axiom of discount monotonicity, introduced by Talmon and Faliszewski [17]. The intuition is that decreasing the cost of a budgeted item does not lead to it being not funded anymore. Using a budgeting method that satisfies this axiom means that there is no incentive to strategize regarding an item's price. Otherwise, one might not take an offer that would reduce the cost of an item, fearing that it could lead to eliminating that item from the winning bundle. This is not desirable, as it would be a waste of resources.

Definition 5. *Given $a_\heartsuit \in A$ and $L, H \in \mathbb{N}_+$ with $L \leq H$, define an alteration function $f_{L,H}$ with $(A, V, c', \ell) \in f_{L,H}(E)$ for every $E = (A, V, c, \ell)$ and $d \in [L, H]$ such that $c'(a_\heartsuit) = d$ and $c'(a) = c(a)$ for all $a \in A \setminus \{a_\heartsuit\}$. The restriction of manipulative interference to such alteration functions, $B_\heartsuit = \{a_\heartsuit\}$, and $k = 1$ will be denoted by $\mathcal{R}_s$-CONTROL-BY-SETTING-AN-ITEM'S-COST ($\mathcal{R}_s$–CSIC).*

With the above defined restrictions, C-$\mathcal{R}_s$-CSIC asks whether the cost of the desired item $a_\heartsuit$ can be changed within the given bounds such that a winning budget contains $a_{\heartsuit}$. In D-$\mathcal{R}_s$-CSIC we ask whether we can obtain a winning budget that does not contain $a_{\heartsuit}$. The complexity of $\mathcal{R}_s$-CSIC in both variants for $\mathcal{R}^m_{|B_v|}$ and $\mathcal{R}^g_{|B_v|}$ follow directly from the results by Talmon and Faliszewski [17] and Baumeister et al. [4]. As both rules satisfy discount monotonicity, the strategy is to set $d = L$ for the constructive variant and $d = H$ for the destructive variant. To see if the control attempt was successful, we can solve the respective winner determination problems, which both are in P.

Corollary 2. C-$\mathcal{R}^m_{|B_v|}$-CSIC, D-$\mathcal{R}^m_{|B_v|}$-CSIC, C-$\mathcal{R}^g_{|B_v|}$-CSIC, and D-$\mathcal{R}^g_{|B_v|}$-CSIC are in P.

We turn to the cost satisfaction function and show that for the maximizing rule the constructive variant of setting an item's cost is complete for Δ^p_2.

Theorem 4. C-$\mathcal{R}^m_{c(B_v)}$-CSIC is Δ^p_2-complete.

Proof. For a lower bound, following Theorem 1, $\mathcal{R}^m_{c(B_v)}$-PB is Δ^p_2-complete. Subsequently, C-$\mathcal{R}^m_{c(B_v)}$-CSIC is at least Δ^p_2-hard, as it coincides with $\mathcal{R}^m_{c(B_v)}$-PB if we choose $f_{L,H}$ such that $L = H = c(a_{\heartsuit})$ for the item $a_{\heartsuit}$ with $B_{\heartsuit} = \{a_{\heartsuit}\}$.

Next, we want to show a matching upper bound. Let $A' = A \setminus \{a_{\heartsuit}\}$, $E' = (A', V, c, \ell)$, and $E_d = (A, V, c', \ell)$ with $c'(a_{\heartsuit}) = d$ and $c'(a) = c(a)$ for all $a \in A'$. First we compute the overall satisfaction t^* of a winning budget for E', which can be done as described in the proof of Lemma 1 by querying an NP-oracle a polynomial number of times. Knowing the optimal score for a winning budget not containing $a_{\heartsuit}$, we can query an NP-oracle to solve C-$\mathcal{R}^m_{c(B_v)}$-CSIC. In particular, we ask whether there exists $d \in [L, H]$, such there exists $B \in \mathcal{B}(E_d)$ with $a_{\heartsuit} \in B$ and $s(V, B) \geq t^*$. Finding an answer to this question is in NP. The answer is yes, if and only if there exists a $d \in [L, H]$, such that there is a budget containing $a_{\heartsuit}$, that yields a satisfaction at least as high as any bundle not containing $a_{\heartsuit}$. $\square$

Note that the above proof does not hold for the destructive control variant, although the lower bound holds for similar reasons. Knowing t^*, does not lead to a bounded number of obvious NP questions. Instead, we still need to determine, whether there exists a $d \in [L, H]$, such that every feasible bundle containing $a_{\heartsuit}$ yields a satisfaction of at most t^*. For the greedy rule and the cost satisfaction function we can again show polynomial-time solvability.

Theorem 5. C-$\mathcal{R}^g_{c(B_v)}$-CSIC and D-$\mathcal{R}^g_{c(B_v)}$-CSIC are in P.

Proof. Consider any budgeting scenario E and a given item $a_{\heartsuit}$, which should be included (or excluded) into the (resolute) final outcome. Further, let $E_d = (A, V, c', \ell)$ denote the modified budgeting scenario with $c'(a_{\heartsuit}) = d$ and $c'(a) = c(a)$ for every $a \in A \setminus \{a_{\heartsuit}\}$. We assume, that there is a linear tie-breaking scheme $\succ$ over the set of items A, which is identical for every E_d.

Note that $s(V, A) = \sum_{v \in V} \sum_{a \in B} c(a)$ is an additive function. Hence, the order in which the greedy rule $\mathcal{R}^g_{c(B_v)}$ determines, whether to add an item or not, is never changing during execution. Yet, the position of $a_{\heartsuit}$ in this order also depends on its cost $c(a_\heartsuit)$. Formally, let the position of $a \in A$ in the processing order with respect to E_d and $\succ$ be denoted by $\text{pos}(a, E_d, \succ)$. To solve C-$\mathcal{R}^g_{c(B_v)}$-CSIC, we compute $\mathcal{R}^g_{c(B_v)}(E_d)$ for at most $|A|$ values $d \in [L, H]$. Initially we set $d = L$ and compute the winning budget. If $a_\heartsuit \in \mathcal{R}^g_s(E_d)$, the input is a yes-instance. Otherwise, we increase d to the minimum value d', such that $\text{pos}(a_\heartsuit, E_{d'}, \succ) < \text{pos}(a_\heartsuit, E_d, \succ)$. Precisely, for the item a with $\text{pos}(a, E_d, \succ) = \text{pos}(a_\heartsuit, E_d, \succ) - 1$ we set $d' = \left\lceil \sum_{v \in V} c(A_v \cap \{a\}) / \sum_{v \in V} |A_v \cap \{a_\heartsuit\}| \right\rceil$. If necessary due to a tie, which is broken favoring a, d' is additionally increased by 1. Again, if $a_\heartsuit \notin \mathcal{R}^g_s(E_{d'})$ holds, we relabel d' to d and repeat the last step, until we cannot increase the cost of item $a_\heartsuit$ without exceeding our upper limit H. If this is the case, we have successfully identified of a no-instance.

To solve D-$\mathcal{R}^g_{c(B_v)}$-CSIC, we use a similar technique. Now, we initialize $d = H$ and decrease d to the highest value, such that the decision, whether to add $a_\heartsuit$, is done one step later in the processing order. We hold and output yes, if for any such d' it holds that $a_\heartsuit \notin \mathcal{R}^g_{c(B_v)}(E_{d'})$, and output no, if d' falls below L.   □

Table 1. Summary of our complexity results for respective control problems.

$\mathcal{R}_s$	Setting-the-Budget-Limit		Setting-an-Item's-Cost	
	Constructive	Destructive	Constructive	Destructive
$\mathcal{R}^g_{|B_v|}$	in P	in P	in P	in P
$\mathcal{R}^g_{c(B_v)}$	in P	in P	in P	in P
$\mathcal{R}^m_{|B_v|}$	in P	in P	in P	in P
$\mathcal{R}^m_{c(B_v)}$	Δ^p_2-h., in Σ^p_2	Δ^p_2-c.	Δ^p_2-h., in Σ^p_2	Δ^p_2-h., in Σ^p_2

5   Conclusions

We extended the study of winner determination problems for the considered budgeting methods, and introduced a general form of manipulative interference. We focussed on two restrictions, the problems of setting the budgeting limit and setting an item's cost. The results are summarized in Table 1. For most of the rules the problems are solvable in P, whereas they are Δ^p_2-hard for the maximizing rule combined with the cost satisfaction function. This correlates with the results obtained for winner determination, where the associated decision problems are complete for coNP and Δ^p_2.

When studying problems of manipulative interference, polynomial-time algorithms are usually undesired, as this does not offer any protection. However, this can also be interpreted from the perspective of robustness. In reality, the budget limit and the cost of an item may both not be perfectly accurate, meaning that there may be some uncertainty about parts of the budget, or that the cost is rather an estimate. Then problems of manipulative interference give insight in how vulnerable the actual solution may be to changes in one of these parameters.

We considered two of the axioms studied by Talmon and Faliszewski [17]. As a task for future research, this should be extended to other axioms and other types of control that are specific for participatory budgeting. Due to our general formulation of manipulative interference, some of our results may still apply. Another task would be, to close the gap between upper and lower bound for the maximizing rule with the cost satisfaction function. The study can also be extended to other budgeting methods. For example, a satisfaction function could also yield dissatisfaction for rejected projects, or a voting rule could measure the overall satisfaction by the minimum voter's satisfaction instead of the sum.

Acknowledgment. We thank all reviewers for their helpful comments. Supported in part by DFG grant BA 6270/1-1 and by the project "Online Participation," funded by the NRW Ministry for Innovation, Science, and Research.

References

1. Arora, S., Barak, B.: Computational Complexity: A Modern Approach. Cambridge University Press, Cambridge (2009)
2. Aziz, H., Lee, B.E., Talmon, N.: Proportionally representative participatory budgeting: axioms and algorithms. In: Proceedings of the 17th International Conference on Autonomous Agents and MultiAgent Systems, pp. 23–31. IFAAMAS (2018)
3. Aziz, H., Shah, N.: Participatory budgeting: models and approaches. In: Rudas, T., Péli, G. (eds.) Pathways Between Social Science and Computational Social Science. CSS, pp. 215–236. Springer, Cham (2021). https://doi.org/10.1007/978-3-030-54936-7_10
4. Baumeister, D., Boes, L., Seeger, T.: Irresolute approval-based budgeting. In: Proceedings of the 19th International Conference on Autonomous Agents and Multi-Agent Systems, pp. 1774–1776. IFAAMAS (2020)
5. Cabannes, Y.: Participatory budgeting: a significant contribution to participatory democracy. Environ. Urbanization **16**(1), 27–46 (2004)
6. Faliszewski, P., Rothe, J.: Control and bribery in voting. In: Brandt, F., Conitzer, V., Endriss, U., Lang, J., Procaccia, A. (eds.) Handbook of Computational Social Choice, chap. 7, pp. 146–168. Cambridge University Press (2016)
7. Garey, M.R., Johnson, D.S.: Computers and Intractability: A Guide to the Theory of NP-Completeness. W. H, Freeman and Company, New York (1979)
8. Goel, A., Krishnaswamy, A.K., Sakshuwong, S., Aitamurto, T.: Knapsack voting for participatory budgeting. ACM Trans. Econ. Comput. **7**(2), 1–27 (2019)
9. Jain, P., Sornat, K., Talmon, N.: Participatory budgeting with project interactions. In: Proceedings of the 29th International Joint Conference on Artificial Intelligence, pp. 386–392. ijcai.org (2020)
10. Kellerer, H., Pferschy, U., Pisinger, D.: Knapsack Problems. Springer, Berlin (2004) https://doi.org/10.1007/978-3-540-24777-7
11. Krentel, M.W.: The Complexity of Optimization Problems. J. Comput. Syst. Sci. **36**(3), 490–509 (1988)
12. Lackner, M., Skowron, P.: Approval-based multi-winner rules and strategic voting. In: Proceedings of the 27th International Joint Conference on Artificial Intelligence, pp. 340–346. ijcai.org (2018)

13. Rey, S., Endriss, U., de Haan, R.: Designing participatory budgeting mechanisms grounded in judgment aggregation. In: Proceedings of the 17th International Conference on Principles of Knowledge Representation and Reasoning, pp. 692–702. ijcai.org (2020)
14. Rey, S., Endriss, U., de Haan, R.: Shortlisting rules and incentives in an end-to-end model for participatory budgeting. In: Proceedings of the 30th International Joint Conference on Artificial Intelligence. ijcai.org (August 2021), to appear
15. Sintomer, Y., Herzberg, C., Röcke, A.: Participatory budgeting in Europe: potentials and challenges. Int. J. Urban Reg. Res. **32**(1), 164–178 (2008)
16. Skowron, P., Faliszewski, P.: Chamberlin-courant rule with approval ballots: approximating the MaxCover problem with bounded frequencies in FPT time. J. Artif. Intell. Res. **60**, 687–716 (2017)
17. Talmon, N., Faliszewski, P.: A Framework for approval-based budgeting methods. In: Proceedings of the 33rd AAAI Conference on Artificial Intelligence, vol. 33, pp. 2181–2188. AAAI Press (2019)

Author Index